SURFACE IMPEDANCE BOUNDARY CONDITIONS

A Comprehensive Approach

SURFACE IMPEDANCE BOUNDARY CONDITIONS

A Comprehensive Approach

Sergey V. Yuferev
Nathan Ida

CRC Press
Taylor & Francis Group
Boca Raton London New York

CRC Press is an imprint of the
Taylor & Francis Group, an **informa** business

CRC Press
Taylor & Francis Group
6000 Broken Sound Parkway NW, Suite 300
Boca Raton, FL 33487-2742

Printed in the United States of America on acid-free paper
10 9 8 7 6 5 4 3 2 1

International Standard Book Number: 978-1-4200-4489-8 (Hardback)

Library of Congress Cataloging-in-Publication Data

Yuferev, Sergey V.
Surface impedance boundary conditions : a comprehensive approach / authors, Sergey V. Yuferev, Nathan Ida.
p. cm.
"A CRC title."
Includes bibliographical references and index.
ISBN 978-1-4200-4489-8 (hardcover : alk. paper)
1. Electromagnetic waves--Scattering--Mathematical models. 2. Radio wave propagation--Mathematical models. 3. Electromagnetic surface waves--Mathematical models. 4. Impedance (Electricity)--Mathematical models. 5. Boundary value problems. I. Ida, Nathan. II. Title.

QC665.S3Y84 2010
537.6'2--dc22 2009026274

Visit the Taylor & Francis Web site at
http://www.taylorandfrancis.com

and the CRC Press Web site at
http://www.crcpress.com

To our parents

Authors

Sergey Yuferev was born in St. Petersburg, Russia, in 1964. He received his MSc in computational fluid mechanics from St. Petersburg Technical University, St. Petersburg, in 1987, and his PhD in computational electromagnetics from the A.F. Ioffe Institute, St. Petersburg, in 1992. From 1987 to 1998, he worked at with the Dense Plasma Dynamics Laboratory, A.F. Ioffe Institute. From 1999 to 2000, he was a visiting associate professor at the University of Akron, Akron, Ohio. Since 2000, he has been with the Nokia Corporation, Tampere, Finland. His current research interests include numerical and analytical methods of computational electromagnetics and their application to electromagnetic compatibility and electromagnetic interference problems of mobile phones.

Nathan Ida is currently a distinguished professor of electrical and computer engineering at the University of Akron, Akron, Ohio. He teaches electromagnetics, antenna theory, electromagnetic compatibility, sensing and actuation, and computational methods and algorithms. His current research interests include numerical modeling of electromagnetic fields, electromagnetic wave propagation, theoretical issues in computation, and nondestructive testing of materials at low and microwave frequencies as well as in communications, especially in low-power remote control and wireless sensing. He has published extensively on electromagnetic field computation, parallel and vector algorithms and computation, nondestructive testing of materials, surface impedance boundary conditions, and other topics. He is the author of three books and co-author of a fourth. Dr. Ida is a fellow of the IEEE and the American Society of Nondestructive Testing.

Preface

The idea for this book grew out of work on surface impedance boundary conditions (SIBCs) over a period of nearly 12 years. We felt that the concept of surface impedance, especially at a low frequency, had been neglected to a large extent. When it was used, it was often viewed as an esoteric issue and incorporated in computation as an ad hoc means of simplifying the treatment for specific problems. In addition, the use of surface impedance was limited almost exclusively to the Leontovich concept—that of a first-order SIBC. No attempt has been made to formalize the concept of SIBC or to extend it to higher orders. This book intends to remedy this state of affairs by providing a comprehensive, consistent, and thorough approach to the subject. The book outlines not only the need for SIBCs, but also offers a simple, systematic method of construction of SIBCs of any order based on a perturbation approach. As one of the first tasks, we develop a so-called surface impedance toolbox, which allows the reader to develop SIBCs for any application. The formulation of the SIBC within common numerical techniques, such as the boundary integral equations method, the finite element method, and the finite difference method, is discussed in detail and is elaborated with specific examples. We felt that one particular reason to shun SIBCs was the fact that their implementation usually required extensive modification of existing software. To mitigate this problem, we developed our own SIBCs so that they could be incorporated within existing software without having to modify the software's structure. The various-order SIBCs are built by adding terms in an expansion so that increasing the order of approximation only requires the addition of terms without modification of the existing terms.

Although the development of the various SIBCs and their formulations is rigorous and detailed, the emphasis in this book is on the practical aspects. Therefore, we include examples for all formulations, and discuss in detail the conditions of applicability and the errors to be expected from the inclusion of SIBCs in formulations. A practical set of guidelines based on the physical dimensions and properties of the problem is outlined to allow the reader to first decide if an SIBC can benefit his or her calculation and then to evaluate, *a priori*, the maximum errors he or she can expect by incorporating the SIBC.

This book has been in the making for over 4 years, but its roots can be traced back to its modest beginnings nearly 12 years ago when the authors first began their association. Over the years, others, including graduate students, have contributed to this book, but we wish to acknowledge the contribution of Professor Luca Di Rienzo from the Polytecnica di Milano, in particular. Many of the results reported here were derived from his work on implementation of the various formulations. The value of this book would have been greatly diminished without his contributions.

Sergey V. Yuferev
Nathan Ida

Contents

Introduction

I.1 Concept of Surface Impedance

In most electromagnetic problems, the domain under consideration consists of several different media. The electromagnetic field governing equations written for each region are linked by boundary conditions that involve values on both sides of the interface. Thus one has to solve the problem for all media simultaneously, even if the main interest is focused only on one of them. However, under certain conditions, the number of regions involved in the solution procedure may be reduced. A classical example is the elimination of a body of infinite conductivity—a perfect electrical conductor (PEC)—from the computational space by enforcing the tangential electric field or normal magnetic flux to be equal to zero at the boundary (the so-called PEC boundary condition):

$$\vec{n} \times \vec{E}\big|_{\text{interface}} = 0; \quad \vec{n} \cdot \vec{B}\big|_{\text{interface}} = 0 \tag{I.1}$$

In practice, any real material has finite conductivity so that the PEC is merely a model of a good conductor in which the skin depth is assumed to be zero. Although the PEC condition is very attractive for implementation, the diffusion of electromagnetic fields into the conductor may be neglected only in a limited number of practical cases. This means that the application of the PEC limit in practical designs is very limited. For example, the electromagnetic penetration depth δ in copper at an incident frequency of 1 MHz is approximately 2×10^{-5} m. Is this skin layer thin or thick? Obviously, the question is meaningful only in relation to the dimensions of the media so that they may be compared. The characteristic size D of the conductor may be used for this comparison. In our example, the penetration depth in the conductor is definitely not small if the conductor's thickness is equal to 5×10^{-5} m (typical thicknesses of conductors in printed circuit boards) and the PEC condition may not be applied in this case. One may expect the use of the PEC condition to lead to errors proportional to δ/D.

Since the PEC is the limiting case of a real conductor, it is only natural to expect that the PEC condition is also a particular case of a more general approximate boundary condition relating electromagnetic quantities at conductor/dielectric interfaces. The existence of approximate boundary conditions of this genre follows directly from Snell's law of refraction; if the electromagnetic wave propagates from a low-conductivity medium to a high-conductivity medium, the refraction angle is about 90° and, in practical terms, does not depend on the incidence angle. Suppose the conducting region is so large that the wave attenuates completely inside the region. Then, the electromagnetic field distribution in the conductor's skin layer can be described as a damped plane wave propagating in the bulk of the conductor normal to its surface. In other words, the behavior of the electromagnetic field in the conducting region may be assumed to be known *a priori*, as in the case of the PEC. The electromagnetic field is continuous across the conductor's surface so that the intrinsic impedance of the wave remains the same at the interface. Therefore, the ratio E_x at the xy-plane of a dielectric/conductor interface is assumed to be equal to the intrinsic impedance of the plane wave propagating in the conductor, in the positive z-direction:

$$\left.\frac{E_x}{H_y}\right|_{\text{interface}} = \left.\sqrt{\frac{j\omega_{\text{source}}\mu_{\text{cond}}}{\sigma_{\text{cond}} + j\omega_{\text{source}}\varepsilon_{\text{cond}}}}\right|_{\sigma \gg \omega\varepsilon} \approx \frac{1+j}{2}\omega_{\text{source}}\mu_{\text{cond}}\delta$$

$$\delta = \sqrt{\frac{2}{\omega_{\text{source}}\sigma_{\text{cond}}\mu_{\text{cond}}}} \qquad \text{(I.2)}$$

The relation in Equation I.2 does not depend on coordinates and, therefore, the surface impedance is assumed to be constant over the conductor's surface.

The "surface" relation in Equation I.2, taking into account parameters of the conductor's material and the source, contains all the necessary information about the field distribution in the conductor's "volume." Thus, it may be used as a boundary condition to the governing equations for the dielectric space and, by doing so, excludes the conductor from the solution region. In contrast to Equation I.1, which is a Dirichlet boundary condition, Equation I.2 can be viewed as an additional equation relating different unknowns at the interface. This is the basic idea of the surface impedance concept.

I.2 Historical Perspective

The roots of the surface impedance concept can be traced to the end of the nineteenth century, a time when the main computational tool of electrical engineers was circuit theory. The lumped circuit representation is based on the approximation that it is acceptable to assign electromagnetic processes such as energy supply, dissipation, and storage to individual components, concentrated virtually at a point (lumped) in space. A lumped circuit is considered to be a group of components (resistors, capacitors, and inductors) interconnected in a certain topology occupying no physical space; a signal propagates from one point to another without time delay. To denote the ratio of amplitudes of an electromotive force V and the alternating current I

produced by it in the circuit comprised of a resistance R and an inductance L, Oliver Lodge introduced the term "impedance" in 1889 [18]. The instantaneous voltage–current relation in such a circuit is described by the following differential equation:

$$L\frac{dI}{dT} + RI = V \tag{I.3}$$

In the steady-state case, Equation I.3 takes the form

$$(R + j\omega L)I = V \tag{I.4}$$

where ω is an angular frequency of the source. Thus, the impedance Z of the electric circuit is represented in the form:

$$Z_{\text{circ}} = V/I = R + j\omega L \tag{I.5}$$

In real physical systems, energy storage and dissipation are mixed together and distributed over relatively large areas. The lumped component representation is applicable if the physical size of the problem is much smaller than the wavelength (electrically small system) and the propagation time of disturbances is negligible compared to their period. However, it is also possible to use a "circuit" approach in devices that do not satisfy this condition, such as transmission lines. The space variation of the complex voltage V across and the electric current I in a transmission line (let it be directed along the x-axis) is described by the following well-known equations:

$$dV/dx = -ZI \tag{I.6}$$

$$dI/dx = -YV \tag{I.7}$$

where the coefficients of proportionality Z and Y are known as the distributed "series impedance" and the distributed "shunt admittance" of the line. In a generalized transmission line, Z and Y may be functions of x and may depend on frequency. If Z and Y are independent of x, they depend upon the distributed series resistance R, the shunt conductance G, the series inductance L, and the shunt capacitance C in the following manner:

$$Z = R + j\omega L; \quad Y = G + j\omega C \tag{I.8}$$

Equations I.6 through I.8 can be combined to solve for $V(x)$ and $I(x)$:

$$\frac{d^2V}{dx^2} = \gamma^2 V \tag{I.9}$$

$$\frac{d^2I}{dx^2} = \gamma^2 I \tag{I.10}$$

where

$$\gamma = \alpha + j\beta = \sqrt{ZY} = \sqrt{(R + j\omega L)(G + j\omega C)} \tag{I.11}$$

is the propagation constant whose real and imaginary parts, α and β, are the attenuation and phase constants of the line, respectively. Solutions of Equations I.9 and I.10 represent waves propagating in the $+x$- and $-x$-directions with a certain finite velocity maintaining a fixed phase between V and I:

$$V(x) = V^+(x) + V^-(x) = V_0^+ \exp(-\gamma x) + V_0^- \exp(\gamma x) \tag{I.12}$$

$$I(x) = I^+(x) + I^-(x) = I_0^+ \exp(-\gamma x) + I_0^- \exp(\gamma x) \tag{I.13}$$

From Equation I.12, one obtains

$$\frac{dV}{dx} = -\gamma V \tag{I.14}$$

Substituting Equation I.14 in Equations I.6 and I.7, the characteristic impedance of the line, that is, the ratio of voltage and current, is obtained:

$$Z_0 = \frac{R + j\omega L}{\gamma} = \frac{\gamma}{G + j\omega C} = \sqrt{\frac{R + j\omega L}{G + j\omega C}} \tag{I.15}$$

Transmission line theory describes wave propagation phenomena. For example, Equations I.6, I.7, I.9, and I.10 are of the same form as the equations governing the propagation of plane electromagnetic waves with $\vec{E} = \vec{a}_y E$ and $\vec{H} = \vec{a}_z H$ in a lossy medium in the x-direction:

$$\frac{dE}{dx} = -j\omega\mu H \tag{I.16}$$

$$\frac{dH}{dx} = -(\sigma + j\omega\varepsilon)H \tag{I.17}$$

$$\frac{d^2E}{dx^2} = \gamma^2 E \tag{I.18}$$

$$\frac{d^2H}{dx^2} = \gamma^2 H \tag{I.19}$$

where the propagation constant γ takes the form:

$$\gamma = \sqrt{j\omega\mu(\sigma + j\omega\varepsilon)} = j\omega\sqrt{\mu\varepsilon\left(1 + \frac{\sigma}{j\omega\varepsilon}\right)} \tag{I.20}$$

Obviously, the solutions of Equations I.18 and I.19 have the same form as Equations I.12 and I.13:

$$E(x) = E_0^+ \exp(-\gamma x) + E_0^- \exp(\gamma x) \tag{I.21}$$

$$H(x) = H_0^+ \exp(-\gamma x) + H_0^- \exp(\gamma x) \tag{I.22}$$

Since even the physical meanings of E and H are closely related to those of V and I (E is V per unit length and H is I per unit length), the terms of impedance are naturally extended to the electromagnetic wave theory and applications. By analogy with Equation I.15, the quantity

$$Z_0 = \frac{E}{H} = \sqrt{\frac{j\omega\mu}{\sigma + j\omega\varepsilon}} = \frac{j\omega\mu}{\gamma} = \frac{\gamma}{\sigma + j\omega\varepsilon} \tag{I.23}$$

has been called the "intrinsic impedance" of the medium. In the general case Z_0 is complex, but in the particular case of a lossless medium it is real. The intrinsic impedance will frequently occur as a multiplier in expressions for the impedances of various types of waves.

Practical applications of impedance in power engineering, transmission lines, shielding, etc., have been analyzed by Schelkunoff, who introduced the term "impedance concept" in 1938 to generalize the idea of impedance [6]. Schelkunoff also recognized analogies between circuit impedance and similar ratios of major quantities in other engineering disciplines where the term "impedance" is also common (such as mechanics) or not so common (such as hydrodynamics and thermodynamics). For example, the well-known one-dimensional equation describing linear oscillations of a mass in a damping medium is written in the form:

$$m\frac{\mathrm{d}(v \exp(iwt))}{\mathrm{d}t} + r(v \exp(iwt)) = F \exp(iwt) \tag{I.24}$$

where

v is the velocity
F is the applied force
m is the mass
r is the resistance coefficient

This equation is similar to Equation I.3, so that mechanical impedance, that is, the ratio "force/velocity," can be represented in the form:

$$Z_{\text{mech}} = r + iwm \tag{I.25}$$

Another example is the propagation of a heat wave that is governed in the one-dimensional case by the following equations:

$$\frac{\partial T}{\partial x} = -\frac{v}{K} \tag{I.26}$$

$$\frac{\partial v}{\partial x} = -c\rho \frac{\partial T}{\partial t} \tag{I.27}$$

where

T is the temperature
v is the rate of heat flow
K is the thermal conductivity
ρ is the density
c is the specific heat

For simple periodic waves, Equations I.26 and I.27 are reduced to

$$\frac{\mathrm{d}T}{\mathrm{d}x} = -\frac{v}{K} \tag{I.28}$$

$$\frac{\mathrm{d}v}{\mathrm{d}x} = -j\omega c\rho T \tag{I.29}$$

Thus, the characteristic impedance and the propagation constant of heat waves are

$$Z_{\text{heat}} = \frac{1}{\sqrt{j\omega c\rho K}}; \quad \gamma = \sqrt{\frac{j\omega c\rho}{K}} \tag{I.30}$$

In the 1930s, the newly evolving radio technology required the development of the theory of propagation of electromagnetic waves of an antenna over the Earth's surface. An analytical solution for the particular case of a vertical dipole radiating over a conducting half-space has been obtained by Sommerfeld [4], but the more general problem involving layered media separated by curved interfaces has not been solved so far. Leontovich [7,8] and Schukin [9] (independently of one another) proposed a different approach, namely, to restrict the general problem by considering only the air region using the surface impedance boundary condition (SIBC) at the air/Earth interface. Of course, the Earth is not a good conductor in the common sense. On the other hand, the characteristic dimensions in this class of electromagnetic problems are sufficiently large for attenuation of the wave in the Earth, so that the surface impedance method may be applied. The considerations of both Leontovich and Schukin are based on the assumption that the variation of the field along the surface is small compared to the variation inside the Earth. Thus the field derivatives in the directions tangential to the surface may be neglected compared to the normal derivative, and the original two-dimensional or even three-dimensional equation of the field diffusion into the conductor is reduced to a one-dimensional problem that yields the relationship in Equation I.2. This is frequently referred to as the skin effect approximation [10]. Note that the same approximation is at the root of the theory of boundary layers in fluid mechanics.

Schukin published this derivation in 1940 [9]. Leontovich published his results without derivation later, in 1944 [7]. However, there is evidence that Leontovich actually came to the idea of approximate boundary conditions at the end of the 1930s, at the same time as Schukin. The first rigorous mathematical analysis was done by Rytov, who wrote the following in his memoirs [2]:

> In 1939 he (Leontovich) posed the following problem to me: calculate the electromagnetic field inside a good conductor near the surface, i.e. develop an approximate, but general theory for strong skin effect. This problem had the following history. Probably in 1938, Leontovich proposed his now well-known boundary conditions for the electromagnetic field on the surface of good conductors. The meaning of these conditions is that they allow solutions to problems of diffraction of electromagnetic waves outside conducting regions without consideration of the field inside the conductors, greatly simplifying the solution of the problem. Although the correctness of the boundary conditions was beyond all doubt, Leontovich was not completely satisfied because the questions concerning accuracy and applicability of the conditions as well as that of acceptable curvature of the conductor's surface were unresolved. In other words, he wanted to have validation of his boundary conditions and estimation of their accuracy. However, when he raised the problem of the skin effect theory, he told me nothing about the approximate boundary conditions (he probably wanted to test my capacity at guesswork). I have developed the theory for the general case of an arbitrarily directed incident wave and double curved surface of the conductor. The theory certainly gave the approximate boundary conditions as the skin depth is reduced, but I missed this fact and so I failed the exam on guesswork.

Rytov's fundamental contribution is the development of the perturbation approach to the problem of the field inside and outside the conductor [12]. He sought a solution in the form of a power series in the small parameter proportional to the ratio δ/D. The first-order terms of the expansions (actually, first nonzero terms) gave Leontovich's condition. Thus, an improvement of the Leontovich condition can be obtained by the inclusion of the next higher-order terms of the expansions. Rytov also stated the problems of calculation of the surface impedance at curved interfaces and nonhomogeneous conductors. Unfortunately, even though [12] has been translated into French [13], Rytov's contribution is not as well known as Leontovich's work.

One may think that the time for approximate analytical approaches passed with the advent of robust numerical methods and fast computers. However, computational electromagnetics is called upon to solve and analyze problems in structures so vast and complex that the cliche "one basic truth to life is that demand will fill up all available resources!" [14] has never sounded truer. Because of this, the surface impedance concept is gaining acceptance as a technique that allows significant savings in computer resources and improved solutions. The Leontovich SIBC is efficiently used in combination with such numerical methods as the finite-element method (FEM) [26–38,79,87–90,108,112,117–182], the boundary-integral equation method [39–56], and the finite-difference time-domain method (FDTD) [57–77]. Among practical applications of the surface impedance concept are transformers [30,33,79–81,166,180], waveguides [82–87], inductive heating devices [88–92], microstrip and transmission lines [44,93–98], electric machines [99–101], electromagnetic scattering [49,65,102–107], electromagnetic compatibility [108–111,160], nondestructive testing analysis [112–115], plasma and magnetic levitation devices [32,116,117], electromagnetic casting [111,112], and many others, including applications outside the sphere of electromagnetics.

I.3 New Developments

The growing number of applications was the main impetus to the further development of the concept in the following basic directions:

1. *Extensions of the planar SIBCs to curved surfaces and edges*. Indeed, the Leontovich SIBC can be applied as long as the smallest radius of curvature of the surface is much larger than the wavelength inside the conductor. Practical important steps have been taken by Mitzner [15], who developed SIBCs, known by his name, for any smooth surface conductors by introducing terms of the order of δ^2 that allow for the conductor's curvature. Although Mitzner has derived the SIBCs in his own way, calculation of the second-order terms in Rytov's expansions leads to the same result. A significantly different situation occurs in the vicinity of a conducting edge where the

magnetic field distribution is singular and the diffusion may not be described by a one-dimensional equation. Semiempirical attempts have been made to modify Leontovich's condition near corners and edges [142–152,194], but the only rigorous solution was obtained using the perturbation approach [153,154]. First, the field distribution in free space surrounding the edge is calculated with the PEC condition (zero-order solution). Then this distribution along the interface is used as a boundary condition to the two- or three-dimensional problem for the field inside the real conductor and its asymptotic solution is used to derive the first-order SIBC near the edge.

2. *Time-domain SIBCs*. The concept of surface impedance can be applied to transient problems, when, for instance, the current pulse duration is so short that the field has no time to diffuse deep into the conductor and remains concentrated near the surface. There are two basic approaches to solve transient problems: (a) by obtaining the solution in the frequency domain for a time-harmonic exciting source and then using the inverse Fourier transform to calculate the required transient data, and (b) by formulating the problem directly in the time domain. The time-domain form of the Leontovich condition can be obtained from Equation I.1 using the Laplace transform. High-order time-domain SIBCs have been obtained using the perturbation approach [21,23–25,55,56,76,77,174]. The choice between time- and frequency-domain forms of the SIBC is often governed by the method used for solution of the problem (for example, time-domain SIBCs are naturally suited for FDTD methods). Due to the frequency-dependent nature of the SIBC, it contains a convolution integral in the time domain. Direct computation of this convolution integral requires large computer resources and, therefore, approximate algorithms have been developed for efficient implementations of SIBCs in the time domain [57–75,155–164].

3. *SIBCs for nonlinear media*. Often, the magnetic or thermal properties of the conductor's material may not be approximated by constants. Traditionally the $B(H)$ or $\sigma(T)$ curves are approximated by an analytical function exploiting features of the problem under consideration and then inserted into the frequency-domain Leontovich condition [90,165–172]. However, in the general case, neither magnetic nor electric fields in the conductor with nonlinear properties are actually time harmonic and the use of a single-frequency SIBC is not feasible. Thus, formulating the problem in the time domain seems to be a more general approach for both linear and nonlinear problems [173,174]. In this case, the SIBC is not obtained as the ratio of the tangential fields at the interface. Instead, it is formulated in the form of a one-dimensional nonlinear equation that must be solved numerically.

On the other hand, time-domain nonlinear SIBCs are usually applicable to any isotropic homogeneous material since they are not restricted to any special approximation of the material property curve.

4. *SIBCs for nonhomogeneous (composite) structures.* The most important case in practice is a conductor coated by a layer of dielectric material. If the nonconducting layer is sufficiently thin compared to the wavelength, the skin layer approximation will still be valid. SIBCs of different orders of approximation for this class of problems have been developed and given the name generalized impedance boundary conditions (GIBCs) [198–207].

As a result of these activities, a wide variety of SIBCs have been developed and are commonly used in computational electromagnetics.

I.4 Purpose and Method of Presentation

The concept of this book grew out of a need to present to the practicing engineer as well as to students a comprehensive, coherent exposition of the ideas, applications, and the general issues involved with the use of SIBCs. In particular, the authors felt that SIBCs are viewed by the computational electromagnetics community as somewhat esoteric and as an ad hoc solution to specific problems. Furthermore, most users of the method unnecessarily restrict themselves to the Leontovich condition. Yet, the concept is much more general and simpler than most potential users suspect, and the authors felt it necessary to expose these aspects of the method. The purpose of this book is to help the practitioner in the understanding, selection, use, and development of impedance boundary conditions for his/her own applications. In the process, the theoretical arguments and mathematical relations needed for the proper definition of SIBCs are developed. However, the main emphasis of the book is on the practical. It provides the engineer with a practical and versatile tool for low-penetration problems in computational electromagnetics. The book provides the following specific benefits:

1. *Decision.* The book provides guidance on how to decide if the surface impedance concept is applicable to a particular problem and if the problem can benefit from the use of the surface impedance concept.
2. *Selection.* The choice of the most appropriate SIBC (if one exists) is important both for accuracy and implementation. This issue is discussed based on performance and accuracy, and simple guidelines to the selection process are provided.

3. *Development.* When an SIBC appropriate for a specific application does not exist, one can be developed based on the methods discussed in this book.
4. *Implementation.* For an SIBC to be truly useful, its implementation in existing code or in new code development should be simple and efficient. A number of formulations of SIBCs in conjunction with well-known numerical techniques are discussed.

Structure of the book

The discussion of SIBCs starts with the Leontovich condition in Chapter 1. The Leontovich condition is derived starting with the electric and magnetic fields, following the classical approach. This serves as an introduction to the more general methods of deriving SIBCs to be taken up in subsequent chapters. It also emphasizes the steps and ideas needed to develop SIBCs for specific applications. The assumptions made in the development of SIBCs and the conditions under which these SIBCs may be used are critical to their successful use. These assumptions and conditions are introduced in Chapter 1, which also discusses the Mitzner and Rytov methods as an introduction to the elaborations and developments to follow.

Chapter 2 introduces the reader to the general idea of building an SIBC as an asymptotic expansion based on a perturbation approach. This method, due to Rytov, is a general tool for the development of SIBCs and is particularly useful for the development of SIBCs of high order of approximation. Following the introduction and discussion of Rytov's formula, a unified time- and frequency-domain approach to the development of SIBCs is introduced. The chapter outlines a general surface impedance "toolbox," that is applicable universally and is later used repeatedly in following chapters.

SIBCs in terms of various formalisms are introduced in Chapter 3. In addition to the electric field–magnetic field formalisms, we also discuss the magnetic scalar potential and the magnetic vector potential formalisms as well as introduce common representations in terms of surface impedance functions. Finally, we discuss the issue of surface impedance near corners and edges.

An important issue in the use of SIBCs is the surface layer itself. Chapter 4 discusses these issues and shows how distributions across the skin layer may be calculated in the context of the formulations of Chapter 3.

Chapter 5 tackles the implementation of SIBCs for nonlinear and nonhomogeneous applications. We treat coupled electromagnetic–thermal formulations as well as magnetic materials.

In Chapter 6, we treat the implementation of SIBCs for the low-frequency boundary equation method. Starting with two-dimensional implementations, we treat in particular the electric field–magnetic field formulation and then introduce the surface current concept followed by the magnetic vector potential–surface current formulation in the context of boundary integral formulations. The concept of common representations discussed first in Chapter 3 is again invoked to extend the implementation to other formulations. Three-dimensional problems are then discussed, followed by a discussion of the surface impedance function and the development of invariant formulations in two and three dimensions.

Chapter 7 continues the subject of implementation of SIBCs for boundary integral equations, but this time we focus on high-frequency problems. Because of the assumption of high frequencies, SIBCs for lossy dielectrics can now be used. Implementations using direct methods and perturbation techniques are discussed.

Although SIBCs are most readily used with boundary integral methods, they can be successfully implemented in volume discretization methods. We take up this subject in Chapter 8 and discuss implementation within the finite-difference time-domain (FDTD), finite-integration technique (FIT), and the finite-element method (FEM).

Chapter 9 presents two important issues. The first is the selection of appropriate SIBCs for a given problem. Starting from the characteristic values of the physical problem, we develop a simple methodology that tells the user the appropriate order of the SIBC and estimates the maximum errors involved. Then we look at experimental validation of SIBCs in two distinct types of applications: one is a simple comparison of computed and measured data; the other is the problem of reconstruction of currents from measured field data. These emphasize not only the accuracy expected from SIBCs, but also the range of applications of the method. Additional applications dealing with multiconductor transmission lines look into the issues of applicability and accuracy of SIBCs.

The book concludes with an appendix discussing representative numerical methods including the FEM, the boundary-element method, and the FDTD method. This appendix only touches on the methods, but it does outline the basic steps so that the user may get a general idea of the issues involved. A concise list of references given may also help in guiding the user to further study.

References

History

1. C. Pelosi and P. Ya. Ufimtsev, The impedance-boundary condition, *IEEE Antennas and Propagation Magazine*, 38(1), 1996, 31–35.
2. S.M. Rytov, In the oscillations laboratory, in *Memoires to Academician Leontovich*, Moscow, Nauka, 1996, pp. 40–66 (in Russian).
3. J.R. Wait, The ancient and modern history of EM ground-wave propagation, *IEEE Antennas and Propagation Magazine*, 40(5), 1998, 7–24.

4. A. Ishimaru, J.D. Rockway, Y. Kuga, and S-W. Lee, Sommerfeld and Zenneck wave propagation for a finitely conducting one-dimensional rough surface, *IEEE Transactions on Antennas and Propagations*, 48(9), 2000, 1475–1484 (also rough surfaces).

General Theory

5. Lord Rayleigh, On the self inductance and resistance of straight conductors, *Philosophical Magazine*, 21, 1886, 381–394.
6. S.A. Shelkunoff, The impedance concept and its application to problems of reflection, radiation, shielding and power absorption, *Bell System Technical Journal*, 17, 1938, 17–48.
7. M.A. Leontovich, On one approach to a problem of the wave propagation along the Earth's surface, *Academy of Science USSR, Series Physics*, 8, 1944, 16–22.
8. M.A. Leontovich, On the approximate boundary conditions for the electromagnetic field on the surface of well conducting bodies, in B.A. Vvedensky, Ed., *Investigations of Radio Waves*, Academy of Sciences of USSR, Moscow, 1948.
9. A.N. Shchukin, *Propagation of Radio Waves*, Svyazizdat, Moscow, 1940.
10. G.S. Smith, On the skin effect approximation, *American Journal of Physics*, 58(10), 1990, 996–1002.
11. H.G. Booker, The elements of wave propagation using the impedance concept, *Journal of the Institute of Electrical Engineers*, 3(94), 1947, 171–184.
12. S.M. Rytov, Calculation of skin effect by perturbation method, *Journal Experimental'noi i Teoreticheskoi Fiziki*, 10(2), 1940, 180–189 (in Russian).
13. S.M. Rytov, Calcul du skin-effet par la méthode des perturbations, *Journal of Physics*, 2(3), 1940, 233–242.
14. C. Brench, TC-9 computational EMC, *IEEE EMC Society Newsletter*, 186, 2000, 13.
15. K.M. Mitzner, An integral equation approach to scattering from a body of finite conductivity, *Radio Science*, 2(12), 1967, 1459–1470.
16. C.A. Balanis, *Advanced Engineering Electromagnetics*, John Wiley & Sons, New York, pp. 210–213.
17. J. van Bladel, *Electromagnetic Fields*, Appendix 2, McGraw-Hill Book Co., Inc., New York, 1964.
18. O. Lodge, On lightning, lightning conductors, and lightning protectors, *Electrical Review*, May 1889, 518.
19. J.R. Wait, The scope of impedance boundary conditions in radio propagation, *IEEE Transactions on Geoscience Remote Sensing*, 28(4), 1990, 721–723.
20. Z. Godzinski, The surface impedance concept and the theory of radio waves over real earth, *Proceedings of the IEE*, 108C, March 1961, 362–373.
21. S. Yuferev and N. Ida, Time domain surface impedance boundary conditions of high order of approximation, *IEEE Transactions on Magnetics*, 34(5), 1998, 2605–2608.
22. S. Yuferev and N. Ida, Selection of the surface impedance boundary conditions for a given problem, *IEEE Transactions on Magnetics*, 35(3), 1999, 1486–1489.
23. S. Yuferev, N. Ida, and L. Kettunen, Invariant BEM-SIBC formulations for time- and frequency-domain eddy current problems, *IEEE Transactions on Magnetics*, 36(4), 2000, 852–855 (also use with numerical methods: integral equations).
24. S. Yuferev and N. Ida, Time domain surface impedance concept for low frequency electromagnetic problems-Part I: Derivation of high order surface impedance

boundary conditions in the time domain, *IEE Proceedings on Science, Measurement and Technology*, 152(4), 2005, 175–185.
25. S. Barmada, L. Di Rienzo, N. Ida, and S. Yuferev, Time domain surface impedance concept for low frequency electromagnetic problems. II. Application to transient skin and proximity effect problems in cylindrical conductors, *IEE Proceedings on Science, Measurement and Technology*, 152(5), 2005, 207–216.

Use of SIBCs with Numerical Methods

FEM

26. E.M. Deeley and J. Xiang, Improved surface impedance methods for 2-D and 3-D problems, *IEEE Transactions on Magnetics*, 24(1), 1988, 209–211.
27. J. Sakellaris, G. Meunier, A. Raizer, and A. Darcherif, The impedance boundary condition applied to the finite element method using the magnetic vector potential as state variable: A rigorous solution for high frequency axisymmetric problems, *IEEE Transactions on Magnetics*, 28(2), 1992, 1643–1646.
28. A.M. El-Sawy Mohamed, Finite-element variational formulation of the impedance boundary condition for solving eddy current problems, *IEE Proceedings on Science, Measurement and Technology*, 142(4), 1995, 293–298.
29. M. Gyimesi, J.D. Lavers, T. Pawlak, and D. Ostergaard, Impedance boundary condition for multiply connected domains with exterior circuit conditions, *IEEE Transactions on Magnetics*, 30(5), 1994, 3056–3059.
30. J. Sakellaris, G. Meunier, X. Brunotte, C. Guerin, and J.C. Sabonnadiere, Application of the impedance boundary condition in a finite element environment using the reduced potential formulation, *IEEE Transactions on Magnetics*, 27(6), 1991, 5022–5024 (also applications: transformers).
31. S. Subramaniam, M. Feliziani, and S.R. Hoole, Open boundary eddy-current problems using edge elements, *IEEE Transactions on Magnetics*, 2(2), 1992, 1499–1503.
32. M. Enokizono and T. Todaka, Approximate boundary element formulation for high-frequency eddy current problems, *IEEE Transactions on Magnetics*, 29(2), 1993, 1504–1507 (also applications: magnetic levitation devices).
33. S.A. Holland, G.P. O'Connel, and L. Haydock, Calculating stray losses in power transformers using surface impedance with finite elements, *IEEE Transactions on Magnetics*, 28(2), 1992, 1355–1358 (also applications: transformers).
34. D. Rodger, P.J. Leonard, H.C. Lai, and P.C. Coles, Finite element modeling of thin skin depth problems using magnetic vector potential, *IEEE Transactions on Magnetics*, 33(2), 1997, 1299–1301.
35. G. Pan and J. Tan, General edge element approach to lossy and dispersive structures in anisotropic media, *IEE Proceedings—Microwave, Antennas, Propagation*, 144(2), 1997, 81–90.
36. I. Mayergoyz and C. Bedrosian, On finite element implementation of impedance boundary conditions, *Journal of Applied Physics*, 75, 1994, 6027–6029.
37. J. Gyselinck, P. Dular, C. Geuzaine, and R.V. Sabariego, Surface-impedance boundary conditions in time-domain finite-element calculations using the magnetic-vector-potential formulation, *IEEE Transactions on Magnetics*, 45(3), 2009, 1280–1283.
38. S. Yuferev and L. Kettunen, Implementation of high order surface impedance boundary conditions using vector approximating functions, *IEEE Transactions on Magnetics*, 36(4), 2000, 1606–1609.

Integral Equations

39. S. Subramaniam and S.R. Hoole, The impedance boundary condition in the boundary element vector potential formulation, *IEEE Transactions on Magnetics*, 4, November 1988, 2503–2505.
40. A. Nicolas, 3D eddy current solution by BIE techniques, *IEEE Transactions on Magnetics*, 24(1), 1988, 130–133.
41. M.T. Ahmed, J.D. Lavers, and P.E. Burke, On the use of the impedance boundary condition with an indirect boundary formulation, *IEEE Transactions on Magnetics*, 24(6), 1988, 2512–2514.
42. M.T. Ahmed, J.D. Lavers, and P.E. Burke, Direct BIE formulation for skin and proximity effect calculation with and without the use of the surface impedance approximation, *IEEE Transactions on Magnetics*, 25, September 1989, 3937–3939.
43. K. Ishibashi, Eddy current analysis by boundary element method utilizing impedance boundary condition, *IEEE Transactions on Magnetics*, 31(3), 1995, 1500–1503.
44. T.E. van Deventer, P. Katehi, and A. Cangellaris, An integral equation method for the evaluation of conductor and dielectric losses in high-frequency interconnects, *IEEE Transactions on Microwave Theory and Techniques*, 37(12), 1989, 1964–1972—also applications: microstrip interconnects (Ohmic losses due to the finite conductivity of the strips constitute the prelevant loss effect at microwave and millimeter-wave frequencies).
45. H. Tsuboi, K. Sue, and K. Kunisue, Surface impedance method using boundary elements for exterior regions, *IEEE Transactions on Magnetics*, 27(5), 1991, 4118–4121.
46. A.A. Kishk and R.K. Gordon, Electromagnetic scattering from conducting bodies of revolution coated with thin magnetic materials, *IEEE Transactions on Magnetics*, 30(2), 1994, 3152–3155.
47. A.H. Chang, K.S. Yee, and J. Prodan, Comparison of different integral equation formulations for bodies of revolution with anisotropic surface impedance boundary conditions, *IEEE Transactions on Antennas and Propagation*, 40(8), 1992, 989–991 (also non-homogeneous structures).
48. A. Bendali, M'B. Fares, and J. Gay, A boundary-element solution of the Leontovich problem, *IEEE Transactions on Antennas and Propagation*, 47(10), 1597–1605, October 1999.
49. P.L. Huddleston, Scattering by finite, open cylinders using approximate boundary conditions, *IEEE Transactions on Antennas and Propagation*, 37(2), 253–257, 1989 (also applications: scattering).
50. O. Sterz and C. Schwab, A scalar boundary integrodifferential equation for eddy current problems using an impedance boundary condition, *Computing and Visualization in Science*, 3, 209–217, 2001.
51. H.C. Jayatilaka and I.R. Ciric, Performance of surface impedance integral equations for quasi-stationary field analysis in axisymmetric systems, *IEEE Transactions on Magnetics*, 40(2), 2004, 1350–1353.
52. G.W. Hanson, On the applicability of the surface impedance integral equation for optical and near infrared copper dipole antennas, *IEEE Transactions on Antennas and Propagation*, 54(12), 2006, 3677–3685.
53. S. Yuferev and V. Yuferev, Calculation of the electromagnetic field diffusion in a system of parallel conductors, *Soviet Physics Technical Physics Journal*, 36, 1991, 1208–1211.

54. S. Yuferev, Generalized BIE-BLA formulation of skin and proximity effect problems for an arbitrary regime of the magnetic field generation, *Second International IEE Conference on Computation in Electromagnetics*, IEE Conference Publication No. 384, 1994, pp. 307–310.
55. S. Yuferev and N. Ida, Application of approximate boundary conditions to electromagnetic transient scattering problems, *Third International IEE Conference on Computation in Electromagnetics*, IEE Conference Publication No. 420, 1996, pp. 51–56.
56. S. Yuferev and N. Ida, Efficient implementation of the time domain surface impedance boundary conditions for the boundary element method, *IEEE Transactions on Magnetics*, 34(2), 1998, 2763–2766.

FDTD

57. J.A. Roden and S.D. Gedney, The efficient implementation of the surface impedance boundary condition in general curvilinear coordinates, *IEEE Transactions on Microwave Theory and Techniques*, 47(10), 1999, 1954–1963.
58. J.G. Maloney and G.S. Smith, The use of surface impedance concepts in the finite-difference time-domain method, *IEEE Transactions on Antennas and Propagation*, 40(1), 1992, 38–48.
59. K.S. Yee, K. Shlager, and A.H. Chang, An algorithm to implement a surface impedance boundary condition for FDTD, *IEEE Transactions on Antennas and Propagation*, 40(7), 1992, 833–837.
60. C.F. Lee, R.T. Shin, and J.A. Kong, Time domain modeling of impedance boundary condition, *IEEE Transactions on Microwave Theory and Techniques*, 40(9), 1992, 1847–1851.
61. S. Kellahi and B. Jecko, Implementation of a surface impedance formalism at oblique incidence in FDTD method, *IEEE Transactions on Electromagnetic Compatibility*, 35(3), 1993, 347–356.
62. K.S. Oh and J.E. Schutt-Aine An efficient implementation of surface impedance boundary conditions for the finite-difference time-domain method, *IEEE Transactions on Antennas and Propagation*, 43(7), 1995, 660–666.
63. C.F. Lee, R.T. Shin, and J.A. Kong, Time domain modeling of impedance boundary condition, *IEEE Transactions on Microwave Theory and Techniques*, 40(9), 1992, 1847–1851.
64. J.H. Beggs, R.J. Lubbers, K.S. Yee, and K.S. Kunz, Finite-difference time-domain implementation of surface impedance boundary conditions, *IEEE Transactions on Antennas and Propagation*, 40(1), 1992, 49–56.
65. C.W. Penney, R.L. Luebbers, and J.W. Schuster, Scattering from coated targets using a frequency-dependent, surface impedance boundary condition in FDTD, *IEEE Transactions on Antennas and Propagation*, 44(4), 1996, 434–443 (also applications: scattering).
66. K.S. Yee and J.S. Chen, Impedance boundary condition simulation in the FDTD/FVTD hybrid, *IEEE Transactions on Antennas and Propagation*, 45(6), 1997, 921–925.
67. J.H. Beggs, A FDTD surface impedance boundary condition using Z-transform, *Applied Computational Electromagnetics Society Journal*, 13(3), 1998, 14–24.

68. M.K. Karkkainen and S.A. Tretiakov, Finite-difference time domain model of interfaces with metals and semiconductors based on a higher order surface impedance boundary conditions, *IEEE Transactions on Antennas and Propagation*, 51(9), 2003, 2448–2455.
69. M.K. Karkkainen, FDTD surface impedance model for coated conductors, *IEEE Transactions on Electromagnetic Compatibility*, 46(2), 2004, 222–233 (also non-homogeneous layered structures).
70. P.G. Petropoulos, Approximating the surface impedance of a homogeneous lossy half-space: An example of "dialable" accuracy, *IEEE Transactions on Antennas and Propagation*, 50(7), 2002, 941–943.
71. R. Makinen, An efficient surface-impedance boundary condition for thin wires of finite conductivity, *IEEE Transactions on Antennas and Propagation*, 52(12), 2004, 3364–3372.
72. Q. Zeng, A method for efficient computation of time domain surface impedances of a lossy half space, *Proceedings 20th Zurich International Symposium on Electromagnetic Compatibility*, 2009, 29–32.
73. T. Kashiwa, O. Chiba, and I. Fukai, A formulation for surface impedance boundary conditions using the finite-difference time-domain method, *Microwave Optical Technology Letter*, 5(10), 1992, 486–490.
74. T. Ohtani, K. Taguchi, T. Kashiwa, Y. Kanai, and J.B. Cole, Scattering analysis of large-scale coated cavity using the complex nonstandard FDTD method with surface impedance boundary condition, *IEEE Transactions on Magnetics*, 45(3), 2009, 1296–1299.
75. M. Feliziani, F. Maradei, and G. Tribellini, Field analysis of penetrable conductive shields by the finite-difference time-domain method with impedance network boundary conditions (INBC's), *IEEE Transactions on Electromagnetic Compatibility*, 41(4), 1999, 307–319.
76. S. Yuferev, N. Farahat, and N. Ida, Use of the perturbation technique for implementation of surface impedance boundary conditions for the FDTD method, *IEEE Transactions on Magnetics*, 36(4), 2000, 942–945.
77. N. Farahat, S. Yuferev, and N. Ida, High order surface impedance boundary conditions for the FDTD method, *IEEE Transactions on Magnetics*, 37(5), 2001, 3242–3245.

Hybrid Formulations

78. H. Wang, M. Xu, C. Wang, and T. Hubing, Impedance boundary conditions in a hybrid FEM/MOM formulation, *IEEE Transactions on Electromagnetic Compatibility*, 45(2), 198–205, 2003.

Applications

Transformers

79. C. Guerin, G. Tanneau, and G. Meunier, 3D eddy current losses calculation in transformer tanks using the finite element method, *IEEE Transactions on Magnetics*, 29(2), 1993, 1419–1422 (also FEM).

80. L. Lin and C. Xiang, Losses calculation in transformer tie plate using the finite element method, *IEEE Transactions on Magnetics*, 34(5), 1998, 3644–3647.
81. Y. Higuchi and M. Koizumi, Integral equation method with surface impedance model for 3D eddy current analysis in transformers, *IEEE Transactions on Magnetics*, 36(4), 2000, 774–779 (also Integral Equations).

Waveguides

82. J.D. Wade and R.H. Macphie, Conservation of complex power technique for waveguide junctions with finite wall conductivity, *IEEE Transactions on Microwave Theory and Techniques*, 38(4), 1990, 373–378.
83. A. Weisshaar, Impedance boundary method of moments for accurate and efficient analysis of planar graded-index optical waveguides, *Journal of Lightwave Technology*, 12(11), 1994, 1943–1951.
84. M.S. Islam, E. Tuncer, and D.P. Neikirk, Calculation of conductor loss in coplanar waveguide using conformal mapping, *Electronic Letters*, 29(13), 24 June 1993, 1189–1191.
85. K. Yoshitomi and H.R. Sharobim, Radiation from a rectangular waveguide with a lossy flange, *IEEE Transactions on Antennas and Propagation*, 42(10), 1994, 1398–1403.
86. K.A. Remley and A. Weisshaar, Impedance boundary method of moments with extended boundary conditions, *Journal of Lightwave Technology*, 13(12), 1995, 2372–2377 (digital-type pulse is considered; looks very interesting).
87. J. Tan and G. Pan, A new edge element analysis of dispersive waveguide structures, *IEEE Transactions on Microwave Theory and Techniques*, 43(11), 1995, 2600–2607 (also FEM).

Induction Heating Systems

88. S.M. Mimoune, J. Fouladgar, A. Chentouf, and G. Develey, A 3D impedance calculation for an induction heating system for materials with poor conductivity, *IEEE Transactions on Magnetics*, 32(3), 1996, 1605–1608 (also coupled electrical/thermal problems and FEM).
89. W. Mai and G. Henneberger, Field and temperature calculations in transverse flux inductive heating devices heating non-parametric materials using surface impedance formulations for non-linear eddy-current problems, *IEEE Transactions on Magnetics*, 35(3), 1999, 1590–1593 (also coupled electrical/thermal problems and FEM).
90. J. Nerg and J. Partanen, A simplified FEM based calculation model for 3-D induction heating problems using surface impedance formulations, *IEEE Transactions on Magnetics*, 37(5), 2001, 3719–3722 (also non-linear materials).
91. F. Bioul and F. Dupret, Application of asymptotic expansions to model two-dimensional induction heating systems. Part I: Calculation of electromagnetic field distribution, *IEEE Transactions on Magnetics*, 41(9), 2005, 2496–2505.
92. F. Bioul and F. Dupret, Application of asymptotic expansions to model two-dimensional induction heating systems. Part II: Calculation of equivalent surface stresses and heat flux, *IEEE Transactions on Magnetics*, 41(9), 2005, 2506–2514.

Transmission Lines

93. W. Thiel, A surface impedance approach for modeling transmission line losses in FDTD, *IEEE Microwave and Guided Wave Letters*, 10(3), 2000, 89–91.
94. S. Kim and D.P. Neikirk, Time domain multiconductor transmission line analysis using effective internal impedance, *IEEE 6th Topical Meeting on Electronic Packaging*, IEEE Cat. Number 97TH8318, San Jose, CA, October 27–29, 255–258 (EPE97_reprint.pdf).

Microstrip Structures

95. E.M. Deeley, Microstrip calculations using an improved minimum-order boundary element method, *IEEE Transactions on Magnetics*, 30(5), 1994, 3753–3756 (also BEM).
96. I.P. Theron and J.H. Cloete, On the surface impedance used to model the conductor losses of microstrip structures, *IEE Proceedings on Microwave Antennas and Propagation*, 142(1), 1995, 35–40.
97. L.L. Lee, W.G. Lyons, T.P. Orlando, S.M. Ali, and R.S. Withers, Full-wave analysis of superconducting microstrip lines on anisotropic substrates using equivalent surface impedance approach, *IEEE Transactions on Microwave Theory and Techniques*, 41(12), 1993, 2359–2367.
98. D. De Zutter and L. Knockaert, Skin effect modeling based on a differential surface admittance operator, *IEEE Transactions on Microwave Theory and Techniques*, 53(8), 2005, 2526–2538.

Electrical Machines

99. K. Adamiak, G.E. Dawson, and A.R. Eastham, Application of impedance boundary conditions in finite element analysis of linear motors, *IEEE Transactions on Magnetics*, 27(6), 1991, 5193–5195 (also FEM).
100. V.C. Silva, Y. Marechal, and A. Foggia, Surface impedance method applied to the prediction of eddy currents in hydrogenerator stator end regions, *IEEE Transactions on Magnetics*, 31(3), 1995, 2072–2075 (also FEM).
101. J.F. Eastham, D. Rodger, H.C. Lai, and H. Nouri, Finite element calculation of fields around the end region of a turbine generator test rig, *IEEE Transactions on Magnetics*, 29(2), 1993, 1415–1418.

Scattering

102. A. Sebak and L. Shafai, Scattering from arbitrarily-shaped objects with impedance boundary conditions, *IEE Proceedings-H*, 136(5), 1989, 371–376.
103. M. Osuda and A. Sebak, Scattering from lossy dielectric cylinders using a multifilament current model with impedance boundary conditions, *IEE Proceedings-H*, 139(5), 1992, 429–434.
104. T.B.A. Senior and J.L. Volakis, Generalized impedance boundary conditions in scattering, *Proceedings of the IEEE*, 79(10), 1991, 1413–1420.
105. A.A. Kishk, Electromagnetic scattering from composite objects using mixture of exact and impedance boundary conditions, *IEEE Transactions on Atennas and Propagation*, 39(6), 826–833, 1991 (also Integral Equations).

106. R.D. Graglia, P.L.E. Uslenghi, R. Vitiello, and U. D'Elia, Electromagnetic scattering for oblique incidence on impedance bodies of revolution, *IEEE Transactions on Antennas and Propagation*, 43(1), 1995, 11–26.
107. L.N. Medgyesi-Mitschang and J.M. Putnam, Integral equation formulations for imperfectly conducting scatters, *IEEE Transaction on Antennas and Propagation*, 33(2), 1985, 206–214 (also integral equations).

EMC

108. A. Darcherif, A. Raizer, J. Sakellaris, and G. Meunier, On the use of the surface impedance concept in shielded and multiconductor cable characterization by the finite element method, *IEEE Transactions on Magnetics*, 2(2), 1992, 1446–1449 (also FEM).
109. M. D'Amore and M.S. Sarto, Theoretical and experimental characterization of the EMP-interaction with composite-metallic enclosures, *IEEE Transactions on Electromagnetic Compatibility*, 42(1), 2000, 152–163—also Time-domain, FDTD (FDTD-SIBC formulation for EMC problems).
110. C. Yang and V. Jandhyala, A time-domain surface integral technique for mixed electromagnetic and circuit simulation, *IEEE Transactions on Advanced Packaging*, 28(4), 2005, 745–753.
111. F. Maradei and M. Raugi, Analysis of upsets and failures due to ESD by the FDTD-INBCs method, *IEEE Transactions on Industry Applications*, 38(4), 2002, 1009–1017.

Inverse Problems and NDT

112. Z. Badics, Y. Matsumoto, K. Aoki, and F. Nakayasu, Effective probe response calculation using impedance boundary condition in eddy current NDE problems with massive conducting regions present, *IEEE Transactions on Magnetics*, 32(3), 1996, 737–740 (also FEM).
113. H. Sahinturk, On the reconstruction of inhomogeneous surface impedance of cylindrical bodies, *IEEE Transactions on Magnetics*, 40(2), 2004, 1152–1155 (also surface impedance measurement and modeling).
114. R. Grimberg, H. Mansir, J.L. Muller, and A. Nicolas, A surface impedance boundary condition for 3-D non destructive testing modeling, *IEEE Transactions on Magnetics*, MAG-22(5), September 1986, 1272–1274.
115. T.H. Fawzi, M.T. Ahmed, and P.E. Burke, On the use of the surface impedance boundary conditions in eddy current problems, *IEEE Transactions on Magnetics*, 21(5), 1985, 1835–1840.

Plasma Devices

116. F.-Z. Louai, D. Benzerga, and M. Feliachi, A 3D finite element analysis coupled to the impedance boundary condition for the magnetodynamic problem in radiofrequency plasma devices, *IEEE Transactions on Magnetics*, 32(3), 1996, 812–815.
117. A. Helaly, E.A. Soliman, and A.A. Megahed, Electromagnetic waves scattering by non-uniform plasma cylinder, *IEE Proceedings of the Microwave Antennas and Propagation*, 144(2), 1997, 61–66.

Electromagnetic Casting

118. M.R. Ahmed and J.D. Lavers, Boundary element analysis of the electromagnet casting mould, *IEEE Transactions on Magnetics*, 25(4), 1989, 2843–2845.
119. B. Dumont and A. Gagnoud, 3D finite element method with impedance boundary condition for the modeling of molten metal shape in electromagnetic casting, *IEEE Transactions on Magnetics*, 36(4), 2000, 1329–1332.

New Developments

Nonhomogeneous Layered Structures

120. S.N. Karp and F.C. Karal, Generalized impedance boundary conditions with applications to surface wave structures, in J. Brown, Ed., *Electromagnetic Wave Theory*, Pt. 1, Pergamon, New York, 1965, pp. 479–483.
121. J.R. Wait, *Electromagnetic Waves in Stratified Media*, Pergamon, New York, 1970.
122. D.J. Hoppe and Y. Rahmat-Samii, Higher order impedance boundary conditions applied to scattering by coated bodies of revolution, *IEEE Transactions on Antennas and Propagation*, 42(12), 1994, 1600–1610.
123. T.B.A. Senior, Approximate boundary conditions, *IEEE Transactions on Antennas and Propagation*, 29, September 1981, 826–829.
124. T.B.A. Senior and J. LO. Volakis, Derivation and application of a class of generalized impedance boundary conditions, *IEEE Transactions on Antennas and Propagation*, 37, December 1989, 1566–1572.
125. D.J. Hoppe and Y. Rahmat-Samii, *Impedance Boundary Conditions in Electromagnetics*, Taylor & Francis, Washington, DC, 1994.
126. T.B.A. Senior and J. LO. Volakis, *Approximate Boundary Conditions in Electromagnetics (IEE Electromagnetic Waves Series)*, IEE, London, 1995.
127. R. Cicchetti, A class of exact and higher order surface boundary conditions for layered structures, *IEEE Transactions on Antennas and Propagation*, 44(2), 1996, 249–259.
128. R. Cicchetti and A. Faraone, Exact impedance/admittance boundary conditions for complex geometries: Theory and applications, *IEEE Transactions on Antennas and Propagation*, 48(2), 2000, 223–230.
129. O. Marceaux and B. Stupfel, High-order impedance boundary conditions for multiplayer coated 3-D objects, *IEEE Transactions on Antennas and Propagation*, 48(3), 2000, 429–436.
130. S.M. Apollonskii and V.T. Erofeenko, Impedance boundary conditions for anisotropic bodies, *IEE Proceedings on Microwave Antennas and Propagation*, 142(2), 1995, 89–92.
131. T.B.A. Senior, J.L. Volakis, and S.R. Lagault, High order impedance and absorbing boundary conditions, *IEEE Transactions on Antennas and Propagation*, 45(1), 1997, 107–114 (links between SIBCs and AbsoBC; also on application of Rytov's approach to inhomogeneous bodies).

Nonplanar Interfaces

Smooth Curved Boundaries

132. K.M. Mitzner, An integral equation approach to scattering from a body of finite conductivity, *Radio Science*, 2(12), 1967, 1459–1470.

133. H. Ammari and S. He, Effective impedance boundary conditions for an inhomogeneous thin layer on a curved metallic surface, *IEEE Transactions on Antennas and Propagation*, 46(5), 1998, 710–715 (also non-homogeneous structures).
134. W.T. Shaw and A.J. Dougan, Curvature corrected impedance boundary conditions in an arbitrary basis, *IEEE Transactions on Antennas and Propagation*, 53(5), 2005, 1699–1705.
135. B.E. Fridman, Skin effect in massive conductors used in pulsed electrical devices: I. Electromagnetic field of massive conductors, *Technical Physics*, 47(9), 2002, 1112–1119.

Rough Surfaces

136. T.B.A. Senior, Impedance boundary conditions for imperfectly conducting surfaces, *Applied Science Research B*, 8, 1960, 418–436.
137. C.L. Holloway and E.F. Kuester, Impedance-type boundary conditions for a periodic interface between a dielectric and a highly conducting medium, *IEEE Transactions on Antennas and Propagation*, 48(10), 2000, 1660–1672 (application of SI concept to 2-D periodic highly conducting rough surface with small scale roughness; new EGIBC improving Leontovich BC for this kind of problems is developed (R)).
138. C.L. Holloway and E.F. Kuester, Power loss associated with conducting and superconducting rough interfaces, *IEEE Transactions on Microwave Theory and Techniques*, 48(10), 2000, 1601–1610 (also applications: power losses in superconductors).
139. A.A. Maradudin, The impedance boundary condition for a one-dimensional, curved, metal surface, *Optics Communications*, 103(3/4), 1993, 227–234 (in this paper the SIBC for a rough conducting surface is derived as function of position; for comments see previous paper by Holloway, etc.).
140. T.T. Ong, V. Celli, and A.A. Maradudin, The impedance of a curved surface, *Optics Communications*, 95(1), 1993, 1–4.
141. J.R. Poirier, A. Bengali, and P. Borderies, Impedance boundary conditions for the scattering of time-harmonic waves by rapidly varying surfaces, *IEEE Transactions on Antennas Propagation*, 54(3), 2006, 995–1005. - *rough surfaces*.

Corners and Edges

142. S.R.H. Hoole and J. Carpenter, Surface impedance models for corners and slots, *IEEE Transactions on Magnetics*, 21(5), 1985, 1841–1843.
143. Y.-G. Park, T.-K. Chung, H.-K. Jung, and S.-Y. Hahn, Three dimensional eddy current computation using the surface impedance method considering geometric singularity, *IEEE Transactions on Magnetics*, 31(3), 1995, 1400–1403.
144. E.M. Deeley, Surface impedance near edges and corners in 3D media, *IEEE Transactions on Magnetics.*, 26(2), March 1990, 712–714.
145. E.M. Deeley, The use of modified surface impedance in a minimum-order boundary element method, *IEEE Transactions on Magnetics*, 26(5), 1990, 2762–2764.
146. J. Wang, J.D. Lavers, and P. Zhou, Modified surface impedance boundary conditions for 3D eddy current problems, *IEEE Transactions on Magnetics*, 29(2), March 1993, 1826–1829.
147. E.M. Deeley, Avoiding surface impedance modification in BE methods by singularity-free representations, *IEEE Transactions on Magnetics*, 28(5), 1992, 2814–2816.

148. E. Tuncer and D.P. Neikirk, Efficient calculation of surface impedance for rectangular conductors, *Electronic Letters*, 29(4), 25 November 1993, 2127–2128.
149. W.E. Boyse and K.D. Paulsen, Accurate solutions of Maxwell's equations around PEC corners and highly curved surfaces using nodal finite elements, *IEEE Transactions on Antennas and Propagation*, 45(12), 1997, 1758–1767.
150. E.M. Deeley, Improved impedance boundary condition for finite elements by cross-coupling at corners, *IEEE Transactions on Magnetics*, 30(5), 1994, 2881–2884.
151. K. Ishibashi, Eddy current analysis by boundary integral equation utilizing edge boundary condition, *IEEE Transactions on Magnetics*, 32, 1996, 832–835.
152. H.T. Anastassiu, D.I. Kaklamani, D.P. Economou, and O. Breinbjerg, Electromagnetic scattering analysis of coated conductors with edges using the method of auxiliary sources (MAS) in conjunction with the standard impedance boundary condition (SIBC), *IEEE Transactions on Antennas and Propagation*, 50(1), 2002, 59–66.
153. S. Yuferev, L. Proekt, and N. Ida, Surface impedance boundary conditions near corners and edges: Rigorous consideration, *IEEE Transactions on Magnetics*, 37(5), 2001, 3465–3468.
154. L. Proekt, S. Yuferev, I. Tsukerman, and N. Ida, Method of overlapping patches for electromagnetic computation near imperfectly conducting cusps and edges, *IEEE Transactions on Magnetics*, 38(2), 2002, 649–652.

Transients

155. F.M. Tesche, On the inclusion of loss in time-domain solutions of electromagnetic interaction problems, *IEEE Transactions on Electromagnetic Compatibility*, 32(1), 1990, 1–4 (also Integral equations).
156. S. Celozzi and M. Feliziani, Time domain finite element formulation of conductive regions, *IEEE Transactions on Magnetics*, 28(2), 1992, 1705–1710 (also FEM).
157. K.R. Davey and L. Turner, Prediction of transient eddy current fields using surface impedance methods, *IEEE Transactions on Magnetics*, 25(5), 1989, 4156–4158.
158. K.R. Davey and L. Turner, Transient eddy current analysis for generalized structures using surface impedances and the Fast Fourier Transform, *IEEE Transactions on Magnetics*, 26(3), 1990, 1164–1170.
159. S. Celozzi and M. Feliziani, Transient scattering problems solution by surface equivalent sources, *IEEE Transactions on Magnetics*, 30(5), 1994, 3148–3151.
160. M.S. Sarto, A new model for the FDTD analysis of the shielding performances of thin composite structures, *IEEE Transactions on Electromagnetic Compatability*, 41(4), 1999, 298–306 (also FDTD, applications: EMC).
161. D. Rodger and H.C. Lai, A surface impedance method for 3D time transient problems, *IEEE Transactions on Magnetics*, 35(3), 1999, 1369–1371 (also FEM).
162. S. Celozzi and M. Feliziani, Analysis of fast transient electromagnetic fields: A frequency dependent 2-D procedure, *IEEE Transactions on Magnetics*, 28(2), 1992, 1146–1149.
163. L. Di Rienzo, N. Ida, and S. Yuferev, Application of surface impedance concept to inverse problems of reconstructing transient currents, *IEEE Transactions on Magnetics*, 39(3), 2003, 1626–1629.
164. Y. O'Keefe, D. Thiel, and K. O'Keefe, Time-dependent surface impedance from sferics, *IEEE Geoscience Remote Sensing Letters*, 2(2), 2005, 104–107.

Nonlinear Materials

165. L. Krahenbuhl, O. Fabregue, S. Wanser, M. De Sousa Dias, and A. Nicolas, Surface impedances, BIEM and FEM coupled with 1D non linear solutions to solve 3D high frequency eddy current problems, *IEEE Transactions on Magnetics*, 33(2), 1997, 1167–1172.
166. C. Guerin, G. Meunier, and G. Tanneau, Surface impedance for 3D non-linear eddy current problems—Application to loss computation in transformers, *IEEE Transactions on Magnetics*, 32(3), 1996, 808–811 (also applications: transformers).
167. F. Azzouz and M. Feliachi, Non-linear surface impedance taking account of thermal effect, *IEEE Transactions on Magnetics*, 37(5), 2001, 3175–3177 (also coupled problems).
168. F. Bertoncini, E. Cardelli, S. Di Fraia, and B. Tellini, Evaluation of surface impedance of hysteretic materials, *IEEE Transactions on Magnetics*, 39(3), 2003, 1369–1372.
169. I. Mayergoyz, *Nonliner Diffusion of Electromagnetic Fields*, Academic Press, New York, 1998.
170. A. Kost, J.P. Bastos, K. Miethner, and L. Janicke, Improvement of nonlinear impedance boundary conditions, *IEEE Transactions on Magnetics*, 38(2), 2002, 573–576.
171. B. Paya and C. Guerin, Magnetic field dependence of nonlinear surface impedance: Which field to choose? *IEEE Transactions on Magnetics*, 38(2), 2002, 585–588.
172. P. Dular, R.V. Sabariego, and L. Krähenbuhl, Subdomain perturbation finite element method for skin and proximity, *IEEE Transactions on Magnetics*, 44(6), 2008, 738–741.
173. S. Yuferev and L. Kettunen, A new boundary element technique to transient nonlinear low penetration problems of multiconductor systems, *IEEE Transactions on Magnetics*, 34(5), 1998, 2613–2616.
174. S.Yuferev and L. Kettunen, A unified time domain surface impedance concept for both linear and non-linear skin effect problems, *IEEE Transactions on Magnetics*, 35(3), 1999, 1454–1457 (also Coupled EM/thermal problems).

Coupled EM/Thermal Problems

175. J.P. Sturgess and T.W. Preston, An economic solution for 3-D electromagnetic and thermal eddy current problems, *IEEE Transactions on Magnetics*, 28(2), 1992, 1267–1270.
176. S. Yuferev and J. Yufereva, A new boundary element formulation of coupled electromagnetic and thermal problems for non-harmonic regimes of current passage, *IEEE Transactions on Magnetics*, 32(3), 1996, 1038–1041.

Thin-Wall Conductors and Shell Elements

177. K.R. Davey, Shell impedance conditions for steady state and transient eddy current problems, *IEEE Transactions on Magnetics*, 25(4), 1989, 3007–3009 (also transients; skin effect approximation).
178. D. Zheng and K.R. Davey, A boundary element formulation for thin shell problems, *IEEE Transactions on Magnetics*, 32(3), 1996, 675–677.

179. D. Rodger, P.J. Leonard, H.C. Lai, and R.J. Hill-Cottingham, Surface elements for modeling eddy currents in high permeability materials, *IEEE Transactions on Magnetics*, 27(6), 1991, 4995–4997 (also FEM).
180. C. Guerin, G. Tanneau, G. Meunier, P. Labie, T. Ngnegueu, and M. Sacotte, A shell element for computing 3D eddy currents—Application to transformers, *IEEE Transactions on Magnetics*, 31(3), 1995, 1360–1363 (also FEM; applications: transformers).
181. M. Feliziani and F. Maradei, Finite-difference time-domain modeling of thin shields, *IEEE Transactions on Magnetics*, 36(4), 2000, 848–851.
182. L. Krahenbuhl and D. Muller, Thin layers in electrical engineering. Example of shell models in analyzing eddy-currents by boundary and finite element methods, *IEEE Transactions on Magnetics*, 29(2), 1993, 1450–1455.

Motion

183. H.-Y. Pao and J.R. Wait, Electromagnetic induction and surface impedance in a half-space from overhead moving current system, *IEEE Transactions on Antennas and Propagation*, 48(9), 2000, 1301–1305.
184. D. Rodger, E. Melgoza, and H.C. Lai, A surface impedance method for moving conductors, *IEEE Transactions on Magnetics*, 36(4), 2000, 736–740.

Surface Impedance Measurements and Modeling

185. C. Wang, J.L. Drewniak, M. Li, J. Fan, J.L. Knighten, N.W. Smith, R. Alexander, and J. Huang, FDTD modeling of skin effect, *Proceedings of 2002 3rd* International Symposium *On Electromagnetic Compatibility*, 2002, 246–249.
186. I. Akduman and A. Yapar, Surface impedance determination of a planar boundary by the use of scattering data, *IEEE Transactions on Antennas and Propagation*, 49(2), 2001, 304–307.
187. D.V. Thiel, On measuring electromagnetic surface impedance—Discussions with professor James R. Wait, *IEEE Transactions on Antennas and Propagation*, 48(10), 2000, 1517–1520.
188. S.R.H. Hoole, T.H. Walsh, and G.H. Stevens, The slot impedance: Experimental verification, *IEEE Transactions on Magnetics*, 24(6), 1988, 3156–3158.
189. D.A. James, S.G. O'Keefe, and D.V. Thiel, Eddy current modeling using the impedance method for surface impedance profiling, *IEEE Transactions on Magnetics*, 35(3), 1999, 1107–1110.
190. D.V. Thiel and R. Mittra, Surface impedance modeling using the finite difference time domain method, *IEEE Transactions on Geoscience Remote Sensing*, 35(5), 1997, 1350–1356.

Limits and Applicability

191. B. Wagner, W. Renhart, and C. Magele, Error evaluation of surface impedance boundary conditions with magnetic vector potential formulation on a cylindrical test problem, *IEEE Transactions on Magnetics*, 44(6), 2008, 734–737.

192. S.R.H. Hoole, Experimental validation of the impedance boundary condition and a review of its limitations, *IEEE Transactions on Magnetics*, 25(4), 1989, 3028–3030 (also experimental validation).
193. D.-S. Wang, Limits and validity of the impedance boundary condition on penetrable surfaces, *IEEE Transactions on Antennas and Propagation*, 35(4), 1987, 453–457.
194. N.G. Alexopoules and G.A. Tadler, Accuracy of the Leontovich boundary condition for continuous and discontinuous surface impedance, *Journal of Applied Physics*, 46, 1975, 3326–3332.
195. N. Aymard, M. Feliachi, and B. Paya, An improved modified surface impedance for transverse electric problems, *IEEE Transactions on Magnetics*, 33(2), 1997, 1267–1270 (also corners and edges).
196. J.R. Wait, Electromagnetic surface impedance of an anisotropic cylinder, *IEE Proceedings A*, 138(3), 1991, 184–186.
197. I. Hanninen and K. Nikoskinen, Implementation of method of moments for numerical analysis of corrugated surfaces with impedance boundary condition, *IEEE Transactions Antennas Propagation*, 56(1), 2008, 278–281.

GBIC

198. T.B.A. Senior and J.L.O. Volakis, *Approximate Boundary Conditions in Electromagnetics (IEE Electromagnetic Waves Series)*, IEE, London, 1995.
199. H.H. Syed and J.L. Volakis, High-frequency scattering by a smooth coated cylinder simulated with generalized impedance boundary conditions, *Radio Science*, 26, 1991, 1305–1314.
200. T.E. van Deventer and L.P.B. Katehi, Generalized boundary conditions with applications to submillimeter and optical waveguides, *Radio Science*, 31(6), 1996, 1407–1416.
201. T.B.A. Senior and J.L. Volakis, Derivation of a class of generalized boundary conditions, *IEEE Transactions on Antennas and Propagation*, AP-37, 1989, 1566–1572.
202. V. Galdi and I.M. Pinto, SDRA approach for higher-order impedance boundary conditions for complex multi-layer coatings on curved conducting bodies, *Progress in Electromagnetics Research*, PIER 24, 1999, pp. 311–335.
203. S.A. Tretyakov, Generalized impedance boundary conditions from isotropic multilayers, *Microwave Optics and Technology Letters*, 17(4), 1998, 262–265.
204. X. Antoine and H. Barucq, Approximation by generalized impedance boundary conditions of a transmission problem in acoustic scattering, *Mathematical Modeling and Numerical Analysis*, 39(5), 2005, 1041–1059.
205. H. Ammari and S. He, Effective impedance boundary conditions for an inhomogeneous thin layer on a curved metallic surface, *IEEE Transactions on Antennas and Propagation*, 46(5), 1998, 710–715.
206. T.B.A. Senior, Generalized boundary conditions for scalar fields, *Journal of the Acoustic Society of America*, 97, 1995, 3473–3477.
207. M.A. Ricoy and J.L Volakis, Generalized impedance boundary conditions for isotropic multilayers, *Radio Science*, 25(4), 1990, 391–405.

1

Classical Surface Impedance Boundary Conditions

1.1 Introduction

Although in recent years, the surface impedance boundary condition (SIBC) concept has taken increasing importance both at low and at high frequencies, its origins are rather modest and are deeply rooted in the concept of skin depth and skin effect. Schelkunoff [1] is credited with coining the term and introducing the concept in the early 1930s, by analogy with circuit theory, as the ratio of the electric and magnetic field intensities. With the rapid advance of the theory of radio waves and propagation, the concept has been further developed for the analysis of wave propagation over the earth's surface. The need for solutions for propagation of waves generated by antennas in the presence of layered, curved surfaces has led to what we refer here as the classical SIBCs. The troubled times before and during World War II seem to have been a catalyst in this development, as the need for these important solutions became more acute. The first systematic attempt to treat SIBCs was undertaken by Rytov [2,3] in 1939. His perturbation method approach allowed calculation of fields inside and outside conductors based on power series expansions, and encompassed the Leontovich condition as the first-order term in the expansion. Inclusion of higher-order terms allows treatment of curved boundaries as well as variations of the field on the surface of conductors. The method is not as well known as others, primarily because Rytov's work was published in Russian and, with the exception of one paper [4] which was translated and published in French [3], none of his work has ever been translated. The work of Leontovich and the SIBC bearing his name are much better known and have found wide acceptance in the past as well as in contemporary work. The first-order SIBC developed by Leontovich [5,6] was published in 1940 but was in existence at least since 1938 [2]. Later, a modified form, which applies to curved surfaces, has been developed [6] and has been discussed (and corrected) by Mitzner [7]. Mitzner developed a second-order SIBC using an integral equations approach solving for the scattered field due to a general scatterer. Although Mitzner's SIBC is equivalent to Rytov's second-order SIBC and can in fact be developed based on the expansion method of Rytov, Mitzner's contribution includes discussion of errors for his own and

Leontovich's conditions as well as limits of applicability and modifications for large skin depths.

The SIBCs due to Leontovich, Mitzner, and Rytov, together with the limiting case of the perfect electric conductor (PEC), form what we call the classical SIBCs and are discussed in this chapter. The purpose is to show how these SIBCs were developed, discuss the limits on their applicability and in so doing to pave the way for additional, new developments to be taken up in subsequent chapters. We start with a short discussion of the skin effect and the skin effect approximation, followed by the Leontovich SIBC. The Mitzner SIBC is introduced as an example of higher-order SIBCs and also to point out to the method by which he arrives at the SIBC. The general approach of Rytov is discussed and used to analyze the conditions of applicability for low- and high-order SIBCs. In the following chapter, we will also show how Rytov's approach can be used to obtain the Leontovich and Mitzner's SIBCs as special cases of the more general higher-order SIBCs.

1.2 Skin Effect Approximation

The basis of the surface impedance concept is intertwined with the concept of skin effect. The idea of skin effect is usually introduced in elementary electromagnetics through the concept of propagation of plane waves from free space into a good conductor, and dates back to 1886 when Lord Rayleigh introduced it [8]. This approach has the advantage of being both simple and, at the same time, quite general even if the approach to derivation hides more than it reveals as to the physical processes and approximations involved. For example, by assuming plane wave propagation, we limit ourselves to planar surfaces and it is not immediately clear how the skin depth applies to curved surfaces or to cylindrical bodies. Similarly, the skin effect is normally understood in the context of time-harmonic fields and its applicability and use in time-domain applications is not obvious. Other questions, including those related to the sources of the fields, will have to be clarified before we can use the idea of skin depth in any general sense. Nevertheless, we shall start with the classical skin depth derivation and then see how the results may be modified and what are the approximations and conditions needed for its universal applicability to derivation of SIBCs. At this point, we limit ourselves to time-harmonic fields. Treatment in the time domain will be taken up later in this book.

Consider a source-free electromagnetic wave propagating in free space, perpendicular to the surface of a conducting half-space as shown in Figure 1.1. The source-free Maxwell's equations inside the conductor may be written as

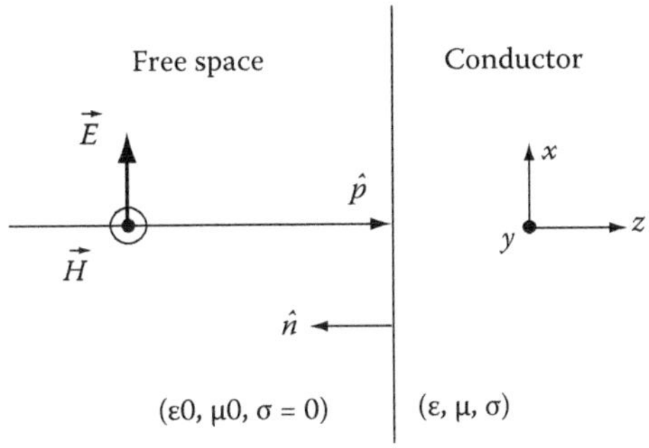

FIGURE 1.1
Conducting half-space used to define skin depth. $\hat{p}$ shows the direction of propagation of the wave.

$$\nabla \times \vec{E} = -j\omega\mu\vec{H} \tag{1.1}$$

$$\nabla \times \vec{H} = j\omega\varepsilon_c\vec{E} \tag{1.2}$$

$$\nabla \cdot \vec{D} = 0 \tag{1.3}$$

$$\nabla \cdot \vec{H} = 0 \tag{1.4}$$

where ε_c is the complex permittivity given by

$$\varepsilon_c = \varepsilon\left(1 - j\frac{\sigma}{\omega\varepsilon}\right) \tag{1.5}$$

where ε is the dielectric constant of the medium (i.e., the "static" value of permittivity).

By direct substitution, we obtain

$$\nabla^2\vec{E} - j\omega\mu(\sigma + j\omega\varepsilon)\vec{E} = 0 \tag{1.6}$$

and

$$\nabla^2\vec{H} - j\omega\mu(\sigma + j\omega\varepsilon)\vec{H} = 0 \tag{1.7}$$

These equations are general and apply equally well to nonconducting media by setting $\sigma = 0$. However, there are two related assumptions that have been used so far.

1. Equations are source free.
2. Plane wave assumption implies the source to be at infinity.

Now, assuming the conductor to be a good conductor, the loss tangent is high and we can write

$$\sigma \gg \omega\varepsilon \tag{1.8}$$

With this, we have

$$\nabla^2\vec{E} - j\omega\mu\sigma\vec{E} = 0 \tag{1.9}$$

$$\nabla^2\vec{H} - j\omega\mu\sigma\vec{H} = 0 \tag{1.10}$$

Under the specific constraints assumed here, the wave is a transverse electromagnetic (TEM) wave whereby both the electric and magnetic field intensities are perpendicular to the direction of propagation and to each other (Figure 1.1). Assuming, arbitrarily that $\vec{E}$ is in the x-direction, and propagation is in the z-direction, we can write

$$\frac{\partial^2 E_x}{\partial z^2} - j\omega\mu\sigma E_x = 0 \tag{1.11}$$

$$\frac{\partial^2 H_y}{\partial z^2} - j\omega\mu\sigma H_y = 0 \tag{1.12}$$

The solution to these equations, assuming only forward propagation, is

$$E_x = E_0 e^{-\gamma z} \tag{1.13}$$

where
E_0 is the electric field in the conductor at $z=0$ (i.e., at the surface)
γ is the propagation constant

$$\gamma = \alpha + j\beta = \sqrt{j\omega\mu\sigma} = (1+j)\sqrt{\frac{\omega\mu\sigma}{2}} \tag{1.14}$$

where
α is the attenuation constant
β is the phase constant

We now define a quantity called the skin depth as

$$\delta = \frac{1}{\alpha} = \sqrt{\frac{2}{\omega\mu\sigma}} \text{ [m]} \tag{1.15}$$

The electric field intensity inside the conductor then may be written as

$$E_x = E_0 e^{-\frac{1+j}{\delta}z} \tag{1.16}$$

We can, of course, write the solution for the magnetic field intensity as

$$H_y = H_0 e^{-\frac{1+j}{\delta}z} \tag{1.17}$$

where H_0 is the magnetic field intensity at $z=0$, since Equations 1.11 and 1.12 are identical in form. However, we can also write from Equation 1.1:

$$\frac{\partial E_x}{\partial z} = -j\omega\mu H_y \quad \rightarrow \quad H_y = \frac{j}{\omega\mu}\frac{\partial E_x}{\partial z} = -\frac{(1+j)}{\delta}\frac{j}{\omega\mu}E_0 e^{-\frac{1+j}{\delta}z} = \frac{(1-j)}{\omega\mu\delta}e^{-\frac{1+j}{\delta}z}E_0 \tag{1.18}$$

Now, from Equation 1.15, we note that $\omega = 2/\mu\sigma\delta^2$ and we have

$$\frac{\partial E_x}{\partial z} = -j\omega\mu H_y \quad \rightarrow \quad H_y = \frac{(1-j)\sigma\delta}{2}e^{-\frac{1+j}{\delta}z}E_0 \tag{1.19}$$

This, together with Equation 1.16, defines a wave impedance in the conductor as

$$\eta_c = \frac{E_x}{H_y} = \frac{E_0 e^{-\frac{1+j}{\delta}z}}{\frac{(1-j)\sigma\delta}{2}e^{-\frac{1+j}{\delta}z}E_0} = \frac{2}{(1-j)\sigma\delta} = \frac{1+j}{\sigma\delta} \quad [\Omega] \tag{1.20}$$

Also, the wave number in the conductor is

$$k_c = \frac{1+j}{\delta} \tag{1.21}$$

We shall see shortly that Equation 1.20 can be, and often is, taken as a surface impedance.

The foregoing development is simple but it also seems very restrictive in that we assumed plane waves, planar surfaces, and semi-infinite thickness. To relax these constraints and therefore to obtain a more universally applicable skin depth approximation, we proceed by assuming a curved surface, finite thickness, and a general field.

Finite thickness: Taking now a conductor of thickness t, we can argue as follows: since the analysis above assumed only a forward propagating wave in the z-direction, the equivalent requirement is that within the conductor there is no reflection from the far surface. For a semi-infinite conductor, this is always satisfied. For practical purposes, t can be finite and not necessarily large as long as it is a few skin depths thick since the field decays to $1/e$ in one skin depth and because any reflection from the opposite surface travels the same distance t to arrive back at the interface between conductor and free space. Thus t must be larger than the skin depth and some have taken this as $t > 2\delta$ [9]. A thickness of 2δ would ensure that the amplitude of the reflected wave at the interface is about 1.8% of the launched wave. A thickness of 3δ will reduce this to about 0.2%. Thus, the minimum thickness required

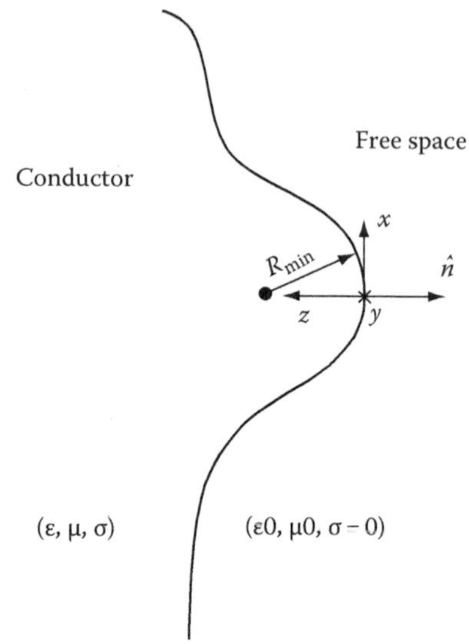

FIGURE 1.2
Local system of coordinates to allow treatment of curved surfaces.

depends on the accuracy one needs, but it is only a very few skin depths for all but the most demanding applications. We can write this as follows:

$$t_{\min} > 2 - 3\delta \tag{1.22}$$

Curved surfaces: If the smallest radius of curvature on a surface is much larger than the skin depth, then we can assume that the surface is locally flat. Assuming now a local system of coordinates as in Figure 1.2, the considerations above apply unchanged. Thus the requirement is

$$R_{\min} \gg \delta \tag{1.23}$$

General field source: Finally, we now assume that a general field impinges on the surface in Figure 1.2, i.e., on a locally flat surface. This brings us back to Equations 1.9 and 1.10. In the local coordinates, we write Equation 1.9 as

$$\nabla^2(\vec{x}\vec{E}_x + \vec{y}\vec{E}_y + \vec{z}\vec{E}_z) - j\omega\mu\sigma(\vec{x}\vec{E}_x + \vec{y}\vec{E}_y + \vec{z}\vec{E}_z) = 0 \tag{1.24}$$

First, it is important to note that the wave propagation inside the conductor is essentially perpendicular to the interface for any angle of incidence. This is best seen from Figure 1.3, where a wave impinges on the interface at angle θ_i and refracts inside the conductor at an angle θ_t. Assuming propagation from free space into a good conductor, we can write for the refraction angle [10]:

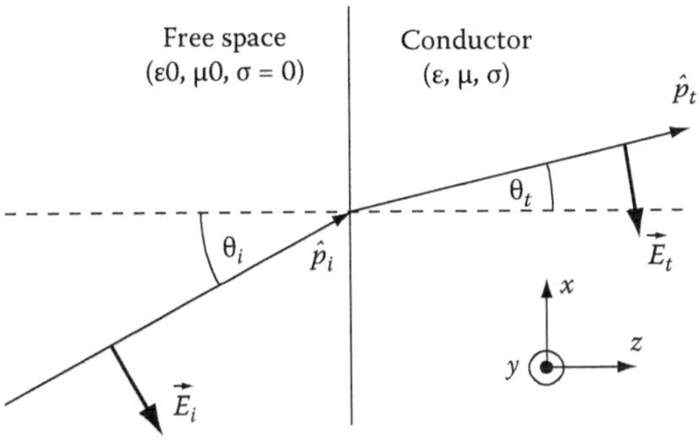

FIGURE 1.3
Incidence and refraction angles at a conducting interface.

$$\sin\theta_t = \frac{n_1}{n_2}\sin\theta_i = \frac{\gamma_1}{\gamma_2}\sin\theta_i = \frac{j\omega\sqrt{\varepsilon_0\mu_0}}{(1+j)\sqrt{\frac{\omega\mu_0\sigma}{2}}}\sin\theta_i \approx 0 \tag{1.25}$$

where
n_1 is the index of refraction in free space
n_2 is the index of refraction in the conductor
γ_1 is the propagation constant in free space
γ_2 is the propagation constant in the conductor

Clearly, Equation 1.25 applies for any good conductor at any frequency for which the condition in Equation 1.8 is satisfied.

Inside the conductor, the variation of the components of the electric and magnetic fields parallel to the surface is assumed to be small compared to the variation of the components perpendicular to the surface. That is, while the wave is quickly attenuated in directions perpendicular to the surface, there is no lateral propagation because of the result in Equation 1.25. This can be written as follows:

$$\left|\frac{\partial^2 E_x}{\partial x^2}\right| \ll \left|\frac{\partial^2 E_x}{\partial z^2}\right|; \quad \left|\frac{\partial^2 E_y}{\partial y^2}\right| \ll \left|\frac{\partial^2 E_y}{\partial z^2}\right|; \quad \left|\frac{\partial^2 E_x}{\partial x \partial y}\right| \ll \left|\frac{\partial^2 E_x}{\partial z^2}\right|; \quad \left|\frac{\partial^2 E_y}{\partial x \partial y}\right| \ll \left|\frac{\partial^2 E_y}{\partial z^2}\right| \tag{1.26}$$

The same considerations apply to the components of the magnetic field intensity:

$$\left|\frac{\partial^2 H_x}{\partial x^2}\right| \ll \left|\frac{\partial^2 H_x}{\partial z^2}\right|; \quad \left|\frac{\partial^2 H_y}{\partial y^2}\right| \ll \left|\frac{\partial^2 H_y}{\partial z^2}\right|; \quad \left|\frac{\partial^2 H_x}{\partial x \partial y}\right| \ll \left|\frac{\partial^2 H_x}{\partial z^2}\right|; \quad \left|\frac{\partial^2 H_y}{\partial x \partial y}\right| \ll \left|\frac{\partial^2 H_y}{\partial z^2}\right| \tag{1.27}$$

Applying these conditions to the general equations in Equations 1.9 and 1.10 results in equations identical to Equations 1.11 and 1.12, which then lead again to the solution in Equation 1.13 and to the skin depth.

Although we have used here the ideas of waves propagating in the conductor, Equations 1.11 and 1.12 are diffusion equations and therefore the dominant physical behavior of the fields in the conductor is one of diffusion perpendicular to the surface with high attenuation, characterized by a small value of the skin depth.

We can now summarize the conditions under which a skin layer may be assumed:

1. Loss tangent is high—conduction effects dominate (Equation 1.8).
2. Conductor is thick with respect to the skin depth (Equation 1.22).
3. Minimum radius of curvature of the conductor is much larger than the skin depth (Equation 1.23).

1.3 SIBCs of the Order of Leontovich's Approximation

Leontovich, while not the first to derive or to use the SIBC that bears his name (it is most often associated with Rytov [3] and Shchukin [11]), was the first to publish its derivation and to discuss its limits of applicability and, since these results were available early on [5,6,11], it was only natural that the SIBC he espoused should be associated with his name. The basic assumptions needed to derive the SIBC are those we assumed in Section 1.2. If the loss tangent of a medium is high (i.e., a high loss dielectric or a conductor), then the wavelength inside the medium is small so that the conditions we discussed in Section 1.2 apply. Specifically, the skin depth is small and Snell's law can be applied at the surface. In addition, the fields are required to vary slowly from point to point on the surface. Under these conditions, the field can be assumed to vary only in the direction normal to the surface of the medium. Although this implies a plane wave in the conducting medium, the wave is only approximately a plane wave in the sense that its radius of curvature is large compared to the skin depth. To summarize, the conditions under which the SIBC is derived are as follows:

1. Loss tangent of the material must be high.
2. As a consequence, the wavelength and skin depth inside the material are small.
3. Fields do not vary or vary slowly on the surface of the medium.

4. As a consequence, the surface of the material must be flat or, equivalently, its smallest radius of curvature must be large compared to the skin depth.
5. Material is sufficiently thick so that no reflected fields across the interface from within the medium exist.
6. There are no sources within the medium.

Although the original SIBC assumes a homogeneous medium, Leontovich also discuses corrections due to inhomogeneities as well as corrections due to finite curvatures. The latter will be discussed in the following section as well as in the more general derivation due to Rytov in Section 1.5.

With these conditions, the fields inside the medium are related to each other as in any plane wave (as a first approximation) and both the electric field intensity and magnetic flux density are parallel to the surface of the conductor. Assuming a system of coordinates as in Figure 1.2, the fields inside the medium are related as

$$E_x = \sqrt{\frac{\mu_c}{\varepsilon_c}} H_y; \quad E_y = -\sqrt{\frac{\mu_c}{\varepsilon_c}} E_x \tag{1.28}$$

where
ε_c is the complex permittivity
μ_c is the complex permeability of the medium

Since the electric and magnetic fields are tangential and tangential fields are continuous across the interface, the electric and magnetic fields also exist in the space outside the lossy medium. Thus, Equation I.2 constitutes the SIBC known as the Leontovich SIBC.

The condition in Equation 1.28 is often written in terms of the electric surface current density, $\vec{K}_e$, and the magnetic surface current density, $\vec{K}_m$, as

$$\vec{K}_m = Z_c \vec{n} \times \vec{K}_e \tag{1.29}$$

where

$$\vec{K}_e = -\vec{n} \times \vec{H}; \quad \vec{K}_m = \vec{n} \times \vec{E} \tag{1.30}$$

and the system of coordinates is the same as in Figure 1.2.

Leontovich [6] also analyzed the errors and conditions of applicability involved in the use of this SIBC. In Section 1.5, we discuss the limits of applicability of the method and, in the next section, also discuss the errors in the Leontovich approximation.

1.4 High-Order SIBCs

1.4.1 Mitzner's Approach

Following the introduction of the Leontovich SIBC and its successful use for a variety of applications, as well as the more general work by Rytov, it became apparent that higher-order boundary conditions are both possible and desirable. Whereas the Leontovich SIBC assumes that the minimum radius of curvature is large compared to the skin depth (Equation 1.23), a correction for smaller radii of curvatures has been implicitly introduced by Rytov [3] and later given by Leontovich [6] and reported, with corrections, by Mitzner [7]. Furthermore, the method by Rytov extends to higher-order SIBCs, as we shall see in Section 1.5. Mitzner's approach is a different way of extending the SIBC to smooth surfaces on which the curvature is not large compared to the skin depth. Unlike Rytov's method, which uses asymptotic expansions, Mitzner [7] derives the SIBC from the magnetic field integral equation (MFIE) and the electric field integral equation (EFIE) on the smooth surface of a homogeneous conductor. We shall show in Chapter 2 that Mitzner's SIBC is in fact identical to Rytov's method in the sense that the same SIBC is obtained from Rytov's method when the order of the expansion is 2. Nevertheless, Mitzner's method is important in that it uses the physical process of scattering by a homogeneous object to derive the SIBC as well as to analyze its accuracy. This method of derivation also indicates the close relationship between SIBCs and integral equation methods and that the use of SIBCs is most natural in boundary integral methods of computation. The latter idea will be used in Chapters 6 and 7 to implement SIBCs for low- and high-frequency applications.

On the other hand, Mitzner's method is ad hoc and does not lend itself to generalization to higher-order SIBCs.

Mitzner started with the general idea of integral equations for scattering by a conducting body but without assuming perfect conductivity. By writing the fields inside and outside the scatterer, he was able to derive impedance boundary conditions (first and second order) as well as to give error estimates to the use of these boundary conditions. The following is a summary of Mitzner's development of his SIBC using, for the most part, his own notation.

By first considering a general homogeneous region V (Figure 1.4), the expressions relating the fields in V and the tangential fields on the surface of the region are given by the EFIE and the MFIE:

$$\vec{E} - L^*(\vec{n} \times \vec{E}) + ZM^*(\vec{n} \times \vec{H}) = \vec{E}^{\text{inc}} \tag{1.31}$$

$$\vec{H} - L^*(\vec{n} \times \vec{H}) + (1/Z)M^*(\vec{n} \times \vec{E}) = \vec{H}^{\text{inc}} \tag{1.32}$$

where $\vec{n}$ is the unit normal vector out of V and the incident fields, $\vec{E}^{\text{inc}}$ and $\vec{H}^{\text{inc}}$, are solely due to sources in V. The operator L^* and M^* are defined as

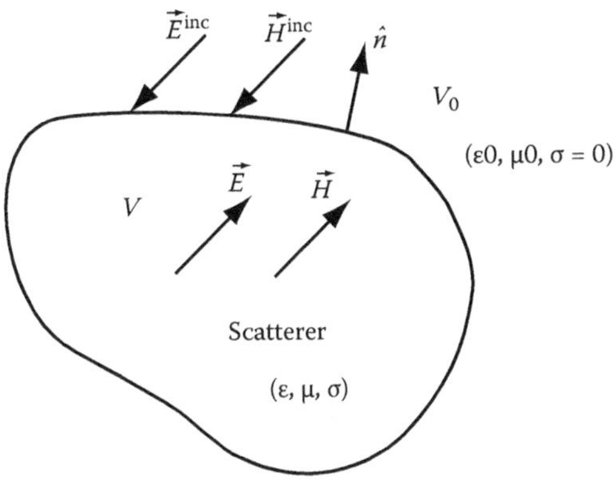

FIGURE 1.4
Scatterer in free space and the fields inside and outside the scatterer. The normal unit vector points out of V.

$$L^*\vec{F} = \int_S \nabla' G \times \vec{F}' \mathrm{d}S' \tag{1.33}$$

$$M^*\vec{F} = \frac{j}{k}\int_S (k^2 GI - \nabla\nabla' G)\cdot \vec{F}' \mathrm{d}S' \tag{1.34}$$

In these relations
$\vec{F}$ represents either $\vec{E}$ or $\vec{H}$
I is the unit dyadic
k is the wave number in V
the symbol "′" denotes the variables of integration
G is the Green function

$$G = \frac{e^{jk|\vec{r}-\vec{r}'|}}{4\pi|\vec{r}-\vec{r}'|}, \quad k = \omega\sqrt{\mu\varepsilon} \tag{1.35}$$

which may be written either inside or outside V by using the appropriate wave number (k_c or k_0).

Equations 1.33 and 1.34 are valid everywhere inside V but not on the boundary s. To obtain the SIBCs, the surface currents $\vec{K}_e$ and $\vec{K}_m$ are defined:

$$\vec{K}_e = -\vec{n}\times\vec{H}; \quad \vec{K}_m = \vec{n}\times\vec{E} \tag{1.36}$$

In Equation 1.36, the normal is assumed to point out of the scatterer V. The surface currents are related through the Leontovich SIBC [6]:

$$\vec{K}_m = Z_c \vec{n}\times\vec{K}_e \tag{1.37}$$

Z_c is the wave impedance in the scatterer:

$$Z_c = \sqrt{\mu_c/\varepsilon_c}; \quad \varepsilon_c = \varepsilon + j\sigma/\omega; \quad \mu_c = \mu + j\tau/\omega \tag{1.38}$$

The SIBC due to Mitzner is developed assuming ε and τ to be negligible (nonferrous, metallic conductors) although the SIBC obtained can be extended to more general media. To derive the condition, Equation 1.32 is first written for the free space, V_0, outside the scatterer, V, by replacing Z_c by Z_0. Next, the cross product between $\vec{n}(\vec{r}_s)$ and Equation 1.32 (where $\vec{r}_s$ is on the surface S) is written

$$\vec{n}(\vec{r}_s) \times \vec{H} - \vec{n}(\vec{r}_s) \times L^*(\vec{n} \times \vec{H}) - (1/Z_0)\vec{n}(\vec{r}_s) \times M^*(\vec{n} \times \vec{E}) = \vec{n}(\vec{r}_s) \times \vec{H}^{\mathrm{inc}} \tag{1.39}$$

Now the following identity is added to the left-hand side of the equation

$$\int_S \left[\frac{j}{k_0 Z_0} (K_m(\vec{r}_s) \cdot \nabla_s')(\vec{n}(\vec{r}_s) \cdot \nabla' G) + \frac{j}{k_0 Z_0} (\vec{n}' \cdot K_m(\vec{r}_s))(\kappa_u' + \kappa_v')(\vec{n}(\vec{r}_s) \times \nabla' G) \right] \mathrm{d}s' = 0 \tag{1.40}$$

where both $\vec{K}_m(\vec{r}_s)$ and $\vec{n}(\vec{r}_s)$ are independent of $\vec{r}$, and κ_u' and κ_v' are the principal curvatures on the surface in the u- and v-directions [12]. Taking the limit as $\vec{r} \to \vec{r}_s$, the MFIE outside the scatterer becomes

$$\frac{1}{2}\vec{K}_e - L_0\vec{K}_e - (1/Z_0)M_0\vec{K}_m = -\vec{n} \times \vec{H}^{\mathrm{inc}} \tag{1.41}$$

Similarly, starting with Equation 1.31, he assumes that the incident electric field intensity inside the scatterer is zero (a conducting scatterer) and, following identical steps, writes the EFIE inside the scatterer as

$$\frac{1}{2}\vec{K}_m + L_c\vec{K}_m - Z_c M_c \vec{K}_e = 0 \tag{1.42}$$

The operators L and M are defined through the following:

$$L\vec{K} = \int_{\bar{S}} \vec{n} \times (\nabla' G \times \vec{K}')\mathrm{d}s' = 0 \tag{1.43}$$

$$M\vec{K} = \frac{1}{jk} \int_{\bar{S}} \vec{n} \times \left[k^2 G\vec{K}' - (\kappa_u' + \kappa_v')(\nabla' G)(\vec{n}' - \vec{n}) \cdot \vec{K} + (\vec{K}' - \vec{K}) \cdot \nabla_s' \nabla' G \right] \mathrm{d}s' \tag{1.44}$$

In these relations, $\bar{S}$ denotes the surface of the scatterer excluding the singularity (punctured surface) and the Green's function is either the free space Green's function or the Green's function in the scatterer as indicated by the

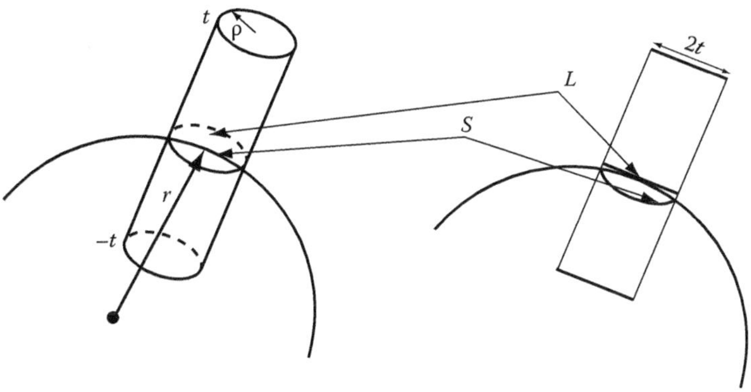

FIGURE 1.5
Local system of coordinates used to derive the SIBC.

use of k_0 or k_c for k in Equation 1.35. ∇_s' denotes the tangential gradient with respect to $\vec{r}'$.

To derive the SIBC, it is necessary here to digress and derive some auxiliary relations on the surface S of the subdomain V, which take into consideration the local geometry of S. First, a local system of coordinates centered at $\vec{r}$ is defined (Figure 1.5) as $(\vec{e}_u, \vec{e}_v, \vec{n})$ so that $(\vec{e}_u \times \vec{e}_v = \vec{n})$ and $\vec{n}$ points into V. A second, cylindrical system (ρ, θ, z), defines a cylinder $-t \leq z \leq t$, $\rho \leq t$, and t is sufficiently small so that the z-coordinate on S, within the cylinder, is single valued. Taking L to be the tangent plane to S at $\vec{r}$, and S_t a disklike subdomain cut by the right cylinder of radius t, then the projection of S_t on L is D and $\mathrm{d}D' = (\vec{n} \cdot \vec{n}')\mathrm{d}S'$. Now, writing a function g with odd periodicity, $g(\rho, \theta, -\pi) = -g(\rho, \theta)$ and a function h with even periodicity, $h(\rho, \theta, -\pi) = h(\rho, \theta)$, Mitzner defines the following integral formulas

$$\int\limits_{\bar{D}} \vec{K}' g/\rho^n \mathrm{d}D' = \lim_{\Delta \to 0} \int\limits_{\Delta}^{t} \mathrm{d}\rho \int\limits_{0}^{\pi} \left[(\vec{K}^+ - \vec{K}^-) g/\rho^{n-1} \right] \mathrm{d}\theta \tag{1.45}$$

$$\int\limits_{\bar{D}} (\vec{K}' - \vec{K}) h/\rho^n \mathrm{d}D' = \lim_{\Delta \to 0} \int\limits_{\Delta}^{t} \mathrm{d}\rho \int\limits_{0}^{\pi} \left[(\vec{K}^+ - 2\vec{K} + \vec{K}^-) g/\rho^{n-1} \right] \mathrm{d}\theta \tag{1.46}$$

where

$$\vec{K}^+ = \vec{K}[\rho, \theta, f(\rho, \theta)]; \quad \vec{K}^- = \vec{K}[\rho, \theta, -\pi, f(\rho, \theta, -\pi)] \tag{1.47}$$

and the integral on $\overline{D}$ indicates the integral over the punctured surface which excludes the singularity at $\vec{r}$.

The surface S_t is required to be sufficiently smooth everywhere so that it can be represented in terms of the principal curvatures at $\vec{r}$ (excluding error terms) as

$$z = f(\rho, \theta) = \frac{\rho^2}{2}\left(\kappa_u \cos^2\theta + \kappa_v \sin^2\theta\right) \tag{1.48}$$

With these results and the assumptions on S_t,

$$\nabla' G_c = \frac{e^{(j-1)R/\delta}(\rho/R) b_1 \vec{s}^*}{4\pi R^2} \tag{1.49}$$

For any vector $\vec{C}$:

$$\vec{n} \times (\vec{C} \cdot \nabla')\nabla' G_c = \frac{e^{(j-1)R/\delta}}{4\pi R^3}\left[b_1 \vec{C}^* \times \vec{n} - b_2(\vec{C}^* \cdot \vec{s}^*)\vec{s}^* \times \vec{n}\right] \tag{1.50}$$

In these relations, $\vec{C}^*$ is the component of $\vec{C}$ normal to $\vec{n}'$: $\vec{C}^* = \vec{n}' \times (\vec{C} \times \vec{n}')$ and

$$b_1 = 1 - (j-1)R/\delta; \quad b_2 = \left[3 - 3(j-1)R/\delta - 2j(R/\delta)^2\right](\rho/R)^2 \tag{1.51}$$

$$\vec{s}^* = \frac{\vec{r} - \vec{r}'}{\rho} = -\left[(\cos\theta \vec{e}_x + \sin\theta \vec{e}_y) + \frac{\rho}{2}(\kappa_u \cos^2\theta + \kappa_v \sin^2\theta)\vec{n}\right] \tag{1.52}$$

Further approximations are possible based on the fact that the skin depth is small, again, excluding error terms. Specifically

$$\frac{\rho}{R} \approx 1 \quad \rightarrow \quad e^{(j-1)R/\delta} = e^{(j-1)\rho/\delta} \tag{1.53}$$

$$(\vec{n} \cdot \vec{n}')^{-1} \approx 1; \quad \vec{n} - \vec{n}' = \rho(\kappa_u \cos\theta \vec{e}_x + \kappa_v \sin\theta \vec{e}_y) \tag{1.54}$$

Now, in a conducting medium, the Green's function is approximated as

$$G = \frac{e^{jk_c R}}{4\pi R^2} = \frac{e^{(j-1)R/\delta}}{4\pi R^2} \quad \rightarrow \quad |G| = \frac{e^{-R/\delta}}{4\pi R^2}$$

With this approximation, the operators L_c and M_c in Equation 1.42 are approximated by the operators L_C^t and M_C^t, in which the integration is over a subarea S_t on S as defined above. S_t has dimensions of the order of δ and, in fact, will be taken to be a smooth curved disk of radius δ.

Once these new operators are evaluated, the two approximations are written as

$$L_C \vec{K}_m \approx L_C^t \vec{K}_m; \quad M_C \vec{K}_e \approx M_C^t \vec{K}_e \tag{1.55}$$

Thus, $L_C\vec{K}_m$ is approximated by $L_C^t\vec{K}_m$ and $M_C\vec{K}_e$ by $M_C^t\vec{K}_e$. These approximations facilitate the evaluation of the various terms needed to obtain the SIBC. These operators are evaluated using the previous relations. Starting with L_C^t:

$$L_C^t\vec{K}_m = -\int_{\tilde{D}} (\vec{n}\cdot\vec{n}')^{-1}\frac{e^{(j-1)R/\delta}}{4\pi\rho^2}b_1(\rho/R)^3$$
$$\left\{(\vec{n}\cdot\vec{s}^*)\vec{K}_m - \left[(\vec{n}-\vec{n}')\vec{K}_m\right]\vec{s}^* + (\vec{n}\cdot\vec{s}^*)\left(\vec{K}_m' - \vec{K}_m\right) - \left[(\vec{n}-\vec{n}')\left(\vec{K}_m' - \vec{K}_m\right)\right]\vec{s}^*\right\}dD' \quad (1.56)$$

To evaluate the integral in Equation 1.56, the various terms are substituted from Equations 1.51 through 1.54 after removing the error terms but, in addition, instead of integrating on D, the integration is performed on the tangent plane L. With these, Equation 1.56 is approximated by

$$L_C\vec{K}_m = \int_L \frac{e^{(j-1)\rho/\delta}}{4\pi\rho^2}[1-(j-1)\rho/\delta]$$
$$\left\{\frac{1}{2}(\kappa_u\cos^2\theta + \kappa_v\sin^2\theta)\vec{K}_m - \left[\kappa_u\cos^2\theta(\vec{K}_m\cdot\vec{e}_u)\vec{e}_u + \kappa_v\sin^2\theta(\vec{K}_m\cdot\vec{e}_v)\vec{e}_v\right]\right\}dD' \quad (1.57)$$

and, from this, on the tangent plane to S

$$L_C\vec{K}_m = \frac{p}{2}\left[(\vec{K}_m\cdot\vec{e}_u)\vec{e}_u - (\vec{K}_m\cdot\vec{e}_v)\vec{e}_v\right] \quad (1.58)$$

where

$$p = \frac{1}{4}(1+j)\delta(\kappa_v - \kappa_u) \quad (1.59)$$

A similar discussion leads to approximation of $M_C\vec{K}_e$, which is

$$M_C\vec{K}_e = (1-j)\int_L \frac{e^{(j-1)\rho/\delta}}{4\pi\rho^2}(\rho/\delta)\vec{n}\times\vec{K}_e dD' = \frac{\vec{n}\times\vec{K}_e}{2} \quad (1.60)$$

Finally, by substituting Equation 1.59 into Equation 1.43, an expression for the SIBC is obtained as

$$(1+p)K_{mu}\vec{e}_u + (1-p)K_{mv}\vec{e}_v = 2Z_c M_c \overleftarrow{K}_e \quad (1.61)$$

where

$$K_{mu} = \vec{K}_{mu} \cdot \vec{e}_u$$
$$K_{mv} = \vec{K}_{mv} \cdot \vec{e}_v$$

Z_c may be written explicitly in terms of the skin depth as

$$Z_c = \frac{(1+j)}{2}\omega\mu\delta \tag{1.62}$$

With this and using Equation 1.60 to evaluate $M_C\vec{K}_e$, Equation 1.61 can be written explicitly in terms of the components of $\vec{E}$ and $\vec{H}$ in the u- and v-directions as

$$E_u = -\frac{(1+j)}{2}\omega\mu\delta\left[1 + \frac{(1-j)}{2}\delta[\kappa_v - \kappa_u]\right]H_v \tag{1.63}$$

$$E_v = \frac{(1+j)}{2}\omega\mu\delta\left[1 - \frac{(1-j)}{2}\delta[\kappa_v - \kappa_u]\right]H_u \tag{1.64}$$

The expression in Equation 1.61 is the SIBC known as the Mitzner boundary condition as published in [7]. Equations 1.63 and 1.64 are in explicit form, useful for implementation. Also, this form is identical to the curvature-dependent SIBC given by Leontovich [6].

It is also interesting to note that the curvature-dependent (first-order) interface condition due to Leontovich as well as the classic (zeroth-order)-Leontovich condition can be obtained from the second-order SIBC in Equation 1.61 by writing

$$(1+p)K_{mu}\vec{e}_u + (1-p)K_{mv}\vec{e}_v = Z_c\vec{n} \times \overleftarrow{K}_e(K_{eu}\vec{e}_v - K_{ev}\vec{e}_u) \tag{1.65}$$

Separating the components

$$(1+p)K_{mu} = -Z_cK_{ev}; \quad (1-p)K_{mv} = Z_cK_{eu} \tag{1.66}$$

Since the correction term, p, is due to the curvature (see Equation 1.59), simply omitting the term p gives the first-order SIBC due to Leontovich:

$$K_{mu} = -Z_cK_{ev}; \quad K_{mv} = Z_cK_{eu} \tag{1.67}$$

Mitzner also analyzed the errors involved in the use of the second-order SIBCs as well as of the first-order SIBC. We will discuss errors of the method in following chapters, including comparison with analytic, numerical, and experimental data, but we note here that the second-order SIBCs are of the order of $O(\delta^2 k_0^2)$, $O(\delta^2/h^2)$, and $O(\delta^2\kappa^2)$, where h is the distance to the nearest source and κ is the larger of $|\kappa_u|$ and $|\kappa_v|$. k_0 is the free space wave number. From the same analysis, he concludes that the curvature-dependent Leontovich SIBC in Equation 1.65 has errors of the order $O(\delta^2 k_0^2)$, $O(\delta^2/h^2)$, and $O(\delta^2\kappa^2)$, just like the Mitzner SIBC. The classical Leontovich SIBC in Equation 1.66 has errors of the order $O(\delta^2 k_0^2)$, $O(\delta^2/h^2)$, and $O(\delta\kappa)$.

1.5 Rytov's Approach

1.5.1 General

Rytov's approach to the idea of SIBCs was much more radical and general than his contemporaries while also preceding most of them. It all started in an attempt to verify and quantify the Leontovich SIBC at the request of Leontovich [2,3]. In the process, Rytov developed a general approach to low- and high-order SIBCs based on the perturbation method [2,3]. This approach also allowed the systematic analysis of the accuracy and applicability of the method and has pointed to the relative value of higher-order SIBCs.

Rytov started with the source-free Maxwell's equations and proceeded to calculate the electromagnetic fields inside and outside the conductor. The electric and magnetic fields were assumed to vary exponentially inside the conductor and were written as a power series expansion in the skin depth δ. The expansion coefficients were evaluated for a system of equations that is obtained by equating terms with equal powers of δ. Equating the external and internal approximations leads to the Rytov boundary conditions. The number of terms retained in the expansion defines the order of the SIBC obtained. Thus, higher-order SIBCs are as easily obtained as lower-order SIBCs. Further, since a higher-order SIBC is obtained by adding terms to the lower-order SIBC, the method is ideally suited for numerical computation since the form of the expansion allows a modular approach whereby addition of terms does not require rewriting software.

In the section that follows, we describe the method and derive the first four terms of the expansion. The zeroth-order term (first term in the expansion) will be shown to be the Leontovich SIBC while the first-order term corresponds to the Mitzner SIBC. The second- and third-order terms lead to higher-order SIBCs, which, in addition to curvature, allow for variations of fields on the surface of the conductor.

Since one of the goals of Rytov's work was the applicability and accuracy of SIBCs, the last part of this section discusses in detail the applicability of the method. The discussion focuses on skin depth as a small parameter and the errors expected are in terms of the skin depth.

1.5.2 Calculation of the Field inside the Conductor

In a conducting medium, under time-harmonic excitation and no source currents, the electric and magnetic field distribution inside the conductor can be described by the following:

$$\nabla \times \vec{E} = -j\omega\mu\vec{H}; \quad \nabla \times \vec{H} = \sigma\vec{E} \tag{1.68}$$

For simplicity, we assume that only one conductor exists and that its shape admits introduction of a system of curvilinear coordinates (ξ_1, ξ_2, η) in which the coordinates ξ_1 and ξ_2 are directed along the surface and coordinate η is normal to it pointing into the conductor.

We seek the solution of Equation 1.68 in the following form

$$\vec{E} = \vec{A}(\xi_1, \xi_2, \eta) \exp(\psi(\eta')); \quad \vec{H} = \vec{B}(\xi_1, \xi_2, \eta) \exp(\psi(\eta')) \tag{1.69}$$

where

$$\eta' = \eta/\delta \tag{1.70}$$

where δ is the skin depth characterizing the electromagnetic penetration and is written in the form

$$\delta = \sqrt{\frac{2}{\omega\sigma\mu}} \tag{1.71}$$

Equation 1.71 can be rewritten in the form:

$$\sigma = \frac{2}{\omega\mu\delta^2} \tag{1.72}$$

Substituting Equation 1.69 into Equation 1.68 and taking into account Equation 1.71, we obtain

$$\delta(\nabla \times \vec{A})_{\xi_1} - \frac{d\psi}{d\eta'} h_\eta A_{\xi_2} = -i\delta B_{\xi_1} \tag{1.73}$$

$$\delta(\nabla \times \vec{A})_{\xi_2} + \frac{d\psi}{d\eta'} h_\eta A_{\xi_1} = -i\delta B_{\xi_2} \tag{1.74}$$

$$\delta(\nabla \times \vec{A})_\eta = -i\delta B_\eta \tag{1.75}$$

$$\delta^2(\nabla \times \vec{B})_{\xi_1} - \delta \frac{d\psi}{d\eta'} h_\eta B_{\xi_2} = 2A_{\xi_1} \tag{1.76}$$

$$\delta^2(\nabla \times \vec{B})_{\xi_2} + \delta \frac{d\psi}{d\eta'} h_\eta B_{\xi_1} = 2A_{\xi_2} \tag{1.77}$$

$$\delta^2(\nabla \times \vec{B})_\eta = 2A_\eta \tag{1.78}$$

where the Lame coefficient, h_η, corresponding to coordinate η was obtained from the following well-known formula:

$$ds^2 = \left(\frac{d\xi_1}{h_{\xi_1}}\right)^2 + \left(\frac{d\xi_2}{h_{\xi_2}}\right)^2 + \left(\frac{d\eta}{h_\eta}\right)^2 \tag{1.79}$$

Representing $\vec{A}$ and $\vec{B}$ in the form of power series in δ gives

$$\vec{A} = \vec{A}_0 + \delta\vec{A}_1 + \delta^2\vec{A}_2 + \cdots; \quad \vec{B} = \vec{B}_0 + \delta\vec{B}_1 + \delta^2\vec{B}_2 + \cdots \tag{1.80}$$

The system of equations for the expansion coefficients is obtained by comparing the terms of equal powers of δ.

The zero-order approximation takes the form ($\psi' = (d\psi/d\eta')$):

$$\psi' h_\eta A_{0\xi_2} = 0; \quad -\psi' h_\eta A_{0\xi_1} = 0$$

$$2A_{0\xi_1} = 0; \quad 2A_{0\xi_2} = 0; \quad 2A_{0\xi_\eta} = 0$$

Therefore,

$$A_{0\xi_1} = A_{0\xi_2} = A_{0\eta} \tag{1.81}$$

The first-order approximation is as follows:

$$\begin{cases} \psi' h_\eta A_{0\xi_2} = iB_{0\xi_1} \\ -\psi' h_\eta A_{0\xi_1} = iB_{0\xi_2}; \\ 0 = iB_{0\eta} \end{cases} \qquad \begin{cases} 2A_{1\xi_1} = -\psi' h_\eta B_{0\xi_2} \\ 2A_{1\xi_1} = \psi' h_\eta B_{0\xi_1} \\ 2A_{1\eta} = 0 \end{cases} \tag{1.82}$$

Therefore,

$$B_{0\eta} = A_{1\eta} = 0 \tag{1.83}$$

The function ψ' can be obtained from the condition of compatibility between the equations for $A_{1\xi_m}$ and $B_{0\xi_m}$, $m = 1, 2$, and is written in the form:

$$\psi' = -\sqrt{2i/h_\eta} \tag{1.84}$$

In the last expression, the negative sign was chosen to reflect the fact that the field decays with increase in the distance from the surface. Performing the integration over the skin layer in Equation 1.84 yields

$$\psi = -\sqrt{2i}\int_0^{\eta'} \frac{d\eta'}{h_\eta} = -\frac{1+i}{\delta}\int_0^{\eta} \frac{d\eta}{h_\eta} \tag{1.85}$$

Using Equation 1.84, the relations between the tangential components of $\vec{A}_1$ and $\vec{B}_0$ can be represented in the form:

$$A_{1\xi_1} = \sqrt{\frac{i}{2}}B_{0\xi_2}; \quad A_{1\xi_2} = -\sqrt{\frac{i}{2}}B_{0\xi_1} \tag{1.86}$$

The second-order approximation looks as follows:

$$\psi' h_\eta A_{2\xi_2} = iB_{1\xi_1} - h_{\xi_2}h_\eta \frac{\partial(A_{1\xi_2}/h_{\xi_2})}{\partial\eta} \tag{1.87}$$

$$-\psi' h_\eta A_{2\xi_1} = iB_{1\xi_2} + h_{\xi_1}h_\eta \frac{\partial(A_{1\xi_1}/h_{\xi_1})}{\partial\eta} \tag{1.88}$$

$$0 = iB_{1\eta} + h_{\xi_1}h_{\xi_2}\left[\frac{\partial(A_{1\xi_2}/h_{\xi_2})}{\partial\xi_1} - \frac{\partial(A_{1\xi_1}/h_{\xi_1})}{\partial\xi_2}\right] \tag{1.89}$$

$$2A_{2\xi_1} = -\psi' h_\eta B_{1\xi_2} - h_{\xi_2}h_\eta \frac{\partial(B_{0\xi_2}/h_{\xi_2})}{\partial\eta} \tag{1.90}$$

$$2A_{2\xi_2} = \psi' h_\eta B_{0\xi_1} + h_{\xi_1}h_\eta \frac{\partial(B_{0\xi_1}/h_{\xi_1})}{\partial\eta} \tag{1.91}$$

$$2A_{2\eta} = h_{\xi_1}h_{\xi_2}\left[\frac{\partial(B_{0\xi_2}/h_{\xi_2})}{\partial\xi_1} - \frac{\partial(B_{0\xi_1}/h_{\xi_1})}{\partial\xi_2}\right] \tag{1.92}$$

Combining Equations 1.86 and 1.92, we obtain

$$B_{1\eta} = \frac{h_{\xi_1}h_{\xi_2}}{\sqrt{2i}}\left[\frac{\partial(B_{0\xi_2}/h_{\xi_2})}{\partial\xi_1} - \frac{\partial(B_{0\xi_1}/h_{\xi_1})}{\partial\xi_2}\right] \tag{1.93}$$

The system of equations in Equations 1.90 through 1.93 can be solved if the following condition is satisfied:

$$\frac{\partial}{\partial\eta}\left(\frac{B_{0\xi_1}^2}{h_{\xi_1}h_{\xi_2}}\right) = \frac{\partial}{\partial\eta}\left(\frac{B_{0\xi_2}^2}{h_{\xi_1}h_{\xi_2}}\right) = 0 \tag{1.94}$$

The pairs of Equations 1.87 and 1.91, and Equations 1.88 and 1.90 are compatible only if Equation 1.94 is satisfied and then the following solutions are obtained:

$$A_{2\xi_1} = \sqrt{\frac{i}{2}} B_{1\xi_2} - \frac{h_\eta}{4} \frac{\partial \ln (h_{\xi_1}/h_{\xi_2})}{\partial \eta} B_{0\xi_2};$$

$$A_{2\xi_2} = \sqrt{\frac{i}{2}} B_{1\xi_1} + \frac{h_\eta}{4} \frac{\partial \ln (h_{\xi_2}/h_{\xi_1})}{\partial \eta} B_{0\xi_1} \tag{1.95}$$

Substituting Equations 1.83, 1.86, and 1.95 into Equations 1.80 and then into Equation 1.69, we obtain

$$E_{\xi_1} = \mu\omega\delta \exp(\psi) \sqrt{\frac{i}{2}} \left\{ B_{0\xi_2} + \delta \left[B_{1\xi_2} - \frac{h_\eta}{2\sqrt{2i}} \frac{\partial \ln (h_{\xi_1}/h_{\xi_2})}{\partial \eta} B_{0\xi_2} \right] + \cdots \right\} \tag{1.96}$$

$$E_{\xi_2} = -\mu\omega\delta \exp(\psi) \sqrt{\frac{i}{2}} \left\{ B_{0\xi_1} + \delta \left[B_{1\xi_1} - \frac{h_\eta}{2\sqrt{2i}} \frac{\partial \ln (h_{\xi_2}/h_{\xi_1})}{\partial \eta} B_{0\xi_1} \right] + \cdots \right\} \tag{1.97}$$

$$E_\eta = \frac{\mu\omega\delta^2}{2} \exp(\psi) \left\{ h_{\xi_1} h_{\xi_2} \left[\frac{\partial (B_{0\xi_2}/h_{\xi_2})}{\partial \xi_1} - \frac{\partial (B_{0\xi_1}/h_{\xi_1})}{\partial \xi_2} \right] + \cdots \right\} \tag{1.98}$$

$$H_{\xi_1} = \exp(\psi) \{ B_{0\xi_1} + \delta B_{1\xi_1} + \cdots \} \tag{1.99}$$

$$H_{\xi_2} = \exp(\psi) \{ B_{0\xi_2} + \delta B_{1\xi_2} + \cdots \} \tag{1.100}$$

$$H_\eta = \delta \exp(\psi) \left\{ \frac{h_{\xi_1} h_{\xi_2}}{\sqrt{2i}} \left[\frac{\partial (B_{0\xi_1}/h_{\xi_2})}{\partial \xi_1} + \frac{\partial (B_{0\xi_2}/h_{\xi_1})}{\partial \xi_2} \right] + \cdots \right\} \tag{1.101}$$

Equations 1.96 through 1.101 lead to the following conclusions about the behavior of the electromagnetic field in the skin layer.

Due to the term, $\exp(\psi)$, all components of the electric and magnetic fields inside the conductor decay with increase in distance from the surface. The tangential components of the magnetic field are quantities of the zero-order magnitude and they tend to be discontinuous at the conductor surface in the limiting case of $\delta \to 0$. The normal magnetic and tangential electric fields decay as quantities of the first order of magnitude, and the normal electric field behaves as a quantity of second-order magnitude.

1.5.3 Boundary Conditions at the Conductor Surface

The expressions in Equations 1.96 through 1.101 giving the electric and magnetic fields inside the conductor must be linked to the external fields E^e and H^e via the following boundary conditions at $\eta = 0$:

$$E^e_{\xi_m} = E_{\xi_m}; \quad \mu_1 H^e_{\xi_m} = \mu H_{\xi_m}; \quad \varepsilon_1 E^e_\eta - \varepsilon E_\eta = \chi; \quad H^e_\eta = H_\eta \tag{1.102}$$

where

χ is the surface charge density

ε_1 and μ_1 are the electric permittivity and magnetic permeability of the dielectric space surrounding the conductor

We now expand the external fields and the surface charge density as power series in the small parameter δ:

$$\vec{E}^{e} = \vec{E}_0^{e} + \delta\vec{E}_1^{e} + \delta^2\vec{E}_2^{e} + \cdots; \quad \vec{H}^{e} = \vec{H}_0^{e} + \delta\vec{H}_1^{e} + \delta^2\vec{H}_2^{e} + \cdots;$$
$$\vec{H}^{e} = \vec{H}_0^{e} + \delta\vec{H}_1^{e} + \delta^2\vec{H}_2^{e} + \cdots \qquad (1.103)$$

Substituting the expansions (Equation 1.103 into Equation 1.102) and equating the coefficients of equal powers of δ, the following relations at the conductor surface ($\eta = 0$) are obtained:

Zero-order approximation:

$$E^{e}_{0\xi_m} = 0; \quad H^{e}_{0\xi_m} = 0; \quad \varepsilon_1 E^{e}_{0\eta} = \chi_0; \quad H^{e}_{0\xi_m} = B_{0\xi_m} \qquad (1.104)$$

These are conditions on the surface of an ideal conductor (PEC condition).

First-order approximation:

$$E^{e}_{1\xi_m} = (-1)^{3-m}\mu\omega\sqrt{\frac{i}{2}}B_{0\xi_{3-m}} = (-1)^{3-m}\mu\omega\sqrt{\frac{i}{2}}H^{e}_{0\xi_{3-m}}, \quad m = 1,2 \qquad (1.105)$$

$$\mu_1 H^{e}_{1\eta} = \mu\frac{h_{\xi_1}h_{\xi_2}}{\sqrt{2i}}\left[\frac{\partial(B_{0\xi_1}/h_{\xi_2})}{\partial\xi_1} + \frac{\partial(B_{0\xi_2}/h_{\xi_1})}{\partial\xi_2}\right] \qquad (1.106)$$

$$\varepsilon_1 E^{e}_{\eta 1} = \chi_1; \quad H^{e}_{1\xi_m} = B_{1\xi_m} \qquad (1.107)$$

The conditions in Equations 1.106 and 1.107 provide corrections to the field distribution around the ideal conductor due to dissipation of energy by the real conductor.

1.5.4 Particular Case of a Planar Interface

Consider the particular case of a planar conductor–dielectric interface. Let the x and y Cartesian coordinates be directed along the surface, and the z-coordinate normal to it inside the conductor. Thus $h_x = h_y = h_z = 1$ and derivations are significantly reduced so that the distribution of the electric and magnetic field components inside the conductor can be represented as functions of the external field at the interface and the z-coordinate:

$$
E_x = \mu\omega\delta \exp\left(-\frac{1+i}{\delta}\right)\sqrt{\frac{i}{2}}\left\{H_{0y}^{e} + \delta\left[H_{1y}^{e} + \frac{z}{2\sqrt{2i}}\left(\frac{\partial^2 H_{0y}^{e}}{\partial x^2} + \frac{\partial^2 H_{0y}^{e}}{\partial y^2}\right)\right]\right.
$$

$$
\left. + \delta^2\left[B_{2y} + \frac{1}{4i}\left(2\frac{\partial^2 H_{0x}^{e}}{\partial x \partial y} - \frac{\partial^2 H_{0y}^{e}}{\partial x^2} + \frac{\partial^2 H_{0y}^{e}}{\partial y^2}\right)\right] + \cdots\right\} \tag{1.108}
$$

$$
E_y = -\mu\omega\delta \exp\left(-\frac{1+i}{\delta}\right)\sqrt{\frac{i}{2}}\left\{H_{0x}^{e} + \delta\left[H_{1x}^{e} + \frac{z}{2\sqrt{2i}}\left(\frac{\partial^2 H_{0x}^{e}}{\partial x^2} + \frac{\partial^2 H_{0x}^{e}}{\partial y^2}\right)\right]\right.
$$

$$
\left. + \delta^2\left[B_{2y} + \frac{1}{4i}\left(2\frac{\partial^2 H_{0x}^{e}}{\partial x \partial y} - \frac{\partial^2 H_{0y}^{e}}{\partial x^2} + \frac{\partial^2 H_{0y}^{e}}{\partial y^2}\right)\right] + \cdots\right\} \tag{1.109}
$$

$$
E_z = \frac{\mu\omega\delta^2}{2} \exp\left(-\frac{1+i}{\delta}\right)\left\{\left(\frac{\partial H_{0y}^{e}}{\partial x} + \frac{\partial H_{0x}^{e}}{\partial y}\right) + \delta\left[\frac{\partial H_{1y}^{e}}{\partial x} - \frac{\partial H_{1x}^{e}}{\partial y}\right.\right.
$$

$$
\left.\left. + \frac{z}{2\sqrt{2i}}\left(\frac{\partial^3 H_{0y}^{e}}{\partial x^3} + \frac{\partial^2 H_{0y}^{e}}{\partial x \partial y^2} - \frac{\partial^3 H_{0x}^{e}}{\partial x^2 \partial y} - \frac{\partial^3 H_{0x}^{e}}{\partial y^3}\right)\right] + \cdots\right\} \tag{1.110}
$$

$$
H_x = \exp\left(-\frac{1+i}{\delta}\right)\left\{H_{0x}^{e} + \delta\left[H_{1x}^{e} + \frac{z}{2\sqrt{2i}}\left(\frac{\partial^2 H_{0x}^{e}}{\partial x^2} + \frac{\partial^2 H_{0x}^{e}}{\partial y^2}\right)\right] + \delta^2 B_{2x} + \cdots\right\} \tag{1.111}
$$

$$
H_y = \exp\left(-\frac{1+i}{\delta}\right)\left\{H_{0y}^{e} + \delta\left[H_{1y}^{e} + \frac{z}{2\sqrt{2i}}\left(\frac{\partial^2 H_{0y}^{e}}{\partial x^2} + \frac{\partial^2 H_{0y}^{e}}{\partial y^2}\right)\right] + \delta^2 B_{2y} + \cdots\right\} \tag{1.112}
$$

$$
H_z = \frac{\delta}{\sqrt{2i}} \exp\left(-\frac{1+i}{\delta}z\right)\left\{\left(\frac{\partial H_{0x}^{e}}{\partial x} + \frac{\partial H_{0y}^{e}}{\partial y}\right) + \delta\left[\frac{\partial H_{1x}^{e}}{\partial x} + \frac{\partial H_{1y}^{e}}{\partial y}\right.\right.
$$

$$
\left.\left. + \frac{z}{2\sqrt{2i}}\left(\frac{\partial^3 H_{0x}^{e}}{\partial x^3} + \frac{\partial^2 H_{0x}^{e}}{\partial x \partial y^2} + \frac{\partial^3 H_{0y}^{e}}{\partial x^2 \partial y} + \frac{\partial^3 H_{0y}^{e}}{\partial y^3}\right)\right] + \cdots\right\} \tag{1.113}
$$

1.5.5 Notes on Applicability of the Method

For validity of the asymptotic behavior of the solution, the first-order terms must be much smaller than the zero-order terms as follows:

$$
\left|H_{0x}^{e}\right| \gg \delta\left|H_{1x}^{e} + \frac{z}{2\sqrt{2i}}\left(\frac{\partial^2 H_{0x}^{e}}{\partial x^2} + \frac{\partial^2 H_{0x}^{e}}{\partial y^2}\right)\right| \tag{1.114}
$$

Equation 1.114 can be written in the following form without consideration of H_{1x}^{e} for now:

$$\left|H_{0x}^{e}\right| \gg \frac{z\delta}{2\sqrt{2i}}\left|\frac{\partial^2 H_{0x}^{e}}{\partial x^2} + \frac{\partial^2 H_{0x}^{e}}{\partial y^2}\right| \tag{1.115}$$

As z increases, the condition in Equation 1.115 will eventually become invalid, but the multiplier $\exp(-z/\delta)$ keeps the solution valid for $z \leq \delta$. Therefore, Equation 1.115 can be represented in the form:

$$\left\{\left|H_{0x}^{e}\right| \gg \delta^2\left|\frac{\partial^2 H_{0x}^{e}}{\partial x^2} + \frac{\partial^2 H_{0x}^{e}}{\partial y^2}\right|\right\} \tag{1.116}$$

The condition in Equation 1.115 means that the field should vary slowly along the interface. In the case of illumination of the conductor by a spherical wave, the distance, R, between the wave source and the interface must be restricted as follows

$$\left\{\begin{array}{ll} \delta^2 \ll R^2 & \text{if} \quad R \ll \lambda \\ \delta^2 \ll \lambda^2 & \text{if} \quad R \gg \lambda \end{array}\right\} \tag{1.117}$$

where λ is the incident wavelength. The second condition in Equation 1.117 can be represented in another form

$$f \ll \mu\sigma \tag{1.118}$$

where $f = \omega/2\pi$ is the source frequency. The condition in Equation 1.118 is always met for metals, but it should be checked when poor conductors are analyzed.

Consider now the first term on the right-hand side of Equation 1.114. The condition

$$\left|H_{0x}^{e}\right| \gg \delta\left|H_{1x}^{e}\right| \tag{1.119}$$

must be satisfied, if we wish to restrict ourselves to zero-order approximation. For spherical waves, Equation 1.119 is met while R is small. In other words, for any R it is possible to estimate a value for δ so that E^{e} and H^{e} will be approximated by E_{0}^{e} and H_{0}^{e} with an accuracy specified *a priori*.

In the general case of a nonplanar surface, an additional condition allowing for the curvature of the interface must be taken into account. Comparing terms of the order of magnitude of δ and δ^2 in Equations 1.96 and 1.97, it is readily found that the following condition must be met in order for terms of the order δ^2 to be neglected:

$$1 \gg \frac{\delta h_\eta}{2\sqrt{2}} \left| \frac{\partial \ln (h_{\xi_1}/h_{\xi_2})}{\partial \eta} \right| \tag{1.120}$$

The right-hand side of Equation 1.120 vanishes in the case of planar interfaces ($h_{\xi_1} = h_x = 1$; $h_{\xi_2} = h_y = 1$), but in the case of cylindrical surfaces ($h_{\xi_1} = h_z = 1$; $h_{\xi_2} = h_\phi = 1/r$; $h_\eta = h_r = 1$; $\eta = r$), Equation 1.120 takes the form:

$$r \gg \frac{\delta}{2\sqrt{2}} \tag{1.121}$$

Application of Equation 1.120 to a sphere ($h_{\xi_1} = h_\theta = 1/R$; $h_{\xi_2} = h_\phi = 1/(R \sin \theta)$; $h_\eta = h_R = 1$; $\eta = R$) yields

$$\frac{\partial \ln (h_{\xi_1}/h_{\xi_2})}{\partial \eta} = \frac{\partial \ln (\sin \theta)}{\partial R} = 0 \tag{1.122}$$

Clearly, this does not mean that the skin depth should not be small as compared to the radius in the case of a sphere. This is a common requirement for all curved surfaces, but it is formulated for each surface in a different way as follows from Equations 1.96 through 1.101.

References

1. S.A. Schelkunoff, The impedance concept and its application to problems of reflection, radiation, shielding and power absorption, *Bell System Technical Journal*, 17, 1938, 17–48.
2. S.M. Rytov, In the oscillations laboratory, in *Memoires to Academician Leontovich*, Nauka, Moscow, 1996, pp. 40–66 (in Russian).
3. S.M. Rytov, Calcul du skin-effet par la méthode des perturbations, *Journal of Physics*, 2(3), 1940, 233–242.
4. S.M. Rytov, Calculation of skin effect by perturbation method, *Journal Experimental'noi i Teoreticheskoi Fiziki*, 10(2), 1940, 180–189 (in Russian).
5. M.A. Leontovich, On one approach to a problem of the wave propagation along the Earth's surface, *Academy of Sciences USSR, Series Physics*, 8, 1944, 16–22.
6. M.A. Leontovich, On the approximate boundary conditions for the electromagnetic field on the surface of well conducting bodies, in B.A. Vvedensky (ed.), *Investigations of Radio Waves*, Academy of Sciences of USSR, Moscow, 1948.
7. K.M. Mitzner, An integral equation approach to scattering from a body of finite conductivity, *Radio Science*, 2(12), 1967, 1459–1470.
8. Lord Rayleigh, On the self inductance and resistance of straight conductors, *Philosophical Magazine*, 21, 1886, 381–394.

9. G.S. Smith, On the skin effect approximation, *American Journal of Physics*, 58(10), 1990, 996–1002.
10. C.A. Balanis, *Advanced Engineering Electromagnetics*, John Wiley & Sons, New York, 1989, pp. 210–213.
11. N. Shchukin, *Propagation of Radio Waves*, Svyazizdat, Moscow, 1940.
12. J. van Bladel, *Electromagnetic Fields, Appendix* 2, McGraw-Hill Book Co., Inc., New York, 1964.

2

General Perturbation Approach to Derivation of Surface Impedance Boundary Conditions

2.1 Introduction

In Chapter 1, we have considered three different approaches (Leontovich's, Mitzner's, and Rytov's) to the development of SIBCs and discussed the historical background. Although Rytov's approach [1] seems to be more robust than those of Mitzner [2] and Leontovich [3,4] and allows one to take into account such high-order effects as variation of the field along the conductor's surface, its application area is restricted to time-harmonic problems with single-frequency excitation. However, Rytov's conditions have been derived in Cartesian, cylindrical, and spherical coordinate systems and therefore, unlike Mitzner's SIBCs, they cannot be directly applied to arbitrarily curved interfaces with variable radii of curvature. Therefore, none of these approaches can be considered a "universal tool" for the development of approximate boundary conditions for skin effect problems.

A general approach should have only those limitations that are common to a wide range of skin effect problems. One such basic limitation is the requirement that the skin depth, δ, be much smaller than the characteristic size, D, of the conductor's surface

$$\delta = \sqrt{\tau/\sigma\mu} \ll D \tag{2.1}$$

where τ is the ratio $2/\omega$ in the case of time-harmonic fields or the incident pulse duration in the case of transient sources.

The condition in Equation 2.1 is related to the well-known fact that the normal magnetic field and tangential electric field in the skin layer of a conductor is much smaller than the tangential magnetic field. In other words, Maxwell's equations for the conducting domain combine terms of different orders of magnitude. In the limiting case of a perfect electrical conductor the skin depth is zero and the normal magnetic and tangential electric fields vanish at the conductor/dielectric interface. It is natural to assume that the equations governing the field distributions in the skin layer "implicitly" include a small parameter proportional to the skin depth. Then the approximate relation between the tangential electric and magnetic fields at the interface can be sought in the form of asymptotic expansions in this small parameter. The question is how to choose the small parameter? Rytov demonstrated that the choice does not look difficult in the particular case of time-harmonic excitation. Consideration of the general case requires a definition suitable for any time dependence of the incident source, even in nonlinear cases, when the one-dimensional problem of the field diffusion into the conducting half-space does not have an analytical solution. It is perhaps better to reformulate the question as follows: Do we need to define the parameter ourselves, in some way, or could it be "extracted" from the governing equations? That is, can the equations be transformed to reveal this parameter? Perturbation theory leads us to favor this possibility. Following this theory, we first introduce the scale factors for the basic variables in the governing equations. Selection of the scale factors is based on knowledge of the characteristic variation of such input data as total current and dimensions of the conductor. Then the equations are rewritten in terms of nondimensional variables so that each is a function of the corresponding dimensional variable and its scale factor. As a result, combination(s) of the scale factors will appear in the governing equations as parameters. Unlike nondimensional variables that are of the same order of magnitude under the definition, the scale factors (as well as the original dimensional variables) are of different orders of magnitude so that one of the combinations of the scale factors is expected to be the desired small parameter of the problem.

The distribution of various electromagnetic quantities in the conducting domain can be described by one specific equation—the equation of diffusion.

$$\nabla \times (\nabla \times \vec{f}) + \sigma\mu \frac{\partial \vec{f}}{\partial t} = 0 \tag{2.2}$$

where the vector function $\vec{f}$ may denote various electromagnetic fields and potentials. The purpose of this chapter is transformation of this equation with the use of perturbation techniques and derivation of a set of approximate relationships between components and derivatives of $\vec{f}$ at the interface. Those results can be applied not only to electromagnetics, but also to any other area where functions under consideration are governed by the diffusion equation.

2.2 Local Coordinates

For simplicity, we first consider the planar case of long parallel conductors. Let us direct the x-coordinate along the conductors and neglect all derivatives with respect to x:

$$\frac{\partial}{\partial x} = 0$$

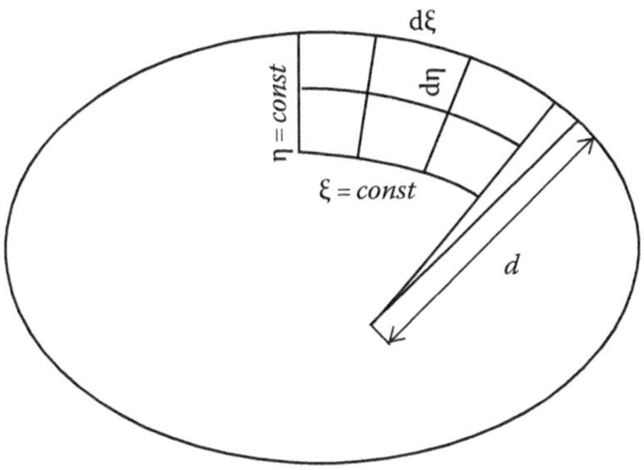

FIGURE 2.1
Local orthogonal curvilinear coordinate system related to the surface. (Adapted from Mestel, A.J., *IMA J. Appl. Math.*, 45, 33, 1990.)

Local to the surface, we define inward normal and tangential orthogonal coordinates, η and ξ, as in Figure 2.1. From Figure 2.1 it is clear that, when the curvature is positive, the scale factor corresponding to ξ in an orthogonal system h_ξ is a decreasing function of η. In fact, when ξ represents the arc length and the contour of the conductor's surface, the Lame coefficients can be written in the form

$$h_\xi = 1 - \eta/d; \quad h_\eta = 1; \quad h_x = 1 \tag{2.3}$$

where $d = d(\xi)$ is the local radius of curvature.

With these new coordinates, the curl, div, and curl curl operators applied to an arbitrary vector function $\vec{f}$ take the form

$$(\nabla\times\vec{f})_x = \frac{1}{h_\xi h_\eta}\left[\frac{\partial(h_\xi f_\xi)}{\partial\eta} - \frac{\partial(h_\eta f_\eta)}{\partial\xi}\right] = \frac{1}{1-\eta/d}\left[\frac{\partial((1-\eta/d)f_\xi)}{\partial\eta} - \frac{\partial f_\eta}{\partial\xi}\right]$$
$$= -\frac{1}{1-\eta/d}\left[(1-\eta/d)\frac{\partial f_\xi}{\partial\eta} - \frac{f_\xi}{d} - \frac{\partial f_\eta}{\partial\xi}\right] = -\left[\frac{\partial f_\xi}{\partial\eta} - \frac{f_\xi}{d-\eta} - \frac{d}{d-\eta}\frac{\partial f_\eta}{\partial\xi}\right] \tag{2.4}$$

$$(\nabla\times\vec{f})_\xi = \frac{1}{h_x h_\eta}\left[\frac{\partial(h_x f_x)}{\partial\eta} - \frac{\partial(h_\eta f_\eta)}{\partial x}\right] = \frac{\partial f_x}{\partial\eta} \tag{2.5}$$

$$(\nabla\times\vec{f})_\eta = \frac{1}{h_x h_\xi}\left[\frac{\partial(h_\xi f_\xi)}{\partial x} - \frac{\partial(h_x f_x)}{\partial\xi}\right] = -\frac{d}{d-\eta}\frac{\partial f_x}{\partial\xi} \tag{2.6}$$

$$\nabla\cdot\vec{f} = \frac{1}{h_x h_\xi h_\eta}\left[\frac{\partial(h_\xi h_\eta f_x)}{\partial x} + \frac{\partial(h_x h_\eta f_\xi)}{\partial\xi} + \frac{\partial(h_x h_\xi f_\eta)}{\partial\eta}\right]$$
$$= \frac{1}{1-\eta/d}\left[\frac{\partial f_\xi}{\partial\xi} + (1-\eta/d)\frac{\partial f_\eta}{\partial\eta} - \frac{f_\eta}{d}\right] = \frac{d}{d-\eta}\frac{\partial f_\xi}{\partial\xi} + \frac{\partial f_\eta}{\partial\eta} - \frac{f_\eta}{d-\eta} \tag{2.7}$$

$$\begin{aligned}
\left[\nabla\times(\nabla\times\vec{f})\right]_x &= \frac{1}{h_\xi h_\eta}\left\{\frac{\partial\xi}{\partial}\left[\frac{h_\eta}{h_x h_\xi}\left(\frac{\partial(h_\xi f_\xi)}{\partial x}-\frac{\partial(h_x f_x)}{\partial\xi}\right)\right]-\frac{\partial}{\partial\eta}\left[\frac{h_\xi}{h_x h_\eta}\left(\frac{\partial(h_x f_x)}{\partial\eta}-\frac{\partial(h_\eta f_\eta)}{\partial x}\right)\right]\right\}\\
&=\frac{1}{h_\xi}\left\{-\frac{\partial}{\partial\xi}\left[\frac{1}{h_\xi}\frac{\partial f_x}{\partial\xi}\right]-\frac{\partial}{\partial\eta}\left[h_\xi\frac{\partial f_x}{\partial\eta}\right]\right\}\\
&=\frac{d}{d-\eta}\left\{\frac{\partial}{\partial\xi}\left(\frac{d}{d-\eta}\right)\frac{\partial f_x}{\partial\xi}-\frac{d}{d-\eta}\frac{\partial^2 f_x}{\partial\xi^2}-\frac{\partial}{\partial\eta}\left(\frac{d-\eta}{d}\right)\frac{\partial f_x}{\partial\eta}-\frac{d-\eta}{d}\frac{\partial^2 f_x}{\partial\eta^2}\right\}\\
&=\frac{d}{d-\eta}\left\{-\frac{\eta d'}{(d-\eta)^2}\frac{\partial f_x}{\partial\xi}-\frac{d}{d-\eta}\frac{\partial^2 f_x}{\partial\xi^2}+\frac{1}{d}\frac{\partial f_x}{\partial\eta}-\frac{d-\eta}{d}\frac{\partial^2 f_x}{\partial\eta^2}\right\}\\
&=-\frac{\eta d d'}{(d-\eta)^3}\frac{\partial f_x}{\partial\xi}-\frac{d^2}{(d-\eta)^2}\frac{\partial^2 f_x}{\partial\xi^2}+\frac{1}{d-\eta}\frac{\partial f_x}{\partial\eta}-\frac{\partial^2 f_x}{\partial\eta^2}
\end{aligned}\tag{2.8}$$

$$\begin{aligned}
\left[\nabla\times(\nabla\times\vec{f})\right]_\xi &= -\frac{1}{h_x h_\eta}\left\{\frac{\partial}{\partial x}\left[\frac{h_\eta}{h_x h_\xi}\left(\frac{\partial(h_\xi f_\xi)}{\partial x}-\frac{\partial(h_x f_x)}{\partial\xi}\right)\right]-\frac{\partial}{\partial\eta}\left[\frac{h_x}{h_\xi h_\eta}\left(\frac{\partial(h_\eta f_\eta)}{\partial\xi}-\frac{\partial(h_\xi f_\xi)}{\partial\eta}\right)\right]\right\}\\
&=\frac{\partial}{\partial\eta}\left[\frac{1}{h_\xi}\left(\frac{\partial f_\eta}{\partial\xi}-h_\xi\frac{\partial f_\xi}{\partial\eta}-f_\xi\frac{\partial h_\xi}{\partial\eta}\right)\right]=\frac{\partial}{\partial\eta}\left[\frac{1}{h_\xi}\frac{\partial f_\eta}{\partial\xi}-\frac{\partial f_\xi}{\partial\eta}-\frac{f_\xi}{h_\xi}\frac{\partial h_\xi}{\partial\eta}\right]\\
&=\frac{\partial}{\partial\eta}\left[\frac{d}{d-\eta}\frac{\partial f_\eta}{\partial\xi}-\frac{\partial f_\xi}{\partial\eta}+\frac{f_\xi}{d-\eta}\right]\\
&=\frac{d}{(d-\eta)^2}\frac{\partial f_\eta}{\partial\xi}+\frac{d}{d-\eta}\frac{\partial^2 f_\eta}{\partial\eta\partial\xi}-\frac{\partial^2 f_\xi}{\partial\eta^2}+\frac{f_\xi}{(d-\eta)^2}+\frac{1}{d-\eta}\frac{\partial f_\xi}{\partial\eta}
\end{aligned}\tag{2.9}$$

$$\begin{aligned}
\left[\nabla\times(\nabla\times\vec{f})\right]_\eta &= \frac{1}{h_x h_\xi}\left\{\frac{\partial}{\partial x}\left[\frac{h_\xi}{h_x h_\eta}\left(\frac{\partial(h_x f_x)}{\partial\eta}-\frac{\partial(h_\eta f_\eta)}{\partial x}\right)\right]-\frac{\partial}{\partial\xi}\left[\frac{h_x}{h_\xi h_\eta}\left(\frac{\partial(h_\eta f_\eta)}{\partial\xi}-\frac{\partial(h_\xi f_\xi)}{\partial\eta}\right)\right]\right\}\\
&=-\frac{1}{h_\xi}\frac{\partial}{\partial\xi}\left[\frac{1}{h_\xi}\left(\frac{\partial f_\eta}{\partial\xi}-f_\xi\frac{\partial h_\xi}{\partial\eta}-h_\xi\frac{\partial f_\xi}{\partial\eta}\right)\right]=-\frac{1}{h_\xi}\frac{\partial}{\partial\xi}\left[\frac{1}{h_\xi}\frac{\partial f_\eta}{\partial\xi}-\frac{f_\xi}{h_\xi}\frac{\partial h_\xi}{\partial\eta}-\frac{\partial f_\xi}{\partial\eta}\right]\\
&=-\frac{d}{d-\eta}\frac{\partial}{\partial\xi}\left[\frac{d}{d-\eta}\frac{\partial f_\eta}{\partial\xi}+\frac{f_\xi}{d-\eta}-\frac{\partial f_\xi}{\partial\eta}\right]\\
&=\frac{\eta d d'}{(d-\eta)^3}\frac{\partial f_\eta}{\partial\xi}-\frac{d^2}{(d-\eta)^2}\frac{\partial^2 f_\eta}{\partial\xi^2}+\frac{d d'}{(d-\eta)^3}f_\xi-\frac{d}{(d-\eta)^2}\frac{\partial f_\xi}{\partial\xi}+\frac{d}{d-\eta}\frac{\partial^2 f_\xi}{\partial\xi\partial\eta}
\end{aligned}\tag{2.10}$$

where d' denotes the derivative of d with respect to ξ.

The characteristic lengths associated with the coordinates ξ and η are the characteristic size D of the conductor's surface and the skin depth δ, respectively. These lengths may be of different orders of magnitude due to the condition in Equation 2.1. Thus, it is natural to introduce nondimensional variables, $\tilde{\xi}$ and $\tilde{\eta}$, that have variation ranges of the same order of magnitude and are related to ξ and η, respectively, as follows:

$$\tilde{\xi}=\xi/D;\quad \tilde{\eta}=\eta/\delta \tag{2.11}$$

Here and below, the sign "∼" denotes nondimensional quantities. Switching to variables $\tilde{\xi}$ and $\tilde{\eta}$ in Equations 2.4 through 2.10, we obtain

$$(\nabla\times\vec{f})_x = -\left[\delta^{-1}\frac{\partial f_\xi}{\partial\tilde{\eta}} - \frac{f_\xi}{d-\delta\tilde{\eta}} - \frac{d}{d-\delta\tilde{\eta}}D^{-1}\frac{\partial f_\eta}{\partial\tilde{\xi}}\right] \tag{2.12}$$

$$(\nabla\times\vec{f})_\xi = \delta^{-1}\frac{\partial f_x}{\partial\tilde{\eta}} \tag{2.13}$$

$$(\nabla\times\vec{f})_\eta = -\frac{d}{d-\delta\tilde{\eta}}D^{-1}\frac{\partial f_x}{\partial\tilde{\xi}} \tag{2.14}$$

$$\nabla\cdot\vec{f} = \frac{d}{d-\delta\tilde{\eta}}D^{-1}\frac{\partial f_\xi}{\partial\tilde{\xi}} + \delta^{-1}\frac{\partial f_\eta}{\partial\tilde{\eta}} - \frac{f_\eta}{d-\delta\tilde{\eta}} \tag{2.15}$$

$$\begin{aligned}\left[\nabla\times(\nabla\times\vec{f})\right]_x = {} & -\delta\frac{\tilde{\eta}dd'}{(d-\delta\tilde{\eta})^3}D^{-1}\frac{\partial f_x}{\partial\tilde{\xi}} - \frac{d^2}{(d-\delta\tilde{\eta})^2}D^{-2}\frac{\partial^2 f_x}{\partial\tilde{\xi}^2}\\ & + \frac{1}{d-\delta\tilde{\eta}}\delta^{-1}\frac{\partial f_x}{\partial\tilde{\eta}} - \delta^{-2}\frac{\partial^2 f_x}{\partial\tilde{\eta}^2}\end{aligned} \tag{2.16}$$

$$\begin{aligned}\left[\nabla\times(\nabla\times\vec{f})\right]_\xi = {} & \frac{d}{(d-\delta\tilde{\eta})^2}D^{-1}\frac{\partial f_\eta}{\partial\tilde{\xi}} + \frac{d}{d-\delta\tilde{\eta}}\delta^{-1}D^{-1}\frac{\partial^2 f_\eta}{\partial\tilde{\eta}\partial\tilde{\xi}} - \delta^{-2}\frac{\partial^2 f_\xi}{\partial\tilde{\eta}^2}\\ & + \frac{f_\xi}{(d-\delta\tilde{\eta})^2} + \frac{1}{d-\delta\tilde{\eta}}\delta^{-1}\frac{\partial f_\xi}{\partial\tilde{\eta}}\end{aligned} \tag{2.17}$$

$$\begin{aligned}\left[\nabla\times(\nabla\times\vec{f})\right]_\eta = {} & \delta\frac{\tilde{\eta}dd'}{(d-\delta\tilde{\eta})^3}D^{-1}\frac{\partial f_\eta}{\partial\tilde{\xi}} - \frac{d^2}{(d-\delta\tilde{\eta})^2}D^{-2}\frac{\partial^2 f_\eta}{\partial\tilde{\xi}^2} + \frac{dd'}{(d-\delta\tilde{\eta})^3}f_\xi\\ & - \frac{d}{(d-\delta\tilde{\eta})^2}D^{-1}\frac{\partial f_\xi}{\partial\tilde{\xi}} + \frac{d}{d-\eta}D^{-1}\delta^{-1}\frac{\partial^2 f_\xi}{\partial\tilde{\xi}\partial\tilde{\eta}}\end{aligned} \tag{2.18}$$

Using Equation 2.1, let us introduce a small parameter $\tilde{p}$ proportional to the ratio of the penetration depth to the characteristic size of the conductor's surface:

$$\tilde{p} = \delta/D = \sqrt{\tau/(\sigma\mu D^2)} \ll 1 \tag{2.19}$$

The second relation in Equation 2.11 can be represented in another form using Equation 2.19:

$$\tilde{\eta} = \eta/(\tilde{p}D) \tag{2.20}$$

The local radius of curvature, d, is directly related to the variation of the function, $\vec{f}$, in the direction tangential to the surface of the conductor. This leads to the following representation:

$$\tilde{d} = d/D \tag{2.21}$$

Substituting Equations 2.20 and 2.21 into Equations 2.12 through 2.18, we get

$$(\nabla \times \vec{f})_x = -\delta^{-1}\left[\frac{\partial f_\xi}{\partial \tilde{\eta}} - \tilde{p}\frac{f_\xi}{\tilde{d}-\tilde{p}\tilde{\eta}} - \tilde{p}\frac{\tilde{d}}{\tilde{d}-\tilde{p}\tilde{\eta}}\frac{\partial f_\eta}{\partial \tilde{\xi}}\right] \tag{2.22}$$

$$(\nabla \times \vec{f})_\xi = \delta^{-1}\frac{\partial f_x}{\partial \tilde{\eta}} \tag{2.23}$$

$$(\nabla \times \vec{f})_\eta = -\delta^{-1}\tilde{p}\frac{\tilde{d}}{\tilde{d}-\tilde{p}\tilde{\eta}}\frac{\partial f_x}{\partial \tilde{\xi}} \tag{2.24}$$

$$\nabla \cdot \vec{f} = \delta^{-1}\left(\tilde{p}\frac{\tilde{d}}{\tilde{d}-\tilde{p}\tilde{\eta}}\frac{\partial f_\xi}{\partial \tilde{\xi}} + \frac{\partial f_\eta}{\partial \tilde{\eta}} - \tilde{p}\frac{f_\eta}{\tilde{d}-\tilde{p}\tilde{\eta}}\right) \tag{2.25}$$

$$\left[\nabla \times (\nabla \times \vec{f})\right]_x = \delta^{-2}\left(-\tilde{p}^3\frac{\tilde{\eta}\tilde{d}\tilde{d}'}{(\tilde{d}-\tilde{p}\tilde{\eta})^3}\frac{\partial f_x}{\partial \tilde{\xi}} - \tilde{p}^2\frac{\tilde{d}^2}{(\tilde{d}-\tilde{p}\tilde{\eta})^2}\frac{\partial^2 f_x}{\partial \tilde{\xi}^2} + \frac{\tilde{p}}{\tilde{d}-\tilde{p}\tilde{\eta}}\frac{\partial f_x}{\partial \tilde{\eta}} - \frac{\partial^2 f_x}{\partial \tilde{\eta}^2}\right) \tag{2.26}$$

$$\left[\nabla \times (\nabla \times \vec{f})\right]_\xi = \delta^{-2}\left(\tilde{p}^2\frac{\tilde{d}}{(\tilde{d}-\tilde{p}\tilde{\eta})^2}\frac{\partial f_\eta}{\partial \tilde{\xi}} + \tilde{p}\frac{\tilde{d}}{\tilde{d}-\tilde{p}\tilde{\eta}}\frac{\partial^2 f_\eta}{\partial \tilde{\eta}\partial \tilde{\xi}} - \frac{\partial^2 f_\xi}{\partial \tilde{\eta}^2} + \tilde{p}^2\frac{f_\xi}{(\tilde{d}-\tilde{p}\tilde{\eta})^2} + \frac{\tilde{p}}{\tilde{d}-\tilde{p}\tilde{\eta}}\frac{\partial f_\xi}{\partial \tilde{\eta}}\right) \tag{2.27}$$

$$\begin{aligned}\left[\nabla \times (\nabla \times \vec{f})\right]_\eta = \delta^{-2}\Bigg(&\tilde{p}^3\frac{\tilde{\eta}\tilde{d}\tilde{d}'}{(\tilde{d}-\tilde{p}\tilde{\eta})^3}\frac{\partial f_\eta}{\partial \tilde{\xi}} - \tilde{p}^2\frac{\tilde{d}^2}{(\tilde{d}-\tilde{p}\tilde{\eta})^2}\frac{\partial^2 f_\eta}{\partial \tilde{\xi}^2} + \tilde{p}^2\frac{\tilde{d}\tilde{d}'}{(\tilde{d}-\tilde{p}\tilde{\eta})^3}f_\xi \\ &-\tilde{p}^2\frac{\tilde{d}}{(\tilde{d}-\tilde{p}\tilde{\eta})^2}\frac{\partial f_\xi}{\partial \tilde{\xi}} + \tilde{p}\frac{\tilde{d}}{\tilde{d}-\tilde{p}\tilde{\eta}}\frac{\partial^2 f_\xi}{\partial \tilde{\xi}\partial \tilde{\eta}}\Bigg)\end{aligned} \tag{2.28}$$

In the general three-dimensional case, both tangential coordinates are curved. We denote them as ξ_1 and ξ_2. Their radii of curvature are d_1 and d_2, respectively, and the Lame coefficients can be written in the form:

$$h_{\xi_k} = 1 - \eta/d_k; \quad h_\eta = 1, \quad k = 1,2 \tag{2.29}$$

The curl, div, and curl curl operators with coordinates (ξ_1, ξ_2, η) are written in the following forms:

$$\begin{aligned}(\nabla \times \vec{f})_{\xi_k} &= (-1)^k\frac{1}{h_{\xi_{3-k}}h_\eta}\left[\frac{\partial(h_{\xi_{3-k}}f_{\xi_{3-k}})}{\partial \eta} - \frac{\partial(h_\eta f_\eta)}{\partial \xi_{3-k}}\right] \\ &= -\frac{1}{1-\eta/d_{3-k}}\left[\frac{\partial\left((1-\eta/d_{3-k})f_{\xi_{3-k}}\right)}{\partial \eta} - \frac{\partial f_\eta}{\partial \xi_{3-k}}\right] \\ &= (-1)^k\frac{1}{1-\eta/d_{3-k}}\left[(1-\eta/d_{3-k})\frac{\partial f_{\xi_{3-k}}}{\partial \eta} - \frac{f_{\xi_{3-k}}}{d} - \frac{\partial f_\eta}{\partial \xi_{3-k}}\right] \\ &= (-1)^k\left[\frac{\partial f_{\xi_{3-k}}}{\partial \eta} - \frac{f_{\xi_{3-k}}}{d_{3-k}-\eta} - \frac{d_{3-k}}{d_{3-k}-\eta}\frac{\partial f_\eta}{\partial \xi_{3-k}}\right], \quad k = 1,2\end{aligned} \tag{2.30}$$

$$(\nabla\times\vec{f})_\eta = \frac{1}{h_{\xi_1}h_{\xi_2}}\left[\frac{\partial(h_{\xi_2}f_{\xi_2})}{\partial\xi_1}-\frac{\partial(h_{\xi_1}f_{\xi_1})}{\partial\xi_2}\right]=\sum_{i=1}^{2}(-1)^i\frac{d_{3-i}}{d_{3-i}-\eta}\frac{\partial f_{\xi_i}}{\partial\xi_{3-i}} \tag{2.31}$$

$$\nabla\cdot\vec{f}=\frac{1}{h_{\xi_1}h_{\xi_2}h_\eta}\left[\sum_{i=1}^{2}\frac{\partial(h_{\xi_i}h_\eta f_{\xi_{3-i}})}{\partial\xi_{3-i}}+\frac{\partial(h_{\xi_1}h_{\xi_2}f_\eta)}{\partial\eta}\right]=\sum_{i=1}^{2}\frac{d_i}{d_i-\eta}\frac{\partial f_{\xi_i}}{\partial\xi_i}+\frac{\partial f_\eta}{\partial\eta}-f_\eta\sum_{l=1}^{2}\frac{1}{d_l-\eta} \tag{2.32}$$

$$\begin{aligned}
&\left[\nabla\times(\nabla\times\vec{f})\right]_{\xi_k}\\
&=\frac{1}{h_{\xi_{3-k}}h_\eta}\left\{\frac{\partial}{\partial\xi_{3-k}}\left[\frac{h_\eta}{h_{\xi_k}h_{\xi_{3-k}}}\left(\frac{\partial(h_{\xi_{3-k}}f_{\xi_{3-k}})}{\partial\xi_k}-\frac{\partial(h_{\xi_k}f_{\xi_k})}{\partial\xi_{3-k}}\right)\right]-\frac{\partial}{\partial\eta}\left[\frac{h_{\xi_{3-k}}}{h_{\xi_k}h_\eta}\left(\frac{\partial(h_{\xi_k}f_{\xi_k})}{\partial\eta}-\frac{\partial(h_\eta f_\eta)}{\partial\xi_k}\right)\right]\right\}\\
&=\frac{1}{h_{\xi_{3-k}}}\left\{\frac{\partial}{\partial\xi_{3-k}}\left[\frac{1}{h_{\xi_k}}\frac{\partial f_{\xi_{3-k}}}{\partial\xi_k}-\frac{1}{h_{\xi_{3-k}}}\frac{\partial f_{\xi_k}}{\partial\xi_{3-k}}\right]-\frac{\partial}{\partial\eta}\left[h_{\xi_{3-k}}\frac{\partial f_{\xi_k}}{\partial\eta}+f_{\xi_k}\frac{h_{\xi_{3-k}}}{h_{\xi_k}}\frac{\partial h_{\xi_k}}{\partial\eta}-\frac{h_{\xi_{3-k}}}{h_{\xi_k}}\frac{\partial f_\eta}{\partial\xi_k}\right]\right\}\\
&=\frac{1}{h_{\xi_{3-k}}}\left\{\left[\frac{1}{h_{\xi_k}}\frac{\partial^2 f_{\xi_{3-k}}}{\partial\xi_k\partial\xi_{3-k}}-\frac{\partial}{\partial\xi_{3-k}}\left(\frac{1}{h_{\xi_{3-k}}}\right)\frac{\partial f_{\xi_k}}{\partial\xi_{3-k}}-\frac{1}{h_{\xi_{3-k}}}\frac{\partial^2 f_{\xi_k}}{\partial\xi_{3-k}^2}\right]\right\}\\
&\quad-\left[\frac{\partial h_{\xi_{3-k}}}{\partial\eta}\frac{\partial f_{\xi_k}}{\partial\eta}+h_{\xi_{3-k}}\frac{\partial^2 f_{\xi_k}}{\partial\eta^2}+f_{\xi_k}\frac{\partial}{\partial\eta}\left(\frac{h_{\xi_{3-k}}}{h_{\xi_k}}\right)\frac{\partial h_{\xi_k}}{\partial\eta}+\frac{h_{\xi_{3-k}}}{h_{\xi_k}}\frac{\partial f_{\xi_k}}{\partial\eta}\frac{\partial h_{\xi_k}}{\partial\eta}-\frac{\partial}{\partial\eta}\left(\frac{h_{\xi_{3-k}}}{h_{\xi_k}}\right)\frac{\partial f_\eta}{\partial\xi_k}-\frac{h_{\xi_{3-k}}}{h_{\xi_k}}\frac{\partial^2 f_\eta}{\partial\eta\partial\xi_k}\right]\\
&=\left\{\frac{d_kd_{3-k}}{(d_k-\eta)(d_{3-k}-\eta)}\frac{\partial^2 f_{\xi_{3-k}}}{\partial\xi_k\partial\xi_{3-k}}+\frac{\eta d'_{3-k}d_{3-k}}{(d_{3-k}-\eta)^3}\frac{\partial f_{\xi_k}}{\partial\xi_{3-k}}-\frac{d_{3-k}^2}{(d_{3-k}-\eta)^2}\frac{\partial^2 f_{\xi_k}}{\partial\xi_{3-k}^2}+\frac{1}{d_{3-k}-\eta}\frac{\partial f_{\xi_k}}{\partial\eta}\right\}\\
&\quad-\frac{\partial^2 f_{\xi_k}}{\partial\eta^2}-f_{\xi_k}\frac{d_k-d_{3-k}}{(d_k-\eta)^2(d_{3-k}-\eta)}+\frac{1}{d_k-\eta}\frac{\partial f_{\xi_k}}{\partial\eta}-\frac{d_k(d_k-d_{3-k})}{(d_k-\eta)^2(d_{3-k}-\eta)}\frac{\partial f_\eta}{\partial\xi_k}+\frac{d_k}{d_k-\eta}\frac{\partial^2 f_\eta}{\partial\eta\partial\xi_k}\\
&=-\frac{\partial^2 f_{\xi_k}}{\partial\eta^2}+\left(\frac{1}{d_{3-k}-\eta}+\frac{1}{d_k-\eta}\right)\frac{\partial f_{\xi_k}}{\partial\eta}+\frac{d_k}{d_k-\eta}\frac{\partial^2 f_\eta}{\partial\eta\partial\xi_k}-f_{\xi_k}\frac{d_k-d_{3-k}}{(d_k-\eta)^2(d_{3-k}-\eta)}\\
&\quad-\frac{d_k(d_k-d_{3-k})}{(d_k-\eta)^2(d_{3-k}-\eta)}\frac{\partial f_\eta}{\partial\xi_k}-\frac{d_{3-k}}{d_{3-k}-\eta}\sum_{i=1}^{2}(-1)^{k+i}\frac{d_{3-i}}{d_{3-i}-\eta}\frac{\partial^2 f_{\xi_i}}{\partial\xi_{3-i}\partial\xi_{3-k}}+\frac{\eta d'_{3-k}d_{3-k}}{(d_{3-k}-\eta)^3}\frac{\partial f_{\xi_k}}{\partial\xi_{3-k}},\quad k=1,2
\end{aligned} \tag{2.33}$$

$$\begin{aligned}
&\left[\nabla\times(\nabla\times\vec{f})\right]_\eta\\
&=\frac{1}{h_{\xi_1}h_{\xi_2}}\left\{\frac{\partial}{\partial\xi_1}\left[\frac{h_{\xi_2}}{h_{\xi_1}h_\eta}\left(\frac{\partial(h_{\xi_1}f_{\xi_1})}{\partial\eta}-\frac{\partial(h_\eta f_\eta)}{\partial\xi_1}\right)\right]-\frac{\partial}{\partial\xi_2}\left[\frac{h_{\xi_1}}{h_{\xi_2}h_\eta}\left(\frac{\partial(h_\eta f_\eta)}{\partial\xi_2}-\frac{\partial(h_{\xi_2}f_{\xi_2})}{\partial\eta}\right)\right]\right\}\\
&=\sum_{i=1}^{2}\frac{1}{h_{\xi_i}h_{\xi_{3-i}}}\frac{\partial}{\partial\xi_i}\left[\frac{h_{\xi_{3-i}}}{h_{\xi_i}}\left(\frac{\partial(h_{\xi_i}f_{\xi_i})}{\partial\eta}-\frac{\partial f_\eta}{\partial\xi_i}\right)\right]=\sum_{i=1}^{2}\frac{1}{h_{\xi_i}}\frac{\partial}{\partial\xi_i}\left[f_{\xi_i}\frac{1}{h_{\xi_i}}\frac{\partial h_{\xi_i}}{\partial\eta}+\frac{\partial f_{\xi_i}}{\partial\eta}-\frac{1}{h_{\xi_i}}\frac{\partial f_\eta}{\partial\xi_i}\right]\\
&=\sum_{i=1}^{2}\frac{1}{h_{\xi_i}}\left[\frac{\partial f_{\xi_i}}{\partial\xi_i}\frac{1}{h_{\xi_i}}\frac{\partial h_{\xi_i}}{\partial\eta}+f_{\xi_i}\frac{\partial}{\partial\xi_i}\left(\frac{1}{h_{\xi_i}}\right)\frac{\partial h_{\xi_i}}{\partial\eta}+\frac{f_{\xi_i}}{h_{\xi_i}}\frac{\partial}{\partial\xi_i}\left(\frac{\partial h_{\xi_i}}{\partial\eta}\right)+\frac{\partial^2 f_{\xi_i}}{\partial\xi_i\partial\eta}-\frac{\partial}{\partial\xi_i}\left(\frac{1}{h_{\xi_i}}\right)\frac{\partial f_\eta}{\partial\xi_i}-\frac{1}{h_{\xi_i}}\frac{\partial^2 f_\eta}{\partial\xi_i^2}\right]\\
&=\sum_{i=1}^{2}\left[-\frac{d_i}{(d_i-\eta)^2}\frac{\partial f_{\xi_i}}{\partial\xi_i}-\frac{\eta d'_i}{(d_i-\eta)^3}f_{\xi_i}+\frac{d'_i}{d_i(d_i-\eta)}f_{\xi_i}+\frac{d_i}{d_i-\eta}\frac{\partial^2 f_{\xi_i}}{\partial\xi_i\partial\eta}+\frac{\eta d'_id_i}{(d_i-\eta)^3}\frac{\partial f_\eta}{\partial\xi_i}-\frac{d_i^2}{(d_i-\eta)^2}\frac{\partial^2 f_\eta}{\partial\xi_i^2}\right]
\end{aligned} \tag{2.34}$$

The scale factor, D, for tangential coordinates is naturally defined as

$$D = \min(d_1, d_2)$$

so that

$$\tilde{\xi}_k = \xi_k / D; \quad \tilde{d}_k = d_k / D, \quad k = 1,2 \tag{2.35}$$

Substituting Equation 2.35 into Equations 2.30 through 2.34, we obtain

$$(\nabla \times \vec{f})_{\xi_k} = (-1)^k \delta^{-1} \left[\frac{\partial f_{\xi_{3-k}}}{\partial \tilde{\eta}} - \tilde{p} \frac{f_{\xi_{3-k}}}{\tilde{d}_{3-k} - \tilde{p}\tilde{\eta}} - \tilde{p} \frac{\tilde{d}_{3-k}}{\tilde{d}_{3-k} - \tilde{p}\tilde{\eta}} \frac{\partial f_\eta}{\partial \tilde{\xi}_{3-k}} \right], \quad k = 1,2 \tag{2.36}$$

$$(\nabla \times \vec{f})_\eta = \delta^{-1} \tilde{p} \sum_{i=1}^{2} (-1)^i \frac{\tilde{d}_{3-i}}{\tilde{d}_{3-i} - \tilde{p}\tilde{\eta}} \frac{\partial f_{\xi_i}}{\partial \tilde{\xi}_{3-i}} \tag{2.37}$$

$$\nabla \cdot \vec{f} = \delta^{-1} \left[\tilde{p} \sum_{i=1}^{2} \frac{\tilde{d}_i}{\tilde{d}_i - \tilde{p}\tilde{\eta}} \frac{\partial f_{\xi_i}}{\partial \tilde{\xi}_i} + \frac{\partial f_\eta}{\partial \tilde{\eta}} - \tilde{p} f_\eta \sum_{l=1}^{2} \frac{1}{\tilde{d}_l - \tilde{p}\tilde{\eta}} \right] \tag{2.38}$$

$$\begin{aligned}
&\left[\nabla \times (\nabla \times \vec{f})\right]_{\xi_k} \\
&= \delta^{-2} \left[-\frac{\partial^2 f_{\xi_k}}{\partial \tilde{\eta}^2} + \tilde{p} \left(\frac{1}{\tilde{d}_{3-k} - \tilde{p}\tilde{\eta}} + \frac{1}{\tilde{d}_k - \tilde{p}\tilde{\eta}} \right) \frac{\partial f_{\xi_k}}{\partial \tilde{\eta}} + \tilde{p} \frac{\tilde{d}_k}{\tilde{d}_k - \tilde{p}\tilde{\eta}} \frac{\partial^2 \tilde{f}_\eta}{\partial \tilde{\eta} \partial \tilde{\xi}_k} \right. \\
&\quad - \tilde{p}^2 \frac{\tilde{d}_k - \tilde{d}_{3-k}}{(\tilde{d}_k - \tilde{p}\tilde{\eta})^2 (\tilde{d}_{3-k} - \tilde{p}\tilde{\eta})} f_{\xi_k} - \tilde{p}^2 \frac{\tilde{d}_k (\tilde{d}_k - \tilde{d}_{3-k})}{(\tilde{d}_k - \tilde{p}\tilde{\eta})^2 (\tilde{d}_{3-k} - \tilde{p}\tilde{\eta})} \frac{\partial f_\eta}{\partial \tilde{\xi}_k} \\
&\quad \left. - \tilde{p}^2 \frac{\tilde{d}_{3-k}}{\tilde{d}_{3-k} - \tilde{p}\tilde{\eta}} \sum_{i=1}^{2} (-1)^{k+i} \frac{\tilde{d}_{3-i}}{\tilde{d}_{3-i} - \tilde{p}\tilde{\eta}} \frac{\partial^2 f_{\xi_i}}{\partial \tilde{\xi}_{3-i} \partial \tilde{\xi}_{3-k}} + \tilde{p}^3 \frac{\eta \tilde{d}'_{3-k} \tilde{d}_{3-k}}{(\tilde{d}_{3-k} - \tilde{p}\tilde{\eta})^3} \frac{\partial f_{\xi_k}}{\partial \tilde{\xi}_{3-k}} \right], \quad k = 1,2
\end{aligned} \tag{2.39}$$

$$\begin{aligned}
\left[\nabla \times (\nabla \times \vec{f})\right]_\eta = \delta^{-2} \tilde{p} \sum_{i=1}^{2} &\left[-\tilde{p} \frac{\tilde{d}_i}{(\tilde{d}_i - \tilde{p}\tilde{\eta})^2} \frac{\partial f_{\xi_i}}{\partial \tilde{\xi}_i} - \tilde{p}^2 \frac{\tilde{\eta} \tilde{d}'_i}{(\tilde{d}_i - \tilde{p}\tilde{\eta})^3} f_{\xi_i} + \tilde{p}^2 \frac{\tilde{d}'_i}{\tilde{d}_i (\tilde{d}_i - \tilde{p}\tilde{\eta})} f_{\xi_i} \right. \\
&\left. + \frac{\tilde{d}_i}{\tilde{d}_i - \tilde{p}\tilde{\eta}} \frac{\partial^2 f_{\xi_i}}{\partial \tilde{\xi}_i \partial \tilde{\eta}} + \tilde{p}^2 \frac{\tilde{\eta} \tilde{d}'_i \tilde{d}_i}{(\tilde{d}_i - \tilde{p}\tilde{\eta})^3} \frac{\partial f_\eta}{\partial \tilde{\xi}_i} - \tilde{p} \frac{\tilde{d}_i^2}{(\tilde{d}_i - \tilde{p}\tilde{\eta})^2} \frac{\partial^2 f_\eta}{\partial \tilde{\xi}_i^2} \right]
\end{aligned} \tag{2.40}$$

Obviously, these derivations are valid only for smooth surfaces (the cases of corners and edges will be considered separately in Section 3.7). In most practically important problems, the radius of curvature varies so slowly along the surface that the terms containing $\tilde{d}'_i$ can be neglected in further derivations, and Equations 2.39 through 2.40 are reduced to the following form:

$$\left[\nabla\times(\nabla\times\vec{f})\right]_{\xi_k}=\delta^{-2}\left[-\frac{\partial^2 f_{\xi_k}}{\partial\tilde{\eta}^2}+\tilde{p}\left(\frac{1}{\tilde{d}_{3-k}-\tilde{p}\tilde{\eta}}+\frac{1}{\tilde{d}_k-\tilde{p}\tilde{\eta}}\right)\frac{\partial f_{\xi_k}}{\partial\tilde{\eta}}+\tilde{p}\frac{\tilde{d}_k}{\tilde{d}_k-\tilde{p}\tilde{\eta}}\frac{\partial^2\tilde{f}_\eta}{\partial\tilde{\eta}\partial\tilde{\xi}_k}\right.$$
$$-\tilde{p}^2\frac{\tilde{d}_k-\tilde{d}_{3-k}}{(\tilde{d}_k-\tilde{p}\tilde{\eta})^2(\tilde{d}_{3-k}-\tilde{p}\tilde{\eta})}f_{\xi_k}-\tilde{p}^2\frac{\tilde{d}_k(\tilde{d}_k-\tilde{d}_{3-k})}{(\tilde{d}_k-\tilde{p}\tilde{\eta})^2(\tilde{d}_{3-k}-\tilde{p}\tilde{\eta})}\frac{\partial f_\eta}{\partial\tilde{\xi}_k}$$
$$\left.-\tilde{p}^2\frac{\tilde{d}_{3-k}}{\tilde{d}_{3-k}-\tilde{p}\tilde{\eta}}\sum_{i=1}^{2}(-1)^{k+i}\frac{\tilde{d}_{3-i}}{\tilde{d}_{3-i}-\tilde{p}\tilde{\eta}}\frac{\partial^2 f_{\xi_i}}{\partial\tilde{\xi}_{3-i}\partial\tilde{\xi}_{3-k}}\right] \tag{2.41}$$

$$\left[\nabla\times(\nabla\times\vec{f})\right]_{\eta}=\delta^{-2}\tilde{p}\sum_{i=1}^{2}\left[-\tilde{p}\frac{\tilde{d}_i}{(\tilde{d}_i-\tilde{p}\tilde{\eta})^2}\frac{\partial f_{\xi_i}}{\partial\tilde{\xi}_i}+\frac{\tilde{d}_i}{\tilde{d}_i-\tilde{p}\tilde{\eta}}\frac{\partial^2 f_{\xi_i}}{\partial\tilde{\xi}_i\partial\tilde{\eta}}-\tilde{p}\frac{\tilde{d}_i^2}{(\tilde{d}_i-\tilde{p}\tilde{\eta})^2}\frac{\partial^2 f_\eta}{\partial\tilde{\xi}_i^2}\right] \tag{2.42}$$

The condition in Equation 2.1 allowed the representation of the vector operators in the form of asymptotic expansions in the small parameter, $\tilde{p}$, that is proportional to the ratio of the depth of penetration to the characteristic size of the conductor's surface.

It is a well-known fact that the tangential derivatives in the conductor's skin layer are much smaller than the normal derivative [5]:

$$\frac{\partial}{\partial\eta}\gg\frac{\partial}{\partial\xi_k};\quad \frac{\partial^2}{\partial\eta^2}\gg\frac{\partial^2}{\partial\xi_k^2} \tag{2.43}$$

Equations 2.36 through 2.38 and Equations 2.41 and 2.42 not only illustrate Equation 2.43, but also facilitate estimation of the tangential and normal derivatives for a given thickness of the skin layer.

Comparison of Equations 2.41 and 2.42 leads us to the following important conclusion: the normal component of the curl of the function $\vec{f}$ in the conductor's skin layer is much smaller than its tangential component. In the limiting case $\tilde{p}=0$ ($\delta=0$), the normal component of curl, $\vec{f}$, vanishes.

2.3 Perturbation Technique

Since the parameter $\tilde{p}$ is small, the following functions can be represented as expansions in the small parameter $\tilde{p}$:

$$\frac{1}{\tilde{d}_k - \tilde{p}\tilde{\eta}} = \frac{1}{\tilde{d}_k} + \tilde{p}\frac{\tilde{\eta}}{\tilde{d}_k^2} + \tilde{p}^2\frac{\tilde{\eta}^2}{\tilde{d}_k^3} + \cdots; \quad k = 1,2 \tag{2.44}$$

$$\frac{\tilde{d}_k}{\tilde{d}_k - \tilde{p}\tilde{\eta}} = 1 + \tilde{p}\frac{\tilde{\eta}}{\tilde{d}_k} + \tilde{p}^2\frac{\tilde{\eta}^2}{\tilde{d}_k^2} + \cdots; \quad k = 1,2 \tag{2.45}$$

Substituting the expansions in Equations 2.44 and 2.45 into Equations 2.36 through 2.38 and Equations 2.41 and 2.42, we obtain

$$(\nabla\times\vec{f})_{\xi_k} = (-1)^k\delta^{-1}\left[\frac{\partial f_{\xi_{3-k}}}{\partial\tilde{\eta}} - \tilde{p}\left(\frac{f_{\xi_{3-k}}}{\tilde{d}_{3-k}} + \frac{\partial f_\eta}{\partial\tilde{\xi}_{3-k}}\right) - \tilde{p}^2\tilde{\eta}\left(\frac{f_{\xi_{3-k}}}{\tilde{d}_{3-k}^2} + \frac{1}{\tilde{d}_{3-k}}\frac{\partial f_\eta}{\partial\tilde{\xi}_{3-k}}\right) + \cdots\right], \quad k=1,2 \tag{2.46}$$

$$(\nabla\times\vec{f})_\eta = \delta^{-1}\tilde{p}\sum_{i=1}^{2}(-1)^i\frac{\partial f_{\xi_i}}{\partial\tilde{\xi}_{3-i}}\left(1 + \tilde{p}\frac{\tilde{\eta}}{\tilde{d}_{3-i}} + \tilde{p}^2\frac{\tilde{\eta}^2}{\tilde{d}_{3-i}^2} + \cdots\right) \tag{2.47}$$

$$\begin{aligned}\nabla\cdot\vec{f} = \delta^{-1}\Bigg[&\frac{\partial f_\eta}{\partial\tilde{\eta}} + \tilde{p}\left(\sum_{i=1}^{2}\frac{\partial f_{\xi_i}}{\partial\tilde{\xi}_i} - f_\eta\sum_{l=1}^{2}\frac{1}{\tilde{d}_l}\right) + \tilde{p}^2\tilde{\eta}\left(\sum_{i=1}^{2}\frac{1}{\tilde{d}_i}\frac{\partial f_{\xi_i}}{\partial\tilde{\xi}_i} - f_\eta\sum_{l=1}^{2}\tilde{d}_l^{-2}\right)\\ &+ \tilde{p}^3\tilde{\eta}^2\left(\sum_{i=1}^{2}\frac{1}{\tilde{d}_i^2}\frac{\partial f_{\xi_i}}{\partial\tilde{\xi}_i} - f_\eta\sum_{l=1}^{2}\tilde{d}_l^{-3}\right)\Bigg]\end{aligned} \tag{2.48}$$

$$\begin{aligned}\left[\nabla\times(\nabla\times\vec{f})\right]_{\xi_k} = \delta^{-2}\Bigg\{&-\frac{\partial^2 f_{\xi_k}}{\partial\tilde{\eta}^2} + \tilde{p}\left[\frac{\partial f_{\xi_k}}{\partial\tilde{\eta}}\sum_{i=1}^{2}\tilde{d}_i^{-1} + \frac{\partial^2 f_\eta}{\partial\tilde{\eta}\partial\tilde{\xi}_k}\right] + \tilde{p}^2\Bigg[\tilde{\eta}\frac{\partial f_{\xi_k}}{\partial\tilde{\eta}}\sum_{i=1}^{2}\tilde{d}_i^{-2} + \tilde{\eta}\tilde{d}_k^{-1}\frac{\partial^2 f_\eta}{\partial\tilde{\eta}\partial\tilde{\xi}_k}\\ &- f_{\xi_k}\frac{\tilde{d}_k - \tilde{d}_{3-k}}{\tilde{d}_k^2\tilde{d}_{3-k}} - \frac{\tilde{d}_k - \tilde{d}_{3-k}}{\tilde{d}_k\tilde{d}_{3-k}}\frac{\partial f_\eta}{\partial\tilde{\xi}_k} - \sum_{i=1}^{2}(-1)^{k+i}\frac{\partial^2 f_{\xi_i}}{\partial\tilde{\xi}_{3-i}\partial\tilde{\xi}_{3-k}}\Bigg] + \cdots\Bigg\}, \quad k=1,2\end{aligned} \tag{2.49}$$

$$\begin{aligned}\left[\nabla\times(\nabla\times\vec{f})\right]_\eta = \delta^{-2}\tilde{p}\sum_{i=1}^{2}\Bigg[&\frac{\partial^2 f_{\xi_i}}{\partial\tilde{\xi}_i\partial\tilde{\eta}} + \tilde{p}\left(\frac{\tilde{\eta}}{\tilde{d}_i}\frac{\partial^2 f_{\xi_i}}{\partial\tilde{\xi}_i\partial\tilde{\eta}} - \frac{1}{\tilde{d}_i}\frac{\partial f_{\xi_i}}{\partial\tilde{\xi}_i} - \frac{\partial^2 f_\eta}{\partial\tilde{\xi}_i^2}\right)\\ &+ \tilde{p}^2\tilde{\eta}\left(\frac{\tilde{\eta}}{\tilde{d}_i^2}\frac{\partial^2 f_{\xi_i}}{\partial\tilde{\xi}_i\partial\tilde{\eta}} - \frac{2}{\tilde{d}_i^2}\frac{\partial f_{\xi_i}}{\partial\tilde{\xi}_i} - \frac{2}{\tilde{d}_i}\frac{\partial^2 f_\eta}{\partial\tilde{\xi}_i^2}\right) + \cdots\Bigg]\end{aligned} \tag{2.50}$$

Consider the practically important case in which the distribution of the function, $\vec{f}$, in a conductor with constant material properties can be described by the diffusion equation (Equation 2.2). Note that the displacement current was neglected in Equation 2.2, therefore its application area is restricted to good conductors. The more general case of lossy dielectrics should be considered using the so-called telegraph equation, which can be represented in the form:

$$\nabla\times(\nabla\times\vec{f}) + \sigma\mu\frac{\partial\vec{f}}{\partial t} + \varepsilon\mu\frac{\partial^2\vec{f}}{\partial t^2} = 0 \tag{2.51}$$

Transformations of Equations 2.2 and 2.51 using the perturbation technique are carried out in a similar fashion. To simplify the description, in this chapter we will consider Equation 2.2 only.

The diffusion equation (Equation 2.2) must be supplemented with the boundary and initial conditions in the form:

$$\eta = 0 \text{ (conductor surface): } f = f^{\mathrm{b}}; \qquad \eta \to \infty \text{ (depth of conductor): } f = 0;$$
$$t = 0\text{: } f = 0 \tag{2.52}$$

Here and below, the superscript "b" denotes values at the interface (boundary). A natural scale factor for time t is the quantity τ defined as the ratio $2/\omega$ in the case of time-harmonic fields or the incident pulse duration in the case of transient sources. Therefore, we can introduce a "nondimensional time" $\tilde{t}$ as follows:

$$\tilde{t} = t/\tau \tag{2.53}$$

Substituting Equation 2.53 into Equation 2.2 and taking into account Equation 2.1, we obtain

$$\nabla \times (\nabla \times \vec{f}) = -\frac{\sigma\mu}{\tau}\frac{\partial \vec{f}}{\partial \tilde{t}} = -\delta^{-2}\frac{\partial \vec{f}}{\partial \tilde{t}} \tag{2.54}$$

Replacing the operator $\nabla \times (\nabla \times \vec{f})$ in Equation 2.54 with Equations 2.49 and 2.50, we get

$$-\frac{\partial^2 f_{\xi_k}}{\partial \tilde{\eta}^2} + \tilde{p}\left[\frac{\partial f_{\xi_k}}{\partial \tilde{\eta}}\sum_{i=1}^{2}\tilde{d}_i^{-1} + \frac{\partial^2 f_\eta}{\partial \tilde{\eta}\partial \tilde{\xi}_k}\right] + \tilde{p}^2\left[\tilde{\eta}\frac{\partial f_{\xi_k}}{\partial \tilde{\eta}}\sum_{i=1}^{2}\tilde{d}_i^{-2} + \tilde{\eta}\tilde{d}_k^{-1}\frac{\partial^2 f_\eta}{\partial \tilde{\eta}\partial \tilde{\xi}_k} - f_{\xi_k}\frac{\tilde{d}_k - \tilde{d}_{3-k}}{\tilde{d}_k^2\tilde{d}_{3-k}}\right.$$
$$\left. - \frac{\tilde{d}_k - \tilde{d}_{3-k}}{\tilde{d}_k\tilde{d}_{3-k}}\frac{\partial f_\eta}{\partial \tilde{\xi}_k} - \sum_{i=1}^{2}(-1)^{k+i}\frac{\partial^2 f_{\xi_i}}{\partial \tilde{\xi}_{3-i}\partial \tilde{\xi}_{3-k}}\right] + O(\tilde{p}^3) = -\frac{\partial f_{\xi_k}}{\partial \tilde{t}}, \quad k = 1,2 \tag{2.55}$$

$$\tilde{p}\sum_{i=1}^{2}\left[\frac{\partial^2 f_{\xi_i}}{\partial \tilde{\xi}_i \partial \tilde{\eta}} + \tilde{p}\left(\frac{\tilde{\eta}}{\tilde{d}_i}\frac{\partial^2 f_{\xi_i}}{\partial \tilde{\xi}_i\partial \tilde{\eta}} - \frac{1}{\tilde{d}_i}\frac{\partial f_{\xi_i}}{\partial \tilde{\xi}_i} - \frac{\partial^2 f_\eta}{\partial \tilde{\xi}_i^2}\right)\right.$$
$$\left. + \tilde{p}^2\tilde{\eta}\left(\frac{\tilde{\eta}}{\tilde{d}_i^2}\frac{\partial^2 f_{\xi_i}}{\partial \tilde{\xi}_i\partial \tilde{\eta}} - \frac{2}{\tilde{d}_i^2}\frac{\partial f_{\xi_i}}{\partial \tilde{\xi}_i} - \frac{2}{\tilde{d}_i}\frac{\partial^2 f_\eta}{\partial \tilde{\xi}_i^2}\right) + O(\tilde{p}^3)\right] = -\frac{\partial f_\eta}{\partial \tilde{t}} \tag{2.56}$$

Since the parameter $\tilde{p}$ is small, we represent the tangential and normal components of $\vec{f}$ in the form of the asymptotic expansions in the parameter $\tilde{p}$:

$$f_{\xi_k} = \sum_{m=0}^{\infty} \tilde{p}^m (\vec{f}_m)_{\xi_k}, \quad k = 1, 2 \tag{2.57}$$

$$f_{\eta} = \sum_{m=0}^{\infty} \tilde{p}^m (\vec{f}_m)_{\eta} \tag{2.58}$$

where $(\vec{f}_m)_{\xi_k}$ and $(\vec{f}_m)_{\eta}$ are unknown coefficients. Substituting Equations 2.57 and 2.58 into Equations 2.55 and 2.56, and equating coefficients of equal powers of $\tilde{p}$, the following equations for the expansion coefficients are obtained

$m = 0$

$$\frac{\partial^2 (\vec{f}_0)_{\xi_k}}{\partial \tilde{\eta}^2} - \frac{\partial (\vec{f}_0)_{\xi_k}}{\partial \tilde{t}} = 0 \tag{2.59}$$

$$\frac{\partial (\vec{f}_0)_{\eta}}{\partial \tilde{t}} = 0 \tag{2.60}$$

$m = 1$

$$\frac{\partial^2 (\vec{f}_1)_{\xi_k}}{\partial \tilde{\eta}^2} - \frac{\partial (\vec{f}_1)_{\xi_k}}{\partial \tilde{t}} = \frac{\partial (\vec{f}_0)_{\xi_k}}{\partial \tilde{\eta}} \sum_{l=1}^{2} \tilde{d}_l^{-1} + \frac{\partial^2 (\vec{f}_0)_{\eta}}{\partial \tilde{\xi}_k \partial \tilde{\eta}} \tag{2.61}$$

$$\frac{\partial (\vec{f}_1)_{\eta}}{\partial \tilde{t}} = -\sum_{i=1}^{2} \frac{\partial^2 (\vec{f}_0)_{\xi_i}}{\partial \tilde{\xi}_i \partial \tilde{\eta}} \tag{2.62}$$

$m = 2$

$$\begin{aligned} \frac{\partial^2 (\vec{f}_2)_{\xi_k}}{\partial \tilde{\eta}^2} - \frac{\partial (\vec{f}_2)_{\xi_k}}{\partial \tilde{t}} = {} & \frac{\partial (\vec{f}_1)_{\xi_k}}{\partial \tilde{\eta}} \sum_{l=1}^{2} \tilde{d}_l^{-1} + \frac{\partial^2 (\vec{f}_1)_{\eta}}{\partial \tilde{\xi}_k \partial \tilde{\eta}} + (\vec{f}_0)_{\xi_k} \left(\tilde{d}_k^{-2} - \tilde{d}_k^{-1} \tilde{d}_{3-k}^{-1} \right) + \frac{\partial (\vec{f}_0)_{\eta}}{\partial \tilde{\xi}_k} \left(\tilde{d}_k^{-1} - d_{3-k}^{-1} \right) \\ & - \sum_{i=1}^{2} (-1)^{k+i} \frac{\partial^2 (\vec{f}_0)_{\xi_i}}{\partial \tilde{\xi}_{3-k} \partial \tilde{\xi}_{3-i}} + \tilde{\eta} \frac{\partial (\vec{f}_0)_{\xi_k}}{\partial \tilde{\eta}} \sum_{i=1}^{2} \tilde{d}_i^{-2} + \tilde{\eta} \tilde{d}_k^{-1} \frac{\partial^2 (\vec{f}_0)_{\eta}}{\partial \tilde{\xi}_k \partial \tilde{\eta}} \end{aligned} \tag{2.63}$$

$$\frac{\partial (\vec{f}_2)_{\eta}}{\partial \tilde{t}} = -\sum_{i=1}^{2} \left[\frac{\partial^2 (\vec{f}_1)_{\xi_i}}{\partial \tilde{\xi}_i \partial \tilde{\eta}} + \tilde{\eta} \tilde{d}_i^{-1} \frac{\partial^2 (\vec{f}_0)_{\xi_i}}{\partial \tilde{\xi}_i \partial \tilde{\eta}} - \tilde{d}_i^{-1} \frac{\partial (\vec{f}_0)_{\xi_i}}{\partial \tilde{\xi}_i} - \frac{\partial^2 (\vec{f}_0)_{\eta}}{\partial \tilde{\xi}_i^2} \right] \tag{2.64}$$

$m = 3$ (for $(\vec{f}_3)_\eta$ only)

$$\frac{\partial(\vec{f}_3)_\eta}{\partial\tilde{t}} = -\sum_{i=1}^{2}\left[\frac{\partial^2(\vec{f}_2)_{\xi_i}}{\partial\tilde{\xi}_i\partial\tilde{\eta}} + \tilde{\eta}\tilde{d}_i^{-1}\frac{\partial^2(\vec{f}_1)_{\xi_i}}{\partial\tilde{\xi}_i\partial\tilde{\eta}} - \tilde{d}_i^{-1}\frac{\partial(\vec{f}_1)_{\xi_i}}{\partial\tilde{\xi}_i} - \frac{\partial^2(\vec{f}_1)_\eta}{\partial\tilde{\xi}_i^2}\right.$$
$$\left. + \tilde{\eta}\left(\frac{\tilde{\eta}}{\tilde{d}_i^2}\frac{\partial^2(\vec{f}_0)_{\xi_i}}{\partial\tilde{\xi}_i\partial\tilde{\eta}} - \frac{2}{\tilde{d}_i^2}\frac{\partial(\vec{f}_0)_{\xi_i}}{\partial\tilde{\xi}_i} - \frac{2}{\tilde{d}_i}\frac{\partial^2(\vec{f}_0)_\eta}{\partial\tilde{\xi}_i^2} + \frac{\tilde{d}_i'}{\tilde{d}_i^2}\frac{\partial(\vec{f}_0)_\eta}{\partial\tilde{\xi}_i}\right)\right] \tag{2.65}$$

Boundary and initial conditions for Equations 2.32, 2.34, and 2.36 are obtained by substituting Equations 2.30 and 2.31 into Equations 2.23 and 2.24:

$$\eta = 0\text{: } (\vec{f}_m)_{\xi_k} = (\vec{f}_m^{\,b})_{\xi_k}; \quad (\vec{f}_m)_\eta = (\vec{f}_m^{\,b})_\eta; \quad \eta \to \infty\text{: } (\vec{f}_m)_{\xi_k} = 0; \quad (\vec{f}_m)_\eta = 0;$$
$$t = 0\text{: } (\vec{f}_m)_{\xi_k} = 0; \quad (\vec{f}_m)_\eta = 0 \tag{2.66}$$

The formulations (Equations 2.59 through 2.65) with the boundary conditions (Equation 2.66) state that the field at any point inside the conductor is determined solely by the field at the geometrically closest point on the surface. Viewed in this way, it is not surprising that the expansions breakdown at $\tilde{\eta} \sim d_k$ when fields due to equidistant points on the surface interfere with each other. Taking into account that higher-order terms of the expansions reduce the approximation error, their derivation is not an easy task. Therefore, we restrict ourselves by keeping the first three nonzero terms in the expansions.

Considering Equation 2.60 with the boundary condition $\eta \to \infty$: $(\vec{f}_m)_\eta = 0$ leads us to the following very important conclusion: "the coefficient $(\vec{f}_0)_\eta$ is equal to zero everywhere inside the conductor including its interface," i.e.

$$(\vec{f}_0)_\eta = 0 \tag{2.67}$$

Equations 2.60 and 2.67 reduce Equations 2.61 through 2.65 to the following form:

$m = 1$

$$\frac{\partial^2(\vec{f}_1)_{\xi_k}}{\partial\tilde{\eta}^2} - \frac{\partial(\vec{f}_1)_{\xi_k}}{\partial\tilde{t}} = \frac{\partial(\vec{f}_0)_{\xi_k}}{\partial\tilde{\eta}}\sum_{l=1}^{2}\tilde{d}_l^{-1} \tag{2.68}$$

$$\frac{\partial(\vec{f}_1)_\eta}{\partial\tilde{t}} = -\sum_{i=1}^{2}\frac{\partial^2(\vec{f}_0)_{\xi_i}}{\partial\tilde{\xi}_i\partial\tilde{\eta}} \tag{2.69}$$

$m=2$

$$\frac{\partial^2(\vec{f}_2)_{\xi_k}}{\partial\tilde{\eta}^2}-\frac{\partial(\vec{f}_2)_{\xi_k}}{\partial\tilde{t}}=\frac{\partial(\vec{f}_1)_{\xi_k}}{\partial\tilde{\eta}}\sum_{l=1}^{2}\tilde{d}_l^{-1}+\frac{\partial^2(\vec{f}_1)_\eta}{\partial\tilde{\xi}_k\partial\tilde{\eta}}+(\vec{f}_0)_{\xi_k}\left(\tilde{d}_k^{-2}-\tilde{d}_k^{-1}\tilde{d}_{3-k}^{-1}\right)$$
$$-\sum_{i=1}^{2}(-1)^{k+i}\frac{\partial^2(\vec{f}_0)_{\xi_i}}{\partial\tilde{\xi}_{3-k}\partial\tilde{\xi}_{3-i}}+\tilde{\eta}\frac{\partial(\vec{f}_0)_{\xi_k}}{\partial\tilde{\eta}}\sum_{i=1}^{2}\tilde{d}_i^{-2} \tag{2.70}$$

$$\frac{\partial(\vec{f}_2)_\eta}{\partial\tilde{t}}=-\sum_{i=1}^{2}\left[\frac{\partial^2(\vec{f}_1)_{\xi_i}}{\partial\tilde{\xi}_i\partial\tilde{\eta}}+\tilde{\eta}\tilde{d}_i^{-1}\frac{\partial^2(\vec{f}_0)_{\xi_i}}{\partial\tilde{\xi}_i\partial\tilde{\eta}}-\tilde{d}_i^{-1}\frac{\partial(\vec{f}_0)_{\xi_i}}{\partial\tilde{\xi}_i}\right] \tag{2.71}$$

$m=3$ (for $(\vec{f}_3)_\eta$ only)

$$\frac{\partial(\vec{f}_3)_\eta}{\partial\tilde{t}}=-\sum_{i=1}^{2}\left[\frac{\partial^2(\vec{f}_2)_{\xi_i}}{\partial\tilde{\xi}_i\partial\tilde{\eta}}+\frac{\tilde{\eta}}{\tilde{d}_i}\frac{\partial^2(\vec{f}_1)_{\xi_i}}{\partial\tilde{\xi}_i\partial\tilde{\eta}}-\frac{1}{\tilde{d}_i}\frac{\partial(\vec{f}_1)_{\xi_i}}{\partial\tilde{\xi}_i}-\frac{\partial^2(\vec{f}_1)_\eta}{\partial\tilde{\xi}_i^2}\right.$$
$$\left.+\tilde{\eta}\left(\frac{\tilde{\eta}}{\tilde{d}_i^2}\frac{\partial^2(\vec{f}_0)_{\xi_i}}{\partial\tilde{\xi}_i\partial\tilde{\eta}}-\frac{2}{\tilde{d}_i^2}\frac{\partial(\vec{f}_0)_{\xi_i}}{\partial\tilde{\xi}_i}\right)\right] \tag{2.72}$$

Substituting Equation 2.69 into Equation 2.70 gives

$$\frac{\partial^2(\vec{f}_2)_{\xi_k}}{\partial\tilde{\eta}^2}-\frac{\partial(\vec{f}_2)_{\xi_k}}{\partial\tilde{t}}=\frac{\partial(\vec{f}_1)_{\xi_k}}{\partial\tilde{\eta}}\sum_{l=1}^{2}\tilde{d}_l^{-1}-\int_0^t\sum_{i=1}^{2}\frac{\partial^4(\vec{f}_0)_{\xi_i}}{\partial\tilde{\xi}_k\partial\tilde{\xi}_i\partial\tilde{\eta}^2}\mathrm{d}t+(\vec{f}_0)_{\xi_k}\left(\tilde{d}_k^{-2}-\tilde{d}_k^{-1}\tilde{d}_{3-k}^{-1}\right)$$
$$-\sum_{i=1}^{2}(-1)^{k+i}\frac{\partial^2(\vec{f}_0)_{\xi_i}}{\partial\tilde{\xi}_{3-k}\partial\tilde{\xi}_{3-i}}+\tilde{\eta}\frac{\partial(\vec{f}_0)_{\xi_k}}{\partial\tilde{\eta}}\sum_{i=1}^{2}\tilde{d}_i^{-2} \tag{2.73}$$

Simplification of Equation 2.73 gives

$$\frac{\partial^2(\vec{f}_2)_{\xi_k}}{\partial\tilde{\eta}^2}-\frac{\partial(\vec{f}_2)_{\xi_k}}{\partial\tilde{t}}=\frac{\partial(\vec{f}_1)_{\xi_k}}{\partial\tilde{\eta}}\sum_{l=1}^{2}\tilde{d}_l^{-1}-\sum_{i=1}^{2}\frac{\partial^2(\vec{f}_0)_{\xi_k}}{\partial\tilde{\xi}_i^2}+\frac{\left(\vec{f}_0\right)_{\xi_k}}{\tilde{d}_k}\frac{\tilde{d}_{3-k}-\tilde{d}_k}{\tilde{d}_k\tilde{d}_{3-k}}+\tilde{\eta}\frac{\partial(\vec{f}_0)_{\xi_k}}{\partial\tilde{\eta}}\sum_{i=1}^{2}\tilde{d}_i^{-2} \tag{2.74}$$

Note that the diffusion equations (Equations 2.59, 2.68, and 2.74) contain only the tangential components $(\vec{f}_m)_{\xi_k}$.

Calculation of the tangential components $(\vec{f}_m)_{\xi_k}$ is a task of primary importance in the surface impedance concept because all other quantities (such as the normal component and tangential and normal derivatives) can be then derived as "postprocessing" operations using the $(\vec{f}_{m-1})_{\xi_k}$ obtained at previous step.

2.4 Tangential Components

We now apply the Laplace transform following the rule:

$$\bar{f}_{\xi_k}(\tilde{s}) = \int_0^\infty f_{\xi_k}(\tilde{t}) \exp(-\tilde{s}\tilde{t}) d\tilde{t} \tag{2.75}$$

In the Laplace domain (LD), Equations 2.59, 2.68, and 2.74 are written in the form:

$m = 0$

$$\frac{\partial^2 \left(\vec{\bar{f}}_0\right)_{\xi_k}}{\partial \tilde{\eta}^2} - \tilde{s}\left(\vec{\bar{f}}_0\right)_{\xi_k} = 0 \tag{2.76}$$

$m = 1$

$$\frac{\partial^2 \left(\vec{\bar{f}}_1\right)_{\xi_k}}{\partial \tilde{\eta}^2} - \tilde{s}\left(\vec{\bar{f}}_1\right)_{\xi_k} = \frac{\partial \left(\vec{\bar{f}}_0\right)_{\xi_k}}{\partial \tilde{\eta}} \sum_{l=1}^{2} \tilde{d}_l^{-1} \tag{2.77}$$

$m = 2$

$$\frac{\partial^2 \left(\vec{\bar{f}}_2\right)_{\xi_k}}{\partial \tilde{\eta}^2} - \tilde{s}\left(\vec{\bar{f}}_2\right)_{\xi_k} = \frac{\partial \left(\vec{\bar{f}}_1\right)_{\xi_k}}{\partial \tilde{\eta}} \sum_{l=1}^{2} \tilde{d}_l^{-1} - \sum_{i=1}^{2} \frac{\partial^2 \left(\vec{\bar{f}}_0\right)_{\xi_k}}{\partial \tilde{\xi}_i^2} + \frac{\left(\vec{\bar{f}}_0\right)_{\xi_k}}{\tilde{d}_k} \frac{\tilde{d}_{3-k} - \tilde{d}_k}{\tilde{d}_k \tilde{d}_{3-k}} + \tilde{\eta} \frac{\partial \left(\vec{\bar{f}}_0\right)_{\xi_k}}{\partial \tilde{\eta}} \sum_{i=1}^{2} \tilde{d}_i^{-2} \tag{2.78}$$

Derivation of $\left(\vec{\bar{f}}_0\right)_{\xi_k}$

The solution of Equation 2.76 with boundary conditions (Equation 2.66) can be written in the form:

$$\left(\vec{\bar{f}}_0\right)_{\xi_k} = \left(\vec{\bar{f}}_0^{\mathrm{b}}\right)_{\xi_k} \exp(-\tilde{\eta}\sqrt{\tilde{s}}) \tag{2.79}$$

Therefore,

$$\frac{\partial \left(\vec{\bar{f}}_0\right)_{\xi_k}}{\partial \tilde{\eta}} = -\sqrt{\tilde{s}}\left(\vec{\bar{f}}_0^{\mathrm{b}}\right)_{\xi_k} \exp(-\tilde{\eta}\sqrt{\tilde{s}}) \tag{2.80}$$

Derivation of $\left(\vec{\vec{f}}_1\right)_{\xi_k}$

Equation 2.77 can be represented in the following form using Equation 2.80:

$$\frac{\partial^2\left(\vec{\vec{f}}_1\right)_{\xi_k}}{\partial\tilde{\eta}^2} - \tilde{s}\left(\vec{\vec{f}}_1\right)_{\xi_k} = -\sqrt{\tilde{s}}\left(\vec{\vec{f}}_0^{\,b}\right)_{\xi_k}\sum_{l=1}^{2}\tilde{d}_l^{-1}\exp\left(-\tilde{\eta}\sqrt{\tilde{s}}\right) \tag{2.81}$$

The solution of Equation 2.81 is

$$\left(\vec{\vec{f}}_1\right)_{\xi_k} = \left(\left(\vec{\vec{f}}_1^{\,b}\right)_{\xi_k} + \tilde{\eta}\frac{\left(\vec{\vec{f}}_0^{\,b}\right)_{\xi_k}}{2}\sum_{l=1}^{2}\tilde{d}_l^{-1}\right)\exp\left(-\tilde{\eta}\sqrt{\tilde{s}}\right) \tag{2.82}$$

Calculation of the normal derivative of $\left(\vec{\vec{f}}_1\right)_{\xi_k}$ yields

$$\frac{\partial\left(\vec{\vec{f}}_1\right)_{\xi_k}}{\partial\tilde{\eta}} = \left[-\sqrt{\tilde{s}}\left(\vec{\vec{f}}_1^{\,b}\right)_{\xi_k} + \left(1-\tilde{\eta}\sqrt{\tilde{s}}\right)\frac{\left(\vec{\vec{f}}_0^{\,b}\right)_{\xi_k}}{2}\sum_{l=1}^{2}\tilde{d}_l^{-1}\right]\exp\left(-\tilde{\eta}\sqrt{\tilde{s}}\right) \tag{2.83}$$

Derivation of $\left(\vec{\vec{f}}_2\right)_{\xi_k}$

Substituting Equations 2.79 and 2.82 into Equation 2.78, we obtain

$$\begin{aligned}\frac{\partial^2\left(\vec{\vec{f}}_2\right)_{\xi_k}}{\partial\tilde{\eta}^2} - \tilde{s}\left(\vec{\vec{f}}_2\right)_{\xi_k} = &\left\{-\sqrt{\tilde{s}}\left(\vec{\vec{f}}_1^{\,b}\right)_{\xi_k}\sum_{l=1}^{2}\tilde{d}_l^{-1} + \left(1-\tilde{\eta}\sqrt{\tilde{s}}\right)\frac{\left(\vec{\vec{f}}_0^{\,b}\right)_{\xi_k}}{2}\left(\sum_{l=1}^{2}\tilde{d}_l^{-1}\right)^2\right.\\ &\left.-\sum_{i=1}^{2}\frac{\partial^2\left(\vec{\vec{f}}_0^{\,b}\right)_{\xi_k}}{\partial\tilde{\xi}_i^2} + \frac{\left(\vec{\vec{f}}_0^{\,b}\right)_{\xi_k}}{\tilde{d}_k}\frac{\tilde{d}_{3-k}-\tilde{d}_k}{\tilde{d}_k\tilde{d}_{3-k}} - \tilde{\eta}\sqrt{\tilde{s}}\left(\vec{\vec{f}}_0^{\,b}\right)_{\xi_k}\sum_{n=1}^{2}\tilde{d}_n^{-2}\right\}\exp\left(-\tilde{\eta}\sqrt{\tilde{s}}\right)\end{aligned} \tag{2.84}$$

The right-hand side of Equation 2.84 is transformed as follows:

$$\begin{aligned}&\left\{-\sqrt{\tilde{s}}\left(\vec{\vec{f}}_1^{\,b}\right)_{\xi_k}\sum_{l=1}^{2}\tilde{d}_l^{-1} + \left(\vec{\vec{f}}_0^{\,b}\right)_{\xi_k}\left[\frac{1}{2}\left(\sum_{l=1}^{2}\tilde{d}_l^{-1}\right)^2 + \frac{\tilde{d}_{3-k}-\tilde{d}_k}{\tilde{d}_k^2\tilde{d}_{3-k}}\right] - \sum_{i=1}^{2}\frac{\partial^2\left(\vec{\vec{f}}_0^{\,b}\right)_{\xi_k}}{\partial\tilde{\xi}_i^2}\right.\\ &\left.-\tilde{\eta}\sqrt{\tilde{s}}\left(\vec{\vec{f}}_0^{\,b}\right)_{\xi_k}\left[\frac{1}{2}\left(\sum_{l=1}^{2}\tilde{d}_l^{-1}\right)^2 + \sum_{n=1}^{2}\tilde{d}_n^{-2}\right]\right\}\exp\left(-\tilde{\eta}\sqrt{\tilde{s}}\right)\\ &= \left\{\left(\vec{W}_1\right)_{\xi_k} + \tilde{\eta}\left(\vec{W}_2\right)_{\xi_k}\right\}\exp\left(-\tilde{\eta}\sqrt{\tilde{s}}\right)\end{aligned} \tag{2.85}$$

where

$$\left(\vec{\bar{W}}_1\right)_{\xi_k} = -\sqrt{\bar{s}}\left(\vec{\bar{f}}_1^{\mathrm{b}}\right)_{\xi_k}\sum_{l=1}^{2}\tilde{d}_l^{-1} + \left(\vec{\bar{f}}_0^{\mathrm{b}}\right)_{\xi_k}\left[\frac{1}{2}\left(\sum_{l=1}^{2}\tilde{d}_l^{-1}\right)^2 + \frac{\tilde{d}_{3-k}-\tilde{d}_k}{\tilde{d}_k^2\tilde{d}_{3-k}}\right] - \sum_{i=1}^{2}\frac{\partial^2\left(\vec{\bar{f}}_0^{\mathrm{b}}\right)_{\xi_k}}{\partial\tilde{\xi}_i^2}$$

$$= -\sqrt{\bar{s}}\left(\vec{\bar{f}}_1^{\mathrm{b}}\right)_{\xi_k}\frac{\tilde{d}_k+\tilde{d}_{3-k}}{\tilde{d}_k\tilde{d}_{3-k}} + \left(\vec{\bar{f}}_0^{\mathrm{b}}\right)_{\xi_k}\frac{\tilde{d}_k^2+3\tilde{d}_{3-k}^2}{2\tilde{d}_k^2\tilde{d}_{3-k}^2} - \sum_{i=1}^{2}\frac{\partial^2\left(\vec{\bar{f}}_0^{\mathrm{b}}\right)_{\xi_k}}{\partial\tilde{\xi}_i^2} \tag{2.86}$$

$$\left(\vec{\bar{W}}_2\right)_{\xi_k} = -\sqrt{\bar{s}}\left(\vec{\bar{f}}_0^{\mathrm{b}}\right)_{\xi_k}\left[\frac{1}{2}\left(\sum_{l=1}^{2}\tilde{d}_l^{-1}\right)^2 + \sum_{n=1}^{2}\tilde{d}_n^{-2}\right] = -\sqrt{\bar{s}}\left(\vec{\bar{f}}_0^{\mathrm{b}}\right)_{\xi_k}\frac{3\tilde{d}_k^2+2\tilde{d}_k\tilde{d}_{3-k}+3\tilde{d}_{3-k}^2}{2\tilde{d}_k^2\tilde{d}_{3-k}^2} \tag{2.87}$$

The solution of Equation 2.84 can be represented in the form:

$$\left(\vec{\bar{f}}_2\right)_{\xi_k} = \left\{\left(\vec{\bar{f}}_2^{\mathrm{b}}\right)_{\xi_k} - \frac{1}{4\sqrt{\bar{s}}}\left[\tilde{\eta}\left(2\left(\vec{\bar{W}}_1\right)_{\xi_k} + \frac{\left(\vec{\bar{W}}_2\right)_{\xi_k}}{\sqrt{\bar{s}}}\right) + \tilde{\eta}^2\left(\vec{\bar{W}}_2\right)_{\xi_k}\right]\right\}\exp\left(-\tilde{\eta}\sqrt{\bar{s}}\right)$$

$$= \left\{\left(\vec{\bar{f}}_2^{\mathrm{b}}\right)_{\xi_k} - \frac{\tilde{\eta}}{2\sqrt{\bar{s}}}\left[-\sqrt{\bar{s}}\left(\vec{\bar{f}}_1^{\mathrm{b}}\right)_{\xi_k}\frac{\tilde{d}_k+\tilde{d}_{3-k}}{\tilde{d}_k\tilde{d}_{3-k}} + \left(\vec{\bar{f}}_0^{\mathrm{b}}\right)_{\xi_k}\frac{\tilde{d}_k^2+3\tilde{d}_{3-k}^2}{2\tilde{d}_k^2\tilde{d}_{3-k}^2} - \sum_{i=1}^{2}\frac{\partial^2\left(\vec{\bar{f}}_0^{\mathrm{b}}\right)_{\xi_k}}{\partial\tilde{\xi}_i^2}\right]\right.$$

$$\left. + \frac{\tilde{\eta}}{4\sqrt{\bar{s}}}\left(\vec{\bar{f}}_0^{\mathrm{b}}\right)_{\xi_k}\frac{3\tilde{d}_k^2+2\tilde{d}_k\tilde{d}_{3-k}+3\tilde{d}_{3-k}^2}{2\tilde{d}_k^2\tilde{d}_{3-k}^2} + \frac{\tilde{\eta}^2}{4}\left(\vec{\bar{f}}_0^{\mathrm{b}}\right)_{\xi_k}\frac{3\tilde{d}_k^2+2\tilde{d}_k\tilde{d}_{3-k}+3\tilde{d}_{3-k}^2}{2\tilde{d}_k^2\tilde{d}_{3-k}^2}\right\}\exp\left(-\tilde{\eta}\sqrt{\bar{s}}\right)$$

$$= \left\{\left(\vec{\bar{f}}_2^{\mathrm{b}}\right)_{\xi_k} + \tilde{\eta}\left(\vec{\bar{f}}_1^{\mathrm{b}}\right)_{\xi_k}\frac{\tilde{d}_k+\tilde{d}_{3-k}}{2\tilde{d}_k\tilde{d}_{3-k}} + \frac{\tilde{\eta}}{2\sqrt{\bar{s}}}\left(\vec{\bar{f}}_0^{\mathrm{b}}\right)_{\xi_k}\left(-\frac{2\tilde{d}_k^2+6\tilde{d}_{3-k}^2}{4\tilde{d}_k^2\tilde{d}_{3-k}^2} + \frac{3\tilde{d}_k^2+2\tilde{d}_k\tilde{d}_{3-k}+3\tilde{d}_{3-k}^2}{4\tilde{d}_k^2\tilde{d}_{3-k}^2}\right)\right.$$

$$\left. + \frac{\tilde{\eta}^2}{4}\left(\vec{\bar{f}}_0^{\mathrm{b}}\right)_{\xi_k}\frac{3\tilde{d}_k^2+2\tilde{d}_k\tilde{d}_{3-k}+3\tilde{d}_{3-k}^2}{2\tilde{d}_k^2\tilde{d}_{3-k}^2} + \frac{\tilde{\eta}}{2\sqrt{\bar{s}}}\sum_{i=1}^{2}\frac{\partial^2\left(\vec{\bar{f}}_0^{\mathrm{b}}\right)_{\xi_k}}{\partial\tilde{\xi}_i^2}\right\}\exp\left(-\tilde{\eta}\sqrt{\bar{s}}\right)$$

$$= \left\{\left(\vec{\bar{f}}_2^{\mathrm{b}}\right)_{\xi_k} + \tilde{\eta}\left(\vec{\bar{f}}_1^{\mathrm{b}}\right)_{\xi_k}\frac{\tilde{d}_k+\tilde{d}_{3-k}}{2\tilde{d}_k\tilde{d}_{3-k}} + \tilde{\eta}^2\left(\vec{\bar{f}}_0^{\mathrm{b}}\right)_{\xi_k}\frac{3\tilde{d}_k^2+2\tilde{d}_k\tilde{d}_{3-k}+3\tilde{d}_{3-k}^2}{8\tilde{d}_k^2\tilde{d}_{3-k}^2}\right.$$

$$\left. + \frac{\tilde{\eta}}{2\sqrt{\bar{s}}}\left[\left(\vec{\bar{f}}_0^{\mathrm{b}}\right)_{\xi_k}\frac{\tilde{d}_k^2+2\tilde{d}_k\tilde{d}_{3-k}-3\tilde{d}_{3-k}^2}{4\tilde{d}_k^2\tilde{d}_{3-k}^2} + \sum_{i=1}^{2}\frac{\partial^2\left(\vec{\bar{f}}_0^{\mathrm{b}}\right)_{\xi_k}}{\partial\tilde{\xi}_i^2}\right]\right\}\exp\left(-\tilde{\eta}\sqrt{\bar{s}}\right) \tag{2.88}$$

Equations 2.79, 2.82, and 2.88 give the distributions of the coefficients $\bar{f}_m$, $m = 0, 1, 2$, in the normal direction inside the conductor in the LD. Substitution of Equations 2.79, 2.82, and 2.88 into Equation 2.57 results in the LD distributions of the tangential components of the function $\bar{f}$ of different orders of approximation:

First-order approximation (only the first nonzero term of the expansions is taken into account)

$$\bar{f}_{\xi_k} = \bar{f}^{\,\mathrm{b}}_{\xi_k} \exp(-\tilde{\eta}\sqrt{\tilde{s}}) \tag{2.89}$$

Second-order approximation (only the first and second nonzero terms of the expansions are taken into account)

$$\bar{f}_{\xi_k} = \left(\bar{f}^{\,\mathrm{b}}_{\xi_k} + \tilde{p}\tilde{\eta}\bar{f}^{\,\mathrm{b}}_{\xi_k}\frac{\tilde{d}_k + \tilde{d}_{3-k}}{2\tilde{d}_k\tilde{d}_{3-k}}\right)\exp(-\tilde{\eta}\sqrt{\tilde{s}}) \tag{2.90}$$

Third-order approximation (three nonzero terms of the expansions are taken into account)

$$\begin{aligned}\bar{f}_{\xi_k} = \Bigg[&\bar{f}^{\,\mathrm{b}}_{\xi_k} + \tilde{p}\tilde{\eta}\bar{f}^{\,\mathrm{b}}_{\xi_k}\frac{\tilde{d}_k + \tilde{d}_{3-k}}{2\tilde{d}_k\tilde{d}_{3-k}} + \tilde{p}^2\tilde{\eta}^2\bar{f}^{\,\mathrm{b}}_{\xi_k}\frac{3\tilde{d}_k^2 + 2\tilde{d}_k\tilde{d}_{3-k} + 3\tilde{d}_{3-k}^2}{8\tilde{d}_k^2\tilde{d}_{3-k}^2}\\ &+\tilde{p}^2\frac{\tilde{\eta}}{2\sqrt{\tilde{s}}}\left(\bar{f}^{\,\mathrm{b}}_{\xi_k}\frac{\tilde{d}_k^2 + 2\tilde{d}_k\tilde{d}_{3-k} - 3\tilde{d}_{3-k}^2}{2\tilde{d}_k^2\tilde{d}_{3-k}^2} + \sum_{i=1}^{2}\frac{\partial^2\bar{f}^{\,\mathrm{b}}_{\xi_k}}{\partial\xi_i^2}\right)\Bigg]\exp(-\tilde{\eta}\sqrt{\tilde{s}})\end{aligned} \tag{2.91}$$

In the case of time-harmonic fields, Equations 2.89 through 2.91 can be represented in the frequency domain (FD) replacing the transform variable $\tilde{s}$ by $2j$ as follows:

First-order approximation

$$\dot{f}_{\xi_k} = \dot{f}^{\mathrm{b}}_{\xi_k}\exp\left(-\tilde{\eta}\sqrt{2j}\right) = \dot{f}^{\mathrm{b}}_{\xi_k}\exp(-(1+j)\tilde{\eta}) \tag{2.92}$$

Second-order approximation

$$\dot{f}_{\xi_k} = \left(\dot{f}^{\mathrm{b}}_{\xi_k} + \tilde{p}\tilde{\eta}\dot{f}^{\mathrm{b}}_{\xi_k}\frac{\tilde{d}_k + \tilde{d}_{3-k}}{2\tilde{d}_k\tilde{d}_{3-k}}\right)\exp(-(1+j)\tilde{\eta}) \tag{2.93}$$

Third-order approximation

$$\begin{aligned}\dot{f}_{\xi_k} = \Bigg[&\dot{f}^{\mathrm{b}}_{\xi_k} + \tilde{p}\tilde{\eta}\dot{f}^{\mathrm{b}}_{\xi_k}\frac{\tilde{d}_k + \tilde{d}_{3-k}}{2\tilde{d}_k\tilde{d}_{3-k}} + \tilde{p}^2\tilde{\eta}^2\dot{f}^{\mathrm{b}}_{\xi_k}\frac{3\tilde{d}_k^2 + 2\tilde{d}_k\tilde{d}_{3-k} + 3\tilde{d}_{3-k}^2}{8\tilde{d}_k^2\tilde{d}_{3-k}^2}\\ &+\tilde{p}^2\frac{1-j}{2}\tilde{\eta}\left(\dot{f}^{\mathrm{b}}_{\xi_k}\frac{\tilde{d}_k^2 + 2\tilde{d}_k\tilde{d}_{3-k} - 3\tilde{d}_{3-k}^2}{8\tilde{d}_k^2\tilde{d}_{3-k}^2} + \frac{1}{2}\sum_{i=1}^{2}\frac{\partial^2\dot{f}^{\mathrm{b}}_{\xi_k}}{\partial\xi_i^2}\right)\Bigg]\exp(-(1+j)\tilde{\eta})\end{aligned} \tag{2.94}$$

The transfer back to the time domain (TD) is performed using the inverse Laplace transform, changing the representation in Equations 2.92 through 2.94 to the following TD form:

First-order approximation

$$f_{\xi_k} = f^{\mathrm{b}}_{\xi_k} * \tilde{T}_1 \tag{2.95}$$

Second-order approximation

$$f_{\xi_k} = \left(1 + \tilde{p}\tilde{\eta}\frac{\tilde{d}_k + \tilde{d}_{3-k}}{2\tilde{d}_k\tilde{d}_{3-k}}\right) f^{\mathrm{b}}_{\xi_k} * \tilde{T}_1 \tag{2.96}$$

Third-order approximation

$$\begin{aligned} f_{\xi_k} = {} & \left(1 + \tilde{p}\tilde{\eta}\frac{\tilde{d}_k + \tilde{d}_{3-k}}{2\tilde{d}_k\tilde{d}_{3-k}} + \tilde{p}^2\tilde{\eta}^2\frac{3\tilde{d}_k^2 + 2\tilde{d}_k\tilde{d}_{3-k} + 3\tilde{d}_{3-k}^2}{8\tilde{d}_k^2\tilde{d}_{3-k}^2}\right) f^{\mathrm{b}}_{\xi_k} * \tilde{T}_1 \\ & + \tilde{p}^2\tilde{\eta}\left(f^{\mathrm{b}}_{\xi_k}\frac{\tilde{d}_k^2 + 2\tilde{d}_k\tilde{d}_{3-k} - 3\tilde{d}_{3-k}^2}{8\tilde{d}_k^2\tilde{d}_{3-k}^2} + \frac{1}{2}\sum_{i=1}^{2}\frac{\partial^2 f^{\mathrm{b}}_{\xi_k}}{\partial\tilde{\xi}_i^2}\right) * \tilde{T}_2 \end{aligned} \tag{2.97}$$

where the sign "*" denotes the nondimensional time convolution product defined as follows:

$$\varphi(\tilde{t}) * \phi(\tilde{t}) = \int_0^{\tilde{t}} \varphi(\tilde{t}')\phi(\tilde{t} - \tilde{t}')\mathrm{d}\tilde{t}'$$

and the nondimensional TD functions, $\tilde{T}_1$ and $\tilde{T}_2$, are obtained using the following Laplace pairs [6]:

$$\exp\left(-\tilde{\eta}\sqrt{\tilde{s}}\right) \quad \Leftrightarrow \quad \frac{1}{2\sqrt{\pi}}\tilde{\eta}\tilde{t}^{-3/2}\exp\left(-\frac{\tilde{\eta}^2}{4\tilde{t}}\right) = \tilde{T}_1(\tilde{\eta}, \tilde{t}) \tag{2.98}$$

$$\frac{1}{\sqrt{\tilde{s}}}\exp\left(-\tilde{\eta}\sqrt{\tilde{s}}\right) \quad \Leftrightarrow \quad \frac{1}{\sqrt{\pi}}\tilde{t}^{-1/2}\exp\left(-\frac{\tilde{\eta}^2}{4\tilde{t}}\right) = \tilde{T}_2(\tilde{\eta}, \tilde{t}) \tag{2.99}$$

2.5 Normal Components

The LD Equations 2.69, 2.71, and 2.72 can be represented in the form:

$$\left(\vec{\tilde{f}}_1\right)_{\eta} = -\tilde{s}^{-1}\sum_{i=1}^{2}\frac{\partial^2\left(\vec{\tilde{f}}_0\right)_{\xi_i}}{\partial\tilde{\xi}_i\partial\tilde{\eta}} \tag{2.100}$$

$$\left(\vec{\bar{f}}_2\right)_\eta = -\tilde{s}^{-1}\sum_{i=1}^{2}\frac{\partial}{\partial\tilde{\xi}_i}\left[\frac{\partial\left(\vec{\bar{f}}_1\right)_{\xi_i}}{\partial\tilde{\eta}} + \frac{\tilde{\eta}}{\tilde{d}_i}\frac{\partial\left(\vec{\bar{f}}_0\right)_{\xi_i}}{\partial\tilde{\eta}} - \frac{\left(\vec{\bar{f}}_0\right)_{\xi_i}}{\tilde{d}_i}\right] \tag{2.101}$$

$$\left(\vec{\bar{f}}_3\right)_\eta = -\tilde{s}^{-1}\sum_{i=1}^{2}\frac{\partial}{\partial\tilde{\xi}_i}\left[\frac{\partial\left(\vec{\bar{f}}_2\right)_{\xi_i}}{\partial\tilde{\eta}} + \frac{\tilde{\eta}}{\tilde{d}_i}\frac{\partial\left(\vec{\bar{f}}_1\right)_{\xi_i}}{\partial\tilde{\eta}} - \frac{\left(\vec{\bar{f}}_1\right)_{\xi_i}}{\tilde{d}_i} - \frac{\partial\left(\vec{\bar{f}}_1\right)_{\eta}}{\partial\tilde{\xi}_i} + \tilde{\eta}\left(\frac{\tilde{\eta}}{\tilde{d}_i^2}\frac{\partial\left(\vec{\bar{f}}_0\right)_{\xi_i}}{\partial\tilde{\eta}} - \frac{2}{\tilde{d}_i^2}\left(\vec{\bar{f}}_0\right)_{\xi_i}\right)\right] \tag{2.102}$$

From Equation 2.100:

$$\left(\vec{\bar{f}}_1\right)_\eta = \tilde{s}^{-1/2}\sum_{i=1}^{2}\frac{\partial\left(\vec{\bar{f}}_0^{\mathrm{b}}\right)_{\xi_i}}{\partial\tilde{\xi}_i}\exp\left(-\tilde{\eta}\sqrt{\tilde{s}}\right) \tag{2.103}$$

From Equation 2.101:

$$\begin{aligned}\left(\vec{\bar{f}}_2\right)_\eta &= -\tilde{s}^{-1}\sum_{i=1}^{2}\frac{\partial}{\partial\tilde{\xi}_i}\left[-\tilde{s}^{1/2}\left(\vec{\bar{f}}_1\right)_{\xi_i} + (1-\tilde{s}^{1/2}\tilde{\eta})\frac{\tilde{d}_i+\tilde{d}_{3-i}}{2\tilde{d}_i\tilde{d}_{3-i}}\left(\vec{\bar{f}}_0\right)_{\xi_i} - \tilde{s}^{1/2}\tilde{\eta}\frac{\left(\vec{\bar{f}}_0\right)_{\xi_i}}{\tilde{d}_i} - \frac{\left(\vec{\bar{f}}_0\right)_{\xi_i}}{\tilde{d}_i}\right]\exp\left(-\tilde{\eta}\sqrt{\tilde{s}}\right)\\ &= \tilde{s}^{-1/2}\sum_{i=1}^{2}\frac{\partial}{\partial\tilde{\xi}_i}\left[\left(\vec{\bar{f}}_1\right)_{\xi_i} + \tilde{\eta}\frac{\tilde{d}_i+3\tilde{d}_{3-i}}{2\tilde{d}_i\tilde{d}_{3-i}}\left(\vec{\bar{f}}_0\right)_{\xi_i} - \tilde{s}^{-1/2}\frac{\tilde{d}_i-\tilde{d}_{3-i}}{2\tilde{d}_i\tilde{d}_{3-i}}\left(\vec{\bar{f}}_0\right)_{\xi_i}\right]\exp\left(-\tilde{\eta}\sqrt{\tilde{s}}\right)\end{aligned} \tag{2.104}$$

From Equation 2.102 (see details in Appendix 2.A.1):

$$\begin{aligned}\left(\vec{\bar{f}}_3\right)_\eta = \sum_{i=1}^{2}\frac{\partial}{\partial\tilde{\xi}_i}\Bigg\{&\tilde{s}^{-1/2}\left[\left(\vec{\bar{f}}_2^{\mathrm{b}}\right)_{\xi_i} + \tilde{\eta}\left(\vec{\bar{f}}_1^{\mathrm{b}}\right)_{\xi_i}\frac{\tilde{d}_i+3\tilde{d}_{3-i}}{2\tilde{d}_i\tilde{d}_{3-i}} + \tilde{\eta}^2\left(\vec{\bar{f}}_0^{\mathrm{b}}\right)_{\xi_i}\frac{3\tilde{d}_i^2+6\tilde{d}_i\tilde{d}_{3-i}+11\tilde{d}_{3-i}^2}{8\tilde{d}_i^2\tilde{d}_{3-i}^2}\right]\\ &-\tilde{s}^{-1}\left[\left(\vec{\bar{f}}_1^{\mathrm{b}}\right)_{\xi_i}\frac{\tilde{d}_i-\tilde{d}_{3-i}}{2\tilde{d}_i\tilde{d}_{3-i}} + \tilde{\eta}\left(\left(\vec{\bar{f}}_0^{\mathrm{b}}\right)_{\xi_i}\frac{5\tilde{d}_i^2-2\tilde{d}_i\tilde{d}_{3-i}-7\tilde{d}_{3-i}^2}{8\tilde{d}_i^2\tilde{d}_{3-i}^2} - \frac{1}{2}\sum_{l=1}^{2}\frac{\partial^2\left(\vec{\bar{f}}_0^{\mathrm{b}}\right)_{\xi_i}}{\partial\tilde{\xi}_l^2}\right)\right]\\ &-\tilde{s}^{-3/2}\left[\left(\vec{\bar{f}}_0^{\mathrm{b}}\right)_{\xi_i}\frac{\tilde{d}_i^2+2\tilde{d}_i\tilde{d}_{3-i}-3\tilde{d}_{3-i}^2}{8\tilde{d}_i^2\tilde{d}_{3-i}^2} - \frac{1}{2}\frac{\partial^2\left(\vec{\bar{f}}_0^{\mathrm{b}}\right)_{\xi_i}}{\partial\tilde{\xi}_i^2} + \frac{1}{2}\frac{\partial^2\left(\vec{\bar{f}}_0^{\mathrm{b}}\right)_{\xi_i}}{\partial\tilde{\xi}_{3-i}^2} - \frac{\partial^2\left(\vec{\bar{f}}_0^{\mathrm{b}}\right)_{\xi_{3-i}}}{\partial\tilde{\xi}_i\partial\tilde{\xi}_{3-i}}\right]\Bigg\}\exp\left(\tilde{\eta}\sqrt{\tilde{s}}\right)\end{aligned} \tag{2.105}$$

Substituting Equations 2.103 through 2.105 into Equation 2.58, we obtain

First-order approximation

$$\bar{f}_\eta = \tilde{p}\sum_{i=1}^{2}\tilde{s}^{-1/2}\frac{\partial\bar{f}_{\xi_i}^{\mathrm{b}}}{\partial\tilde{\xi}_i}\exp\left(-\tilde{\eta}\sqrt{\tilde{s}}\right) \tag{2.106}$$

Second-order approximation

$$\bar{f}_\eta = \tilde{s}^{-1/2}\tilde{p}\sum_{i=1}^{2}\frac{\partial}{\partial\tilde{\xi}_i}\left[\bar{f}^{\mathrm{b}}_{\xi_i} + \tilde{p}\tilde{\eta}\,\bar{f}^{\mathrm{b}}_{\xi_i}\frac{\tilde{d}_i + 3\tilde{d}_{3-i}}{2\tilde{d}_i\tilde{d}_{3-i}} - \tilde{p}\tilde{s}^{-1/2}\bar{f}^{\mathrm{b}}_{\xi_i}\frac{\tilde{d}_i - \tilde{d}_{3-i}}{2\tilde{d}_i\tilde{d}_{3-i}}\right]\exp\left(-\tilde{\eta}\sqrt{\tilde{s}}\right) \tag{2.107}$$

Third-order approximation

$$\begin{aligned}\bar{f}_\eta = \tilde{s}^{-1/2}\tilde{p}\sum_{i=1}^{2}\frac{\partial}{\partial\tilde{\xi}_i}\Bigg\{&\bar{f}^{\mathrm{b}}_{\xi_i} + \tilde{p}\tilde{\eta}\,\bar{f}^{\mathrm{b}}_{\xi_i}\frac{\tilde{d}_i + 3\tilde{d}_{3-i}}{2\tilde{d}_i\tilde{d}_{3-i}} + \tilde{p}^2\tilde{\eta}^2\bar{f}^{\mathrm{b}}_{\xi_i}\frac{3\tilde{d}_i^2 + 6\tilde{d}_i\tilde{d}_{3-i} + 11\tilde{d}_{3-i}^2}{8\tilde{d}_i^2\tilde{d}_{3-i}^2}\\ &- \tilde{p}\tilde{s}^{-1/2}\left[\bar{f}^{\mathrm{b}}_{\xi_i}\frac{\tilde{d}_i - \tilde{d}_{3-i}}{2\tilde{d}_i\tilde{d}_{3-i}} + \tilde{p}\tilde{\eta}\left(\bar{f}^{\mathrm{b}}_{\xi_i}\frac{5\tilde{d}_i^2 + 2\tilde{d}_i\tilde{d}_{3-i} - 7\tilde{d}_{3-i}^2}{8\tilde{d}_i^2\tilde{d}_{3-i}^2} - \frac{1}{2}\sum_{l=1}^{2}\frac{\partial^2\bar{f}^{\mathrm{b}}_{\xi_i}}{\partial\tilde{\xi}_l^2}\right)\right]\\ &-\tilde{p}^2\tilde{s}^{-1}\left[\bar{f}^{\mathrm{b}}_{\xi_i}\frac{\tilde{d}_i^2 + 2\tilde{d}_i\tilde{d}_{3-i} - 3\tilde{d}_{3-i}^2}{8\tilde{d}_i^2\tilde{d}_{3-i}^2} - \frac{1}{2}\frac{\partial^2\bar{f}^{\mathrm{b}}_{\xi_i}}{\partial\tilde{\xi}_i^2} + \frac{1}{2}\frac{\partial^2\bar{f}^{\mathrm{b}}_{\xi_i}}{\partial\tilde{\xi}_{3-i}^2} - \frac{\partial^2\bar{f}^{\mathrm{b}}_{\xi_{3-i}}}{\partial\tilde{\xi}_i\partial\tilde{\xi}_{3-i}}\right]\Bigg\}\exp\left(\tilde{\eta}\sqrt{\tilde{s}}\right)\end{aligned} \tag{2.108}$$

The FD and TD versions of Equations 2.106 through 2.108 can be represented in the form:

First-order approximation

$$\dot{f}_\eta = \tilde{p}\frac{1-j}{2}\sum_{i=1}^{2}\frac{\partial\dot{f}^{\mathrm{b}}_{\xi_i}}{\partial\tilde{\xi}_i}\exp(-(1+j)\tilde{\eta}) \tag{2.109}$$

$$f_\eta = \tilde{p}\sum_{i=1}^{2}\frac{\partial f^{\mathrm{b}}_{\xi_i}}{\partial\tilde{\xi}_i} * \tilde{T}_2 \tag{2.110}$$

Second-order approximation

$$\dot{f}_\eta = \tilde{p}\frac{1-j}{2}\sum_{i=1}^{2}\frac{\partial}{\partial\tilde{\xi}_i}\left(\dot{f}^{\mathrm{b}}_{\xi_i} + \tilde{p}\tilde{\eta}\,\dot{f}^{\mathrm{b}}_{\xi_i}\frac{\tilde{d}_i + 3\tilde{d}_{3-i}}{2\tilde{d}_i\tilde{d}_{3-i}} - \tilde{p}\frac{1-j}{2}\dot{f}^{\mathrm{b}}_{\xi_i}\frac{\tilde{d}_i - \tilde{d}_{3-i}}{2\tilde{d}_i\tilde{d}_{3-i}}\right)\exp(-(1+j)\tilde{\eta}) \tag{2.111}$$

$$f_\eta = \tilde{p}\sum_{i=1}^{2}\frac{\partial}{\partial\tilde{\xi}_i}\left[\left(1 + \tilde{p}\tilde{\eta}\frac{\tilde{d}_i + 3\tilde{d}_{3-i}}{2\tilde{d}_i\tilde{d}_{3-i}}\right)f^{\mathrm{b}}_{\xi_i} * \tilde{T}_2 - \tilde{p}\frac{\tilde{d}_i - \tilde{d}_{3-i}}{2\tilde{d}_i\tilde{d}_{3-i}}f^{\mathrm{b}}_{\xi_i} * \tilde{T}_3\right] \tag{2.112}$$

Third-order approximation

$$\dot{f}_\eta = \tilde{p}\frac{1-j}{2}\sum_{i=1}^{2}\frac{\partial}{\partial\tilde{\xi}_i}\left\{\dot{f}^{\text{b}}_{\xi_i} + \tilde{p}\tilde{\eta}\dot{f}^{\text{b}}_{\xi_i}\frac{\tilde{d}_i + 3\tilde{d}_{3-i}}{2\tilde{d}_i\tilde{d}_{3-i}} + \tilde{p}^2\tilde{\eta}^2\dot{f}^{\text{b}}_{\xi_i}\frac{3\tilde{d}_i^2 + 6\tilde{d}_i\tilde{d}_{3-i} + 11\tilde{d}_{3-i}^2}{8\tilde{d}_i^2\tilde{d}_{3-i}^2}\right.$$
$$-\tilde{p}\frac{1-j}{2}\left[\dot{f}^{\text{b}}_{\xi_i}\frac{\tilde{d}_i - \tilde{d}_{3-i}}{2\tilde{d}_i\tilde{d}_{3-i}} + \tilde{p}\tilde{\eta}\left(\dot{f}^{\text{b}}_{\xi_i}\frac{5\tilde{d}_i^2 + 2\tilde{d}_i\tilde{d}_{3-i} - 7\tilde{d}_{3-i}^2}{8\tilde{d}_i^2\tilde{d}_{3-i}^2} - \frac{1}{2}\sum_{l=1}^{2}\frac{\partial^2\dot{f}^{\text{b}}_{\xi_i}}{\partial\tilde{\xi}_l^2}\right)\right]$$
$$\left.+\tilde{p}^2\frac{j}{2}\left[\dot{f}^{\text{b}}_{\xi_i}\frac{\tilde{d}_i^2 + 2\tilde{d}_i\tilde{d}_{3-i} - 3\tilde{d}_{3-i}^2}{8\tilde{d}_i^2\tilde{d}_{3-i}^2} - \frac{1}{2}\frac{\partial^2\dot{f}^{\text{b}}_{\xi_i}}{\partial\tilde{\xi}_i^2} + \frac{1}{2}\frac{\partial^2\dot{f}^{\text{b}}_{\xi_i}}{\partial\tilde{\xi}_{3-i}^2} - \frac{\partial^2\dot{f}^{\text{b}}_{\xi_{3-i}}}{\partial\tilde{\xi}_i\partial\tilde{\xi}_{3-i}}\right]\right\}\exp((1+j)\tilde{\eta}) \quad (2.113)$$

$$f_\eta = \tilde{p}\sum_{i=1}^{2}\frac{\partial}{\partial\tilde{\xi}_i}\left\{\left(1 + \tilde{p}\tilde{\eta}\frac{\tilde{d}_i + 3\tilde{d}_{3-i}}{2\tilde{d}_i\tilde{d}_{3-i}} + \tilde{p}^2\tilde{\eta}^2\frac{3\tilde{d}_i^2 + 6\tilde{d}_i\tilde{d}_{3-i} + 11\tilde{d}_{3-i}^2}{8\tilde{d}_i^2\tilde{d}_{3-i}^2}\right)f^{\text{b}}_{\xi_i} * \tilde{T}_2\right.$$
$$-\tilde{p}\left[f^{\text{b}}_{\xi_i}\frac{\tilde{d}_i - \tilde{d}_{3-i}}{2\tilde{d}_i\tilde{d}_{3-i}} + \tilde{p}\tilde{\eta}\left(f^{\text{b}}_{\xi_i}\frac{5\tilde{d}_i^2 + 2\tilde{d}_i\tilde{d}_{3-i} - 7\tilde{d}_{3-i}^2}{8\tilde{d}_i^2\tilde{d}_{3-i}^2} - \frac{1}{2}\sum_{l=1}^{2}\frac{\partial^2 f^{\text{b}}_{\xi_i}}{\partial\tilde{\xi}_l^2}\right)\right] * \tilde{T}_3$$
$$\left.-\tilde{p}^2\left(f^{\text{b}}_{\xi_i}\frac{\tilde{d}_i^2 + 2\tilde{d}_i\tilde{d}_{3-i} - 3\tilde{d}_{3-i}^2}{8\tilde{d}_i^2\tilde{d}_{3-i}^2} - \frac{1}{2}\frac{\partial^2 f^{\text{b}}_{\xi_i}}{\partial\tilde{\xi}_i^2} + \frac{1}{2}\frac{\partial^2 f^{\text{b}}_{\xi_i}}{\partial\tilde{\xi}_{3-i}^2} - \frac{\partial^2 f^{\text{b}}_{\xi_{3-i}}}{\partial\tilde{\xi}_i\partial\tilde{\xi}_{3-i}}\right) * \tilde{T}_4\right\} \quad (2.114)$$

Here the functions $\tilde{T}_3$ and $\tilde{T}_4$ have been obtained using the inverse Laplace transform:

$$\tilde{s}^{-1}\exp(-\tilde{\eta}\sqrt{\tilde{s}}) \quad \Leftrightarrow \quad (\pi\tilde{t})^{-1}\text{erfc}\left(\frac{\tilde{\eta}}{\sqrt{2\tilde{t}}}\right) = \tilde{T}_3(\tilde{\eta},\tilde{t}) \quad (2.115)$$

$$\tilde{s}^{-3/2}\exp(-\tilde{\eta}\sqrt{\tilde{s}}) \quad \Leftrightarrow \quad 2\tilde{t}^{1/2}\pi^{-1/2}\exp\left(-\frac{\tilde{\eta}^2}{4\tilde{t}}\right) - \tilde{\eta}\text{erfc}\left(\frac{\tilde{\eta}}{\sqrt{2\tilde{t}}}\right) = \tilde{T}_4(\tilde{\eta},\tilde{t}) \quad (2.116)$$

and

$$\text{erfc}(x) = 1 - \text{erf}(x) = \frac{2}{\sqrt{\pi}}\int_x^\infty \exp(-u^2)\text{d}u \quad (2.117)$$

As follows from Equations 2.106 through 2.108, the normal component of a function governed by the equation of diffusion in the conductor's skin layer (including surface) is much smaller than the tangential components and can be neglected in the zero-order approximation.

2.6 Normal Derivatives

The normal derivatives of the tangential components in the LD can be obtained directly from Equations 2.79, 2.82, and 2.88 and are written in the following forms:

$$\frac{\partial\left(\vec{f}_0\right)_{\xi_k}}{\partial\tilde{\eta}} = -\tilde{s}^{1/2}\left(\vec{f}_0^{\mathrm{b}}\right)_{\xi_k}\exp\left(-\tilde{\eta}\sqrt{\tilde{s}}\right) \tag{2.118}$$

$$\begin{aligned}\frac{\partial\left(\vec{f}_1\right)_{\xi_k}}{\partial\tilde{\eta}} &= \left[-\tilde{s}^{1/2}\left(\vec{f}_1^{\mathrm{b}}\right)_{\xi_k} + (1-\tilde{\eta}\tilde{s}^{1/2})\left(\vec{f}_0^{\mathrm{b}}\right)_{\xi_k}\frac{\tilde{d}_k+\tilde{d}_{3-k}}{2\tilde{d}_k\tilde{d}_{3-k}}\right]\exp\left(-\tilde{\eta}\sqrt{\tilde{s}}\right)\\ &= -\tilde{s}^{1/2}\left[\left(\vec{f}_1^{\mathrm{b}}\right)_{\xi_k} + \tilde{\eta}\left(\vec{f}_0^{\mathrm{b}}\right)_{\xi_k}\frac{\tilde{d}_k+\tilde{d}_{3-k}}{2\tilde{d}_k\tilde{d}_{3-k}} - \tilde{s}^{-1/2}\left(\vec{f}_0^{\mathrm{b}}\right)_{\xi_k}\frac{\tilde{d}_k+\tilde{d}_{3-k}}{2\tilde{d}_k\tilde{d}_{3-k}}\right]\exp\left(-\tilde{\eta}\sqrt{\tilde{s}}\right)\end{aligned} \tag{2.119}$$

$$\begin{aligned}\frac{\partial\left(\vec{f}_2\right)_{\xi_k}}{\partial\tilde{\eta}} &= \Bigg\{-\sqrt{\tilde{s}}\left(\vec{f}_2^{\mathrm{b}}\right)_{\xi_k} + (1-\tilde{\eta}\sqrt{\tilde{s}})\left(\vec{f}_1^{\mathrm{b}}\right)_{\xi_k}\frac{\tilde{d}_k+\tilde{d}_{3-k}}{2\tilde{d}_k\tilde{d}_{3-k}} + (2\tilde{\eta}-\tilde{\eta}^2\sqrt{\tilde{s}})\left(\vec{f}_0^{\mathrm{b}}\right)_{\xi_k}\frac{3\tilde{d}_k^2+2\tilde{d}_k\tilde{d}_{3-k}+3\tilde{d}_{3-k}^2}{8\tilde{d}_k^2\tilde{d}_{3-k}^2}\\ &\quad + \left(\frac{1}{2\sqrt{\tilde{s}}}-\frac{\tilde{\eta}}{2}\right)\left[\left(\vec{f}_0^{\mathrm{b}}\right)_{\xi_k}\frac{\tilde{d}_k^2+2\tilde{d}_k\tilde{d}_{3-k}-3\tilde{d}_{3-k}^2}{4\tilde{d}_k^2\tilde{d}_{3-k}^2} + \sum_{i=1}^{2}\frac{\partial^2\left(\vec{f}_0^{\mathrm{b}}\right)_{\xi_k}}{\partial\tilde{\xi}_i^2}\right]\Bigg\}\exp\left(-\tilde{\eta}\sqrt{\tilde{s}}\right)\\ &= -\tilde{s}^{1/2}\Bigg\{\left(\vec{f}_2^{\mathrm{b}}\right)_{\xi_k} + \tilde{\eta}\left(\vec{f}_1^{\mathrm{b}}\right)_{\xi_k}\frac{\tilde{d}_k+\tilde{d}_{3-k}}{2\tilde{d}_k\tilde{d}_{3-k}} + \tilde{\eta}^2\left(\vec{f}_0^{\mathrm{b}}\right)_{\xi_k}\frac{3\tilde{d}_k^2+2\tilde{d}_k\tilde{d}_{3-k}+3\tilde{d}_{3-k}^2}{8\tilde{d}_k^2\tilde{d}_{3-k}^2}\\ &\quad - \tilde{s}^{-1/2}\left(\left(\vec{f}_1^{\mathrm{b}}\right)_{\xi_k}\frac{\tilde{d}_k+\tilde{d}_{3-k}}{2\tilde{d}_k\tilde{d}_{3-k}} + \tilde{\eta}\left(\vec{f}_0^{\mathrm{b}}\right)_{\xi_k}\frac{5\tilde{d}_k^2+2\tilde{d}_k\tilde{d}_{3-k}+9\tilde{d}_{3-k}^2}{8\tilde{d}_k^2\tilde{d}_{3-k}^2} - \frac{\tilde{\eta}}{2}\sum_{i=1}^{2}\frac{\partial^2\left(\vec{f}_0^{\mathrm{b}}\right)_{\xi_k}}{\partial\tilde{\xi}_i^2}\right)\\ &\quad - \tilde{s}^{-1}\left(\left(\vec{f}_0^{\mathrm{b}}\right)_{\xi_k}\frac{\tilde{d}_k^2+2\tilde{d}_k\tilde{d}_{3-k}-3\tilde{d}_{3-k}^2}{8\tilde{d}_k^2\tilde{d}_{3-k}^2} + \frac{1}{2}\sum_{i=1}^{2}\frac{\partial^2\left(\vec{f}_0^{\mathrm{b}}\right)_{\xi_k}}{\partial\tilde{\xi}_i^2}\right)\Bigg\}\exp\left(-\tilde{\eta}\sqrt{\tilde{s}}\right)\end{aligned} \tag{2.120}$$

Similarly, we obtain the normal derivatives of the normal component in the LD from Equations 2.103 through 2.105:

$$\frac{\partial\left(\vec{f}_1\right)_{\eta}}{\partial\tilde{\eta}} = -\sum_{i=1}^{2}\frac{\partial\left(\vec{f}_0^{\mathrm{b}}\right)_{\xi_i}}{\partial\tilde{\xi}_i}\exp\left(-\tilde{\eta}\sqrt{\tilde{s}}\right) \tag{2.121}$$

$$\begin{aligned}\frac{\partial\left(\vec{f}_2\right)_{\eta}}{\partial\tilde{\eta}} &= -\sum_{i=1}^{2}\frac{\partial}{\partial\tilde{\xi}_i}\left[\left(\vec{f}_1\right)_{\xi_i} + (\tilde{\eta}-\tilde{s}^{-1/2})\left(\vec{f}_0\right)_{\xi_i}\frac{\tilde{d}_i+3\tilde{d}_{3-i}}{2\tilde{d}_i\tilde{d}_{3-i}} - \tilde{s}^{-1/2}\left(\vec{f}_0\right)_{\xi_i}\frac{\tilde{d}_i-\tilde{d}_{3-i}}{2\tilde{d}_i\tilde{d}_{3-i}}\right]\exp\left(-\tilde{\eta}\sqrt{\tilde{s}}\right)\\ &= -\sum_{i=1}^{2}\frac{\partial}{\partial\tilde{\xi}_i}\left[\left(\vec{f}_1^{\mathrm{b}}\right)_{\xi_i} + \tilde{\eta}\left(\vec{f}_0^{\mathrm{b}}\right)_{\xi_i}\frac{\tilde{d}_i+3\tilde{d}_{3-i}}{2\tilde{d}_i\tilde{d}_{3-i}} - \tilde{s}^{-1/2}\left(\vec{f}_0\right)_{\xi_i}\frac{\tilde{d}_i+\tilde{d}_{3-i}}{\tilde{d}_i\tilde{d}_{3-i}}\right]\exp\left(-\tilde{\eta}\sqrt{\tilde{s}}\right)\end{aligned} \tag{2.122}$$

$$\frac{\partial\left(\vec{f}_3\right)_{\eta}}{\partial\tilde{\eta}} = -\sum_{i=1}^{2}\frac{\partial}{\partial\tilde{\xi}_i}\left\{\left[\left(\vec{f}_2^{\,b}\right)_{\xi_i} + (\tilde{\eta}-\tilde{s}^{-1/2})\left(\vec{f}_1^{\,b}\right)_{\xi_i}\frac{\tilde{d}_i+3\tilde{d}_{3-i}}{2\tilde{d}_i\tilde{d}_{3-i}}\right.\right.$$
$$\left. + (\tilde{\eta}^2 - 2\tilde{s}^{-1/2}\tilde{\eta})\left(\vec{f}_0^{\,b}\right)_{\xi_i}\frac{3\tilde{d}_i^2+6\tilde{d}_i\tilde{d}_{3-i}+11\tilde{d}_{3-i}^2}{8\tilde{d}_i^2\tilde{d}_{3-i}^2}\right]$$
$$-\tilde{s}^{-1/2}\left[\left(\vec{f}_1^{\,b}\right)_{\xi_i}\frac{\tilde{d}_i-\tilde{d}_{3-i}}{2\tilde{d}_i\tilde{d}_{3-i}} + (\tilde{\eta}-\tilde{s}^{-1/2})\left(\left(\vec{f}_0^{\,b}\right)_{\xi_i}\frac{5\tilde{d}_i^2+2\tilde{d}_i\tilde{d}_{3-i}-7\tilde{d}_{3-i}^2}{8\tilde{d}_i^2\tilde{d}_{3-i}^2} - \frac{1}{2}\sum_{l=1}^{2}\frac{\partial^2\left(\vec{f}_0^{\,b}\right)_{\xi_i}}{\partial\tilde{\xi}_l^2}\right)\right]$$
$$\left.-\tilde{s}^{-1}\left[\left(\vec{f}_0^{\,b}\right)_{\xi_i}\frac{\tilde{d}_i^2+2\tilde{d}_i\tilde{d}_{3-i}-3\tilde{d}_{3-i}^2}{8\tilde{d}_i^2\tilde{d}_{3-i}^2} - \frac{1}{2}\frac{\partial^2\left(\vec{f}_0^{\,b}\right)_{\xi_i}}{\partial\tilde{\xi}_i^2} + \frac{1}{2}\frac{\partial^2\left(\vec{f}_0^{\,b}\right)_{\xi_i}}{\partial\tilde{\xi}_{3-i}^2} - \frac{\partial^2\left(\vec{f}_0^{\,b}\right)_{\xi_{3-i}}}{\partial\tilde{\xi}_i\partial\tilde{\xi}_{3-i}}\right]\right\}\exp(\tilde{\eta}\sqrt{\tilde{s}})$$
$$= -\sum_{i=1}^{2}\frac{\partial}{\partial\tilde{\xi}_i}\left\{\left[\left(\vec{f}_2^{\,b}\right)_{\xi_i} + \tilde{\eta}\left(\vec{f}_1^{\,b}\right)_{\xi_i}\frac{\tilde{d}_i+3\tilde{d}_{3-i}}{2\tilde{d}_i\tilde{d}_{3-i}} + \tilde{\eta}^2\left(\vec{f}_0^{\,b}\right)_{\xi_i}\frac{3\tilde{d}_i^2+6\tilde{d}_i\tilde{d}_{3-i}+11\tilde{d}_{3-i}^2}{8\tilde{d}_i^2\tilde{d}_{3-i}^2}\right]\right.$$
$$-\tilde{s}^{-1/2}\left[\left(\vec{f}_1^{\,b}\right)_{\xi_i}\frac{\tilde{d}_i+\tilde{d}_{3-i}}{\tilde{d}_i\tilde{d}_{3-i}} + \tilde{\eta}\left(\left(\vec{f}_0^{\,b}\right)_{\xi_i}\frac{11\tilde{d}_i^2+14\tilde{d}_i\tilde{d}_{3-i}+15\tilde{d}_{3-i}^2}{8\tilde{d}_i^2\tilde{d}_{3-i}^2} - \frac{1}{2}\sum_{l=1}^{2}\frac{\partial^2\left(\vec{f}_0^{\,b}\right)_{\xi_i}}{\partial\tilde{\xi}_l^2}\right)\right]$$
$$\left.-\tilde{s}^{-1}\left[\left(\vec{f}_0^{\,b}\right)_{\xi_i}\frac{-\tilde{d}_i^2+\tilde{d}_{3-i}^2}{2\tilde{d}_i^2\tilde{d}_{3-i}^2} + \frac{\partial^2\left(\vec{f}_0^{\,b}\right)_{\xi_i}}{\partial\tilde{\xi}_{3-i}^2} - \frac{\partial^2\left(\vec{f}_0^{\,b}\right)_{\xi_{3-i}}}{\partial\tilde{\xi}_i\partial\tilde{\xi}_{3-i}}\right]\right\}\exp(\tilde{\eta}\sqrt{\tilde{s}}) \tag{2.123}$$

Substituting Equations 2.121 through 2.123 into Equation 2.57 and representing the results in the FD and TD, we obtain

First-order approximation

$$\frac{\partial\dot{f}_{\xi k}}{\partial\tilde{\eta}} = -(1+j)\dot{f}_{\xi k}^{b}\exp\left(-(1+j)\tilde{\eta}\right) \quad \text{(FD)} \tag{2.124}$$

$$\frac{\partial\dot{f}_{\eta}}{\partial\tilde{\eta}} = -\tilde{p}\sum_{i=1}^{2}\frac{\partial\dot{f}_{\xi_i}^{b}}{\partial\tilde{\xi}_i}\exp(-(1+j)\tilde{\eta}) \quad \text{(FD)} \tag{2.125}$$

$$\frac{\partial f_{\xi_k}}{\partial\tilde{\eta}} = -\frac{\mathrm{d}}{\mathrm{d}\tilde{t}}\left(f_{\xi_k}^{b} * \tilde{T}_2\right) \quad \text{(TD)} \tag{2.126}$$

$$\frac{\partial f_{\eta}}{\partial\tilde{\eta}} = -\tilde{p}\sum_{i=1}^{2}\frac{\partial f_{\xi_i}^{b}}{\partial\tilde{\xi}_i} * \tilde{T}_1 \quad \text{(TD)} \tag{2.127}$$

Second-order approximation

$$\frac{\partial\dot{f}_{\xi_k}}{\partial\tilde{\eta}} = -(1+j)\left(\dot{f}_{\xi_k}^{b} + \tilde{p}\tilde{\eta}\dot{f}_{\xi_k}^{b}\frac{\tilde{d}_k+\tilde{d}_{3-k}}{2\tilde{d}_k\tilde{d}_{3-k}} - \tilde{p}\frac{1-j}{2}\dot{f}_{\xi_k}^{b}\frac{\tilde{d}_k+\tilde{d}_{3-k}}{2\tilde{d}_k\tilde{d}_{3-k}}\right)\exp\left(-(1+j)\tilde{\eta}\right) \tag{2.128}$$

$$\frac{\partial \dot{f}_\eta}{\partial \tilde{\eta}} = -\tilde{p}\sum_{i=1}^{2}\frac{\partial}{\partial \tilde{\xi}_i}\left(\dot{f}^{\mathrm{b}}_{\xi_i} + \tilde{p}\tilde{\eta}\dot{f}^{\mathrm{b}}_{\xi_i}\frac{\tilde{d}_i + 3\tilde{d}_{3-i}}{2\tilde{d}_i\tilde{d}_{3-i}} - \tilde{p}\frac{1-j}{2}\dot{f}^{\mathrm{b}}_{\xi_i}\frac{\tilde{d}_i + \tilde{d}_{3-i}}{\tilde{d}_i\tilde{d}_{3-i}}\right)\exp\left(-(1+j)\tilde{\eta}\right) \tag{2.129}$$

$$\frac{\partial f_{\xi_k}}{\partial \tilde{\eta}} = -\frac{\mathrm{d}}{\mathrm{d}\tilde{t}}\left[\left(1 + \tilde{p}\tilde{\eta}\frac{\tilde{d}_k + \tilde{d}_{3-k}}{2\tilde{d}_k\tilde{d}_{3-k}}\right)f^{\mathrm{b}}_{\xi_k} * \tilde{T}_2\right] + \tilde{p}\frac{\tilde{d}_k + \tilde{d}_{3-k}}{2\tilde{d}_k\tilde{d}_{3-k}}f^{\mathrm{b}}_{\xi_k} * \tilde{T}_1 \tag{2.130}$$

$$\frac{\partial f_\eta}{\partial \tilde{\eta}} = -\tilde{p}\sum_{i=1}^{2}\frac{\partial}{\partial \tilde{\xi}_i}\left[\left(1 + \tilde{p}\tilde{\eta}\frac{\tilde{d}_i + 3\tilde{d}_{3-i}}{2\tilde{d}_i\tilde{d}_{3-i}}\right)f^{\mathrm{b}}_{\xi_i} * \tilde{T}_1 - \tilde{p}\frac{\tilde{d}_i + \tilde{d}_{3-i}}{\tilde{d}_i\tilde{d}_{3-i}}f^{\mathrm{b}}_{\xi_i} * \tilde{T}_2\right] \tag{2.131}$$

Third-order approximation

$$\begin{aligned}\frac{\partial \dot{f}_{\xi_k}}{\partial \tilde{\eta}} = -(1+j)\Bigg[&\dot{f}^{\mathrm{b}}_{\xi_k} + \tilde{p}\tilde{\eta}\dot{f}^{\mathrm{b}}_{\xi_k}\frac{\tilde{d}_k + \tilde{d}_{3-k}}{2\tilde{d}_k\tilde{d}_{3-k}} + \tilde{p}^2\tilde{\eta}^2\dot{f}^{\mathrm{b}}_{\xi_k}\frac{3\tilde{d}_k^2 + 2\tilde{d}_k\tilde{d}_{3-k} + 3\tilde{d}_{3-k}^2}{8\tilde{d}_k^2\tilde{d}_{3-k}^2}\\ &-\tilde{p}\frac{1-j}{2}\left(\dot{f}^{\mathrm{b}}_{\xi_k}\frac{\tilde{d}_k + \tilde{d}_{3-k}}{2\tilde{d}_k\tilde{d}_{3-k}} + \tilde{p}\tilde{\eta}\dot{f}^{\mathrm{b}}_{\xi_k}\frac{5\tilde{d}_k^2 + 2\tilde{d}_k\tilde{d}_{3-k} + 9\tilde{d}_{3-k}^2}{8\tilde{d}_k^2\tilde{d}_{3-k}^2} - \tilde{p}\frac{\tilde{\eta}}{2}\sum_{i=1}^{2}\frac{\partial^2 \dot{f}^{\mathrm{b}}_{\xi_k}}{\partial \tilde{\xi}_i^2}\right)\\ &+\tilde{p}^2\frac{j}{2}\left(\dot{f}^{\mathrm{b}}_{\xi_k}\frac{\tilde{d}_k^2 + 2\tilde{d}_k\tilde{d}_{3-k} - 3\tilde{d}_{3-k}^2}{8\tilde{d}_k^2\tilde{d}_{3-k}^2} + \frac{1}{2}\sum_{i=1}^{2}\frac{\partial^2 \dot{f}^{\mathrm{b}}_{\xi_k}}{\partial \tilde{\xi}_i^2}\right)\Bigg]\exp(-(1+j)\tilde{\eta})\end{aligned} \tag{2.132}$$

$$\begin{aligned}\frac{\partial \dot{f}_\eta}{\partial \tilde{\eta}} = -\tilde{p}\sum_{i=1}^{2}\frac{\partial}{\partial \tilde{\xi}_i}\Bigg[&\dot{f}^{\mathrm{b}}_{\xi_i} + \tilde{p}\tilde{\eta}\dot{f}^{\mathrm{b}}_{\xi_i}\frac{\tilde{d}_i + 3\tilde{d}_{3-i}}{2\tilde{d}_i\tilde{d}_{3-i}} + \tilde{p}^2\tilde{\eta}^2\dot{f}^{\mathrm{b}}_{\xi_i}\frac{3\tilde{d}_i^2 + 6\tilde{d}_i\tilde{d}_{3-i} + 11\tilde{d}_{3-i}^2}{8\tilde{d}_i^2\tilde{d}_{3-i}^2}\\ &-\tilde{p}\frac{1-j}{2}\left(\dot{f}^{\mathrm{b}}_{\xi_i}\frac{\tilde{d}_i + \tilde{d}_{3-i}}{\tilde{d}_i\tilde{d}_{3-i}} + \tilde{p}\tilde{\eta}\dot{f}^{\mathrm{b}}_{\xi_i}\frac{11\tilde{d}_i^2 + 14\tilde{d}_i\tilde{d}_{3-i} + 15\tilde{d}_{3-i}^2}{8\tilde{d}_i^2\tilde{d}_{3-i}^2} - \tilde{p}\tilde{\eta}\frac{1}{2}\sum_{l=1}^{2}\frac{\partial^2 \dot{f}^{\mathrm{b}}_{\xi_i}}{\partial \tilde{\xi}_l^2}\right)\\ &+\tilde{p}^2\frac{j}{2}\left(\dot{f}^{\mathrm{b}}_{\xi_i}\frac{-\tilde{d}_i^2 + \tilde{d}_{3-i}^2}{2\tilde{d}_i^2\tilde{d}_{3-i}^2} + \frac{\partial^2 \dot{f}^{\mathrm{b}}_{\xi_i}}{\partial \tilde{\xi}_{3-i}^2} - \frac{\partial^2 \dot{f}^{\mathrm{b}}_{\xi_{3-i}}}{\partial \tilde{\xi}_i\partial \tilde{\xi}_{3-i}}\right)\Bigg]\exp(-(1+j)\tilde{\eta})\end{aligned} \tag{2.133}$$

$$\begin{aligned}\frac{\partial f_{\xi_k}}{\partial \tilde{\eta}} = -\frac{\mathrm{d}}{\mathrm{d}\tilde{t}}\Bigg[&\left(1 + \tilde{p}\tilde{\eta}\frac{\tilde{d}_k + \tilde{d}_{3-k}}{2\tilde{d}_k\tilde{d}_{3-k}} + \tilde{p}^2\tilde{\eta}^2\frac{3\tilde{d}_k^2 + 2\tilde{d}_k\tilde{d}_{3-k} + 3\tilde{d}_{3-k}^2}{8\tilde{d}_k^2\tilde{d}_{3-k}^2}\right)f^{\mathrm{b}}_{\xi_k} * \tilde{T}_2\Bigg]\\ &+\tilde{p}\left(f^{\mathrm{b}}_{\xi_k}\frac{\tilde{d}_k + \tilde{d}_{3-k}}{2\tilde{d}_k\tilde{d}_{3-k}} + \tilde{p}\tilde{\eta}f^{\mathrm{b}}_{\xi_k}\frac{5\tilde{d}_k^2 + 2\tilde{d}_k\tilde{d}_{3-k} + 9\tilde{d}_{3-k}^2}{8\tilde{d}_k^2\tilde{d}_{3-k}^2} - \tilde{p}\tilde{\eta}\frac{1}{2}\sum_{i=1}^{2}\frac{\partial^2 f^{\mathrm{b}}_{\xi_k}}{\partial \tilde{\xi}_i^2}\right) * \tilde{T}_1\\ &+\tilde{p}^2\left(f^{\mathrm{b}}_{\xi_k}\frac{\tilde{d}_k^2 + 2\tilde{d}_k\tilde{d}_{3-k} - 3\tilde{d}_{3-k}^2}{8\tilde{d}_k^2\tilde{d}_{3-k}^2} + \frac{1}{2}\sum_{i=1}^{2}\frac{\partial^2 f^{\mathrm{b}}_{\xi_k}}{\partial \tilde{\xi}_i^2}\right) * \tilde{T}_2\Bigg]\end{aligned} \tag{2.134}$$

$$\frac{\partial f_\eta}{\partial\tilde{\eta}} = -\tilde{p}\sum_{i=1}^{2}\frac{\partial}{\partial\tilde{\xi}_i}\left[\left(1+\tilde{p}\tilde{\eta}\frac{\tilde{d}_i+3\tilde{d}_{3-i}}{2\tilde{d}_i\tilde{d}_{3-i}}+\tilde{p}^2\tilde{\eta}^2\frac{3\tilde{d}_i^2+6\tilde{d}_i\tilde{d}_{3-i}+11\tilde{d}_{3-i}^2}{8\tilde{d}_i^2\tilde{d}_{3-i}^2}\right)f_{\xi_i}^{\mathrm{b}} * \tilde{T}_1\right.$$
$$-\tilde{p}\left(f_{\xi_i}^{\mathrm{b}}\frac{\tilde{d}_i+\tilde{d}_{3-i}}{\tilde{d}_i\tilde{d}_{3-i}}+\tilde{p}\tilde{\eta}f_{\xi_i}^{\mathrm{b}}\frac{11\tilde{d}_i^2+14\tilde{d}_i\tilde{d}_{3-i}+15\tilde{d}_{3-i}^2}{8\tilde{d}_i^2\tilde{d}_{3-i}^2}-\tilde{p}\tilde{\eta}f_{\xi_i}^{\mathrm{b}}\frac{1}{2}\sum_{l=1}^{2}\frac{\partial^2 f_{\xi_i}^{\mathrm{b}}}{\partial\tilde{\xi}_l^2}\right) * \tilde{T}_2$$
$$\left.-\tilde{p}^2\left(f_{\xi_i}^{\mathrm{b}}\frac{-\tilde{d}_i^2+\tilde{d}_{3-i}^2}{2\tilde{d}_i^2\tilde{d}_{3-i}^2}+\frac{\partial^2 f_{\xi_i}^{\mathrm{b}}}{\partial\tilde{\xi}_{3-i}^2}-\frac{\partial^2 f_{\xi_{3-i}}^{\mathrm{b}}}{\partial\tilde{\xi}_i\partial\tilde{\xi}_{3-i}}\right) * \tilde{T}_3\right] \quad (2.135)$$

Here the nondimensional time-dependent functions, $\tilde{T}_1$, $\tilde{T}_2$, and $\tilde{T}_3$, are given in Equations 2.99, 2.98, and 2.115, respectively.

2.7 Components of the *Curl* Operator

To simplify notations in the following derivations we denote curl $\vec{f}$ as $\vec{g}$:

$$\nabla\times\vec{f} = \vec{g} \quad (2.136)$$

Rewriting Equation 2.136 in local nondimensional coordinates, $\tilde{\xi}_1$, $\tilde{\xi}_2$, and $\tilde{\eta}$, using Equations 2.46 and 2.47, we obtain

$$g_{\xi_{3-k}} = (-1)^{3-k}\delta^{-1}\left[\frac{\partial f_{\xi_k}}{\partial\tilde{\eta}}-\tilde{p}\left(\frac{f_{\xi_k}}{\tilde{d}_k}+\frac{\partial f_\eta}{\partial\tilde{\xi}_k}\right)-\tilde{p}^2\tilde{\eta}\left(\frac{f_{\xi_k}}{\tilde{d}_k^2}+\frac{1}{\tilde{d}_k}\frac{\partial f_\eta}{\partial\tilde{\xi}_k}\right)+\cdots\right],\quad k=1,2 \quad (2.137)$$

$$g_\eta = \delta^{-1}\sum_{i=1}^{2}(-1)^i\frac{\partial f_{\xi_i}}{\partial\tilde{\xi}_{3-i}}\left(\tilde{p}+\tilde{p}^2\frac{\tilde{\eta}}{\tilde{d}_{3-i}}+\cdots\right) \quad (2.138)$$

The tangential and normal components of the function $\vec{g}$ are now represented as expansions in the small parameter $\tilde{p}$:

$$g_{\xi_k} = \sum_{m=0}^{\infty}\tilde{p}^m(\vec{g}_m)_{\xi_k};\quad g_\eta = \sum_{m=0}^{\infty}\tilde{p}^m(\vec{g}_m)_\eta,\quad k=1,2 \quad (2.139)$$

Substituting Equations 2.139 into Equations 2.137 and 2.138 and equating coefficients of equal powers of $\tilde{p}$, the following equations are obtained (we restrict ourselves to the first three nonzero terms of the expansions):

First-order approximation

$$(\vec{g}_0)_{\xi_{3-k}} = (-1)^{3-k}\frac{\partial(\vec{f}_0)_{\xi_k}}{\partial\tilde{\eta}}, \quad k = 1,2 \tag{2.140}$$

$$(\vec{g}_1)_{\eta} = \sum_{i=1}^{2}(-1)^i\frac{\partial(\vec{f}_0)_{\xi_i}}{\partial\tilde{\xi}_{3-i}} \tag{2.141}$$

Second-order approximation

$$(\vec{g}_1)_{\xi_{3-k}} = (-1)^{3-k}\left(\frac{\partial(\vec{f}_1)_{\xi_k}}{\partial\tilde{\eta}} - \frac{(\vec{f}_0)_{\xi_k}}{\tilde{d}_k} - \frac{\partial(\vec{f}_0)_{\eta}}{\partial\tilde{\xi}_k}\right) = (-1)^{3-k}\left(\frac{\partial(\vec{f}_1)_{\xi_k}}{\partial\tilde{\eta}} - \frac{(\vec{f}_0)_{\xi_k}}{\tilde{d}_k}\right), \quad k = 1,2 \tag{2.142}$$

$$(\vec{g}_2)_{\eta} = \sum_{i=1}^{2}(-1)^i\left(\frac{\partial(\vec{f}_1)_{\xi_i}}{\partial\tilde{\xi}_{3-i}} + \frac{\tilde{\eta}}{\tilde{d}_{3-i}}\frac{\partial(\vec{f}_0)_{\xi_i}}{\partial\tilde{\xi}_{3-i}}\right) \tag{2.143}$$

Third-order approximation

$$\begin{aligned}(\vec{g}_2)_{\xi_{3-k}} &= (-1)^{3-k}\left[\frac{\partial(\vec{f}_2)_{\xi_k}}{\partial\tilde{\eta}} - \frac{(\vec{f}_1)_{\xi_k}}{\tilde{d}_k} - \frac{\partial(\vec{f}_1)_{\eta}}{\partial\tilde{\xi}_k} - \tilde{\eta}\left(\frac{(\vec{f}_0)_{\xi_k}}{\tilde{d}_k^2} + \frac{1}{\tilde{d}_k}\frac{\partial(\vec{f}_0)_{\eta}}{\partial\tilde{\xi}_k}\right)\right] \\ &= (-1)^{3-k}\left[\frac{\partial(\vec{f}_2)_{\xi_k}}{\partial\tilde{\eta}} - \frac{(\vec{f}_1)_{\xi_k}}{\tilde{d}_k} - \frac{\partial(\vec{f}_1)_{\eta}}{\partial\tilde{\xi}_k} - \tilde{\eta}\frac{(\vec{f}_0)_{\xi_k}}{\tilde{d}_k^2}\right], \quad k = 1,2\end{aligned} \tag{2.144}$$

$$(\vec{g}_3)_{\eta} = \sum_{i=1}^{2}(-1)^i\left(\frac{\partial(\vec{f}_2)_{\xi_i}}{\partial\tilde{\xi}_{3-i}} + \frac{\tilde{\eta}}{\tilde{d}_{3-i}}\frac{\partial(\vec{f}_1)_{\xi_i}}{\partial\tilde{\xi}_{3-i}} + \frac{\tilde{\eta}^2}{\tilde{d}_{3-i}^2}\frac{\partial(\vec{f}_0)_{\xi_i}}{\partial\tilde{\xi}_{3-i}}\right) \tag{2.145}$$

Representing Equation 2.140 through 2.145 in the LD and substituting Equations 2.79, 2.82, 2.88, and 2.103, we obtain the following expressions:

$$\begin{aligned}\left(\bar{\vec{g}}_0\right)_{\xi_{3-k}} &= -(-1)^{3-k}\tilde{s}^{1/2}\left(\bar{\vec{f}}_0^{\,b}\right)_{\xi_k}\exp\left(-\tilde{\eta}\sqrt{\tilde{s}}\right) \\ &= (-1)^k\tilde{s}^{1/2}\left(\bar{\vec{f}}_0^{\,b}\right)_{\xi_k}\exp\left(-\tilde{\eta}\sqrt{\tilde{s}}\right), \quad k = 1,2\end{aligned} \tag{2.146}$$

$$
\begin{aligned}
(\vec{\bar{g}}_1)_{\xi_{3-k}} &= (-1)^{3-k}\left[-\tilde{s}^{1/2}\left(\vec{f}_1^{\rm b}\right)_{\xi_k} + (1-\tilde{s}^{1/2}\tilde{\eta})\left(\vec{f}_0^{\rm b}\right)_{\xi_k}\frac{\tilde{d}_k+\tilde{d}_{3-k}}{2\tilde{d}_k\tilde{d}_{3-k}} - \frac{\left(\vec{f}_0^{\rm b}\right)_{\xi_k}}{\tilde{d}_k}\right]\exp(-\tilde{\eta}\sqrt{\tilde{s}}) \\
&= (-1)^{3-k}\left[-\tilde{s}^{1/2}\left(\vec{f}_1^{\rm b}\right)_{\xi_k} + \left(\vec{f}_0^{\rm b}\right)_{\xi_k}\frac{\tilde{d}_k-\tilde{d}_{3-k}}{2\tilde{d}_k\tilde{d}_{3-k}} - \tilde{s}^{1/2}\tilde{\eta}\left(\vec{f}_0^{\rm b}\right)_{\xi_k}\frac{\tilde{d}_k+\tilde{d}_{3-k}}{2\tilde{d}_k\tilde{d}_{3-k}}\right]\exp(-\tilde{\eta}\sqrt{\tilde{s}}) \\
&= (-1)^k\tilde{s}^{1/2}\left(\left(\vec{f}_1^{\rm b}\right)_{\xi_k} + \tilde{\eta}\left(\vec{f}_0^{\rm b}\right)_{\xi_k}\frac{\tilde{d}_k+\tilde{d}_{3-k}}{2\tilde{d}_k\tilde{d}_{3-k}} - \tilde{s}^{-1/2}\left(\vec{f}_0^{\rm b}\right)_{\xi_k}\frac{\tilde{d}_k-\tilde{d}_{3-k}}{2\tilde{d}_{k0}\tilde{d}_{3-k}}\right)\exp(-\tilde{\eta}\sqrt{\tilde{s}}),\quad k=1,2
\end{aligned}
\tag{2.147}
$$

$$
\begin{aligned}
(\vec{\bar{g}}_2)_{\xi_{3-k}} = (-1)^k\tilde{s}^{1/2}\Bigg\{&\left(\vec{f}_2^{\rm b}\right)_{\xi_k} + \tilde{\eta}\left(\vec{f}_1^{\rm b}\right)_{\xi_k}\frac{\tilde{d}_k+\tilde{d}_{3-k}}{2\tilde{d}_k\tilde{d}_{3-k}} + \tilde{\eta}^2\left(\vec{f}_0^{\rm b}\right)_{\xi_k}\frac{3\tilde{d}_k^2+2\tilde{d}_k\tilde{d}_{3-k}+3\tilde{d}_{3-k}^2}{8\tilde{d}_k^2\tilde{d}_{3-k}^2} \\
&-\tilde{s}^{-1/2}\left[\frac{\tilde{d}_k-\tilde{d}_{3-k}}{2\tilde{d}_k\tilde{d}_{3-k}}\left(\vec{f}_1^{\rm b}\right)_{\xi_k} - \tilde{\eta}\left(\left(\vec{f}_0^{\rm b}\right)_{\xi_k}\frac{-5\tilde{d}_k^2+2\tilde{d}_k\tilde{d}_{3-k}+3\tilde{d}_{3-k}^2}{2\tilde{d}_k^2\tilde{d}_{3-k}^2} + \frac{1}{2}\sum_{l=1}^{2}\frac{\partial^2\left(\vec{f}_0^{\rm b}\right)_{\xi_l}}{\partial\tilde{\xi}_l^2}\right)\right] \\
&-\tilde{s}^{-1}\left(\left(\vec{f}_0^{\rm b}\right)_{\xi_k}\frac{\tilde{d}_k^2+2\tilde{d}_k\tilde{d}_{3-k}-3\tilde{d}_{3-k}^2}{8\tilde{d}_k^2\tilde{d}_{3-k}^2} - \frac{1}{2}\frac{\partial^2\left(\vec{f}_0^{\rm b}\right)_{\xi_k}}{\partial\tilde{\xi}_k^2} + \frac{1}{2}\frac{\partial^2\left(\vec{f}_0^{\rm b}\right)_{\xi_k}}{\partial\tilde{\xi}_{3-k}^2} - \frac{\partial^2\left(\vec{f}_0^{\rm b}\right)_{\xi_{3-k}}}{\partial\tilde{\xi}_k\partial\tilde{\xi}_{3-k}}\right)\Bigg\} \\
&\times\exp(-\tilde{\eta}\sqrt{\tilde{s}}),\quad k=1,2
\end{aligned}
\tag{2.148}
$$

$$
(\vec{\bar{g}}_1)_\eta = \sum_{i=1}^{2}(-1)^i\frac{\partial\left(\vec{f}_0^{\rm b}\right)_{\xi_i}}{\partial\tilde{\xi}_{3-i}}\exp(-\tilde{\eta}\sqrt{\tilde{s}}) \tag{2.149}
$$

$$
\begin{aligned}
(\vec{\bar{g}}_2)_\eta &= \sum_{i=1}^{2}(-1)^i\left(\frac{\partial\left(\vec{f}_1^{\rm b}\right)_{\xi_i}}{\partial\tilde{\xi}_{3-i}} + \tilde{\eta}\frac{\tilde{d}_i+\tilde{d}_{3-i}}{2\tilde{d}_i\tilde{d}_{3-i}}\frac{\partial\left(\vec{f}_0^{\rm b}\right)_{\xi_i}}{\partial\tilde{\xi}_{3-i}} + \frac{\tilde{\eta}}{\tilde{d}_{3-i}}\frac{\partial\left(\vec{f}_0\right)_{\xi_i}}{\partial\tilde{\xi}_{3-i}}\right)\exp(-\tilde{\eta}\sqrt{\tilde{s}}) \\
&= \sum_{i=1}^{2}(-1)^i\left(\frac{\partial\left(\vec{f}_1^{\rm b}\right)_{\xi_i}}{\partial\tilde{\xi}_{3-i}} + \tilde{\eta}\frac{3\tilde{d}_i+\tilde{d}_{3-i}}{2\tilde{d}_i\tilde{d}_{3-i}}\frac{\partial\left(\vec{f}_0^{\rm b}\right)_{\xi_i}}{\partial\tilde{\xi}_{3-i}}\right)\exp(-\tilde{\eta}\sqrt{\tilde{s}})
\end{aligned}
\tag{2.150}
$$

$$
\begin{aligned}
(\vec{\bar{g}}_3)_\eta = \sum_{i=1}^{2}(-1)^i\frac{\partial}{\partial\tilde{\xi}_{3-i}}\Bigg[&\left(\vec{f}_2^{\rm b}\right)_{\xi_i} + \tilde{\eta}\left(\vec{f}_1^{\rm b}\right)_{\xi_i}\frac{3\tilde{d}_i+\tilde{d}_{3-i}}{2\tilde{d}_i\tilde{d}_{3-i}} + \tilde{\eta}^2\left(\vec{f}_0^{\rm b}\right)_{\xi_i}\frac{11\tilde{d}_i^2+6\tilde{d}_i\tilde{d}_{3-i}+3\tilde{d}_{3-i}^2}{8\tilde{d}_i^2\tilde{d}_{3-i}^2} \\
&+\tilde{\eta}\tilde{s}^{-1/2}\left(\left(\vec{f}_0^{\rm b}\right)_{\xi_i}\frac{\tilde{d}_i^2+2\tilde{d}_i\tilde{d}_{3-i}-3\tilde{d}_{3-i}^2}{8\tilde{d}_i^2\tilde{d}_{3-i}^2} + \frac{1}{2}\sum_{l=1}^{2}\frac{\partial^2\left(\vec{f}_0^{\rm b}\right)_{\xi_i}}{\partial\tilde{\xi}_l^2}\right)\Bigg]\exp(-\tilde{\eta}\sqrt{\tilde{s}})
\end{aligned}
\tag{2.151}
$$

Details of derivation of Equations 2.148 and 2.151 are given in Appendix 2.A.1.

Substituting the coefficients $(\vec{\bar{g}}_m)_{\xi_k}$ given in Equations 2.146 through 2.148, and $(\vec{\bar{g}}_m)_\eta$ given in Equations 2.149 through 2.151 into the expansions in

Equation 2.139, we obtain the tangential and normal components of $\bar{\vec{g}}$ of various orders of approximation:

First-order approximation

$$\bar{g}_{\xi_{3-k}} = (-1)^k \tilde{s}^{1/2} \bar{f}^{\mathrm{b}}_{\xi_k} \exp(-\tilde{\eta}\sqrt{\tilde{s}}), \quad k = 1,2 \tag{2.152}$$

$$\bar{g}_{\eta} = \tilde{p} \sum_{i=1}^{2} (-1)^i \frac{\partial \bar{f}^{\mathrm{b}}_{\xi_i}}{\partial \tilde{\xi}_{3-i}} \exp(-\tilde{\eta}\sqrt{\tilde{s}}) \tag{2.153}$$

Second-order approximation

$$\bar{g}_{\xi_{3-k}} = (-1)^k \tilde{s}^{1/2} \left(\bar{f}^{\mathrm{b}}_{\xi_k} + \tilde{p}\tilde{\eta} \bar{f}^{\mathrm{b}}_{\xi_k} \frac{\tilde{d}_k + \tilde{d}_{3-k}}{2\tilde{d}_k \tilde{d}_{3-k}} - \tilde{p}\tilde{s}^{-1/2} \bar{f}^{\mathrm{b}}_{\xi_k} \frac{\tilde{d}_k - \tilde{d}_{3-k}}{2\tilde{d}_k \tilde{d}_{3-k}} \right) \exp(-\tilde{\eta}\sqrt{\tilde{s}}) \tag{2.154}$$

$$\bar{g}_{\eta} = \tilde{p} \sum_{i=1}^{2} (-1)^i \frac{\partial}{\partial \tilde{\xi}_{3-i}} \left(\bar{f}^{\mathrm{b}}_{\xi_i} + \tilde{p}\tilde{\eta} \frac{3\tilde{d}_i + \tilde{d}_{3-i}}{2\tilde{d}_i \tilde{d}_{3-i}} \bar{f}^{\mathrm{b}}_{\xi_i} \right) \exp(-\tilde{\eta}\sqrt{\tilde{s}}) \tag{2.155}$$

Third-order approximation

$$\begin{aligned} \bar{g}_{\xi_{3-k}} = (-1)^k \tilde{s}^{1/2} \Bigg\{ & \bar{f}^{\mathrm{b}}_{\xi_k} + \tilde{p}\tilde{\eta} \bar{f}^{\mathrm{b}}_{\xi_k} \frac{\tilde{d}_k + \tilde{d}_{3-k}}{2\tilde{d}_k \tilde{d}_{3-k}} + \tilde{p}^2 \tilde{\eta}^2 \bar{f}^{\mathrm{b}}_{\xi_k} \frac{3\tilde{d}_k^2 + 2\tilde{d}_k \tilde{d}_{3-k} + 3\tilde{d}_{3-k}^2}{8\tilde{d}_k^2 \tilde{d}_{3-k}^2} \\ & - \tilde{p}\tilde{s}^{-1/2} \left[\frac{\tilde{d}_k - \tilde{d}_{3-k}}{2\tilde{d}_k \tilde{d}_{3-k}} \bar{f}^{\mathrm{b}}_{\xi_k} - \tilde{\eta} \left(\bar{f}^{\mathrm{b}}_{\xi_k} \frac{-5\tilde{d}_k^2 + 2\tilde{d}_k \tilde{d}_{3-k} + 3\tilde{d}_{3-k}^2}{2\tilde{d}_k^2 \tilde{d}_{3-k}^2} + \frac{1}{2} \sum_{l=1}^{2} \frac{\partial^2 \bar{f}^{\mathrm{b}}_{\xi_k}}{\partial \tilde{\xi}_l^2} \right) \right] \\ & - \tilde{p}^2 \tilde{s}^{-1} \left(\bar{f}^{\mathrm{b}}_{\xi_k} \frac{\tilde{d}_k^2 + 2\tilde{d}_k \tilde{d}_{3-k} - 3\tilde{d}_{3-k}^2}{8\tilde{d}_k^2 \tilde{d}_{3-k}^2} - \frac{1}{2} \frac{\partial^2 \bar{f}^{\mathrm{b}}_{\xi_k}}{\partial \tilde{\xi}_k^2} + \frac{1}{2} \frac{\partial^2 \bar{f}^{\mathrm{b}}_{\xi_k}}{\partial \tilde{\xi}_{3-k}^2} - \frac{\partial^2 \bar{f}^{\mathrm{b}}_{\xi_{3-k}}}{\partial \tilde{\xi}_k \partial \tilde{\xi}_{3-k}} \right) \Bigg\} \exp\left(-\tilde{\eta}\sqrt{\tilde{s}}\right), \quad k = 1,2 \end{aligned} \tag{2.156}$$

$$\begin{aligned} \bar{g}_{\eta} = \tilde{p} \sum_{i=1}^{2} (-1)^i \frac{\partial}{\partial \tilde{\xi}_{3-i}} \Bigg[& \bar{f}^{\mathrm{b}}_{\xi_i} + \tilde{p}\tilde{\eta} \bar{f}^{\mathrm{b}}_{\xi_i} \frac{3\tilde{d}_i + \tilde{d}_{3-i}}{2\tilde{d}_i \tilde{d}_{3-i}} + \tilde{p}^2 \tilde{\eta}^2 \bar{f}^{\mathrm{b}}_{\xi_i} \frac{11\tilde{d}_i^2 + 6\tilde{d}_i \tilde{d}_{3-i} + 3\tilde{d}_{3-i}^2}{8\tilde{d}_i^2 \tilde{d}_{3-i}^2} \\ & + \tilde{p}^2 \tilde{\eta} \tilde{s}^{-1/2} \left(\bar{f}^{\mathrm{b}}_{\xi_i} \frac{\tilde{d}_i^2 + 2\tilde{d}_i \tilde{d}_{3-i} - 3\tilde{d}_{3-i}^2}{8\tilde{d}_i^2 \tilde{d}_{3-i}^2} + \frac{1}{2} \sum_{l=1}^{2} \frac{\partial^2 \bar{f}^{\mathrm{b}}_{\xi_i}}{\partial \tilde{\xi}_l^2} \right) \Bigg] \exp\left(-\tilde{\eta}\sqrt{\tilde{s}}\right) \end{aligned} \tag{2.157}$$

Equations 2.152 through 2.157 give the tangential and normal components of $\vec{g}$ in the LD as functions of the tangential components, $f^{\mathrm{b}}_{\xi_k}$, at the conductor's surface.

Finally, let us represent Equations 2.152 through 2.157 in the FD and TD:

First-order approximation

$$\dot{g}_{\xi_{3-k}} = (-1)^k (1 + j) \dot{f}^{\mathrm{b}}_{\xi_k} \exp(-(1 + j)\tilde{\eta}) \tag{2.158}$$

$$\dot{g}_\eta = \tilde{p}\sum_{i=1}^{2}(-1)^i \frac{\partial \dot{f}^{\rm b}_{\xi_i}}{\partial \tilde{\xi}_{3-i}} \exp\left(-(1+j)\tilde{\eta}\right) \tag{2.159}$$

$$g_{\xi_{3-k}} = (-1)^k \frac{\rm d}{{\rm d}\tilde{t}}\left(f^{\rm b}_{\xi_k} * \tilde{T}_2\right) \tag{2.160}$$

$$g_\eta = \tilde{p}\sum_{i=1}^{2}(-1)^i \frac{\partial f^{\rm b}_{\xi_i}}{\partial \tilde{\xi}_{3-i}} * \tilde{T}_1 \tag{2.161}$$

Second-order approximation

$$\dot{g}_{\xi_{3-k}} = (-1)^k(1+j)\left[1 + \tilde{p}\tilde{\eta}\frac{\tilde{d}_k + \tilde{d}_{3-k}}{2\tilde{d}_k\tilde{d}_{3-k}} - \tilde{p}\frac{1-j}{2}\frac{\tilde{d}_k - \tilde{d}_{3-k}}{2\tilde{d}_k\tilde{d}_{3-k}}\right]\dot{f}^{\rm b}_{\xi_k} \exp\left(-(1+j)\tilde{\eta}\right) \tag{2.162}$$

$$\dot{g}_\eta = \tilde{p}\sum_{i=1}^{2}(-1)^i\left(1 + \tilde{p}\tilde{\eta}\frac{3\tilde{d}_i + \tilde{d}_{3-i}}{2\tilde{d}_i\tilde{d}_{3-i}}\right)\frac{\partial \dot{f}^{\rm b}_{\xi_i}}{\partial \tilde{\xi}_{3-i}} \exp\left(-(1+j)\tilde{\eta}\right) \tag{2.163}$$

$$g_{\xi_k} = (-1)^k\left[\left(1 + \tilde{p}\tilde{\eta}\frac{\tilde{d}_k + \tilde{d}_{3-k}}{2\tilde{d}_k\tilde{d}_{3-k}}\right)\frac{\rm d}{{\rm d}\tilde{t}}\left(f^{\rm b}_{\xi_k} * \tilde{T}_2\right) - \tilde{p}\frac{\tilde{d}_k - \tilde{d}_{3-k}}{2\tilde{d}_k\tilde{d}_{3-k}} f^{\rm b}_{\xi_k} * \tilde{T}_1\right] \tag{2.164}$$

$$g_\eta = \tilde{p}\sum_{i=1}^{2}(-1)^i \frac{\partial}{\partial \tilde{\xi}_{3-i}}\left(1 + \tilde{p}\tilde{\eta}\frac{3\tilde{d}_i + \tilde{d}_{3-i}}{2\tilde{d}_i\tilde{d}_{3-i}}\right) f^{\rm b}_{\xi_i} * \tilde{T}_1 \tag{2.165}$$

Third-order approximation

$$\begin{aligned}
\dot{g}_{\xi_{3-k}} = (-1)^k(1+j)\Bigg[&\dot{f}^{\rm b}_{\xi_k}\left(1 + \tilde{p}\tilde{\eta}\frac{\tilde{d}_k + \tilde{d}_{3-k}}{2\tilde{d}_k\tilde{d}_{3-k}} + \tilde{p}^2\tilde{\eta}^2\frac{3\tilde{d}_k^2 + 2\tilde{d}_k\tilde{d}_{3-k} + 3\tilde{d}_{3-k}^2}{8\tilde{d}_k^2\tilde{d}_{3-k}^2}\right)\\
&- \tilde{p}\frac{1-j}{2}\left(\dot{f}^{\rm b}_{\xi_k}\frac{\tilde{d}_k - \tilde{d}_{3-k}}{2\tilde{d}_k\tilde{d}_{3-k}} + \tilde{p}\tilde{\eta}\dot{f}^{\rm b}_{\xi_k}\frac{5\tilde{d}_k^2 - 2\tilde{d}_k\tilde{d}_{3-k} - 3\tilde{d}_{3-k}^2}{2\tilde{d}_k^2\tilde{d}_{3-k}^2} - \tilde{p}\tilde{\eta}\frac{1}{2}\sum_{l=1}^{2}\frac{\partial^2 \dot{f}^{\rm b}_{\xi_k}}{\partial \tilde{\xi}_l^2}\right)\\
&+ \tilde{p}^2\frac{j}{2}\left(\dot{f}^{\rm b}_{\xi_k}\frac{\tilde{d}_k^2 + 2\tilde{d}_k\tilde{d}_{3-k} - 3\tilde{d}_{3-k}^2}{8\tilde{d}_k^2\tilde{d}_{3-k}^2} - \frac{1}{2}\frac{\partial^2 \dot{f}^{\rm b}_{\xi_k}}{\partial \tilde{\xi}_k^2} + \frac{\partial^2 \dot{f}^{\rm b}_{\xi_k}}{\partial \tilde{\xi}_{3-k}^2} - \frac{\partial^2 \dot{f}^{\rm b}_{\xi_{3-k}}}{\partial \tilde{\xi}_k \partial \tilde{\xi}_{3-k}}\right)\Bigg]\\
&\times \exp\left(-(1+j)\tilde{\eta}\right), \quad k = 1, 2
\end{aligned} \tag{2.166}$$

$$\begin{aligned}
\dot{g}_\eta = \tilde{p}\sum_{i=1}^{2}(-1)^i \frac{\partial}{\partial \tilde{\xi}_{3-i}}\Bigg[&\dot{f}^{\rm b}_{\xi_i} + \tilde{p}\tilde{\eta}\dot{f}^{\rm b}_{\xi_i}\frac{3\tilde{d}_i + \tilde{d}_{3-i}}{2\tilde{d}_i\tilde{d}_{3-i}} + \tilde{p}^2\tilde{\eta}^2\dot{f}^{\rm b}_{\xi_i}\frac{11\tilde{d}_i^2 + 6\tilde{d}_i\tilde{d}_{3-i} + 3\tilde{d}_{3-i}^2}{8\tilde{d}_i^2\tilde{d}_{3-i}^2}\\
&+ \tilde{p}^2\frac{1-j}{2}\tilde{\eta}\left(\dot{f}^{\rm b}_{\xi_i}\frac{\tilde{d}_i^2 + 2\tilde{d}_i\tilde{d}_{3-i} - 3\tilde{d}_{3-i}^2}{8\tilde{d}_i^2\tilde{d}_{3-i}^2} + \frac{1}{2}\sum_{l=1}^{2}\frac{\partial^2 \dot{f}^{\rm b}_{\xi_i}}{\partial \tilde{\xi}_l^2}\right)\Bigg]\exp\left(-(1+j)\tilde{\eta}\right)
\end{aligned} \tag{2.167}$$

$$g_{\xi_{3-k}} = (-1)^k \left[\left(1 + \tilde{p}\tilde{\eta}\frac{\tilde{d}_k + \tilde{d}_{3-k}}{2\tilde{d}_k\tilde{d}_{3-k}} + \tilde{p}^2\tilde{\eta}^2\frac{3\tilde{d}_k^2 + 2\tilde{d}_k\tilde{d}_{3-k} + 3\tilde{d}_{3-k}^2}{8\tilde{d}_k^2\tilde{d}_{3-k}^2} \right) \frac{\mathrm{d}}{\mathrm{d}\tilde{t}}\left(f_{\xi_k}^{\mathrm{b}} * \tilde{T}_2 \right) \right.$$
$$- \tilde{p}\left(\frac{\tilde{d}_k - \tilde{d}_{3-k}}{2\tilde{d}_k\tilde{d}_{3-k}} f_{\xi_k}^{\mathrm{b}} + \tilde{p}\tilde{\eta}\frac{5\tilde{d}_k^2 - 2\tilde{d}_k\tilde{d}_{3-k} - 3\tilde{d}_{3-k}^2}{2\tilde{d}_k^2\tilde{d}_{3-k}^2} f_{\xi_k}^{\mathrm{b}} - \tilde{p}\tilde{\eta}\frac{1}{2}\sum_{l=1}^{2}\frac{\partial^2 f_{\xi_k}^{\mathrm{b}}}{\partial\tilde{\xi}_l^2} \right) * \tilde{T}_1$$
$$\left. - \tilde{p}^2\left(f_{\xi_k}^{\mathrm{b}}\frac{\tilde{d}_k^2 + 2\tilde{d}_k\tilde{d}_{3-k} - 3\tilde{d}_{3-k}^2}{8\tilde{d}_k^2\tilde{d}_{3-k}^2} - \frac{1}{2}\frac{\partial^2 f_{\xi_k}^{\mathrm{b}}}{\partial\tilde{\xi}_k^2} + \frac{\partial^2 f_{\xi_k}^{\mathrm{b}}}{\partial\tilde{\xi}_{3-k}^2} - \frac{\partial^2 f_{\xi_{3-k}}^{\mathrm{b}}}{\partial\tilde{\xi}_k\partial\tilde{\xi}_{3-k}} \right) * \tilde{T}_2, \quad k = 1,2 \right] \tag{2.168}$$

$$g_{\eta} = \tilde{p}\sum_{i=1}^{2}(-1)^i\frac{\partial}{\partial\tilde{\xi}_{3-i}}\left[\left(1 + \tilde{p}\tilde{\eta}\frac{3\tilde{d}_i + \tilde{d}_{3-i}}{2\tilde{d}_i\tilde{d}_{3-i}} + \tilde{p}^2\tilde{\eta}^2\frac{11\tilde{d}_i^2 + 6\tilde{d}_i\tilde{d}_{3-i} + 3\tilde{d}_{3-i}^2}{8\tilde{d}_i^2\tilde{d}_{3-i}^2} \right) f_{\xi_i}^{\mathrm{b}} * \tilde{T}_1 \right.$$
$$\left. + \tilde{p}^2\tilde{\eta}\left(f_{\xi_i}^{\mathrm{b}}\frac{\tilde{d}_i^2 + 2\tilde{d}_i\tilde{d}_{3-i} - 3\tilde{d}_{3-i}^2}{8\tilde{d}_i^2\tilde{d}_{3-i}^2} + \frac{1}{2}\sum_{l=1}^{2}\frac{\partial^2 f_{\xi_i}^{\mathrm{b}}}{\partial\tilde{\xi}_l^2} \right) * \tilde{T}_2 \right] \tag{2.169}$$

2.8 Surface Impedance "Toolbox" Concept

The results obtained in the previous sections can be summarized into a set of formulae that can be then used for derivation of appropriate SIBCs for a given problem. Suppose the function $\vec{f}$ obeys the equation of diffusion (Equation 2.2). Then its normal component, derivatives of tangential and normal components in direction normal to the surface, and tangential and normal components of curl $\vec{f}$ at the interface ($\eta = 0$) can be written in a form depending on the order of approximation in the LD, FD, and TD:

Zero-order approximation (diffusion into the body is neglected)

$$f_{\eta} = 0 \tag{2.170}$$

$$\left.\frac{\partial f_{\eta}}{\partial\tilde{\eta}}\right|^{\mathrm{b}} = 0 \tag{2.171}$$

$$(\nabla \times \vec{f})_{\eta}^{\mathrm{b}} = 0 \tag{2.172}$$

First-order approximation (only diffusion in the normal direction is considered; the surface of the body is assumed to be planar)

$$f_{\eta}^{\mathrm{b}} = \begin{cases} \tilde{p}\tilde{s}^{-1/2}\displaystyle\sum_{i=1}^{2}\frac{\partial\bar{f}_{\xi_i}^{\mathrm{b}}}{\partial\tilde{\xi}_i} & \text{LD} \\ \tilde{p}\dfrac{1-j}{2}\displaystyle\sum_{i=1}^{2}\frac{\partial\dot{f}_{\xi_i}^{\mathrm{b}}}{\partial\tilde{\xi}_i} & \text{FD} \\ \tilde{p}\displaystyle\sum_{i=1}^{2}\frac{\partial f_{\xi_i}^{\mathrm{b}}}{\partial\tilde{\xi}_i} * \tilde{T}_2^{\mathrm{b}} & \text{TD} \end{cases} \tag{2.173}$$

$$\left.\frac{\partial f_{\xi_k}}{\partial\tilde{\eta}}\right|^{\mathrm{b}} = \begin{cases} -\tilde{s}^{1/2}\bar{f}^{\mathrm{b}}_{\xi_k} & \text{LD} \\ -(1+j)\dot{f}^{\mathrm{b}}_{\xi_k} & \text{FD}, \quad k=1,2 \\ -\frac{\mathrm{d}}{\mathrm{d}\tilde{t}}\left(f^{\mathrm{b}}_{\xi_k} * \tilde{T}^{\mathrm{b}}_2\right) & \text{TD} \end{cases} \tag{2.174}$$

$$\left.\frac{\partial f_{\eta}}{\partial\tilde{\eta}}\right|^{\mathrm{b}} = \begin{cases} -\tilde{p}\sum_{i=1}^{2}\frac{\partial\bar{f}^{\mathrm{b}}_{\xi_i}}{\partial\tilde{\xi}_i} & \text{LD} \\ -\tilde{p}\sum_{i=1}^{2}\frac{\partial\dot{f}^{\mathrm{b}}_{\xi_i}}{\partial\tilde{\xi}_i} & \text{FD} \\ -\tilde{p}\sum_{i=1}^{2}e\frac{\partial f^{\mathrm{b}}_{\xi_i}}{\partial\tilde{\xi}_i} * \tilde{T}^{\mathrm{b}}_1 & \text{TD} \end{cases} \tag{2.175}$$

$$(\tilde{\nabla}\times\vec{f})^{\mathrm{b}}_{\xi_{3-k}} = \begin{cases} (-1)^k\tilde{s}^{1/2}\bar{f}^{\mathrm{b}}_{\xi_k} & \text{LD} \\ (-1)^k(1+j)\dot{f}^{\mathrm{b}}_{\xi_k} & \text{FD}, \quad k=1,2 \\ (-1)^k\frac{\mathrm{d}}{\mathrm{d}\tilde{t}}\left(f^{\mathrm{b}}_{\xi_k} * \tilde{T}^{\mathrm{b}}_2\right) & \text{TD} \end{cases} \tag{2.176}$$

$$(\tilde{\nabla}\times\vec{f})^{\mathrm{b}}_{\eta} = \begin{cases} \tilde{p}\sum_{i=1}^{2}(-1)^i\frac{\partial\bar{f}^{\mathrm{b}}_{\xi_i}}{\partial\tilde{\xi}_{3-i}} & \text{LD} \\ \tilde{p}\sum_{i=1}^{2}(-1)^i\frac{\partial\dot{f}^{\mathrm{b}}_{\xi_i}}{\partial\tilde{\xi}_{3-i}} & \text{FD} \\ \tilde{p}\sum_{i=1}^{2}(-1)^i\frac{\partial f^{\mathrm{b}}_{\xi_i}}{\partial\tilde{\xi}_{3-i}} * \tilde{T}^{\mathrm{b}}_1 & \text{TD} \end{cases} \tag{2.177}$$

Second-order approximation (only diffusion in the normal direction is considered, the curvature of the surface of the body is taken into account)

$$f^{\mathrm{b}}_{\eta} = \begin{cases} \tilde{p}\tilde{s}^{-1/2}\sum_{i=1}^{2}\frac{\partial}{\partial\tilde{\xi}_i}\left(\bar{f}^{\mathrm{b}}_{\xi_i} - \tilde{p}\tilde{s}^{-1/2}\bar{f}^{\mathrm{b}}_{\xi_i}\frac{\tilde{d}_i-\tilde{d}_{3-i}}{2\tilde{d}_i\tilde{d}_{3-i}}\right) & \text{LD} \\ \tilde{p}\frac{1-j}{2}\sum_{i=1}^{2}\frac{\partial}{\partial\tilde{\xi}_i}\left(\dot{f}^{\mathrm{b}}_{\xi_i} - \tilde{p}\frac{1-j}{2}\dot{f}^{\mathrm{b}}_{\xi_i}\frac{\tilde{d}_i-\tilde{d}_{3-i}}{2\tilde{d}_i\tilde{d}_{3-i}}\right) & \text{FD} \\ \tilde{p}\sum_{i=1}^{2}\frac{\partial}{\partial\tilde{\xi}_i}\left(f^{\mathrm{b}}_{\xi_i} * \tilde{T}^{\mathrm{b}}_2 - \tilde{p}\frac{\tilde{d}_i-\tilde{d}_{3-i}}{2\tilde{d}_i\tilde{d}_{3-i}}f^{\mathrm{b}}_{\xi_i} * \tilde{T}^{\mathrm{b}}_3\right) & \text{TD} \end{cases} \tag{2.178}$$

$$\left.\frac{\partial f_{\xi_k}}{\partial\tilde{\eta}}\right|^{\mathrm{b}} = \begin{cases} -\tilde{s}^{1/2}\left(\bar{f}^{\mathrm{b}}_{\xi_k} - \tilde{p}\tilde{s}^{-1/2}\bar{f}^{\mathrm{b}}_{\xi_k}\frac{\tilde{d}_k+\tilde{d}_{3-k}}{2\tilde{d}_k\tilde{d}_{3-k}}\right) & \text{LD} \\ -(1+j)\left(\dot{f}^{\mathrm{b}}_{\xi_k} - \tilde{p}\frac{1-j}{2}\dot{f}^{\mathrm{b}}_{\xi_k}\frac{\tilde{d}_k+\tilde{d}_{3-k}}{2\tilde{d}_k\tilde{d}_{3-k}}\right) & \text{FD}, \quad k=1,2 \\ -\frac{\mathrm{d}}{\mathrm{d}\tilde{t}}\left(f^{\mathrm{b}}_{\xi_k} * \tilde{T}^{\mathrm{b}}_2\right) + \tilde{p}\frac{\tilde{d}_k+\tilde{d}_{3-k}}{2\tilde{d}_k\tilde{d}_{3-k}}f^{\mathrm{b}}_{\xi_k} * \tilde{T}^{\mathrm{b}}_1 & \text{TD} \end{cases} \tag{2.179}$$

$$\left.\frac{\partial f_\eta}{\partial\tilde{\eta}}\right|^{\mathrm{b}} = \begin{cases} -\tilde{p}\displaystyle\sum_{i=1}^{2}\frac{\partial}{\partial\tilde{\xi}_i}\left(\bar{f}^{\mathrm{b}}_{\xi_i} - \tilde{p}\tilde{s}^{-1/2}\bar{f}^{\mathrm{b}}_{\xi_i}\frac{\tilde{d}_i+\tilde{d}_{3-i}}{\tilde{d}_i\tilde{d}_{3-i}}\right) & \text{LD} \\ -\tilde{p}\displaystyle\sum_{i=1}^{2}\frac{\partial}{\partial\tilde{\xi}_i}\left(\dot{f}^{\mathrm{b}}_{\xi_i} - \tilde{p}\frac{1-j}{2}\dot{f}^{\mathrm{b}}_{\xi_i}\frac{\tilde{d}_i+\tilde{d}_{3-i}}{\tilde{d}_i\tilde{d}_{3-i}}\right) & \text{FD} \\ -\tilde{p}\displaystyle\sum_{i=1}^{2}\frac{\partial}{\partial\tilde{\xi}_i}\left(f^{\mathrm{b}}_{\xi_i}*\tilde{T}^{\mathrm{b}}_1 - \tilde{p}\frac{\tilde{d}_i+\tilde{d}_{3-i}}{\tilde{d}_i\tilde{d}_{3-i}}f^{\mathrm{b}}_{\xi_i}*\tilde{T}^{\mathrm{b}}_2\right) & \text{TD} \end{cases} \tag{2.180}$$

$$(\tilde{\nabla}\times\vec{f})^{\mathrm{b}}_{\xi_{3-k}} = \begin{cases} (-1)^k\tilde{s}^{1/2}\left(\bar{f}^{\mathrm{b}}_{\xi_k} - \tilde{p}\tilde{s}^{-1/2}\bar{f}^{\mathrm{b}}_{\xi_k}\dfrac{\tilde{d}_k-\tilde{d}_{3-k}}{2\tilde{d}_k\tilde{d}_{3-k}}\right) & \text{LD} \\ (-1)^k(1+j)\left(\dot{f}^{\mathrm{b}}_{\xi_k} - \tilde{p}\frac{1-j}{2}\dot{f}^{\mathrm{b}}_{\xi_k}\dfrac{\tilde{d}_k-\tilde{d}_{3-k}}{2\tilde{d}_k\tilde{d}_{3-k}}\right) & \text{FD}, \quad k=1,2 \\ (-1)^k\left[\dfrac{\mathrm{d}}{\mathrm{d}\tilde{t}}\left(f^{\mathrm{b}}_{\xi_k}*\tilde{T}^{\mathrm{b}}_2\right) - \tilde{p}\dfrac{\tilde{d}_k-\tilde{d}_{3-k}}{2\tilde{d}_k\tilde{d}_{3-k}}f^{\mathrm{b}}_{\xi_k}*\tilde{T}^{\mathrm{b}}_1\right] & \text{TD} \end{cases} \tag{2.181}$$

$$(\tilde{\nabla}\times\vec{f})^{\mathrm{b}}_{\eta} = \begin{cases} \tilde{p}\displaystyle\sum_{i=1}^{2}(-1)^i\frac{\partial\bar{f}^{\mathrm{b}}_{\xi_i}}{\partial\tilde{\xi}_{3-i}} & \text{LD} \\ \tilde{p}\displaystyle\sum_{i=1}^{2}(-1)^i\frac{\partial\dot{f}^{\mathrm{b}}_{\xi_i}}{\partial\tilde{\xi}_{3-i}} & \text{FD} \\ \tilde{p}\displaystyle\sum_{i=1}^{2}(-1)^i\frac{\partial f^{\mathrm{b}}_{\xi_i}}{\partial\tilde{\xi}_{3-i}}*\tilde{T}^{\mathrm{b}}_1 & \text{TD} \end{cases} \tag{2.182}$$

Third-order approximation (diffusion in the normal and tangential directions is considered)

$$f^{\mathrm{b}}_{\eta} = \begin{cases} \tilde{p}\tilde{s}^{-1/2}\displaystyle\sum_{i=1}^{2}\frac{\partial}{\partial\tilde{\xi}_i}\left[\bar{f}^{\mathrm{b}}_{\xi_i} - \tilde{p}\tilde{s}^{-1/2}\bar{f}^{\mathrm{b}}_{\xi_i}\frac{\tilde{d}_i-\tilde{d}_{3-i}}{2\tilde{d}_i\tilde{d}_{3-i}}\right. \\ \quad\left. -\tilde{p}^2\tilde{s}^{-1}\left(\bar{f}^{\mathrm{b}}_{\xi_i}\dfrac{\tilde{d}_i^2+2\tilde{d}_i\tilde{d}_{3-i}-3\tilde{d}_{3-i}^2}{8\tilde{d}_i^2\tilde{d}_{3-i}^2} - \dfrac{1}{2}\dfrac{\partial^2\bar{f}^{\mathrm{b}}_{\xi_i}}{\partial\tilde{\xi}_i^2} + \dfrac{1}{2}\dfrac{\partial^2\bar{f}^{\mathrm{b}}_{\xi_i}}{\partial\tilde{\xi}_{3-i}^2} - \dfrac{\partial^2\bar{f}^{\mathrm{b}}_{\xi_{3-i}}}{\partial\tilde{\xi}_i\partial\tilde{\xi}_{3-i}}\right)\right] & \text{LD} \\ \tilde{p}\dfrac{1-j}{2}\displaystyle\sum_{i=1}^{2}\frac{\partial}{\partial\tilde{\xi}_i}\left[\dot{f}^{\mathrm{b}}_{\xi_i} - \tilde{p}\frac{1-j}{2}\dot{f}^{\mathrm{b}}_{\xi_i}\frac{\tilde{d}_i-\tilde{d}_{3-i}}{2\tilde{d}_i\tilde{d}_{3-i}}\right. \\ \quad\left. +\tilde{p}^2\dfrac{j}{2}\left(\dot{f}^{\mathrm{b}}_{\xi_i}\dfrac{\tilde{d}_i^2+2\tilde{d}_i\tilde{d}_{3-i}-3\tilde{d}_{3-i}^2}{8\tilde{d}_i^2\tilde{d}_{3-i}^2} - \dfrac{1}{2}\dfrac{\partial^2\dot{f}^{\mathrm{b}}_{\xi_i}}{\partial\tilde{\xi}_i^2} + \dfrac{1}{2}\dfrac{\partial^2\dot{f}^{\mathrm{b}}_{\xi_i}}{\partial\tilde{\xi}_{3-i}^2} - \dfrac{\partial^2\dot{f}^{\mathrm{b}}_{\xi_{3-i}}}{\partial\tilde{\xi}_i\partial\tilde{\xi}_{3-i}}\right)\right] & \text{FD} \\ \tilde{p}\displaystyle\sum_{i=1}^{2}\frac{\partial}{\partial\tilde{\xi}_i}\left[f^{\mathrm{b}}_{\xi_i}*\tilde{T}^{\mathrm{b}}_2 - \tilde{p}\frac{\tilde{d}_i-\tilde{d}_{3-i}}{2\tilde{d}_i\tilde{d}_{3-i}}f^{\mathrm{b}}_{\xi_i}*\tilde{T}^{\mathrm{b}}_3\right. \\ \quad\left. -\tilde{p}^2\left(f^{\mathrm{b}}_{\xi_i}\dfrac{\tilde{d}_i^2+2\tilde{d}_i\tilde{d}_{3-i}-3\tilde{d}_{3-i}^2}{8\tilde{d}_i^2\tilde{d}_{3-i}^2} - \dfrac{1}{2}\dfrac{\partial^2 f^{\mathrm{b}}_{\xi_i}}{\partial\tilde{\xi}_i^2} + \dfrac{1}{2}\dfrac{\partial^2 f^{\mathrm{b}}_{\xi_i}}{\partial\tilde{\xi}_{3-i}^2} - \dfrac{\partial^2 f^{\mathrm{b}}_{\xi_{3-i}}}{\partial\tilde{\xi}_i\partial\tilde{\xi}_{3-i}}\right)*\tilde{T}^{\mathrm{b}}_4\right] & \text{TD} \end{cases} \tag{2.183}$$

$$
\left.\frac{\partial f_{\xi_k}}{\partial \tilde{\eta}}\right|^{\mathrm{b}} \begin{cases}
-\tilde{s}^{1/2}\Bigg[\bar{f}^{\mathrm{b}}_{\xi_k}-\tilde{p}\tilde{s}^{-1/2}\bar{f}^{\mathrm{b}}_{\xi_k}\dfrac{\tilde{d}_k+\tilde{d}_{3-k}}{2\tilde{d}_k\tilde{d}_{3-k}} \\
\qquad -\tilde{p}^2\tilde{s}^{-1}\left(\bar{f}^{\mathrm{b}}_{\xi_k}\dfrac{\tilde{d}_k^2+2\tilde{d}_k\tilde{d}_{3-k}-3\tilde{d}_{3-k}^2}{8\tilde{d}_k^2\tilde{d}_{3-k}^2}+\dfrac{1}{2}\displaystyle\sum_{i=1}^{2}\dfrac{\partial^2\bar{f}^{\mathrm{b}}_{\xi_k}}{\partial\tilde{\xi}_i^2}\right)\Bigg] & \text{LD} \\
-(1+j)\Bigg[\dot{f}^{\mathrm{b}}_{\xi_k}-\tilde{p}\dfrac{1-j}{2}\dot{f}^{\mathrm{b}}_{\xi_k}\dfrac{\tilde{d}_k+\tilde{d}_{3-k}}{2\tilde{d}_k\tilde{d}_{3-k}} \\
\qquad +\tilde{p}^2\dfrac{j}{4}\left(\dot{f}^{\mathrm{b}}_{\xi_k}\dfrac{\tilde{d}_k^2+2\tilde{d}_k\tilde{d}_{3-k}-3\tilde{d}_{3-k}^2}{8\tilde{d}_k^2\tilde{d}_{3-k}^2}+\dfrac{1}{2}\displaystyle\sum_{i=1}^{2}\dfrac{\partial^2\dot{f}^{\mathrm{b}}_{\xi_k}}{\partial\tilde{\xi}_i^2}\right)\Bigg] & \text{FD}, \quad k=1,2 \\
-\dfrac{\mathrm{d}}{\mathrm{d}\tilde{t}}\left(f^{\mathrm{b}}_{\xi_k}*\tilde{T}^{\mathrm{b}}_2\right)+\tilde{p}\dfrac{\tilde{d}_k+\tilde{d}_{3-k}}{2\tilde{d}_k\tilde{d}_{3-k}}f^{\mathrm{b}}_{\xi_k}*\tilde{T}^{\mathrm{b}}_1 \\
\qquad +\left(f^{\mathrm{b}}_{\xi_k}\dfrac{\tilde{d}_k^2+2\tilde{d}_k\tilde{d}_{3-k}-3\tilde{d}_{3-k}^2}{8\tilde{d}_k^2\tilde{d}_{3-k}^2}+\dfrac{1}{2}\displaystyle\sum_{i=1}^{2}\dfrac{\partial^2 f^{\mathrm{b}}_{\xi_k}}{\partial\tilde{\xi}_i^2}\right)*\tilde{T}^{\mathrm{b}}_2 & \text{TD}
\end{cases} \tag{2.184}
$$

$$
\left.\frac{\partial f_{\eta}}{\partial \tilde{\eta}}\right|^{\mathrm{b}} = \begin{cases}
-\tilde{p}\displaystyle\sum_{i=1}^{2}\dfrac{\partial}{\partial\tilde{\xi}_i}\Bigg[\bar{f}^{\mathrm{b}}_{\xi_i}-\tilde{p}\tilde{s}^{-1/2}\bar{f}^{\mathrm{b}}_{\xi_i}\dfrac{\tilde{d}_i+\tilde{d}_{3-i}}{\tilde{d}_i\tilde{d}_{3-i}} \\
\qquad -\tilde{p}^2\tilde{s}^{-1}\left(\bar{f}^{\mathrm{b}}_{\xi_i}\dfrac{-3\tilde{d}_i^2+2\tilde{d}_i\tilde{d}_{3-i}+\tilde{d}_{3-i}^2}{8\tilde{d}_i^2\tilde{d}_{3-i}^2}+\dfrac{\partial^2\bar{f}^{\mathrm{b}}_{\xi_i}}{\partial\tilde{\xi}_{3-i}^2}-\dfrac{\partial^2\bar{f}^{\mathrm{b}}_{\xi_{3-i}}}{\partial\tilde{\xi}_i\partial\tilde{\xi}_{3-i}}\right)\Bigg] & \text{LD} \\
-\tilde{p}\displaystyle\sum_{i=1}^{2}\dfrac{\partial}{\partial\tilde{\xi}_i}\Bigg[\dot{f}^{\mathrm{b}}_{\xi_i}-\tilde{p}\dfrac{1-j}{2}\dot{f}^{\mathrm{b}}_{\xi_i}\dfrac{\tilde{d}_i+\tilde{d}_{3-i}}{\tilde{d}_i\tilde{d}_{3-i}} \\
\qquad +\tilde{p}^2\dfrac{j}{2}\left(\dot{f}^{\mathrm{b}}_{\xi_i}\dfrac{-3\tilde{d}_i^2+2\tilde{d}_i\tilde{d}_{3-i}+\tilde{d}_{3-i}^2}{8\tilde{d}_i^2\tilde{d}_{3-i}^2}+\dfrac{\partial^2\dot{f}^{\mathrm{b}}_{\xi_i}}{\partial\tilde{\xi}_{3-i}^2}-\dfrac{\partial^2\dot{f}^{\mathrm{b}}_{\xi_{3-i}}}{\partial\tilde{\xi}_i\partial\tilde{\xi}_{3-i}}\right)\Bigg] & \text{FD} \\
-\tilde{p}\displaystyle\sum_{i=1}^{2}\dfrac{\partial}{\partial\tilde{\xi}_i}\Bigg[f^{\mathrm{b}}_{\xi_i}*\tilde{T}^{\mathrm{b}}_1-\tilde{p}\dfrac{\tilde{d}_i+\tilde{d}_{3-i}}{\tilde{d}_i\tilde{d}_{3-i}}f^{\mathrm{b}}_{\xi_i}*\tilde{T}^{\mathrm{b}}_2 \\
\qquad -\tilde{p}^2\left(f^{\mathrm{b}}_{\xi_i}\dfrac{-3\tilde{d}_i^2+2\tilde{d}_i\tilde{d}_{3-i}+\tilde{d}_{3-i}^2}{8\tilde{d}_i^2\tilde{d}_{3-i}^2}+\dfrac{\partial^2 f^{\mathrm{b}}_{\xi_i}}{\partial\tilde{\xi}_{3-i}^2}-\dfrac{\partial^2 f^{\mathrm{b}}_{\xi_{3-i}}}{\partial\tilde{\xi}_i\partial\tilde{\xi}_{3-i}}\right)*\tilde{T}^{\mathrm{b}}_3\Bigg] & \text{TD}
\end{cases} \tag{2.185}
$$

$$
\left(\tilde{\nabla}\times\vec{f}\right)^{\mathrm{b}}_{\xi_{3-k}} = \begin{cases}
(-1)^k\tilde{s}^{1/2}\Bigg[\bar{f}^{\mathrm{b}}_{\xi_k}-\tilde{p}\tilde{s}^{-1/2}\bar{f}^{\mathrm{b}}_{\xi_k}\dfrac{\tilde{d}_k-\tilde{d}_{3-k}}{2\tilde{d}_k\tilde{d}_{3-k}} \\
\qquad -\tilde{p}^2\tilde{s}^{-1}\left(\bar{f}^{\mathrm{b}}_{\xi_k}\dfrac{\tilde{d}_k^2+2\tilde{d}_k\tilde{d}_{3-k}-3\tilde{d}_{3-k}^2}{8\tilde{d}_k^2\tilde{d}_{3-k}^2}-\dfrac{1}{2}\dfrac{\partial^2\bar{f}^{\mathrm{b}}_{\xi_k}}{\partial\tilde{\xi}_k^2}+\dfrac{1}{2}\dfrac{\partial^2\bar{f}^{\mathrm{b}}_{\xi_k}}{\partial\tilde{\xi}_{3-k}^2}-\dfrac{\partial^2\bar{f}^{\mathrm{b}}_{\xi_{3-k}}}{\partial\tilde{\xi}_k\partial\tilde{\xi}_{3-k}}\right)\Bigg] & \text{LD} \\
(-1)^k(1+j)\Bigg[\dot{f}^{\mathrm{b}}_{\xi_k}-\tilde{p}\dfrac{1-j}{2}\dot{f}^{\mathrm{b}}_{\xi_k}\dfrac{\tilde{d}_k-\tilde{d}_{3-k}}{2\tilde{d}_k\tilde{d}_{3-k}} \\
\qquad +\tilde{p}^2\dfrac{j}{2}\left(\dot{f}^{\mathrm{b}}_{\xi_k}\dfrac{\tilde{d}_k^2+2\tilde{d}_k\tilde{d}_{3-k}-3\tilde{d}_{3-k}^2}{8\tilde{d}_k^2\tilde{d}_{3-k}^2}-\dfrac{1}{2}\dfrac{\partial^2\dot{f}^{\mathrm{b}}_{\xi_k}}{\partial\tilde{\xi}_k^2}+\dfrac{1}{2}\dfrac{\partial^2\dot{f}^{\mathrm{b}}_{\xi_k}}{\partial\tilde{\xi}_{3-k}^2}-\dfrac{\partial^2\dot{f}^{\mathrm{b}}_{\xi_{3-k}}}{\partial\tilde{\xi}_k\partial\tilde{\xi}_{3-k}}\right)\Bigg] & \text{FD}, \\
(-1)^k\Bigg[\dfrac{\mathrm{d}}{\mathrm{d}\tilde{t}}\left(f^{\mathrm{b}}_{\xi_k}*\tilde{T}^{\mathrm{b}}_2\right)-\tilde{p}\dfrac{\tilde{d}_k-\tilde{d}_{3-k}}{2\tilde{d}_k\tilde{d}_{3-k}}f^{\mathrm{b}}_{\xi_k}*\tilde{T}^{\mathrm{b}}_1 \\
\qquad -\tilde{p}^2\left(f^{\mathrm{b}}_{\xi_k}\dfrac{\tilde{d}_k^2+2\tilde{d}_k\tilde{d}_{3-k}-3\tilde{d}_{3-k}^2}{8\tilde{d}_k^2\tilde{d}_{3-k}^2}-\dfrac{1}{2}\dfrac{\partial^2 f^{\mathrm{b}}_{\xi_k}}{\partial\tilde{\xi}_k^2}+\dfrac{1}{2}\dfrac{\partial^2 f^{\mathrm{b}}_{\xi_k}}{\partial\tilde{\xi}_{3-k}^2}-\dfrac{\partial^2 f^{\mathrm{b}}_{\xi_{3-k}}}{\partial\tilde{\xi}_k\partial\tilde{\xi}_{3-k}}\right)*\tilde{T}^{\mathrm{b}}_2\Bigg] & \text{TD}
\end{cases}
$$

$$k=1,2 \tag{2.186}$$

$$(\tilde{\nabla}\times\vec{f})^{\mathrm{b}}_{\eta}=\begin{cases}\tilde{p}\sum_{i=1}^{2}(-1)^{i}\dfrac{\partial\bar{f}^{\mathrm{b}}_{\xi_i}}{\partial\tilde{\xi}_{3-i}} & \text{LD}\\ \tilde{p}\sum_{i=1}^{2}(-1)^{i}\dfrac{\partial\dot{f}^{\mathrm{b}}_{\xi_i}}{\partial\tilde{\xi}_{3-i}} & \text{FD}\\ \tilde{p}\sum_{i=1}^{2}(-1)^{i}\dfrac{\partial f^{\mathrm{b}}_{\xi_i}}{\partial\tilde{\xi}_{3-i}}*\tilde{T}^{\mathrm{b}}_{1} & \text{TD}\end{cases} \tag{2.187}$$

Here we used the following:

$$\tilde{T}^{\mathrm{b}}_{1}(\tilde{t})=|\tilde{T}_{1}(\tilde{\eta},\tilde{t})|_{\tilde{\eta}=0}=\tilde{U}'(\tilde{t}) \tag{2.188}$$

$$\tilde{T}^{\mathrm{b}}_{2}(\tilde{t})=|\tilde{T}_{2}(\tilde{\eta},\tilde{t})|_{\tilde{\eta}=0}=(\pi\tilde{t})^{-1/2} \tag{2.189}$$

$$\tilde{T}^{\mathrm{b}}_{3}(\tilde{t})=|\tilde{T}_{3}(\tilde{\eta},\tilde{t})|_{\tilde{\eta}=0}=\tilde{U}(\tilde{t}) \tag{2.190}$$

$$\tilde{T}^{\mathrm{b}}_{4}(\tilde{t})=|\tilde{T}_{4}(\tilde{\eta},\tilde{t})|_{\tilde{\eta}=0}=2\tilde{t}^{1/2}\pi^{-1/2} \tag{2.191}$$

where $U(t)$ and $U'(t)$ denote the Heaviside function and its derivative (the Dirac function), respectively.

The return to dimensional variables can be performed by substituting Equations 2.19, 2.20, 2.35, and 2.53 into Equations 2.173 through 2.191. For example, the FD and TD versions of Equation 2.173 are

$$\dot{f}^{\mathrm{b}}_{\eta}=(2j)^{-1/2}\frac{\delta}{D}\sum_{i=1}^{2}\frac{\partial\dot{f}^{\mathrm{b}}_{\xi_i}}{\partial(\xi_i D^{-1})}=(2j)^{-1/2}\delta\sum_{i=1}^{2}\frac{\partial\dot{f}^{\mathrm{b}}_{\xi_i}}{\partial\xi_i} \tag{2.192}$$

$$\begin{aligned}f^{\mathrm{b}}_{\eta}&=\frac{\delta}{D}\sum_{i=1}^{2}\frac{\partial}{\partial(\xi_i D^{-1})}\int_{0}^{t}f^{\mathrm{b}}_{\xi_i}(t')\left(\pi\frac{t-t'}{\tau}\right)^{-1/2}\mathrm{d}(t'\tau^{-1})=\delta\tau^{-1/2}\sum_{i=1}^{2}\frac{\partial}{\partial\xi_i}\int_{0}^{t}f^{\mathrm{b}}_{\xi_i}(t')(\pi(t-t'))^{-1/2}\mathrm{d}t'\\&=\delta\tau^{-1/2}\sum_{i=1}^{2}\frac{\partial f^{\mathrm{b}}_{\xi_i}}{\partial\xi_i}*T^{\mathrm{b}}_{2};\qquad T^{\mathrm{b}}_{2}(t)=(\pi t)^{-1/2}\end{aligned} \tag{2.193}$$

Taking into account the definition of skin depth in the FD and TD

$$\delta=\begin{cases}\sqrt{\dfrac{2}{\omega\sigma\mu}} & \text{FD}\\ \sqrt{\dfrac{\tau}{\sigma\mu}} & \text{TD}\end{cases} \tag{2.194}$$

casts Equations 2.193 and 2.194 into the following form:

$$\dot{f}_{\eta}^{\mathrm{b}} = (j\omega\sigma\mu)^{-1/2}\sum_{i=1}^{2}\frac{\partial \dot{f}_{\xi_i}^{\mathrm{b}}}{\partial \xi_i} \tag{2.195}$$

$$f_{\eta}^{\mathrm{b}} = (\sigma\mu)^{-1/2}\sum_{i=1}^{2}\frac{\partial f_{\xi_i}^{\mathrm{b}}}{\partial \xi_i} * T_2^{\mathrm{b}} \tag{2.196}$$

Transforming Equations 2.174 through 2.191 in a similar way, and taking into account that

$$\int_0^t f(t')T_1^{\mathrm{b}}(t-t')\mathrm{d}t' = \int_0^t f(t')U'(t-t')\mathrm{d}t' = f(t) \tag{2.197}$$

and

$$\frac{\mathrm{d}}{\mathrm{d}t}\left(T_2^{\mathrm{b}}(t) * f(\vec{r},t)\right) = \frac{\mathrm{d}T_2^{\mathrm{b}}(t)}{dt} * f(\vec{r},t) + T_2^{\mathrm{b}}(0)f(\vec{r},t) = \hat{T}_2^{\mathrm{b}}(t) * f(\vec{r},t) \quad \text{(Duhamels theorem)} \tag{2.198}$$

the following approximate boundary conditions are finally obtained:

First-order approximation

$$f_{\eta}^{\mathrm{b}} = \begin{cases} (j\omega\sigma\mu)^{-1/2}\displaystyle\sum_{i=1}^{2}\frac{\partial \dot{f}_{\xi_i}^{\mathrm{b}}}{\partial \xi_i} & \text{FD} \\ (\sigma\mu)^{-1/2}\displaystyle\sum_{i=1}^{2}\frac{\partial f_{\xi_i}^{\mathrm{b}}}{\partial \xi_i} * T_2^{\mathrm{b}} & \text{TD} \end{cases} \tag{2.199}$$

$$\left.\frac{\partial f_{\xi_k}}{\partial \eta}\right|^{\mathrm{b}} = \begin{cases} -(j\omega\sigma\mu)^{1/2}\dot{f}_{\xi_k}^{\mathrm{b}} & \text{FD} \\ -(\sigma\mu)^{1/2} f_{\xi_k}^{\mathrm{b}} * \hat{T}_2^{\mathrm{b}} & \text{TD} \end{cases}, \quad k = 1, 2 \tag{2.200}$$

$$\left.\frac{\partial f_{\eta}}{\partial \eta}\right|^{\mathrm{b}} = \begin{cases} -\displaystyle\sum_{i=1}^{2}\frac{\partial \dot{f}_{\xi_i}^{\mathrm{b}}}{\partial \xi_i} & \text{FD} \\ -\displaystyle\sum_{i=1}^{2}\frac{\partial f_{\xi_i}^{\mathrm{b}}}{\partial \xi_i} & \text{TD} \end{cases} \tag{2.201}$$

$$(\nabla \times \vec{f})^{\mathrm{b}}_{\xi_{3-k}} = \begin{cases} (-1)^k (j\omega\sigma\mu)^{1/2} \dot{f}^{\mathrm{b}}_{\xi_k} & \text{FD} \\ (-1)^k (\sigma\mu)^{1/2} f^{\mathrm{b}}_{\xi_k} * \hat{T}^{\mathrm{b}}_2 & \text{TD} \end{cases}, \quad k = 1, 2 \tag{2.202}$$

$$(\nabla \times \vec{f})^{\mathrm{b}}_{\eta} = \begin{cases} \displaystyle\sum_{i=1}^{2} (-1)^i \frac{\partial \dot{f}^{\mathrm{b}}_{\xi_i}}{\partial \xi_{3-i}} & \text{FD} \\ \displaystyle\sum_{i=1}^{2} (-1)^i \frac{\partial f^{\mathrm{b}}_{\xi_i}}{\partial \xi_{3-i}} & \text{TD} \end{cases} \tag{2.203}$$

Second-order approximation

$$f^{\mathrm{b}}_{\eta} = \begin{cases} (j\omega\sigma\mu)^{-1/2} \displaystyle\sum_{i=1}^{2} \frac{\partial}{\partial \xi_i} \left(\dot{f}^{\mathrm{b}}_{\xi_i} - (j\omega\sigma\mu)^{-1/2} \frac{d_i - d_{3-i}}{2 d_i d_{3-i}} \dot{f}^{\mathrm{b}}_{\xi_i} \right) & \text{FD} \\ (\sigma\mu)^{-1/2} \displaystyle\sum_{i=1}^{2} \frac{\partial}{\partial \xi_i} \left(f^{\mathrm{b}}_{\xi_i} * T^{\mathrm{b}}_2 - (\sigma\mu)^{-1/2} \frac{d_i - d_{3-i}}{2 d_i d_{3-i}} f^{\mathrm{b}}_{\xi_i} * T^{\mathrm{b}}_3 \right) & \text{TD} \end{cases} \tag{2.204}$$

$$\left.\frac{\partial f_{\xi_k}}{\partial \eta}\right|^{\mathrm{b}} = \begin{cases} -(j\omega\sigma\mu)^{1/2} \dot{f}^{\mathrm{b}}_{\xi_k} + \dfrac{d_k + d_{3-k}}{2 d_k d_{3-k}} \dot{f}^{\mathrm{b}}_{\xi_k} & \text{FD} \\ -(\sigma\mu)^{1/2} f^{\mathrm{b}}_{\xi_k} * \hat{T}^{\mathrm{b}}_2 + \dfrac{d_k + d_{3-k}}{2 d_k d_{3-k}} f^{\mathrm{b}}_{\xi_k} & \text{TD} \end{cases}, \quad k = 1, 2 \tag{2.205}$$

$$\left.\frac{\partial f_{\eta}}{\partial \eta}\right|^{\mathrm{b}} = \begin{cases} -\displaystyle\sum_{i=1}^{2} \frac{\partial}{\partial \tilde{\xi}_i} \left(\dot{f}^{\mathrm{b}}_{\xi_i} - (j\omega\sigma\mu)^{-1/2} \frac{d_i + d_{3-i}}{d_i d_{3-i}} \dot{f}^{\mathrm{b}}_{\xi_i} \right) & \text{FD} \\ -\displaystyle\sum_{i=1}^{2} \frac{\partial}{\partial \tilde{\xi}_i} \left(f^{\mathrm{b}}_{\xi_i} - (\sigma\mu)^{-1/2} \frac{d_i + d_{3-i}}{d_i d_{3-i}} f^{\mathrm{b}}_{\xi_i} * T^{\mathrm{b}}_2 \right) & \text{TD} \end{cases} \tag{2.206}$$

$$(\nabla \times \vec{f})^{\mathrm{b}}_{\xi_{3-k}} = \begin{cases} (-1)^k \left((j\omega\sigma\mu)^{1/2} \dot{f}^{\mathrm{b}}_{\xi_k} - \dfrac{d_k - d_{3-k}}{2 d_k d_{3-k}} \dot{f}^{\mathrm{b}}_{\xi_k} \right) & \text{FD} \\ (-1)^k \left((\sigma\mu)^{1/2} f^{\mathrm{b}}_{\xi_k} * \hat{T}^{\mathrm{b}}_2 - \dfrac{d_k - d_{3-k}}{2 d_k d_{3-k}} f^{\mathrm{b}}_{\xi_k} \right) & \text{TD} \end{cases}, \quad k = 1, 2 \tag{2.207}$$

$$(\nabla \times \vec{f})^{\mathrm{b}}_{\eta} = \begin{cases} \displaystyle\sum_{i=1}^{2} (-1)^i \frac{\partial \dot{f}^{\mathrm{b}}_{\xi_i}}{\partial \xi_{3-i}} & \text{FD} \\ \displaystyle\sum_{i=1}^{2} (-1)^i \frac{\partial f^{\mathrm{b}}_{\xi_i}}{\partial \xi_{3-i}} & \text{TD} \end{cases} \tag{2.208}$$

Third-order approximation

$$f_\eta^{\mathrm{b}} = \begin{cases} (j\omega\sigma\mu)^{-1/2} \displaystyle\sum_{i=1}^{2} \frac{\partial}{\partial \xi_i} \left[\dot{f}_{\xi_i}^{\mathrm{b}} - (j\omega\sigma\mu)^{-1/2} \frac{d_i - d_{3-i}}{2d_i d_{3-i}} \dot{f}_{\xi_i}^{\mathrm{b}} \right. \\ \left. \quad - (j\omega\sigma\mu)^{-1} \left(\dot{f}_{\xi_i}^{\mathrm{b}} \dfrac{d_i^2 + 2d_i d_{3-i} - 3d_{3-i}^2}{8d_i^2 d_{3-i}^2} - \dfrac{1}{2} \dfrac{\partial^2 \dot{f}_{\xi_i}^{\mathrm{b}}}{\partial \xi_i^2} + \dfrac{1}{2} \dfrac{\partial^2 \dot{f}_{\xi_i}^{\mathrm{b}}}{\partial \xi_{3-i}^2} - \dfrac{\partial^2 \dot{f}_{\xi_{3-i}}^{\mathrm{b}}}{\partial \xi_i \partial \xi_{3-i}} \right) \right] & \text{FD} \\ (\sigma\mu)^{-1/2} \displaystyle\sum_{i=1}^{2} \frac{\partial}{\partial \xi_i} \left[f_{\xi_i}^{\mathrm{b}} * T_2^{\mathrm{b}} - (\sigma\mu)^{-1/2} \frac{d_i - d_{3-i}}{2d_i d_{3-i}} f_{\xi_i}^{\mathrm{b}} * T_3^{\mathrm{b}} \right. \\ \left. \quad - (\sigma\mu)^{-1} \left(f_{\xi_i}^{\mathrm{b}} \dfrac{d_i^2 + 2d_i d_{3-i} - 3d_{3-i}^2}{8d_i^2 d_{3-i}^2} - \dfrac{1}{2} \dfrac{\partial^2 f_{\xi_i}^{\mathrm{b}}}{\tilde{\xi}_i^2} + \dfrac{1}{2} \dfrac{\partial^2 f_{\xi_i}^{\mathrm{b}}}{\partial \xi_{3-i}^2} - \dfrac{\partial^2 f_{\xi_{3-i}}^{\mathrm{b}}}{\partial \xi_i \partial \xi_{3-i}} \right) * T_4^{\mathrm{b}} \right] & \text{TD} \end{cases} \tag{2.209}$$

$$\left.\frac{\partial f_{\xi_k}}{\partial \eta}\right|^{\mathrm{b}} = \begin{cases} -(j\omega\sigma\mu)^{1/2} \dot{f}_{\xi_k}^{\mathrm{b}} + \dfrac{d_k + d_{3-k}}{2d_k d_{3-k}} \dot{f}_{\xi_k}^{\mathrm{b}} \\ \quad + (j\omega\sigma\mu)^{-1/2} \left(\dot{f}_{\xi_k}^{\mathrm{b}} \dfrac{d_k^2 + 2d_k d_{3-k} - 3d_{3-k}^2}{8d_k^2 d_{3-k}^2} + \dfrac{1}{2} \displaystyle\sum_{i=1}^{2} \dfrac{\partial^2 \dot{f}_{\xi_k}^{\mathrm{b}}}{\partial \xi_i^2} \right) \Bigg] & \text{FD} \\ -(\sigma\mu)^{1/2} f_{\xi_k}^{\mathrm{b}} * \hat{T}_2^{\mathrm{b}} + \dfrac{d_k + d_{3-k}}{2d_k d_{3-k}} f_{\xi_k}^{\mathrm{b}} \\ \quad + (\sigma\mu)^{-1/2} \left(f_{\xi_k}^{\mathrm{b}} \dfrac{d_k^2 + 2d_k d_{3-k} - 3d_{3-k}^2}{8d_k^2 d_{3-k}^2} + \dfrac{1}{2} \displaystyle\sum_{i=1}^{2} \dfrac{\partial^2 f_{\xi_k}^{\mathrm{b}}}{\partial \xi_i^2} \right) * T_2^{\mathrm{b}} & \text{TD} \end{cases}, \quad k = 1,2 \tag{2.210}$$

$$\left.\frac{\partial f_{\xi_k}}{\partial \eta}\right|^{\mathrm{b}} = \begin{cases} -(j\omega\sigma\mu)^{1/2} \dot{f}_{\xi_k}^{\mathrm{b}} + \dfrac{d_k + d_{3-k}}{2d_k d_{3-k}} \dot{f}_{\xi_k}^{\mathrm{b}} \\ \quad + (j\omega\sigma\mu)^{-1/2} \left(\dot{f}_{\xi_k}^{\mathrm{b}} \dfrac{d_k^2 + 2d_k d_{3-k} - 3d_{3-k}^2}{8d_k^2 d_{3-k}^2} + \dfrac{1}{2} \displaystyle\sum_{i=1}^{2} \dfrac{\partial^2 \dot{f}_{\xi_k}^{\mathrm{b}}}{\partial \xi_i^2} \right) \Bigg] & \text{FD} \\ -(\sigma\mu)^{1/2} f_{\xi_k}^{\mathrm{b}} * \hat{T}_2^{\mathrm{b}} + \dfrac{d_k + d_{3-k}}{2d_k d_{3-k}} f_{\xi_k}^{\mathrm{b}} \\ \quad + (\sigma\mu)^{-1/2} \left(f_{\xi_k}^{\mathrm{b}} \dfrac{d_k^2 + 2d_k d_{3-k} - 3d_{3-k}^2}{8d_k^2 d_{3-k}^2} + \dfrac{1}{2} \displaystyle\sum_{i=1}^{2} \dfrac{\partial^2 f_{\xi_k}^{\mathrm{b}}}{\partial \xi_i^2} \right) * T_2^{\mathrm{b}} & \text{TD} \end{cases} \tag{2.211}$$

$$(\nabla \times \vec{f})_{\xi_{3-k}}^{\mathrm{b}} = \begin{cases} (-1)^k \left[(j\omega\sigma\mu)^{1/2} \dot{f}_{\xi k}^{\mathrm{b}} - \dfrac{d_k - d_{3-k}}{2d_k d_{3-k}} \dot{f}_{\xi k}^{\mathrm{b}} \right. \\ \left. \quad - (j\omega\sigma\mu)^{-1/2} \left(\dot{f}_{\xi_k}^{\mathrm{b}} \dfrac{d_k^2 + 2d_k d_{3-k} - 3d_{3-k}^2}{8d_k^2 d_{3-k}^2} - \dfrac{1}{2} \dfrac{\partial^2 \dot{f}_{\xi k}^{\mathrm{b}}}{\partial \xi_k^2} + \dfrac{1}{2} \dfrac{\partial^2 \dot{f}_{\xi k}^{\mathrm{b}}}{\partial \xi_{3-k}^2} - \dfrac{\partial^2 \dot{f}_{\xi_{3-k}}^{\mathrm{b}}}{\partial \xi_k \partial \xi_{3-k}} \right) \right] & \text{FD} \\ (-1)^k \left[(\sigma\mu)^{1/2} f_{\xi k}^{\mathrm{b}} * \hat{T}_2^{\mathrm{b}} - \dfrac{d_k - d_{3-k}}{2d_k d_{3-k}} f_{\xi k}^{\mathrm{b}} \right. \\ \left. \quad - (\sigma\mu)^{-1/2} \left(f_{\xi k}^{\mathrm{b}} \dfrac{d_k^2 + 2d_k d_{3-k} - 3\tilde{d}_{3-k}^2}{8d_k^2 d_{3-k}^2} - \dfrac{1}{2} \dfrac{\partial^2 f_{\xi k}^{\mathrm{b}}}{\partial \xi_k^2} + \dfrac{1}{2} \dfrac{\partial^2 f_{\xi k}^{\mathrm{b}}}{\partial \xi_{3-k}^2} - \dfrac{\partial^2 f_{\xi_{3-k}}^{\mathrm{b}}}{\partial \xi_k \partial \xi_{3-k}} \right) * T_2^{\mathrm{b}} \right] & \text{TD} \end{cases}, \quad k = 1,2 \tag{2.212}$$

$$(\nabla \times \vec{f})_{\eta}^{\mathrm{b}} = \begin{cases} \displaystyle\sum_{i=1}^{2} (-1)^i \frac{\partial \dot{f}_{\xi_i}^{\mathrm{b}}}{\partial \xi_{3-i}} & \text{FD} \\ \displaystyle\sum_{i=1}^{2} (-1)^i \frac{\partial f_{\xi_i}^{\mathrm{b}}}{\partial \xi_{3-i}} & \text{TD} \end{cases} \tag{2.213}$$

Here we used the following:

$$T_2^{\mathrm{b}}(t) = (\pi t)^{-1/2} \tag{2.214}$$

$$T_3^{\mathrm{b}}(t) = U(t) \tag{2.215}$$

$$T_4^{\mathrm{b}}(t) = 2t^{1/2}\pi^{-1/2} \tag{2.216}$$

$$\hat{T}_2^{\mathrm{b}} = \frac{\mathrm{d}T_2^{\mathrm{b}}}{\mathrm{d}t} = -(4\pi t^3)^{-1/2} \tag{2.217}$$

and the nondimensional convolution product was replaced with

$$\tau^{-1} \int_0^t \varphi(t')\phi(t - t')\mathrm{d}t' \tag{2.218}$$

The formulae we derived are applicable to such functions as the magnetic field, the electric field, and the magnetic vector potential (with Coulomb's gauge) since they obey the diffusion equation (Equation 2.2). We emphasize that the application of "toolbox" formulae requires that the equations be represented in local nondimensional coordinates.

2.9 Numerical Example

Consider an infinitely long straight conducting cylinder of circular cross section. In this case the coordinate ξ_1 is directed along the cylinder as shown in Figure 2.2 so that $d_1 = \infty$. The radius $d = d_2$ of the cylinder cross section is constant and is used as the characteristic scale D. Let the distribution of the time-harmonic function $\vec{f}$ in the cylinder be described by the diffusion equation and assume no variation of $\vec{f}$ in the ξ_1-direction, i.e.

$$\frac{\partial}{\partial \xi_1} = 0 \tag{2.219}$$

The problem is therefore considered one-dimensional in cylindrical coordinates.

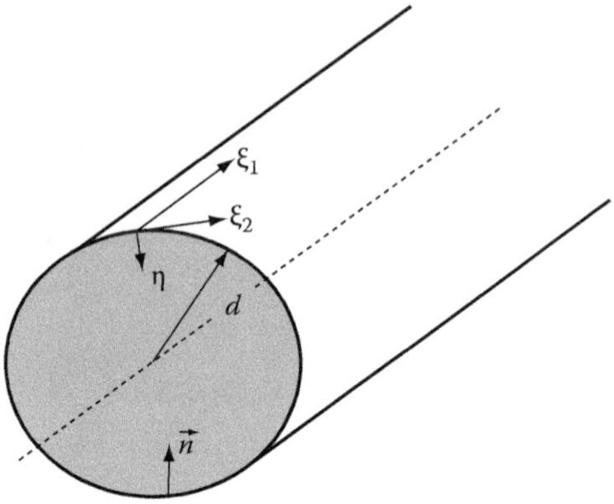

FIGURE 2.2
The test problem analyzed: infinitely long straight conducting cylinder of circular cross section.

Suppose that the function $\vec{f}$ has only a circumferential component (analogous to the magnetic field in our current carrying infinitely long straight cylinder):

$$\vec{f} = f_{\xi_2}\vec{e}_{\xi_2} = f\vec{e}_{\xi_2} \tag{2.220}$$

Then the governing equation of diffusion is

$$\frac{\partial^2 \dot{f}}{\partial r^2} + \frac{1}{r}\frac{\partial \dot{f}}{\partial r} - \frac{\dot{f}}{r^2} + \vartheta^2 \dot{f} = 0; \quad \dot{f}\Big|_{r=d} = \dot{f}^{\text{b}} \tag{2.221}$$

where

$$\vartheta^2 = -j\omega\sigma\mu = -2j/\delta^2; \quad \vartheta = \sqrt{-2j}/\delta = (1-j)/\delta \tag{2.222}$$

and r is the radial coordinate in the cylindrical coordinate system. It varies from 0 (center of the cylinder) to d (radius of the cylinder).

The exact solution of Equation 2.221 can be written in the form [7]

$$\dot{f}(r) = \dot{f}^{\text{b}}\frac{J_1(\vartheta r)}{J_1(\vartheta d)} \tag{2.223}$$

where J_1 is the Bessel function of the first kind and first order. Note that

$$\vartheta d = (1-j)\frac{d}{\delta} = \frac{1-j}{\tilde{p}} \tag{2.224}$$

Turning to the dimensionless variable

$$\tilde{r} = r/d \tag{2.225}$$

we represent Equation 2.223 in the form:

$$\dot{f}(d\tilde{r}) = \dot{f}^{\mathrm{b}} \frac{J_1\left[\frac{1-j}{\tilde{p}}\tilde{r}\right]}{J_1\left[\frac{1-j}{\tilde{p}}\right]} \tag{2.226}$$

where $\tilde{r}$ varies from 0 (center of the cylinder) to 1 (surface of the cylinder).

An approximate solution under the conditions of skin effect is given by Equations 2.92 through 2.94 which, in this case, reduce to the following forms:

First-order approximation

$$\dot{f}(\eta) = \dot{f}^{\mathrm{b}} \exp\left(-\eta\sqrt{j\omega\sigma\mu}\right) = \dot{f}^{\mathrm{b}} \exp\left(-\frac{1+j}{\delta}\eta\right) \tag{2.227}$$

Second-order approximation

$$\dot{f}(\eta) = \dot{f}^{\mathrm{b}}\left(1+\frac{\eta}{2d}\right)\exp\left(-\frac{1+j}{\delta}\eta\right) \tag{2.228}$$

Third-order approximation

$$\dot{f}(\eta) = \dot{f}^{\mathrm{b}}\left(1+\frac{\eta}{2d}+\frac{3\eta^2}{8d^2}+\frac{\delta}{\sqrt{2j}}\frac{\eta}{8d^2}\right)\exp\left(-\frac{1+j}{\delta}\eta\right) \tag{2.229}$$

Introducing d as a scale factor for η

$$\tilde{\eta} = \eta/d \tag{2.230}$$

we represent Equations 2.227 through 2.229 in the form:

First-order approximation

$$\dot{f}(\tilde{\eta}) = \dot{f}^{\mathrm{b}} \exp\left(-\frac{1+j}{\tilde{p}}\tilde{\eta}\right) \tag{2.231}$$

Second-order approximation

$$\dot{f}(\tilde{\eta}) = \dot{f}^{\mathrm{b}}\left(1+\frac{\tilde{\eta}}{2}\right)\exp\left(-\frac{1+j}{\tilde{p}}\tilde{\eta}\right) \tag{2.232}$$

Third-order approximation

$$\dot{f}(\tilde{\eta}) = \dot{f}^{\mathrm{b}}\left(1+\frac{1}{2}\tilde{\eta}+\frac{3}{8}\tilde{\eta}^2+\frac{\tilde{p}}{8\sqrt{2j}}\tilde{\eta}\right)\exp\left(-\frac{1+j}{\tilde{p}}\tilde{\eta}\right) \tag{2.233}$$

Note that the variables $\tilde{\eta}$ and $\tilde{r}$ have the same scale factor, d, and are related as follows:

$$\tilde{r} = 1 - \tilde{\eta} \tag{2.234}$$

Using Equation 2.234, we recast Equation 2.226 into the form:

$$\dot{f}(\tilde{\eta}) = \dot{f}^{b} \frac{J_1\left[\frac{1-j}{\tilde{p}}(1-\tilde{\eta})\right]}{J_1\left[\frac{1-j}{\tilde{p}}\right]} \tag{2.235}$$

Figures 2.3 through 2.11 show distributions of the real part, imaginary part, and amplitude of the function $\dot{f}$ given by the approximate formulae (Equations 2.231 through 2.233) and the exact solution (Equation 2.235) for the parameter $\tilde{p}$ equal to 0.1, 0.3, and 0.5. The nondimensional variable, $\tilde{\eta}$, varies from 0 (surface) to 0.5 (half the distance from the surface to the center of the cylinder). The boundary value, $\dot{f}^{b}$, is equal to 1. It is clearly seen that

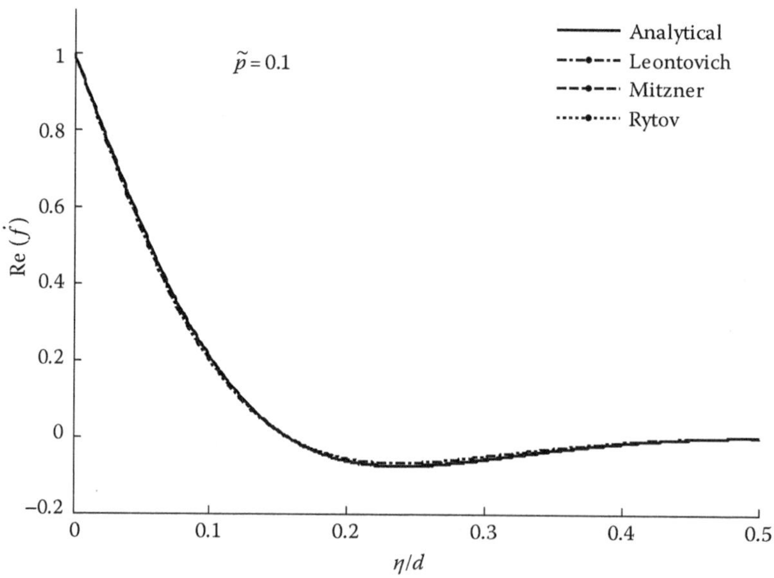

FIGURE 2.3
Real part of $\dot{f}$ as a function of the distance from the cylinder surface, for the case $\tilde{p} = 0.1$ and $\dot{f}^{b} = 1$.

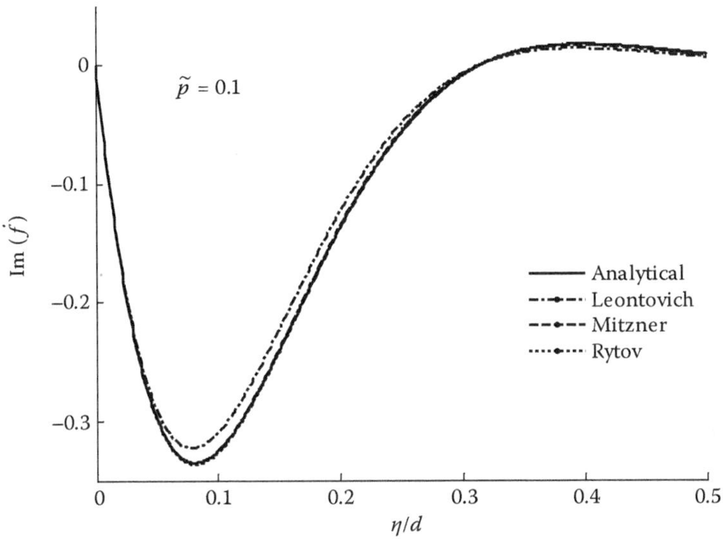

FIGURE. 2.4
Imaginary part of $\dot{f}$ as a function of the distance from the cylinder surface, for the case $\tilde{p} = 0.1$ and $\dot{f}^b = 1$.

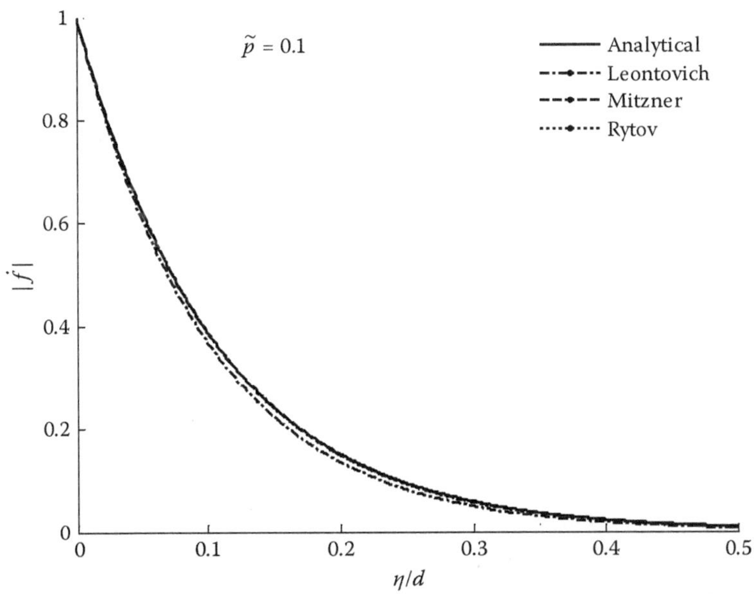

FIGURE 2.5
Amplitude of $\dot{f}$ as a function of the distance from the cylinder surface, for the case $\tilde{p} = 0.1$ and $\dot{f}^b = 1$.

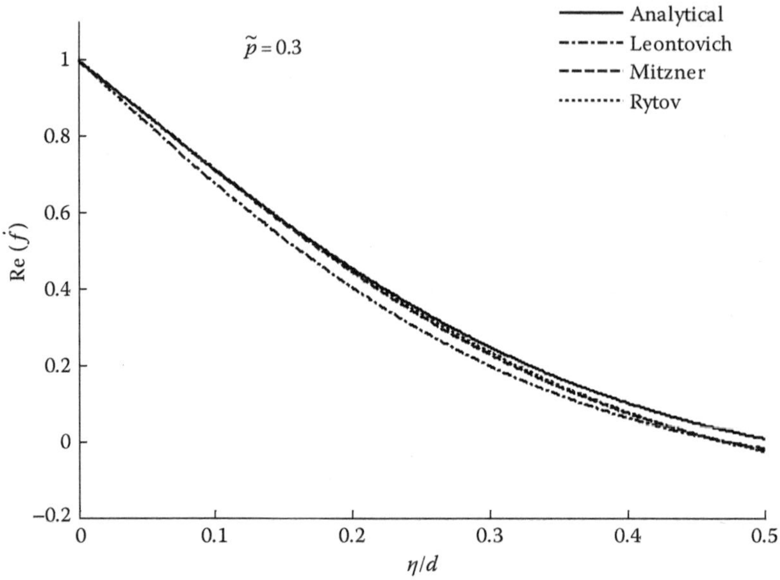

FIGURE 2.6
Real part of $\dot{f}$ as a function of the distance from the cylinder surface, for the case $\tilde{p} = 0.3$ and $\dot{f}^{b} = 1$.

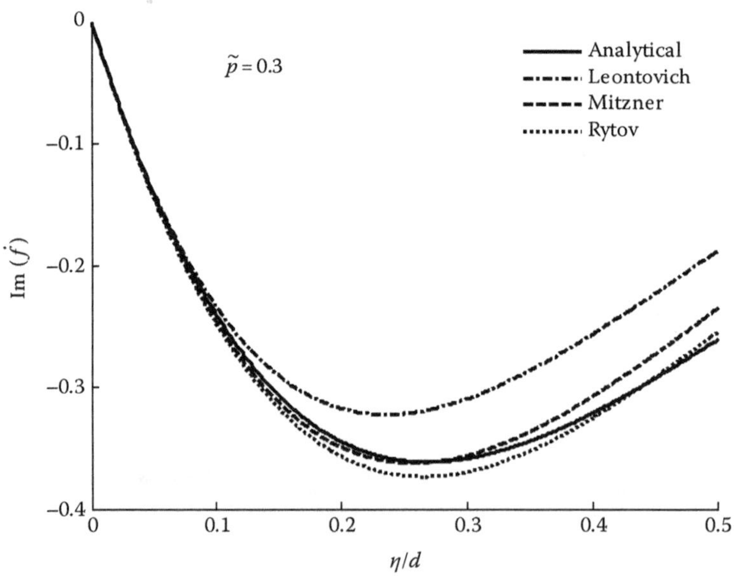

FIGURE 2.7
Imaginary part of $\dot{f}$ as a function of the distance from the cylinder surface, for the case $\tilde{p} = 0.3$ and $\dot{f}^{b} = 1$.

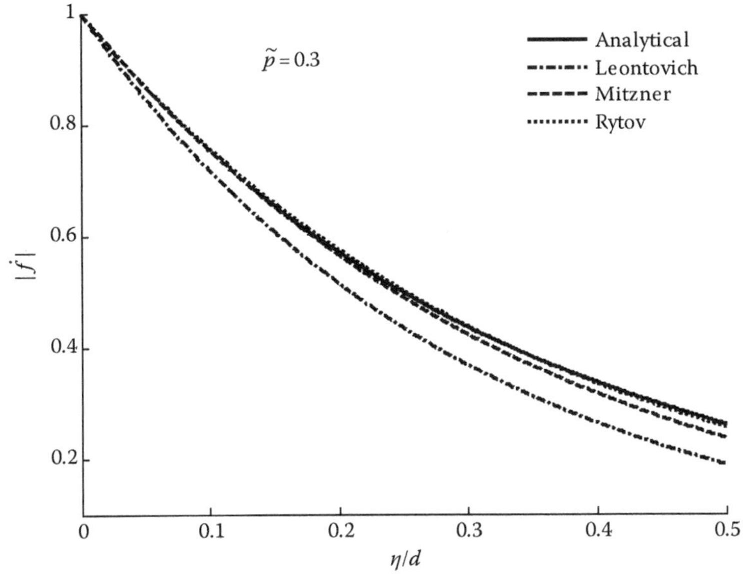

FIGURE 2.8
Amplitude of $\dot{f}$ as a function of the distance from the cylinder surface, for the case $\tilde{p} = 0.3$ and $\dot{f}^b = 1$.

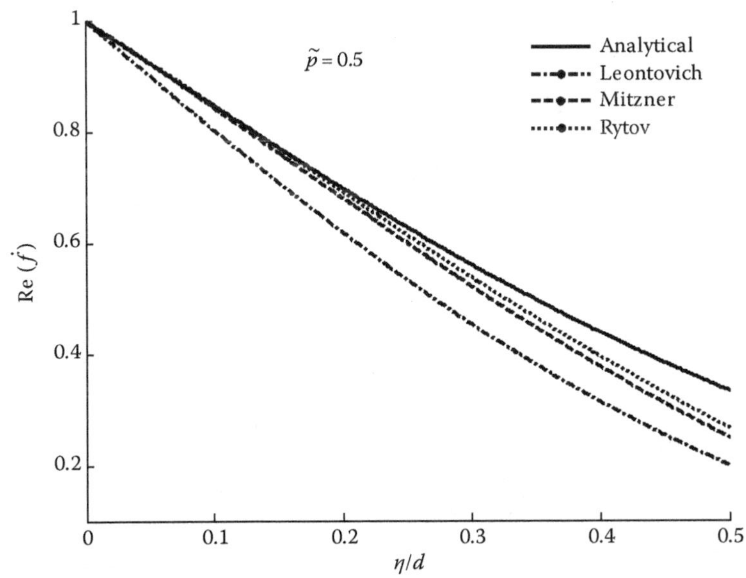

FIGURE 2.9
Real part of $\dot{f}$ as a function of the distance from the cylinder surface, for the case $\tilde{p} = 0.5$ and $\dot{f}^b = 1$.

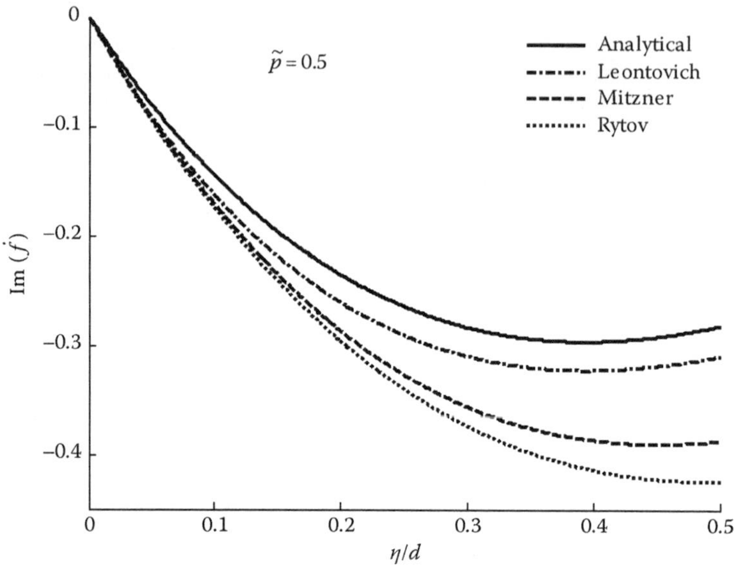

FIGURE 2.10
Imaginary part of $\dot{f}$ as a function of the distance from the cylinder surface, for the case $\tilde{p} = 0.5$ and $\dot{f}^{b} = 1$.

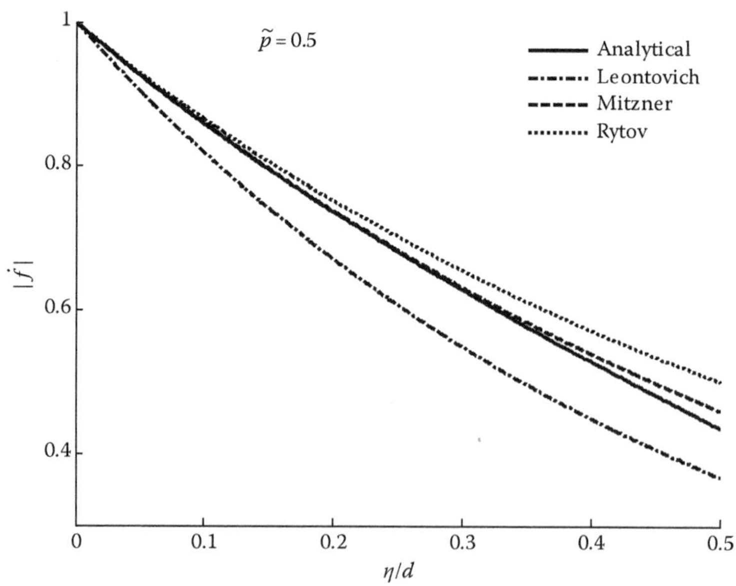

FIGURE 2.11
Amplitude of $\dot{f}$ as a function of the distance from the cylinder surface, for the case $\tilde{p} = 0.5$ and $\dot{f}^{b} = 1$.

in the case of $\tilde{p}=0.1$, the Leontovich's SIBC provides suitable accuracy; however with the increase in thickness of the skin layer ($\tilde{p}=0.3$), higher-order SIBCs (Miztner's and Rytov's approximations) should be applied. When the skin depth equals half the cylinder radius ($\tilde{p}=0.5$), the surface impedance concept yields significant error.

Appendix 2.A.1: Calculation of $\left(\vec{f}_3\right)_\eta$, $\left(\vec{\bar{g}}_2\right)_{\xi k}$, and $\left(\vec{\bar{g}}_2\right)_\eta$

The following notations are introduced to simplify further derivations:

$$\left(\vec{f}_2\right)_{\xi_i} = \left(a_{20} + \tilde{\eta}a_{21} + \tilde{\eta}^2 a_{22} + \tilde{\eta}\tilde{s}^{-1/2}a_{23}\right)\exp\left(-\tilde{\eta}\sqrt{\tilde{s}}\right) \tag{2.A.1}$$

$$\left(\vec{f}_1\right)_{\xi_i} = (a_{10} + \tilde{\eta}a_{11})\exp(-\tilde{\eta}\sqrt{\tilde{s}}) \tag{2.A.2}$$

$$\left(\vec{f}_0\right)_{\xi_i} = a_{00}\exp(-\tilde{\eta}\sqrt{\tilde{s}}) \tag{2.A.3}$$

$$\left(\vec{f}_2\right)_\eta = \tilde{s}^{-1/2}\sum_{i=1}^{2}\frac{\partial}{\partial\tilde{\xi}_i}\left[a_{10} + \tilde{\eta}a_{24} - \tilde{s}^{-1/2}a_{25}\right]\exp(-\tilde{\eta}\sqrt{\tilde{s}}) \tag{2.A.4}$$

$$\left(\vec{f}_1\right)_\eta = \tilde{s}^{-1/2}\sum_{i=1}^{2}\frac{\partial a_{00}}{\partial\tilde{\xi}_i}\exp(-\tilde{\eta}\sqrt{\tilde{s}}) \tag{2.A.5}$$

where

$$a_{20} = \left(\vec{f}_2^{\,b}\right)_{\xi_i} \tag{2.A.6}$$

$$a_{21} = \left(\vec{f}_1^{\,b}\right)_{\xi_i}\frac{\tilde{d}_i + \tilde{d}_{3-i}}{2\tilde{d}_i\tilde{d}_{3-i}} \tag{2.A.7}$$

$$a_{22} = \left(\vec{f}_0^{\,b}\right)_{\xi_i}\frac{3\tilde{d}_i^2 + 2\tilde{d}_i\tilde{d}_{3-i} + 3\tilde{d}_{3-i}^2}{8\tilde{d}_i^2\tilde{d}_{3-i}^2} \tag{2.A.8}$$

$$a_{23} = \left(\vec{f}_0^{\,b}\right)_{\xi_i}\frac{\tilde{d}_i^2 + 2\tilde{d}_i\tilde{d}_{3-i} - 3\tilde{d}_{3-i}^2}{8\tilde{d}_i^2\tilde{d}_{3-i}^2} + \frac{1}{2}\sum_{l=1}^{2}\frac{\partial^2\left(\vec{f}_0^{\,b}\right)_{\xi_i}}{\partial\tilde{\xi}_l^2} \tag{2.A.9}$$

$$a_{24} = \left(\vec{f}_0\right)_{\xi_i}\frac{\tilde{d}_i + 3\tilde{d}_{3-i}}{2\tilde{d}_i\tilde{d}_{3-i}} \tag{2.A.10}$$

$$a_{25} = \left(\vec{\vec{f}}_0\right)_{\xi_i} \frac{\tilde{d}_i - \tilde{d}_{3-i}}{2\tilde{d}_i\tilde{d}_{3-i}} \tag{2.A.11}$$

$$a_{10} = \left(\vec{\vec{f}}_1^{\,b}\right)_{\xi_i} \tag{2.A.12}$$

$$a_{11} = \left(\vec{\vec{f}}_0^{\,b}\right)_{\xi_i} \frac{\tilde{d}_i + \tilde{d}_{3-i}}{2\tilde{d}_i\tilde{d}_{3-i}} \tag{2.A.13}$$

$$a_{00} = \left(\vec{\vec{f}}_0^{\,b}\right)_{\xi_i} \tag{2.A.14}$$

Equations 2.102, 2.144, and 2.145 can now be represented in the forms:

$$\begin{aligned}
\left(\vec{\vec{f}}_3\right)_{\eta} &= -\tilde{s}^{-1}\sum_{i=1}^{2}\frac{\partial}{\partial\tilde{\xi}_i}\left[\frac{\partial\left(\vec{\vec{f}}_2\right)_{\xi_i}}{\partial\tilde{\eta}} + \frac{\tilde{\eta}}{\tilde{d}_i}\frac{\partial\left(\vec{\vec{f}}_1\right)_{\xi_i}}{\partial\tilde{\eta}} - \frac{\left(\vec{\vec{f}}_1\right)_{\xi_i}}{\tilde{d}_i} - \frac{\partial\left(\vec{\vec{f}}_1\right)_{\eta}}{\partial\tilde{\xi}_i} + \tilde{\eta}\left(\frac{\tilde{\eta}}{\tilde{d}_i^2}\frac{\partial\left(\vec{\vec{f}}_0\right)_{\xi_i}}{\partial\tilde{\eta}} - \frac{2}{\tilde{d}_i^2}\left(\vec{\vec{f}}_0\right)_{\xi_i}\right)\right] \\
&= -\tilde{s}^{-1}\sum_{i=1}^{2}\frac{\partial}{\partial\tilde{\xi}_i}\left\{\left[-\tilde{s}^{1/2}\left(a_{20} + \tilde{\eta}a_{21} + \tilde{\eta}^2 a_{22} + \tilde{\eta}\tilde{s}^{-1/2}a_{23}\right) + a_{21} + 2\tilde{\eta}a_{22} + \tilde{s}^{-1/2}a_{23}\right]\right. \\
&\quad + \frac{\tilde{\eta}}{\tilde{d}_i}\left[-\tilde{s}^{1/2}(a_{10} + \tilde{\eta}a_{11}) + a_{11}\right] - \frac{1}{\tilde{d}_i}[a_{10} + \tilde{\eta}a_{11}] - \tilde{s}^{-1/2}\sum_{l=1}^{2}\frac{\partial^2 a_{00}}{\partial\tilde{\xi}_i\partial\tilde{\xi}_l} \\
&\quad \left. + \tilde{\eta}a_{00}\left[-\tilde{s}^{1/2}\frac{\tilde{\eta}}{\tilde{d}_i^2} - \frac{2}{\tilde{d}_i^2}\right]\right\}\exp\left(-\tilde{\eta}\sqrt{\tilde{s}}\right) \\
&= -\tilde{s}^{-1}\sum_{i=1}^{2}\frac{\partial}{\partial\tilde{\xi}_i}\left\{\tilde{s}^{-1/2}\left[a_{23} - \sum_{l=1}^{2}\frac{\partial^2 a_{00}}{\partial\tilde{\xi}_i\partial\tilde{\xi}_l}\right]\right. \\
&\quad + \left[-\tilde{\eta}a_{23} + a_{21} + 2\tilde{\eta}a_{22} + \tilde{\eta}\frac{a_{11}}{\tilde{d}_i} - \frac{a_{10}}{\tilde{d}_i} - \tilde{\eta}\frac{a_{11}}{\tilde{d}_i} - \tilde{\eta}\frac{2a_{00}}{\tilde{d}_i^2}\right] \\
&\quad \left. + \tilde{s}^{1/2}\left[-a_{20} - \tilde{\eta}a_{21} - \tilde{\eta}^2 a_{22} - \tilde{\eta}\frac{a_{10}}{\tilde{d}_i} - \tilde{\eta}^2\frac{a_{11}}{\tilde{d}_i} - \tilde{\eta}^2\frac{a_{00}}{\tilde{d}_i^2}\right]\right\}\exp\left(\tilde{\eta}\sqrt{\tilde{s}}\right) \\
&= -\tilde{s}^{-1}\sum_{i=1}^{2}\frac{\partial}{\partial\tilde{\xi}_i}\left\{\tilde{s}^{-1/2}\left[a_{23} - \sum_{l=1}^{2}\frac{\partial^2 a_{00}}{\partial\tilde{\xi}_i\partial\tilde{\xi}_l}\right]\right. \\
&\quad + \left[a_{21} - \frac{a_{10}}{\tilde{d}_i} + \tilde{\eta}\left(-a_{23} + 2a_{22} - \frac{2a_{00}}{\tilde{d}_i^2}\right)\right] \\
&\quad \left. + \tilde{s}^{1/2}\left[-a_{20} - \tilde{\eta}\left(a_{21} + \frac{a_{10}}{\tilde{d}_i}\right) - \tilde{\eta}^2\left(a_{22} + \frac{a_{11}}{\tilde{d}_i} + \frac{a_{00}}{\tilde{d}_i^2}\right)\right]\right\}\exp\left(\tilde{\eta}\sqrt{\tilde{s}}\right)
\end{aligned} \tag{2.A.15}$$

$$(\vec{\bar{g}}_2)_{\xi_{3-k}} = (-1)^{3-k}\left[\frac{\partial\left(\vec{\bar{f}}_2\right)_{\xi_k}}{\partial\tilde{\eta}} - \frac{\left(\vec{\bar{f}}_1\right)_{\xi_k}}{\tilde{d}_k} - \frac{\partial\left(\vec{\bar{f}}_1\right)_{\eta}}{\partial\tilde{\xi}_k} - \tilde{\eta}\frac{\left(\vec{\bar{f}}_0\right)_{\xi_k}}{\tilde{d}_k^2}\right]$$

$$= (-1)^{3-k}\left[-\tilde{s}^{1/2}\left(a_{20} + \tilde{\eta}a_{21} + \tilde{\eta}^2 a_{22} + \tilde{\eta}\tilde{s}^{-1/2}a_{23}\right) + a_{21} + 2\tilde{\eta}a_{22}\right.$$

$$\left. + \tilde{s}^{-1/2}a_{23} - \frac{1}{\tilde{d}_k}(a_{10} + \tilde{\eta}a_{11}) - \tilde{s}^{-1/2}\sum_{l=1}^{2}\frac{\partial^2 a_{00}}{\partial\tilde{\xi}_k\partial\tilde{\xi}_l} - \tilde{\eta}\frac{a_{00}}{\tilde{d}_k^2}\right]\exp(-\tilde{\eta}\sqrt{\tilde{s}})$$

$$= (-1)^{3-k}\left[-\tilde{s}^{1/2}(a_{20} + \tilde{\eta}a_{21} + \tilde{\eta}^2 a_{22}) + a_{21} - \frac{a_{10}}{\tilde{d}_k} - \tilde{\eta}\left(a_{23} + 2a_{22} + \frac{a_{11}}{\tilde{d}_k} + \frac{a_{00}}{\tilde{d}_k^2}\right)\right.$$

$$\left. + \tilde{s}^{-1/2}\left(a_{23} - \sum_{l=1}^{2}\frac{\partial^2 a_{00}}{\partial\tilde{\xi}_k\partial\tilde{\xi}_l}\right)\right]\exp(-\tilde{\eta}\sqrt{\tilde{s}}) \qquad (2.A.16)$$

$$(\vec{\bar{g}}_3)_\eta = \sum_{i=1}^{2}(-1)^i\left(\frac{\partial\left(\vec{\bar{f}}_2\right)_{\xi_i}}{\partial\tilde{\xi}_{3-i}} + \frac{\tilde{\eta}}{\tilde{d}_{3-i}}\frac{\partial\left(\vec{\bar{f}}_1\right)_{\xi_i}}{\partial\tilde{\xi}_{3-i}} + \frac{\tilde{\eta}^2}{\tilde{d}_{3-i}^2}\frac{\partial\left(\vec{\bar{f}}_0\right)_{\xi_i}}{\partial\tilde{\xi}_{3-i}}\right)$$

$$= \sum_{i=1}^{2}(-1)^i\frac{\partial}{\partial\tilde{\xi}_{3-i}}\left[a_{20} + \tilde{\eta}a_{21} + \tilde{\eta}^2 a_{22} + \tilde{\eta}\tilde{s}^{-1/2}a_{23} + \frac{\tilde{\eta}}{\tilde{d}_{3-i}}(a_{10} + \tilde{\eta}a_{11}) + \frac{\tilde{\eta}^2}{\tilde{d}_{3-i}^2}a_{00}\right]$$

$$= \sum_{i=1}^{2}(-1)^i\frac{\partial}{\partial\tilde{\xi}_{3-i}}\left[a_{20} + \tilde{\eta}\left(a_{21} + \frac{a_{10}}{\tilde{d}_{3-i}}\right) + \tilde{\eta}^2\left(a_{22} + \frac{a_{11}}{\tilde{d}_{3-i}} + \frac{a_{00}}{\tilde{d}_{3-i}^2}\right) + \tilde{\eta}\tilde{s}^{-1/2}a_{23}\right]$$

(2.A.17)

Substitution of Equations 2.A.6 through 2.A.14 into Equations 2.A.15 through 2.A.17 leads to the following results:

$$\left(\vec{\bar{f}}_3\right)_\eta = -\sum_{i=1}^{2}\frac{\partial}{\partial\tilde{\xi}_i}\left\{-\tilde{s}^{-1/2}\left[\left(\vec{\bar{f}}_2^{\,b}\right)_{\xi_i} + \tilde{\eta}\left(\vec{\bar{f}}_1^{\,b}\right)_{\xi_i}\frac{\tilde{d}_i + 3\tilde{d}_{3-i}}{2\tilde{d}_i\tilde{d}_{3-i}} + \tilde{\eta}^2\left(\vec{\bar{f}}_0^{\,b}\right)_{\xi_i}\frac{3\tilde{d}_i^2 + 6\tilde{d}_i\tilde{d}_{3-i} + 11\tilde{d}_{3-i}^2}{8\tilde{d}_i^2\tilde{d}_{3-i}^2}\right]\right.$$

$$+ \tilde{s}^{-1}\left[\left(\vec{\bar{f}}_1^{\,b}\right)_{\xi_i}\frac{\tilde{d}_i - \tilde{d}_{3-i}}{2\tilde{d}_i\tilde{d}_{3-i}} + \tilde{\eta}\left(\left(\vec{\bar{f}}_0^{\,b}\right)_{\xi_i}\frac{5\tilde{d}_i^2 + 2\tilde{d}_i\tilde{d}_{3-i} - 7\tilde{d}_{3-i}^2}{8\tilde{d}_i^2\tilde{d}_{3-i}^2} - \frac{1}{2}\sum_{l=1}^{2}\frac{\partial^2\left(\vec{\bar{f}}_0^{\,b}\right)_{\xi_i}}{\partial\tilde{\xi}_l^2}\right)\right]$$

$$\left. + \tilde{s}^{-3/2}\left[\left(\vec{\bar{f}}_0^{\,b}\right)_{\xi_i}\frac{\tilde{d}_i^2 + 2\tilde{d}_i\tilde{d}_{3-i} - 3\tilde{d}_{3-i}^2}{8\tilde{d}_i^2\tilde{d}_{3-i}^2} - \frac{1}{2}\frac{\partial^2\left(\vec{\bar{f}}_0^{\,b}\right)_{\xi_i}}{\partial\tilde{\xi}_i^2} + \frac{1}{2}\frac{\partial^2\left(\vec{\bar{f}}_0^{\,b}\right)_{\xi_i}}{\partial\tilde{\xi}_{3-i}^2} - \frac{\partial^2\left(\vec{\bar{f}}_0^{\,b}\right)_{\xi_{3-i}}}{\partial\tilde{\xi}_i\partial\tilde{\xi}_{3-i}}\right]\right\}\exp(\tilde{\eta}\sqrt{\tilde{s}})$$

(2.A.18)

$$
\begin{aligned}
(\vec{\tilde{g}}_2)_{\xi_{3-k}} = (-1)^{3-k}\Bigg[& -\tilde{s}^{1/2}\left(\left(\vec{\tilde{f}}{}^{\mathrm{b}}_2\right)_{\xi_k} + \tilde{\eta}\left(\vec{\tilde{f}}{}^{\mathrm{b}}_1\right)_{\xi_k}\frac{\tilde{d}_k+\tilde{d}_{3-k}}{2\tilde{d}_k\tilde{d}_{3-k}} + \tilde{\eta}^2\left(\vec{\tilde{f}}{}^{\mathrm{b}}_0\right)_{\xi_k}\frac{3\tilde{d}_k^2+2\tilde{d}_k\tilde{d}_{3-k}+3\tilde{d}_{3-k}^2}{8\tilde{d}_k^2\tilde{d}_{3-k}^2}\right) \\
& + \frac{\tilde{d}_k-\tilde{d}_{3-k}}{2\tilde{d}_k\tilde{d}_{3-k}}\left(\vec{\tilde{f}}{}^{\mathrm{b}}_1\right)_{\xi_k} - \tilde{\eta}\left(\left(\vec{\tilde{f}}{}^{\mathrm{b}}_0\right)_{\xi_k}\frac{-5\tilde{d}_k^2+2\tilde{d}_k\tilde{d}_{3-k}+3\tilde{d}_{3-k}^2}{2\tilde{d}_k^2\tilde{d}_{3-k}^2} + \frac{1}{2}\sum_{l=1}^{2}\frac{\partial^2\left(\vec{\tilde{f}}{}^{\mathrm{b}}_0\right)_{\xi_i}}{\partial\tilde{\xi}_l^2}\right) \\
& + \tilde{s}^{-1/2}\left(\left(\vec{\tilde{f}}{}^{\mathrm{b}}_0\right)_{\xi_k}\frac{\tilde{d}_k^2+2\tilde{d}_k\tilde{d}_{3-k}-3\tilde{d}_{3-k}^2}{8\tilde{d}_k^2\tilde{d}_{3-k}^2} - \frac{1}{2}\frac{\partial^2\left(\vec{\tilde{f}}{}^{\mathrm{b}}_0\right)_{\xi_k}}{\partial\tilde{\xi}_k^2} + \frac{1}{2}\frac{\partial^2\left(\vec{\tilde{f}}{}^{\mathrm{b}}_0\right)_{\xi_k}}{\partial\tilde{\xi}_{3-k}^2} - \frac{\partial^2\left(\vec{\tilde{f}}{}^{\mathrm{b}}_0\right)_{\xi_{3-k}}}{\partial\tilde{\xi}_k\partial\tilde{\xi}_{3-k}}\right)\Bigg]\exp\left(-\tilde{\eta}\sqrt{\tilde{s}}\right)
\end{aligned}
\tag{2.A.19}
$$

$$
\begin{aligned}
(\vec{\tilde{g}}_3)_{\eta} = \sum_{i=1}^{2}(-1)^i\frac{\partial}{\partial\tilde{\xi}_{3-i}}\Bigg[& \left(\vec{\tilde{f}}{}^{\mathrm{b}}_2\right)_{\xi_i} + \tilde{\eta}\left(\vec{\tilde{f}}{}^{\mathrm{b}}_1\right)_{\xi_i}\frac{3\tilde{d}_i+\tilde{d}_{3-i}}{2\tilde{d}_i\tilde{d}_{3-i}} + \tilde{\eta}^2\left(\vec{\tilde{f}}{}^{\mathrm{b}}_0\right)_{\xi_i}\frac{11\tilde{d}_i^{?}+6\tilde{d}_i\tilde{d}_{3-\iota}+3\tilde{d}_{3-i}^2}{8\tilde{d}_i^2\tilde{d}_{3-i}^2} \\
& + \tilde{\eta}\tilde{s}^{-1/2}\left(\left(\vec{\tilde{f}}{}^{\mathrm{b}}_0\right)_{\xi_i}\frac{\tilde{d}_i^2+2\tilde{d}_i\tilde{d}_{3-i}-3\tilde{d}_{3-i}^2}{8\tilde{d}_i^2\tilde{d}_{3-i}^2} + \frac{1}{2}\sum_{l=1}^{2}\frac{\partial^2\left(\vec{\tilde{f}}{}^{\mathrm{b}}_0\right)_{\xi_i}}{\partial\tilde{\xi}_l^2}\right)\Bigg]\exp\left(-\tilde{\eta}\sqrt{\tilde{s}}\right)
\end{aligned}
\tag{2.A.20}
$$

At the conductor's surface ($\tilde{\eta}=0$), Equations 2.A.18 through 2.A.20 reduce to the following formulae:

$$
\begin{aligned}
\left(\vec{\tilde{f}}_3\right)_{\eta}\Big|_{\tilde{\eta}=0} = -\tilde{s}^{-1}\sum_{i=1}^{2}\frac{\partial}{\partial\tilde{\xi}_i}\Bigg[& -\tilde{s}^{1/2}\left(\vec{\tilde{f}}{}^{\mathrm{b}}_2\right)_{\xi_i} + \left(\vec{\tilde{f}}{}^{\mathrm{b}}_1\right)_{\xi_i}\frac{\tilde{d}_i-\tilde{d}_{3-i}}{2\tilde{d}_i\tilde{d}_{3-i}} \\
& + \tilde{s}^{-1/2}\left(\left(\vec{\tilde{f}}{}^{\mathrm{b}}_0\right)_{\xi_i}\frac{\tilde{d}_i^2+2\tilde{d}_i\tilde{d}_{3-i}-3\tilde{d}_{3-i}^2}{8\tilde{d}_i^2\tilde{d}_{3-i}^2} - \frac{1}{2}\frac{\partial^2\left(\vec{\tilde{f}}{}^{\mathrm{b}}_0\right)_{\xi_i}}{\partial\tilde{\xi}_i^2} + \frac{1}{2}\frac{\partial^2\left(\vec{\tilde{f}}{}^{\mathrm{b}}_0\right)_{\xi_i}}{\partial\tilde{\xi}_{3-i}^2} - \frac{\partial^2\left(\vec{\tilde{f}}{}^{\mathrm{b}}_0\right)_{\xi_{3-i}}}{\partial\tilde{\xi}\partial\tilde{\xi}_{3-i}}\right)\Bigg]
\end{aligned}
\tag{2.A.21}
$$

$$
\begin{aligned}
(\vec{\tilde{g}}_2)_{\xi_{3-k}}\Big|_{\tilde{\eta}=0} = (-1)^{3-k}\Bigg[& -\tilde{s}^{1/2}\left(\vec{\tilde{f}}{}^{\mathrm{b}}_2\right)_{\xi_k} + \frac{\tilde{d}_k-\tilde{d}_{3-k}}{2\tilde{d}_k\tilde{d}_{3-k}}\left(\vec{\tilde{f}}{}^{\mathrm{b}}_1\right)_{\xi_k} \\
& + \tilde{s}^{-1/2}\left(\left(\vec{\tilde{f}}{}^{\mathrm{b}}_0\right)_{\xi_k}\frac{\tilde{d}_k^2+2\tilde{d}_k\tilde{d}_{3-k}-3\tilde{d}_{3-k}^2}{8\tilde{d}_k^2\tilde{d}_{3-k}^2} - \frac{1}{2}\frac{\partial^2\left(\vec{\tilde{f}}{}^{\mathrm{b}}_0\right)_{\xi_k}}{\partial\tilde{\xi}_k^2} + \frac{1}{2}\frac{\partial^2\left(\vec{\tilde{f}}{}^{\mathrm{b}}_0\right)_{\xi_k}}{\partial\tilde{\xi}_{3-k}^2} - \frac{\partial^2\left(\vec{\tilde{f}}{}^{\mathrm{b}}_0\right)_{\xi_{3-k}}}{\partial\tilde{\xi}_k\partial\tilde{\xi}_{3-k}}\right)\Bigg]
\end{aligned}
\tag{2.A.22}
$$

$$
(\vec{\tilde{g}}_3)_{\eta}\Big|_{\tilde{\eta}=0} = \sum_{i=1}^{2}(-1)^i\frac{\partial\left(\vec{\tilde{f}}{}^{\mathrm{b}}_2\right)_{\xi_i}}{\partial\tilde{\xi}_{3-i}}
\tag{2.A.23}
$$

References

1. S.M. Rytov, Calcul du skin-effet par la méthode des perturbations, *Journal of Physics*, 2(3), 1940, 233–242.
2. K.M. Mitzner, An integral equation approach to scattering from a body of finite, *Radio Science*, 2(12), 1967, 1459–1470.
3. M.A. Leontovich, On one approach to a problem of the wave propagation along the Earth's surface, *Academy of Science USSR, Series Physics*, 8, 1944, 16–22.
4. M.A. Leontovich, On the approximate boundary conditions for the electromagnetic field on the surface of well conducting bodies, in B.A. Vvedensky (ed.), *Investigations of Radio Waves*, Academy of Sciences of the USSR, Moscow, 1948.
5. G.S. Smith, On the skin effect approximation, *American Journal of Physics*, 58(10), 1990, 996–1002.
6. M. Abramowitz and I.A. Stegun (Eds.), *Handbook of Mathematical Functions with Formulas, Graphs and Mathematical Tables*, Applied Math Series 55, National Bureau of Standards, Washington, DC, 1964.
7. K. Simonyi, *Theoretische Elektrotechnik*, Ver Deutscher Verlag, Berlin, 1956.
8. A.J. Mestel, More accurate skin-depth approximations, *IMA Journal of Applied Mathematics*, 45, 1990, 33–48.

3

SIBCs in Terms of Various Formalisms

3.1 Introduction

Surface impedance boundary conditions are usually used in combination with a numerical method. Therefore, both the formulation of the numerical method and the SIBC should be represented in terms of the same variables. It is rather natural that boundary conditions should not dictate the selection of the numerical technique. Instead, the numerical method is selected first and boundary conditions must then be derived in the state variables of the numerical method. Numerical methods used in practice frequently operate in terms other than the electric and/or magnetic fields (for example, the magnetic vector or scalar potential formalisms are commonly employed). The method may require quantities such as normal and tangential components of a function or the function and its normal derivative being related at the interface. These types of relationships are obtained in this chapter using the "surface impedance toolbox."

Since most electromagnetic quantities in conductors are governed by the diffusion equation, it is natural to suppose they are mutually connected. Thus, another goal of this chapter is to derive general relations suitable for representation of various SIBCs in terms of different formalisms. In this sense, the discussion in this chapter forms the cornerstone of implementations that follow in Chapters 5 through 8.

3.2 Basic Equations

Consider a homogeneous body of finite conductivity ($\mu_1 = \mu$, $\sigma_1 = \sigma$), surrounded by a nonconductive medium ($\varepsilon_2 = \varepsilon_0$, $\mu_2 = \mu_0$, $\sigma_2 = 0$). The parameters of the conducting material are assumed to be constant. In this chapter, we restrict ourselves to good conductors so that displacement currents inside the body as well as surface charges can be neglected. The more general case of a lossy dielectric materials will be considered in Chapter 8. The electric field, $\vec{E}$, and magnetic field, $\vec{H}$, inside the body can be described by Maxwell's equations:

$$\nabla \times \vec{E} = -\mu \frac{\partial \vec{H}}{\partial t} \tag{3.1}$$

$$\nabla \times \vec{H} = \sigma \vec{E} \tag{3.2}$$

The exact boundary conditions at the surface of the conductor are

$$\vec{n} \times \vec{H}\Big|_{\text{cond}} = \vec{n} \times \vec{H}\Big|_{\text{diel}}; \quad \mu_1 \vec{n} \cdot \vec{H}\Big|_{\text{cond}} = \mu_2 \vec{n} \cdot \vec{H}\Big|_{\text{diel}}; \quad \vec{n} \times \vec{E}\Big|_{\text{cond}} = \vec{n} \times \vec{E}\Big|_{\text{diel}}; \quad \vec{n} \cdot \vec{E}\Big|_{\text{cond}} = \vec{n} \cdot \vec{E}\Big|_{\text{diel}} \tag{3.3}$$

Let the time variation of the incident field be such that the electromagnetic penetration depth, δ, into the body remains small compared with the characteristic dimension, D, of the surface of the body so that the condition in Equation 2.1 is satisfied.

3.3 Electric Field–Magnetic Field Formalism

Equations 3.1 and 3.2 with the local coordinates $\tilde{\xi}_1$, $\tilde{\xi}_2$, and $\tilde{\eta}$, introduced in Section 2.2, are written in the following form using Equations 2.46 and 2.47:

$$\delta^{-1}\left[\frac{\partial E_{\xi_k}}{\partial \tilde{\eta}} - \tilde{p}\left(\frac{E_{\xi_k}}{\tilde{d}_k} + \frac{\partial E_{\eta}}{\partial \tilde{\xi}_k}\right) - \tilde{p}^2\tilde{\eta}\left(\frac{E_{\xi_k}}{\tilde{d}_k^2} + \frac{1}{\tilde{d}_k}\frac{\partial E_{\eta}}{\partial \tilde{\xi}_k}\right) + \cdots\right] = (-1)^k \mu \frac{\partial H_{\xi_{3-k}}}{\partial t}, \quad k = 1,2 \tag{3.4}$$

$$\delta^{-1}\sum_{i=1}^{2}(-1)^i \frac{\partial E_{\xi_i}}{\partial \tilde{\xi}_{3-i}}\left(\tilde{p} + \tilde{p}^2\frac{\tilde{\eta}}{\tilde{d}_{3-i}} + \cdots\right) = \mu \frac{\partial H_{\eta}}{\partial t} \tag{3.5}$$

$$\delta^{-1}\left[\frac{\partial H_{\xi_{3-k}}}{\partial \tilde{\eta}} - \tilde{p}\left(\frac{H_{\xi_{3-k}}}{\tilde{d}_{3-k}} + \frac{\partial H_{\eta}}{\partial \tilde{\xi}_{3-k}}\right) - \tilde{p}^2\tilde{\eta}\left(\frac{H_{\xi_{3-k}}}{\tilde{d}_{3-k}^2} + \frac{1}{\tilde{d}_{3-k}}\frac{\partial H_{\eta}}{\partial \tilde{\xi}_{3-k}}\right) + \cdots\right] = (-1)^k \sigma E_{\xi_k}, \quad k = 1,2 \tag{3.6}$$

$$\delta^{-1}\sum_{i=1}^{2}(-1)^{3-i} \frac{\partial H_{\xi_{3-i}}}{\partial \tilde{\xi}_i}\left(\tilde{p} + \tilde{p}^2\frac{\tilde{\eta}}{\tilde{d}_i} + \cdots\right) = \sigma E_{\eta} \tag{3.7}$$

Solution of Equations 3.4 through 3.7 using perturbation theory requires switching to nondimensional electric and magnetic fields. We denote the scale factors for the electric and magnetic fields as $\boldsymbol{E}$ and $\boldsymbol{H}$, respectively, and replace the dimensional $\vec{E}$ and $\vec{H}$ by the nondimensional quantities $\vec{\tilde{E}}$ and $\vec{\tilde{H}}$ multiplied by the appropriate scale factors:

$$\vec{E} = \vec{\tilde{E}}\boldsymbol{E}; \quad \vec{H} = \vec{\tilde{H}}\boldsymbol{H} \tag{3.8}$$

The relation between **E** and **H** follows directly from Equation 3.1:

$$\boldsymbol{E} = \frac{\mu D}{\tau}\boldsymbol{H} \tag{3.9}$$

The relation in Equation 3.9 can be represented in another form:

$$\boldsymbol{E} = \frac{\mu\sigma}{\tau}\frac{D}{\sigma}\boldsymbol{H} = \delta^{-2}\frac{D}{\sigma}\boldsymbol{H} = \delta^{-2}\frac{D}{\sigma}\boldsymbol{H} = \tilde{p}^{-1}(\sigma\delta)^{-1}\boldsymbol{H} \tag{3.10}$$

In many practical problems, the total current is used as input data. In such cases it is only natural to use the characteristic current **I** as one of the scale factors and express **E** and **H** in terms of **I**. The relation between **I** and **H** can be obtained using the Biot–Savart law:

$$\vec{H} = (4\pi)^{-1}\int_L \vec{I} \times (\vec{R}/R^3)\mathrm{d}l \tag{3.11}$$

where L is any closed contour surrounding the conductor carrying the current $\vec{I}$. Transfer to nondimensional variables in Equation 3.11 yields

$$\boldsymbol{H}\vec{\tilde{H}} = \boldsymbol{I}(4\pi D)^{-1}\int_L \vec{\tilde{I}} \times \left(\vec{\tilde{R}}/\tilde{R}^3\right)\mathrm{d}\tilde{l} \tag{3.12}$$

The latter equation can be rewritten in the form:

$$\vec{\tilde{H}} = (4\pi)^{-1}\int_L \vec{\tilde{I}} \times \left(\vec{\tilde{R}}/\tilde{R}^3\right)\mathrm{d}\tilde{l} \tag{3.13}$$

if **H** is chosen as

$$\boldsymbol{H} = \boldsymbol{I}D^{-1} \tag{3.14}$$

Substitution of Equation 3.14 in Equation 3.9 yields the new definition of **E**:

$$\boldsymbol{E} = \frac{\mu}{\tau}\boldsymbol{I} \tag{3.15}$$

Scale factors for all quantities are collected in Table 3.1. Note that there are three "basic" scale factors: **I**, τ, and D. The other scale factors are combinations of the basic scale factors.

TABLE 3.1

Quantities and Appropriate Scale Factors

Quantity	Scale Factor	Unit
ξ_2, ξ_2	$D = \min(d_1, d_2)$	m
$\partial/\partial\xi_1$, $\partial/\partial\xi_2$	D^{-1}	m^{-1}
η	$\delta = \tilde{p}D$	m
$\partial/\partial\eta$	$\delta^{-1} = (\tilde{p}D)^{-1}$	m^{-1}
T	τ	s
I	I	A
E	$I \cdot \tau^{-1}\mu$	$\mathrm{V\,m}^{-1}$
H	$I \cdot D^{-1}$	$\mathrm{A\,m}^{-1}$
A	I	A
"*" (in time convolution products)	τ	s

Substituting Equation 3.8 into Equations 3.4 through 3.7 and using Equations 3.9 and 3.10, we obtain

$$\frac{\partial \tilde{E}_{\xi_k}}{\partial \tilde{\eta}} - \tilde{p}\left(\frac{\tilde{E}_{\xi_k}}{\tilde{d}_k} + \frac{\partial \tilde{E}_\eta}{\partial \tilde{\xi}_k}\right) - \tilde{p}^2\tilde{\eta}\left(\frac{\tilde{E}_{\xi_k}}{\tilde{d}_k^2} + \frac{1}{\tilde{d}_k}\frac{\partial \tilde{E}_\eta}{\partial \tilde{\xi}_k}\right) = (-1)^k \tilde{p}\frac{\partial \tilde{H}_{\xi_{3-k}}}{\partial \tilde{t}}, \quad k=1,2 \tag{3.16}$$

$$\sum_{i=1}^{2}(-1)^i \frac{\partial \tilde{E}_{\xi_i}}{\partial \tilde{\xi}_{3-i}}\left(\tilde{p} + \tilde{p}^2\frac{\tilde{\eta}}{\tilde{d}_{3-i}}\right) = \tilde{p}\frac{\partial \tilde{H}_\eta}{\partial \tilde{t}} \tag{3.17}$$

$$\frac{\partial \tilde{H}_{\xi_{3-k}}}{\partial \tilde{\eta}} - \tilde{p}\left(\frac{\tilde{H}_{\xi_{3-k}}}{\tilde{d}_{3-k}} + \frac{\partial \tilde{H}_\eta}{\partial \tilde{\xi}_{3-k}}\right) - \tilde{p}^2\tilde{\eta}\left(\frac{\tilde{H}_{\xi_{3-k}}}{\tilde{d}_{3-k}^2} + \frac{1}{\tilde{d}_{3-k}}\frac{\partial \tilde{H}_\eta}{\partial \tilde{\xi}_{3-k}}\right) = (-1)^k \tilde{p}^{-1}\tilde{E}_{\xi_k}, \quad k=1,2 \tag{3.18}$$

$$\sum_{i=1}^{2}(-1)^{3-i} \frac{\partial \tilde{H}_{\xi_{3-i}}}{\partial \tilde{\xi}_i}\left(\tilde{p} + \tilde{p}^2\frac{\tilde{\eta}}{\tilde{d}_i}\right) = \tilde{p}^{-1}\tilde{E}_\eta \tag{3.19}$$

Equations 3.16 through 3.19 do not contain the scale factors for the electric and magnetic fields any longer.

We now apply the Laplace transform following the rule:

$$\vec{\bar{\tilde{E}}}(\tilde{s}) = \int_0^\infty \vec{\tilde{E}}(\tilde{t})\exp(-\tilde{s}\tilde{t})\mathrm{d}\tilde{t}, \quad \vec{\bar{\tilde{H}}}(\tilde{s}) = \int_0^\infty \vec{\tilde{H}}(\tilde{t})\exp(-\tilde{s}\tilde{t})\mathrm{d}\tilde{t} \tag{3.20}$$

In the Laplace domain, Equations 3.16 through 3.19 are written in the form:

$$\frac{\partial \bar{\tilde{E}}_{\xi_k}}{\partial \tilde{\eta}} - \tilde{p}\left(\frac{\bar{\tilde{E}}_{\xi_k}}{\tilde{d}_k} + \frac{\partial \bar{\tilde{E}}_\eta}{\partial \tilde{\xi}_k}\right) - \tilde{p}^2\tilde{\eta}\left(\frac{\bar{\tilde{E}}_{\xi_k}}{\tilde{d}_k^2} + \frac{1}{\tilde{d}_k}\frac{\partial \bar{\tilde{E}}_\eta}{\partial \tilde{\xi}_k}\right) = (-1)^k \tilde{p}\tilde{s}\bar{\tilde{H}}_{\xi_{3-k}}, \quad k=1,2 \tag{3.21}$$

$$\sum_{i=1}^{2}(-1)^i \frac{\partial \bar{\tilde{E}}_{\xi_i}}{\partial \tilde{\xi}_{3-i}}\left(\tilde{p}+\tilde{p}^2\frac{\tilde{\eta}}{\tilde{d}_{3-i}}\right) = \tilde{p}\tilde{s}\bar{\tilde{H}}_\eta \tag{3.22}$$

$$\frac{\partial \bar{\tilde{H}}_{\xi_{3-k}}}{\partial \tilde{\eta}} - \tilde{p}\left(\frac{\bar{\tilde{H}}_{\xi_{3-k}}}{\tilde{d}_{3-k}}+\frac{\partial \bar{\tilde{H}}_\eta}{\partial \tilde{\xi}_{3-k}}\right) - \tilde{p}^2\tilde{\eta}\left(\frac{\bar{\tilde{H}}_{\xi_{3-k}}}{\tilde{d}_{3-k}^2}+\frac{1}{\tilde{d}_{3-k}}\frac{\partial \bar{\tilde{H}}_\eta}{\partial \tilde{\xi}_{3-k}}\right) = (-1)^k\tilde{p}^{-1}\bar{\tilde{E}}_{\xi_k},\quad k=1,2 \tag{3.23}$$

$$\sum_{i=1}^{2}(-1)^{3-i} \frac{\partial \bar{\tilde{H}}_{\xi_{3-i}}}{\partial \tilde{\xi}_i}\left(\tilde{p}+\tilde{p}^2\frac{\tilde{\eta}}{\tilde{d}_i}\right) = \tilde{p}^{-1}\bar{\tilde{E}}_\eta \tag{3.24}$$

Since the parameter $\tilde{p}$ is small, we represent the functions for which the solutions are sought in the form of asymptotic expansions in the parameter $\tilde{p}$:

$$\vec{\bar{\tilde{E}}} = \sum_{m=0}^{\infty}\tilde{p}^m\vec{\bar{\tilde{E}}}_m \tag{3.25}$$

$$\vec{\bar{\tilde{H}}} = \sum_{m=0}^{\infty}\tilde{p}^m\vec{\bar{\tilde{H}}}_m \tag{3.26}$$

Substituting the expansions (Equations 3.25 and 3.26) into Equations 3.21 through 3.24 and equating coefficients of equal powers of $\tilde{p}$, we obtain the following equations for the expansion coefficients:

$m=0$:

$$\left(\vec{\bar{\tilde{E}}}_0\right)_{\xi_1} = \left(\vec{\bar{\tilde{E}}}_0\right)_{\xi_2} = \left(\vec{\bar{\tilde{E}}}_0\right)_{\eta} = \left(\vec{\bar{\tilde{H}}}_0\right)_{\eta} = 0 \tag{3.27}$$

$m=1$:

$$\frac{\partial\left(\vec{\bar{\tilde{E}}}_1\right)_{\xi_k}}{\partial\tilde{\eta}} = (-1)^k\tilde{s}\left(\vec{\bar{\tilde{H}}}_0\right)_{\xi_{3-k}},\quad k=1,2 \tag{3.28}$$

$$\sum_{i=1}^{2}(-1)^i\frac{\partial\left(\vec{\bar{\tilde{E}}}_1\right)_{\xi_i}}{\partial\tilde{\xi}_{3-i}} = \tilde{s}\left(\vec{\bar{\tilde{H}}}_1\right)_{\eta} \tag{3.29}$$

$$\frac{\partial\left(\vec{\bar{\tilde{H}}}_0\right)_{\xi_{3-k}}}{\partial\tilde{\eta}} = (-1)^k\left(\vec{\bar{\tilde{E}}}_1\right)_{\xi_k},\quad k=1,2 \tag{3.30}$$

$$\left(\vec{\tilde{\tilde{E}}}_1\right)_{\eta} = 0 \tag{3.31}$$

$m = 2$:

$$\frac{\partial\left(\vec{\tilde{\tilde{E}}}_2\right)_{\xi_k}}{\partial\tilde{\eta}} - \frac{\left(\vec{\tilde{\tilde{E}}}_1\right)_{\xi_k}}{\tilde{d}_k} = (-1)^k \tilde{s}\left(\vec{\tilde{\tilde{H}}}_1\right)_{\xi_{3-k}}, \quad k = 1,2 \tag{3.32}$$

$$\sum_{i=1}^{2}(-1)^i\left(\frac{\partial\left(\vec{\tilde{\tilde{E}}}_2\right)_{\xi_i}}{\partial\tilde{\xi}_{3-i}} + \frac{\tilde{\eta}}{\tilde{d}_{3-i}}\frac{\partial\left(\vec{\tilde{\tilde{E}}}_1\right)_{\xi_i}}{\partial\tilde{\xi}_{3-i}}\right) = \tilde{s}\left(\vec{\tilde{\tilde{H}}}_2\right)_{\eta} \tag{3.33}$$

$$\frac{\partial\left(\vec{\tilde{\tilde{H}}}_1\right)_{\xi_{3-k}}}{\partial\tilde{\eta}} - \frac{\left(\vec{\tilde{\tilde{H}}}_0\right)_{\xi_{3-k}}}{\tilde{d}_{3-k}} = (-1)^k\left(\vec{\tilde{\tilde{E}}}_2\right)_{\xi_k}, \quad k = 1,2 \tag{3.34}$$

$$\sum_{i=1}^{2}(-1)^{3-i}\frac{\partial\left(\vec{\tilde{\tilde{H}}}_0\right)_{\xi_{3-i}}}{\partial\tilde{\xi}_i} = \left(\vec{\tilde{\tilde{E}}}_2\right)_{\eta} \tag{3.35}$$

$m = 3$:

$$\frac{\partial\left(\vec{\tilde{\tilde{E}}}_3\right)_{\xi_k}}{\partial\tilde{\eta}} - \frac{\left(\vec{\tilde{\tilde{E}}}_2\right)_{\xi_k}}{\tilde{d}_k} - \frac{\partial\left(\vec{\tilde{\tilde{E}}}_2\right)_{\eta}}{\partial\tilde{\xi}_k} - \tilde{\eta}\frac{\left(\vec{\tilde{\tilde{E}}}_1\right)_{\xi_k}}{\tilde{d}_k^2} = (-1)^k\tilde{s}\left(\vec{\tilde{\tilde{H}}}_2\right)_{\xi_{3-k}}, \quad k = 1,2 \tag{3.36}$$

$$\sum_{i=1}^{2}(-1)^i\left[\frac{\partial\left(\vec{\tilde{\tilde{E}}}_3\right)_{\xi_i}}{\partial\tilde{\xi}_{3-i}} + \frac{\tilde{\eta}}{\tilde{d}_{3-i}}\frac{\partial\left(\vec{\tilde{\tilde{E}}}_2\right)_{\xi_i}}{\partial\tilde{\xi}_{3-i}} - \frac{\tilde{\eta}^2}{2\tilde{d}_{3-i}^2}\frac{\partial\left(\vec{\tilde{\tilde{E}}}_1\right)_{\xi_i}}{\partial\tilde{\xi}_{3-i}}\right] = \tilde{s}\left(\vec{\tilde{\tilde{H}}}_3\right)_{\eta} \tag{3.37}$$

$$\frac{\partial\left(\vec{\tilde{\tilde{H}}}_2\right)_{\xi_{3-k}}}{\partial\tilde{\eta}} - \frac{\left(\vec{\tilde{\tilde{H}}}_1\right)_{\xi_{3-k}}}{\tilde{d}_{3-k}} - \frac{\partial\left(\vec{\tilde{\tilde{H}}}_1\right)_{\eta}}{\partial\tilde{\xi}_{3-k}} - \tilde{\eta}\frac{\left(\vec{\tilde{\tilde{H}}}_0\right)_{\xi_{3-k}}}{\tilde{d}_{3-k}^2} = (-1)^k\left(\vec{\tilde{\tilde{E}}}_3\right)_{\xi_k}, \quad k = 1,2 \tag{3.38}$$

$$\sum_{i=1}^{2}(-1)^{3-i}\left[\frac{\partial\left(\vec{\tilde{\tilde{H}}}_1\right)_{\xi_{3-i}}}{\partial\tilde{\xi}_i} + \frac{\tilde{\eta}}{\tilde{d}_i}\frac{\partial\left(\vec{\tilde{\tilde{H}}}_0\right)_{\xi_{3-i}}}{\partial\tilde{\xi}_i}\right] = \left(\vec{\tilde{E}}_3\right)_{\eta} \tag{3.39}$$

From Equation 3.27 it follows that only the magnetic field at the interface does not vanish at $\delta \to 0$. The tangential electric field and the normal magnetic field tend to zero as $\tilde{p}$ whereas the normal electric field tends to zero as $\tilde{p}^2$.

The representation in Equations 3.25 and 3.26 has a clear physical meaning, namely

1. Zero-order terms of the Equations 3.25 and 3.26 give the solution of the problem in the perfect electrical conductor (PEC) limit, in which the electromagnetic field diffusion into the body is neglected.
2. First-order terms describe the diffusion in the well-known Leontovich approximation, in which the surface of the body is considered as a plane and the field is assumed to be penetrating into the body only in the direction normal to the surface of the body.
3. Second-order terms yield a correction by taking into account the curvature of the body's surface, but the diffusion is assumed to be only in the direction normal to the surface as in the Leontovich approximation. This is Mitzner's approximation.
4. Third-order terms and higher allow for electromagnetic field diffusion in directions tangential to the surface of the body. This approximation is referred to as Rytov's approximation.

Equations 3.28 through 3.39 can be solved sequentially to derive the first-, second-, and third-order terms of expansion of the tangential electric and magnetic fields in the conductor's skin layer. However, the shortest way to obtain the mentioned quantities is application of the surface impedance "toolbox" since the magnetic field obeys the diffusion equation (Equation 2.2). As a first step, we represent Equation 3.2 with nondimensional variables in the following form:

$$\bar{\tilde{E}}^{\mathrm{b}}_{\xi_k} = \tilde{p}\left(\tilde{\nabla} \times \bar{\tilde{\vec{H}}}\right)^{\mathrm{b}}_{\xi_k} \tag{3.40}$$

Equation 3.40 is similar to Equation 2.136. Therefore, treating the magnetic field as $\vec{f}$ and applying Equations 2.176, 2.181, and 2.186 to the interface ($\tilde{\eta} = 0$), we obtain relations between the coefficients of expansions directly in the Laplace domain:

First-order (Leontovich) approximation:

$$\left(\bar{\tilde{\vec{E}}}_1\right)^{\mathrm{b}}_{\xi_{3-k}} = (-1)^k \tilde{p}\tilde{s}^{1/2}\left(\bar{\tilde{\vec{H}}}_0\right)^{\mathrm{b}}_{\xi_k} \tag{3.41}$$

Second-order (Mitzner) approximation:

$$\left(\tilde{\bar{\bar{E}}}_2\right)^{\mathrm{b}}_{\xi_{3-k}} = (-1)^k \tilde{p}\tilde{s}^{1/2}\left(\left(\tilde{\bar{\bar{H}}}_1\right)^{\mathrm{b}}_{\xi_k} - \tilde{p}\tilde{s}^{-1/2}\left(\tilde{\bar{\bar{H}}}_0\right)^{\mathrm{b}}_{\xi_k}\frac{\tilde{d}_k - \tilde{d}_{3-k}}{2\tilde{d}_k\tilde{d}_{3-k}}\right) \tag{3.42}$$

Third-order (Rytov) approximation:

$$\left(\tilde{\bar{\bar{E}}}_3\right)^{\mathrm{b}}_{\xi_{3-k}} = (-1)^k \tilde{s}^{1/2}\tilde{p}\left[\left(\tilde{\bar{\bar{H}}}_2\right)^{\mathrm{b}}_{\xi_k} - \tilde{p}\tilde{s}^{-1/2}\left(\tilde{\bar{\bar{H}}}_1\right)^{\mathrm{b}}_{\xi_k}\frac{\tilde{d}_k - \tilde{d}_{3-k}}{2\tilde{d}_k\tilde{d}_{3-k}}\right.$$
$$\left. -\tilde{p}^2\tilde{s}^{-1}\left(\left(\tilde{\bar{\bar{H}}}_0\right)^{\mathrm{b}}_{\xi_k}\frac{\tilde{d}_k^2 + 2\tilde{d}_k\tilde{d}_{3-k} - 3\tilde{d}_{3-k}^2}{8\tilde{d}_k^2\tilde{d}_{3-k}^2} - \frac{1}{2}\frac{\partial^2\left(\tilde{\bar{\bar{H}}}_0\right)^{\mathrm{b}}_{\xi_k}}{\partial\tilde{\xi}_k^2} + \frac{1}{2}\frac{\partial^2\left(\tilde{\bar{\bar{H}}}_0\right)^{\mathrm{b}}_{\xi_k}}{\partial\tilde{\xi}_{3-k}^2} - \frac{\partial^2\left(\tilde{\bar{\bar{H}}}_0\right)^{\mathrm{b}}_{\xi_{3-k}}}{\partial\tilde{\xi}_k\partial\tilde{\xi}_{3-k}}\right)\right], \quad k=1,2 \tag{3.43}$$

Substituting Equations 3.41 through 3.43 into the expansions in Equations 3.25 and 3.26 and using Equations 2.176, 2.181, and 2.186, we obtain

Zero-order approximation (PEC limit):

$$\tilde{E}^{\mathrm{b}}_{\xi_{3-k}} = 0 \tag{3.44}$$

First-order (Leontovich) approximation:

$$\tilde{E}^{\mathrm{b}}_{\xi_{3-k}} = \begin{cases} (-1)^k\tilde{p}\tilde{s}^{1/2}\tilde{\bar{\bar{H}}}^{\mathrm{b}}_{\xi_k} & \text{LD} \\ (-1)^k\tilde{p}(1+j)\dot{\tilde{H}}^{\mathrm{b}}_{\xi_k} & \text{FD}, \quad k=1,2 \\ (-1)^k\tilde{p}\dfrac{\mathrm{d}}{\mathrm{d}\tilde{t}}\left(\tilde{H}^{\mathrm{b}}_{\xi_k} * \tilde{T}^{\mathrm{b}}_2\right) & \text{TD} \end{cases} \tag{3.45}$$

Second-order (Mitzner) approximation:

$$\tilde{E}^{\mathrm{b}}_{\xi_{3-k}} = \begin{cases} (-1)^k\tilde{s}^{1/2}\tilde{p}\left(\tilde{\bar{\bar{H}}}^{\mathrm{b}}_{\xi_k} - \tilde{p}\tilde{s}^{-1/2}\tilde{\bar{\bar{H}}}^{\mathrm{b}}_{\xi_k}\dfrac{\tilde{d}_k - \tilde{d}_{3-k}}{2\tilde{d}_k\tilde{d}_{3-k}}\right) & \text{LD} \\ (-1)^k(1+j)\tilde{p}\left(\dot{\tilde{H}}^{\mathrm{b}}_{\xi_k} - \tilde{p}\dfrac{1-j}{2}\dot{\tilde{H}}^{\mathrm{b}}_{\xi_k}\dfrac{\tilde{d}_k - \tilde{d}_{3-k}}{2\tilde{d}_k\tilde{d}_{3-k}}\right) & \text{FD}, \quad k=1,2 \\ (-1)^k\tilde{p}\left[\dfrac{\mathrm{d}}{\mathrm{d}\tilde{t}}\left(f^{\mathrm{b}}_{\xi_k} * \tilde{T}^{\mathrm{b}}_2\right) - \tilde{p}\dfrac{\tilde{d}_k - \tilde{d}_{3-k}}{2\tilde{d}_k\tilde{d}_{3-k}}f^{\mathrm{b}}_{\xi_k} * \tilde{T}^{\mathrm{b}}_1\right] & \text{TD} \end{cases} \tag{3.46}$$

Third-order (Rytov) approximation:

$$
\tilde{E}^{\mathrm{b}}_{\xi_{3-k}} = \begin{cases}
(-1)^k \tilde{s}^{1/2}\tilde{p}\left[\bar{\tilde{H}}^{\mathrm{b}}_{\xi_k} - \tilde{p}\tilde{s}^{-1/2}\bar{\tilde{H}}^{\mathrm{b}}_{\xi_k}\dfrac{\tilde{d}_k - \tilde{d}_{3-k}}{2\tilde{d}_k\tilde{d}_{3-k}}\right. \\
\quad\left. -\tilde{p}^2\tilde{s}^{-1}\left(\bar{\tilde{H}}^{\mathrm{b}}_{\xi_k}\dfrac{\tilde{d}_k^2 + 2\tilde{d}_k\tilde{d}_{3-k} - 3\tilde{d}_{3-k}^2}{8\tilde{d}_k^2\tilde{d}_{3-k}^2} - \dfrac{1}{2}\dfrac{\partial^2\bar{\tilde{H}}^{\mathrm{b}}_{\xi_k}}{\partial\tilde{\xi}_k^2} + \dfrac{1}{2}\dfrac{\partial^2\bar{\tilde{H}}^{\mathrm{b}}_{\xi_k}}{\partial\tilde{\xi}_{3-k}^2} - \dfrac{\partial^2\bar{\tilde{H}}^{\mathrm{b}}_{\xi_{3-k}}}{\partial\tilde{\xi}_k\partial\tilde{\xi}_{3-k}}\right)\right] & \text{LD} \\
(-1)^k(1+j)\tilde{p}\left[\dot{\tilde{H}}^{\mathrm{b}}_{\xi_k} - \tilde{p}\dfrac{1-j}{2}\dot{\tilde{H}}^{\mathrm{b}}_{\xi_k}\dfrac{\tilde{d}_k - \tilde{d}_{3-k}}{2\tilde{d}_k\tilde{d}_{3-k}}\right. \\
\quad\left. +\tilde{p}^2\dfrac{j}{2}\left(\dot{\tilde{H}}^{\mathrm{b}}_{\xi_k}\dfrac{\tilde{d}_k^2 + 2\tilde{d}_k\tilde{d}_{3-k} - 3\tilde{d}_{3-k}^2}{8\tilde{d}_k^2\tilde{d}_{3-k}^2} - \dfrac{1}{2}\dfrac{\partial^2\dot{\tilde{H}}^{\mathrm{b}}_{\xi_k}}{\partial\tilde{\xi}_k^2} + \dfrac{1}{2}\dfrac{\partial^2\dot{\tilde{H}}^{\mathrm{b}}_{\xi_k}}{\partial\tilde{\xi}_{3-k}^2} - \dfrac{\partial^2\dot{\tilde{H}}^{\mathrm{b}}_{\xi_{3-k}}}{\partial\tilde{\xi}_k\partial\tilde{\xi}_{3-k}}\right)\right] & \text{FD}, \quad k=1,2 \\
(-1)^k\tilde{p}\left[\dfrac{\mathrm{d}}{\mathrm{d}\tilde{t}}\left(\tilde{H}^{\mathrm{b}}_{\xi_k} * \tilde{T}^{\mathrm{b}}_2\right) - \tilde{p}\dfrac{\tilde{d}_k - \tilde{d}_{3-k}}{2\tilde{d}_k\tilde{d}_{3-k}}\tilde{H}^{\mathrm{b}}_{\xi_k} * \tilde{T}^{\mathrm{b}}_1\right. \\
\quad\left. -\tilde{p}^2\left(\tilde{H}^{\mathrm{b}}_{\xi_k}\dfrac{\tilde{d}_k^2 + 2\tilde{d}_k\tilde{d}_{3-k} - 3\tilde{d}_{3-k}^2}{8\tilde{d}_k^2\tilde{d}_{3-k}^2} - \dfrac{1}{2}\dfrac{\partial^2\tilde{H}^{\mathrm{b}}_{\xi_k}}{\partial\tilde{\xi}_k^2} + \dfrac{1}{2}\dfrac{\partial^2\tilde{H}^{\mathrm{b}}_{\xi_k}}{\partial\tilde{\xi}_{3-k}^2} - \dfrac{\partial^2\tilde{H}^{\mathrm{b}}_{\xi_{3-k}}}{\partial\tilde{\xi}_k\partial\tilde{\xi}_{3-k}}\right) * \tilde{T}^{\mathrm{b}}_2\right] & \text{TD}
\end{cases}
\tag{3.47}
$$

In Equations 3.45 through 3.47, LD, FD, and TD stands for the Laplace domain, frequency domain, and time domain respectively.

Note that the electric field also obeys Equation 2.2. Thus, the technique can be applied to Equation 3.1, which is written with nondimensional variables in the Laplace domain as

$$
\bar{\tilde{H}}^{\mathrm{b}}_{\xi_k} = -\tilde{s}^{-1}\tilde{p}^{-1}\left(\tilde{\nabla} \times \vec{\bar{\tilde{E}}}\right)^{\mathrm{b}}_{\xi_k} \tag{3.48}
$$

Therefore, we get

Leontovich's approximation:

$$
\bar{\tilde{H}}^{\mathrm{b}}_{\xi_{3-k}} = -\tilde{s}^{-1}\tilde{p}^{-1}(-1)^k\sqrt{\tilde{s}}\bar{\tilde{E}}^{\mathrm{b}}_{\xi_k} = (-1)^{3-k}\tilde{p}^{-1}\tilde{s}^{-1/2}\bar{\tilde{E}}^{\mathrm{b}}_{\xi_k}, \quad k = 1,2 \tag{3.49}
$$

Mitzner's approximation:

$$
\bar{\tilde{H}}^{\mathrm{b}}_{\xi_{3-k}} = (-1)^{3-k}\tilde{p}^{-1}\left(\tilde{s}^{-1/2}\bar{\tilde{E}}^{\mathrm{b}}_{\xi_k} - \tilde{p}\tilde{s}^{-1}\bar{\tilde{E}}^{\mathrm{b}}_{\xi_k}\frac{\tilde{d}_k - \tilde{d}_{3-k}}{2\tilde{d}_k\tilde{d}_{3-k}}\right) \tag{3.50}
$$

Rytov's approximation:

$$
\begin{aligned}
\bar{\tilde{H}}^{\mathrm{b}}_{\xi_{3-k}} = (-1)^{3-k}\tilde{p}^{-1}&\left[\tilde{s}^{-1/2}\bar{\tilde{E}}^{\mathrm{b}}_{\xi_k} - \tilde{p}\tilde{s}^{-1}\bar{\tilde{E}}^{\mathrm{b}}_{\xi_k}\frac{\tilde{d}_k - \tilde{d}_{3-k}}{2\tilde{d}_k\tilde{d}_{3-k}}\right. \\
&\left. -\tilde{p}^2\tilde{s}^{-3/2}\left(\bar{\tilde{E}}^{\mathrm{b}}_{\xi_k}\frac{\tilde{d}_k^2 + 2\tilde{d}_k\tilde{d}_{3-k} - 3\tilde{d}_{3-k}^2}{8\tilde{d}_k^2\tilde{d}_{3-k}^2} - \frac{1}{2}\frac{\partial^2\bar{\tilde{E}}^{\mathrm{b}}_{\xi_k}}{\partial\tilde{\xi}_k^2} + \frac{1}{2}\frac{\partial^2\bar{\tilde{E}}^{\mathrm{b}}_{\xi_k}}{\partial\tilde{\xi}_{3-k}^2} - \frac{\partial^2\bar{\tilde{E}}^{\mathrm{b}}_{\xi_{3-k}}}{\partial\tilde{\xi}_k\partial\tilde{\xi}_{3-k}}\right)\right]
\end{aligned}
\tag{3.51}
$$

Both sets of equations (Equations 3.45 through 3.47 and Equations 3.49 through 3.51) give relationships between the tangential electric and magnetic fields at the interface. However, the presence of $\tilde{p}^{-1}$ in Equations 3.49 through 3.51 makes their use less practical as $\tilde{p} \to 0$ compared with Equations 3.45 through 3.47.

The return to dimensional variables in Equations 3.45 through 3.47 can be performed by representing the nondimensional variables in the following form using the scale factors given in Table 3.1:

$$\tilde{t} = t/\tau; \quad \tilde{\xi}_k = \xi_k/D; \quad \tilde{\eta} = \eta/\delta; \quad \tilde{E} = E\tau/(\mu \boldsymbol{I}); \quad \tilde{H} = HD/\boldsymbol{I} \tag{3.52}$$

Substituting Equation 3.52 into Equation 3.45, we obtain the Leontovich approximation:

$$\dot{E}^{\mathrm{b}}_{\xi_{3-k}} 2(\mu\omega \boldsymbol{I})^{-1} = (-1)^k \delta D^{-1}(1+j)\dot{H}^{\mathrm{b}}_{\xi_k} D\boldsymbol{I}^{-1} \tag{3.53}$$

$$E^{\mathrm{b}}_{\xi_{3-k}} \tau(\mu \boldsymbol{I})^{-1} = (-1)^k \delta D^{-1} D\boldsymbol{I}^{-1} \frac{\partial}{\partial t}\left[H^{\mathrm{b}}_{\xi_k} * \left(-\frac{1}{\sqrt{\pi t}}\right)\right]\tau\frac{\tau^{1/2}}{\tau} \tag{3.54}$$

Taking into account that

$$\frac{\tau^{1/2}}{\delta} = \sqrt{\sigma\mu}$$

we get Equations 3.53 and 3.54 in the following form:

Leontovich's approximation:

$$\dot{E}^{\mathrm{b}}_{\xi_{3-k}} = (-1)^k(1+j)\frac{\mu\omega\delta}{2}\dot{H}^{\mathrm{b}}_{\xi_k} = (-1)^k\sqrt{\frac{j\omega\mu}{\sigma}}\dot{H}^{\mathrm{b}}_{\xi_k} \tag{3.55}$$

$$E^{\mathrm{b}}_{\xi_{3-k}} = (-1)^k\sqrt{\frac{\mu}{\sigma}}\frac{\partial}{\partial t}\left(H^{\mathrm{b}}_{\xi_k} * T^{\mathrm{b}}_2\right) \tag{3.56}$$

Here T^{b}_2 is given by Equation 2.113.

The transfer to dimensional variables in Equations 3.50 and 3.51 is done in a similar way. We give below the final results:

Mitzner's approximation:

$$\dot{E}^{\mathrm{b}}_{\xi_{3-k}} = (-1)^k\frac{1}{\sigma}\left(\sqrt{j\omega\sigma\mu}\dot{H}^{\mathrm{b}}_{\xi_k} - \frac{d_k - d_{3-k}}{2d_k d_{3-k}}\dot{H}^{\mathrm{b}}_{\xi_k}\right) \tag{3.57}$$

$$E^{\mathrm{b}}_{\xi_{3-k}} = (-1)^k \frac{1}{\sigma}\left(\sqrt{\sigma\mu}\frac{\partial}{\partial t}\left(H^{\mathrm{b}}_{\xi_k} * T^{\mathrm{b}}_2\right) - \frac{d_k - d_{3-k}}{2d_k d_{3-k}} H^{\mathrm{b}}_{\xi_k}\right) \tag{3.58}$$

Rytov's approximation:

$$\dot{E}^{\mathrm{b}}_{\xi_{3-k}} = (-1)^k \frac{1}{\sigma}\left[\sqrt{j\omega\sigma\mu}\dot{H}^{\mathrm{b}}_{\xi_k} - \frac{d_k - d_{3-k}}{2d_k d_{3-k}}\dot{H}^{\mathrm{b}}_{\xi_k}\right.$$
$$\left. - \frac{1}{\sqrt{j\omega\sigma\mu}}\left(\frac{d_k^2 + 2d_k d_{3-k} - 3d_{3-k}^2}{8d_k^2 d_{3-k}^2}\dot{H}^{\mathrm{b}}_{\xi_{3-k}} - \frac{1}{2}\frac{\partial^2 \dot{H}^{\mathrm{b}}_{\xi_k}}{\partial \xi_k^2} + \frac{1}{2}\frac{\partial^2 \dot{H}^{\mathrm{b}}_{\xi_k}}{\partial \xi_{3-k}^2} - \frac{\partial^2 \dot{H}^{\mathrm{b}}_{\xi_{3-k}}}{\partial \xi_k \partial \xi_{3-k}}\right)\right] \tag{3.59}$$

$$E^{\mathrm{b}}_{\xi_{3-k}} = (-1)^k \frac{1}{\sigma}\left[\sqrt{\sigma\mu}\frac{\partial}{\partial t}\left(H^{\mathrm{b}}_{\xi_k} * T^{\mathrm{b}}_2\right) - \frac{d_k - d_{3-k}}{2d_k d_{3-k}} H^{\mathrm{b}}_{\xi_k}\right.$$
$$\left. - \frac{1}{\sqrt{\sigma\mu}}\left(\frac{d_k^2 + 2d_k d_{3-k} - 3d_{3-k}^2}{8d_k^2 d_{3-k}^2}H^{\mathrm{b}}_{\xi_{3-k}} - \frac{1}{2}\frac{\partial^2 H^{\mathrm{b}}_{\xi_k}}{\partial \xi_k^2} + \frac{1}{2}\frac{\partial^2 H^{\mathrm{b}}_{\xi_k}}{\partial \xi_{3-k}^2} - \frac{\partial^2 H^{\mathrm{b}}_{\xi_{3-k}}}{\partial \xi_k \partial \xi_{3-k}}\right) * T^{\mathrm{b}}_2\right] \tag{3.60}$$

It is a simple matter to show that by neglecting the terms containing δ^2, Equations 3.59 and 3.60 reduce to the Mitzner condition in Equations 1.63 and 1.64. Similarly, by setting $d_1 \to \infty$ and $d_2 \to \infty$, the Rytov condition at the interface in Equations 1.108 and 1.109 is obtained. Note that the third term in Equations 3.59 and 3.60 is absent in both Mitzner's and Rytov's conditions since Equations 3.59 and 3.60 cannot, naturally, be obtained by adding Equations 1.63 and 1.64 and Equations 1.108 and 1.109. Neglecting all terms except for the very first leads to the original Leontovich SIBC as given in Section I.2.

3.4 Magnetic Scalar Potential Formalism

We first restrict ourselves to a source-free region where any closed path, c, does not enclose any current so that the following condition is satisfied:

$$\oint_c \vec{H} \cdot \mathrm{d}\vec{l} = 0 \tag{3.61}$$

Then the magnetic scalar potential, ϕ, can be introduced as follows:

$$\vec{H} = -\nabla\phi \tag{3.62}$$

Substituting Equation 3.62 into Equation 3.3, we obtain the Laplace equation governing the scalar potential distribution in free space:

$$\nabla^2\phi = 0 \tag{3.63}$$

The exact boundary conditions at the dielectric/conductor interface follow directly from Equation 3.62:

$$\vec{n}\times\nabla\phi|_{\text{diel}} = \vec{n}\times\vec{H}\Big|_{\text{cond}} \tag{3.64}$$

$$\left.\frac{\partial\phi}{\partial\vec{n}}\right|_{\text{diel}} = \vec{n}\cdot\vec{H}\Big|_{\text{cond}} \tag{3.65}$$

The use of the conditions in Equations 3.64 and 3.65 requires that the problem inside the conductor in terms of the magnetic field be considered together with Equation 3.63. To eliminate the conducting region from the numerical procedure, another relation between the magnetic scalar potential and its normal derivative at the conductor surface, and taking into account material properties of the conductor, must be derived and used instead of Equations 3.64 and 3.65. Obviously, this is equivalent to derivation of an equation relating the tangential and normal components of the magnetic field at the interface. Since the magnetic field in the conductor obeys the diffusion equation, Equations 2.170, 2.173, 2.178, and 2.183 of the surface impedance "toolbox" developed in Chapter 2 can be applied and the results written in the form:

PEC limit:

$$H_\eta = 0 \tag{3.66}$$

Leontovich's approximation:

$$H_\eta^{\text{b}} = \begin{cases} \tilde{p}\tilde{s}^{-1/2}\displaystyle\sum_{i=1}^{2}\frac{\partial\bar{H}_{\xi_i}^{\text{b}}}{\partial\tilde{\xi}_i} & \text{LD} \\ \tilde{p}\dfrac{1-j}{2}\displaystyle\sum_{i=1}^{2}\frac{\partial\dot{H}_{\xi_i}^{\text{b}}}{\partial\tilde{\xi}_i} & \text{FD} \\ \tilde{p}\displaystyle\sum_{i=1}^{2}\frac{\partial H_{\xi_i}^{\text{b}}}{\partial\tilde{\xi}_i} * \tilde{T}_2^{\text{b}} & \text{TD} \end{cases} \tag{3.67}$$

Mitzner's approximation:

$$H_\eta^b = \begin{cases} \tilde{p}\tilde{s}^{-1/2}\sum_{i=1}^{2}\dfrac{\partial}{\partial\tilde{\xi}_i}\left(\bar{H}_{\xi_i}^b - \tilde{p}\tilde{s}^{-1/2}\bar{H}_{\xi_i}^b\dfrac{\tilde{d}_i - \tilde{d}_{3-i}}{2\tilde{d}_i\tilde{d}_{3-i}}\right) & \text{LD} \\ \tilde{p}\dfrac{1-j}{2}\sum_{i=1}^{2}\dfrac{\partial}{\partial\tilde{\xi}_i}\left(\dot{H}_{\xi_i}^b - \tilde{p}\dfrac{1-j}{2}\dot{H}_{\xi_i}^b\dfrac{\tilde{d}_i - \tilde{d}_{3-i}}{2\tilde{d}_i\tilde{d}_{3-i}}\right) & \text{FD} \\ \tilde{p}\sum_{i=1}^{2}\dfrac{\partial}{\partial\tilde{\xi}_i}\left(H_{\xi_i}^b * \tilde{T}_2^b - \tilde{p}\dfrac{\tilde{d}_i - \tilde{d}_{3-i}}{2\tilde{d}_i\tilde{d}_{3-i}}H_{\xi_i}^b * \tilde{T}_3^b\right) & \text{TD} \end{cases} \tag{3.68}$$

Rytov's approximation:

$$H_\eta^b = \begin{cases} \tilde{p}\tilde{s}^{-1/2}\sum_{i=1}^{2}\dfrac{\partial}{\partial\tilde{\xi}_i}\Bigg[\bar{H}_{\xi_i}^b - \tilde{p}\tilde{s}^{-1/2}\bar{H}_{\xi_i}^b\dfrac{\tilde{d}_i - \tilde{d}_{3-i}}{2\tilde{d}_i\tilde{d}_{3-i}} \\ \quad -\tilde{p}^2\tilde{s}^{-1}\left(\bar{H}_{\xi_i}^b\dfrac{\tilde{d}_i^2 + 2\tilde{d}_i\tilde{d}_{3-i} - 3\tilde{d}_{3-i}^2}{8\tilde{d}_i^2\tilde{d}_{3-i}^2} - \dfrac{1}{2}\dfrac{\partial^2\bar{H}_{\xi_i}^b}{\partial\tilde{\xi}_i^2} + \dfrac{1}{2}\dfrac{\partial^2\bar{H}_{\xi_i}^b}{\partial\tilde{\xi}_{3-i}^2} - \dfrac{\partial^2\bar{H}_{\xi_{3-i}}^b}{\partial\tilde{\xi}_i\partial\tilde{\xi}_{3-i}}\right)\Bigg] & \text{LD} \\ \tilde{p}\dfrac{1-j}{2}\sum_{i=1}^{2}\dfrac{\partial}{\partial\tilde{\xi}_i}\Bigg[\dot{H}_{\xi_i}^b - \tilde{p}\dfrac{1-j}{2}\dot{H}_{\xi_i}^b\dfrac{\tilde{d}_i - \tilde{d}_{3-i}}{2\tilde{d}_i\tilde{d}_{3-i}} \\ \quad +\tilde{p}^2\dfrac{j}{2}\left(\dot{H}_{\xi_i}^b\dfrac{\tilde{d}_i^2 + 2\tilde{d}_i\tilde{d}_{3-i} - 3\tilde{d}_{3-i}^2}{8\tilde{d}_i^2\tilde{d}_{3-i}^2} - \dfrac{1}{2}\dfrac{\partial^2\dot{H}_{\xi_i}^b}{\partial\tilde{\xi}_i^2} + \dfrac{1}{2}\dfrac{\partial^2\dot{H}_{\xi_i}^b}{\partial\tilde{\xi}_{3-i}^2} - \dfrac{\partial^2\dot{H}_{\xi_{3-i}}^b}{\partial\tilde{\xi}_i\partial\tilde{\xi}_{3-i}}\right)\Bigg] & \text{FD} \\ \tilde{p}\sum_{i=1}^{2}\dfrac{\partial}{\partial\tilde{\xi}_i}\Bigg[H_{\xi_i}^b * \tilde{T}_2^b - \tilde{p}\dfrac{\tilde{d}_i - \tilde{d}_{3-i}}{2\tilde{d}_i\tilde{d}_{3-i}}H_{\xi_i}^b * \tilde{T}_3^b \\ \quad -\tilde{p}^2\left(H_{\xi_i}^b\dfrac{\tilde{d}_i^2 + 2\tilde{d}_i\tilde{d}_{3-i} - 3\tilde{d}_{3-i}^2}{8\tilde{d}_i^2\tilde{d}_{3-i}^2} - \dfrac{1}{2}\dfrac{\partial^2 H_{\xi_i}^b}{\partial\tilde{\xi}_i^2} + \dfrac{1}{2}\dfrac{\partial^2 H_{\xi_i}^b}{\partial\tilde{\xi}_{3-i}^2} - \dfrac{\partial^2 H_{\xi_{3-i}}^b}{\partial\tilde{\xi}_i\partial\tilde{\xi}_{3-i}}\right) * \tilde{T}_4^b\Bigg] & \text{TD} \end{cases} \tag{3.69}$$

The shortest way to write Equations 3.67 through 3.69 with dimensional variables is to use Equations 2.198, 2.203, and 2.208:

Leontovich's approximation:

$$H_\eta^b = \begin{cases} (j\omega\sigma\mu)^{-1/2}\sum_{i=1}^{2}\dfrac{\partial\dot{H}_{\xi_i}^b}{\partial\xi_i} & \text{FD} \\ (\sigma\mu)^{-1/2}\sum_{i=1}^{2}\dfrac{\partial H_{\xi_i}^b}{\partial\xi_i} * T_2^b & \text{TD} \end{cases} \tag{3.70}$$

Mitzner's approximation:

$$H_\eta^b = \begin{cases} (j\omega\sigma\mu)^{-1/2}\sum_{i=1}^{2}\dfrac{\partial}{\partial\xi_i}\left(\dot{H}_{\xi_i}^b - (j\omega\sigma\mu)^{-1/2}\dfrac{d_i - d_{3-i}}{2d_id_{3-i}}\dot{H}_{\xi_i}^b\right) & \text{FD} \\ (\sigma\mu)^{-1/2}\sum_{i=1}^{2}\dfrac{\partial}{\partial\xi_i}\left(H_{\xi_i}^b * T_2^b - (\sigma\mu)^{-1/2}\dfrac{d_i - d_{3-i}}{2d_id_{3-i}}H_{\xi_i}^b * T_3^b\right) & \text{TD} \end{cases} \tag{3.71}$$

Rytov's approximation:

$$H_\eta^{\mathrm{b}} = \begin{cases} (j\omega\sigma\mu)^{-1/2} \sum_{i=1}^{2} \frac{\partial}{\partial \xi_i} \left[\dot{H}_{\xi_i}^{\mathrm{b}} - (j\omega\sigma\mu)^{-1/2} \frac{d_i - d_{3-i}}{2d_i d_{3-i}} \dot{H}_{\xi_i}^{\mathrm{b}} \right. \\ \left. -(j\omega\sigma\mu)^{-1} \left(\dot{H}_{\xi_i}^{\mathrm{b}} \frac{d_i^2 + 2d_i d_{3-i} - 3d_{3-i}^2}{8d_i^2 d_{3-i}^2} - \frac{1}{2}\frac{\partial^2 \dot{H}_{\xi_i}^{\mathrm{b}}}{\partial \xi_i^2} + \frac{1}{2}\frac{\partial^2 \dot{H}_{\xi_i}^{\mathrm{b}}}{\partial \xi_{3-i}^2} - \frac{\partial^2 \dot{H}_{\xi_{3-i}}^{\mathrm{b}}}{\partial \xi_i \partial \xi_{3-i}} \right) \right] & \text{FD} \\ (\sigma\mu)^{-1/2} \sum_{i=1}^{2} \frac{\partial}{\partial \xi_i} \left[H_{\xi_i}^{\mathrm{b}} * T_2^{\mathrm{b}} - (\sigma\mu)^{-1/2} \frac{d_i - d_{3-i}}{2d_i d_{3-i}} H_{\xi_i}^{\mathrm{b}} * T_3^{\mathrm{b}} \right. \\ \left. -(\sigma\mu)^{-1} \left(H_{\xi_i}^{\mathrm{b}} \frac{d_i^2 + 2d_i d_{3-i} - 3d_{3-i}^2}{8d_i^2 d_{3-i}^2} - \frac{1}{2}\frac{\partial^2 H_{\xi_i}^{\mathrm{b}}}{\partial \xi_i^2} + \frac{1}{2}\frac{\partial^2 H_{\xi_i}^{\mathrm{b}}}{\partial \xi_{3-i}^2} - \frac{\partial^2 H_{\xi_{3-i}}^{\mathrm{b}}}{\partial \xi_i \partial \xi_{3-i}} \right) * T_4^{\mathrm{b}} \right] & \text{TD} \end{cases} \tag{3.72}$$

From Equation 3.62, it directly follows that

$$H_{\xi_i} = -\frac{\partial \phi}{\partial \xi_i}; \quad H_\eta = -\frac{\partial \phi}{\partial \eta} \tag{3.73}$$

So that

$$\frac{\partial^2 H_{\xi_i}^{\mathrm{b}}}{\partial \xi_{3-i}^2} = \frac{\partial^2}{\partial \xi_{3-i}^2}\left(\frac{\partial \phi}{\partial \xi_i}\right) = \frac{\partial^3 \phi}{\partial \xi_{3-i}^2 \partial \xi_i} \quad \text{and} \quad \frac{\partial^2 H_{\xi_{3-i}}^{\mathrm{b}}}{\partial \xi_i \partial \xi_{3-i}} = \frac{\partial^2}{\partial \xi_i \partial \xi_{3-i}}\left(\frac{\partial \phi}{\partial \xi_{3-i}}\right) = \frac{\partial^3 \phi}{\partial \xi_{3-i}^2 \partial \xi_i} \tag{3.74}$$

Substituting Equation 3.73 into Equations 3.70 through 3.72 and taking into account Equation 3.74 yields the boundary relations between the normal and tangential derivatives of the scalar potential in the following form:

Leontovich's approximation:

$$\left(\frac{\partial \phi}{\partial \vec{n}}\right)^{\mathrm{b}} = \begin{cases} (j\omega\sigma\mu)^{-1/2} \sum_{i=1}^{2} \frac{\partial^2 \dot{\phi}^{\mathrm{b}}}{\partial \xi_i^2} & \text{FD} \\ (\sigma\mu)^{-1/2} \sum_{i=1}^{2} \frac{\partial^2 \phi^{\mathrm{b}}}{\partial \xi_i^2} * T_2^{\mathrm{b}} & \text{TD} \end{cases} \tag{3.75}$$

Mitzner's approximation:

$$\left(\frac{\partial \phi}{\partial \vec{n}}\right)^{\mathrm{b}} = \begin{cases} (j\omega\sigma\mu)^{-1/2} \sum_{i=1}^{2} \frac{\partial^2}{\partial \xi_i^2} \left(\dot{\phi}^{\mathrm{b}} - (j\omega\sigma\mu)^{-1/2} \frac{d_i - d_{3-i}}{2d_i d_{3-i}} \dot{\phi}^{\mathrm{b}} \right) & \text{FD} \\ (\sigma\mu)^{-1/2} \sum_{i=1}^{2} \frac{\partial^2}{\partial \xi_i^2} \left(\phi^{\mathrm{b}} * T_2^{\mathrm{b}} - (\sigma\mu)^{-1/2} \frac{d_i - d_{3-i}}{2d_i d_{3-i}} \phi^{\mathrm{b}} * T_3^{\mathrm{b}} \right) & \text{TD} \end{cases} \tag{3.76}$$

Rytov's approximation:

$$\left(\frac{\partial \phi}{\partial \vec{n}}\right)^{\mathrm{b}} = \begin{cases} (j\omega\sigma\mu)^{-1/2} \sum_{i=1}^{2} \frac{\partial^2}{\partial \xi_i^2} \left[\dot{\phi}^{\mathrm{b}} - (j\omega\sigma\mu)^{-1/2} \frac{d_i - d_{3-i}}{2 d_i d_{3-i}} \dot{\phi}^{\mathrm{b}} \right. \\ \left. -(j\omega\sigma\mu)^{-1} \left(\dot{\phi}^{b} \frac{d_i^2 + 2 d_i d_{3-i} - 3 d_{3-i}^2}{8 d_i^2 d_{3-i}^2} - \frac{1}{2} \frac{\partial^2 \dot{\phi}^{\mathrm{b}}}{\partial \xi_i^2} - \frac{1}{2} \frac{\partial^2 \dot{\phi}^{\mathrm{b}}}{\partial \xi_{3-i}^2} \right) \right] & \text{FD} \\ (\sigma\mu)^{-1/2} \sum_{i=1}^{2} \left[\frac{\partial^2}{\partial \xi_i^2} \phi^{\mathrm{b}} * T_2^{\mathrm{b}} - (\sigma\mu)^{-1/2} \frac{d_i - d_{3-i}}{2 d_i d_{3-i}} \phi^{\mathrm{b}} * T_3^{\mathrm{b}} \right. \\ \left. -(\sigma\mu)^{-1} \left(\phi^{\mathrm{b}} \frac{d_i^2 + 2 d_i d_{3-i} - 3 d_{3-i}^2}{8 d_i^2 d_{3-i}^2} - \frac{1}{2} \frac{\partial^2 \phi^{\mathrm{b}}}{\partial \xi_i^2} - \frac{1}{2} \frac{\partial^2 \phi^{\mathrm{b}}}{\partial \xi_{3-i}^2} \right) * T_4^{\mathrm{b}} \right] & \text{TD} \end{cases} \tag{3.77}$$

Since the tangential derivative of a function can be approximated using the function calculated at nodes located along the conductor's surface, Equations 3.75 through 3.77 are SIBCs of various orders of approximation relating the scalar potential and its normal derivative at the surface of the conductor.

Now let a source current $\vec{I}(t)$ flow in the conductor so that Equation 3.61 takes the form:

$$\oint_{\mathrm{c}} \vec{H} \cdot \mathrm{d}\vec{l} = I \tag{3.78}$$

To ensure uniqueness of the solution under the condition of Equation 3.78, the scalar magnetic potential in nonconducting regions can be introduced as follows

$$\vec{H} = \vec{H}^{\mathrm{s}} - \nabla \phi \tag{3.79}$$

where $\vec{H}^{\mathrm{s}}$ is the "source" magnetic field satisfying only the following requirement:

$$\oint_{\mathrm{c}} \vec{H}^{\mathrm{s}} \cdot \mathrm{d}\vec{l} = I \tag{3.80}$$

Details of selection of $\vec{H}^{\mathrm{s}}$ will be discussed in Chapter 6. Applying Equation 3.79 to the interface and using Equations 3.75 through 3.77, the following boundary conditions are obtained and written in the form:

$$\left(\frac{\partial \phi}{\partial \eta}\right) = \vec{n} \cdot (\vec{H}^{\mathrm{s}} - \vec{H}) = H_\eta^{\mathrm{s}} - H_\eta = H_\eta^{\mathrm{s}} - \sum_{i=1}^{2} \frac{\partial}{\partial \xi_i} F\left[H_{\xi_i}^{\mathrm{s}}\right] + \sum_{i=1}^{2} \frac{\partial}{\partial \xi_i} F[\partial \phi / \partial \xi_i] \tag{3.81}$$

where the operator-function F takes the following forms depending on the order of approximation and excitation applied:

$$F^{\text{Leontovich}}\left[f_{\xi_k}\right] = \begin{cases} (j\omega\sigma\mu)^{-1/2}\dot{f}^{\text{b}}_{\xi_k} & \text{FD} \\ (\sigma\mu)^{-1/2}f^{\text{b}}_{\xi_k} * T^{\text{b}}_2 & \text{TD} \end{cases}, \quad k = 1,2 \tag{3.82}$$

$$F^{\text{Mitzner}}\left[f_{\xi_k}\right] = \begin{cases} (j\omega\sigma\mu)^{-1/2}\dot{f}^{\text{b}}_{\xi_k} - (j\omega\sigma\mu)^{-1}\dfrac{d_k - d_{3-k}}{2d_k d_{3-k}}\dot{f}^{\text{b}}_{\xi_k} & \text{FD} \\ (\sigma\mu)^{-1/2}f^{\text{b}}_{\xi_k} * T^{\text{b}}_2 - (\sigma\mu)^{-1}\dfrac{d_k - d_{3-k}}{2d_k d_{3-k}}f^{\text{b}}_{\xi_k} * T^{\text{b}}_3 & \text{TD} \end{cases}, \quad k = 1,2 \tag{3.83}$$

$$F^{\text{Rytov}}[f_{\xi_k}] = \begin{cases} (j\omega\sigma\mu)^{-1/2}\dot{f}^{\text{b}}_{\xi_k} - (j\omega\sigma\mu)^{-1}\dfrac{d_k - d_{3-k}}{2d_k d_{3-k}}\dot{f}^{\text{b}}_{\xi_k} \\ \quad -(j\omega\sigma\mu)^{-3/2}\left(\dot{f}^{\text{b}}_{\xi_k}\dfrac{d_k^2 + 2d_k d_{3-k} - 3d_{3-k}^2}{8d_k^2 d_{3-k}^2} - \dfrac{1}{2}\dfrac{\partial^2 \dot{f}^{\text{b}}_{\xi_k}}{\partial \xi_k^2} + \dfrac{1}{2}\dfrac{\partial^2 \dot{f}^{\text{b}}_{\xi_k}}{\partial \xi_{3-k}^2} - \dfrac{\partial^2 \dot{f}^{\text{b}}_{\xi_{3-k}}}{\partial \xi_k \partial \xi_{3-k}}\right) & \text{FD} \\ (\sigma\mu)^{-1/2}f^{\text{b}}_{\xi_k} * T^{\text{b}}_2 - (\sigma\mu)^{-1}\dfrac{d_k - d_{3-k}}{2d_k d_{3-k}}f^{\text{b}}_{\xi_k} * T^{\text{b}}_3 \\ \quad -(\sigma\mu)^{-3/2}\left(f^{\text{b}}_{\xi_k}\dfrac{d_k^2 + 2d_k d_{3-k} - 3d_{3-k}^2}{8d_k^2 d_{3-k}^2} - \dfrac{1}{2}\dfrac{\partial^2 f^{\text{b}}_{\xi_k}}{\partial \xi_k^2} + \dfrac{1}{2}\dfrac{\partial^2 f^{\text{b}}_{\xi_k}}{\partial \xi_{3-k}^2} - \dfrac{\partial^2 f^{\text{b}}_{\xi_{3-k}}}{\partial \xi_k \partial \xi_{3-k}}\right) * T^{\text{b}}_4 & \text{TD} \end{cases}, \quad k=1,2 \tag{3.84}$$

The product $F[f]$ has dimensions of fD where f is the scale factor of the function f. Let us also represent F in nondimensional form in the frequency, time, and Laplace domains as follows (we will need these representations later for implementation of SIBCs into numerical methods)

$$F[\tilde{f}] = D\tilde{F}[\tilde{f}] \tag{3.85}$$

where the nondimensional product $\tilde{F}[\tilde{f}]$ takes the following form depending on the order of approximation:

$$\tilde{F}^{\text{Leontovich}}[\tilde{f}_{\xi_k}] = \begin{cases} \tilde{p}\tilde{s}^{-1/2}\bar{\tilde{f}}^{\text{b}}_{\xi_k} & \text{LD} \\ \tilde{p}\frac{1-j}{2}\dot{\tilde{f}}^{\text{b}}_{\xi_k} & \text{FD}, \quad k = 1,2 \\ \tilde{p}\tilde{f}^{\text{s}}_{\xi_k} * \tilde{T}^{\text{b}}_2 & \text{TD} \end{cases} \tag{3.86}$$

$$\tilde{F}^{\text{Mitzner}}[\tilde{f}_{\xi_k}] = \begin{cases} \tilde{s}^{-1/2}\tilde{p}\bar{\tilde{f}}^{\text{b}}_{\xi_k} - \tilde{p}^2\tilde{s}^{-1}\bar{\tilde{f}}^{\text{b}}_{\xi_k}\dfrac{\tilde{d}_k - \tilde{d}_{3-k}}{2\tilde{d}_k\tilde{d}_{3-k}} & \text{LD} \\ \tilde{p}\frac{1-j}{2}\dot{\tilde{f}}^{\text{b}}_{\xi_k} + \tilde{p}^2\dfrac{j}{2}\dfrac{\tilde{d}_k - \tilde{d}_{3-k}}{2\tilde{d}_k\tilde{d}_{3-k}}\dot{\tilde{f}}^{\text{b}}_{\xi_k} & \text{FD}, \quad k = 1,2 \\ \tilde{p}\tilde{f}^{\text{b}}_{\xi_k} * \tilde{T}^{\text{b}}_2 - \tilde{p}^2\dfrac{\tilde{d}_k - \tilde{d}_{3-k}}{2\tilde{d}_k\tilde{d}_{3-k}}\tilde{f}^{\text{b}}_{\xi_k} * \tilde{T}^{\text{b}}_3 & \text{TD} \end{cases} \tag{3.87}$$

$$\tilde{F}^{\text{Rytov}}[\tilde{f}_{\xi_k}] = \begin{cases} \tilde{p}\tilde{s}^{-1/2}\bar{\tilde{f}}^{\text{b}}_{\xi_k} - \tilde{p}^2\tilde{s}^{-1}\bar{\tilde{f}}^{\text{b}}_{\xi_k}\dfrac{\tilde{d}_k - \tilde{d}_{3-k}}{2\tilde{d}_k\tilde{d}_{3-k}} \\ \quad -\tilde{p}^3\tilde{s}^{-3/2}\left(\bar{\tilde{f}}^{\text{b}}_{\xi_k}\dfrac{\tilde{d}_k^2 + 2\tilde{d}_k\tilde{d}_{3-k} - 3\tilde{d}_{3-k}^2}{8\tilde{d}_k^2\tilde{d}_{3-k}^2} - \dfrac{1}{2}\dfrac{\partial^2\bar{\tilde{f}}^{\text{b}}_{\xi_k}}{\partial\tilde{\xi}_k^2} + \dfrac{1}{2}\dfrac{\partial^2\bar{\tilde{f}}^{\text{b}}_{\xi_k}}{\partial\tilde{\xi}_{3-k}^2} - \dfrac{\partial^2\bar{\tilde{f}}^{\text{b}}_{\xi_{3-k}}}{\partial\tilde{\xi}_k\partial\tilde{\xi}_{3-k}}\right) & \text{LD} \\ \tilde{p}\frac{1-j}{2}\dot{\tilde{f}}^{\text{b}}_{\xi_k} + \tilde{p}^2\frac{j}{2}\dot{\tilde{f}}^{\text{b}}_{\xi_k}\dfrac{\tilde{d}_k - \tilde{d}_{3-k}}{2\tilde{d}_k\tilde{d}_{3-k}} \\ \quad -\tilde{p}^3\frac{1+j}{4}\left(\dot{\tilde{f}}^{\text{b}}_{\xi_k}\dfrac{\tilde{d}_k^2 + 2\tilde{d}_k\tilde{d}_{3-k} - 3\tilde{d}_{3-k}^2}{8\tilde{d}_k^2\tilde{d}_{3-k}^2} - \dfrac{1}{2}\dfrac{\partial^2\dot{\tilde{f}}^{\text{b}}_{\xi_k}}{\partial\tilde{\xi}_k^2} + \dfrac{1}{2}\dfrac{\partial^2\dot{\tilde{f}}^{\text{b}}_{\xi_k}}{\partial\tilde{\xi}_{3-k}^2} - \dfrac{\partial^2\dot{\tilde{f}}^{\text{b}}_{\xi_{3-k}}}{\partial\tilde{\xi}_k\partial\tilde{\xi}_{3-k}}\right) & \text{FD} \\ \tilde{p}\tilde{f}^{\text{b}}_{\xi_k} * \tilde{T}^{\text{b}}_2 - \tilde{p}^2\dfrac{\tilde{d}_k - \tilde{d}_{3-k}}{2\tilde{d}_k\tilde{d}_{3-k}}\tilde{f}^{\text{b}}_{\xi_k} * \tilde{T}^{\text{b}}_3 \\ \quad -\tilde{p}^3\left(\tilde{f}^{\text{b}}_{\xi_k}\dfrac{\tilde{d}_k^2 + 2\tilde{d}_k\tilde{d}_{3-k} - 3\tilde{d}_{3-k}^2}{8\tilde{d}_k^2\tilde{d}_{3-k}^2} - \dfrac{1}{2}\dfrac{\partial^2\tilde{f}^{\text{b}}_{\xi_k}}{\partial\tilde{\xi}_k^2} + \dfrac{1}{2}\dfrac{\partial^2\tilde{f}^{\text{b}}_{\xi_k}}{\partial\tilde{\xi}_{3-k}^2} - \dfrac{\partial^2\tilde{f}^{\text{b}}_{\xi_{3-k}}}{\partial\tilde{\xi}_k\partial\tilde{\xi}_{3-k}}\right) * \tilde{T}^{\text{b}}_4 & \text{TD} \end{cases} \tag{3.88}$$

The physical meaning of the operator-function F will be explained in Section 3.6.

3.5 Magnetic Vector Potential Formalism

The magnetic vector potential $\vec{A}$ is defined as follows:

$$\vec{H} = \nabla \times \vec{A} \tag{3.89}$$

Substituting Equation 3.89 into Equations 3.1 and 3.2, the distribution of the vector potential in the conductor can be written in the form:

$$\nabla \times \vec{E} = -\nabla \times \left(\mu\frac{\partial\vec{A}}{\partial t}\right) \tag{3.90}$$

$$\nabla \times \nabla \times \vec{A} = \sigma\vec{E} \tag{3.91}$$

From Equation 3.90 we obtain

$$\vec{E} = -\mu\frac{\partial\vec{A}}{\partial t} + \nabla V \tag{3.92}$$

where V is the electric scalar potential. Substitution of Equation 3.92 into Equation 3.91 gives

$$\nabla \times \nabla \times \vec{A} = -\sigma\mu \frac{\partial \vec{A}}{\partial t} + \sigma \nabla V \tag{3.93}$$

Following the approach developed by Emson and Simkin [1], we split the vector potential into "source" and "eddy" components:

$$\vec{A} = \vec{A}^{\mathrm{s}} + \vec{A}^{\mathrm{e}} \tag{3.94}$$

where the source component $\vec{A}^{\mathrm{s}}$ can be represented in the form:

$$\vec{A}^{\mathrm{s}} = \frac{1}{\mu} \int \nabla V \mathrm{d}t \tag{3.95}$$

Substitution of Equation 3.95 into Equation 3.92 yields the relation between the electric field and the eddy component of the magnetic vector potential:

$$\vec{E} = -\mu \frac{\partial \vec{A}^{\mathrm{e}}}{\partial t} \tag{3.96}$$

Substitution of Equation 3.94 into Equation 3.93 leads to elimination of the electric scalar potential for the equation of diffusion of the eddy component of the magnetic vector potential:

$$\nabla \times \nabla \times \vec{A}^{\mathrm{e}} = -\sigma\mu \frac{\partial \vec{A}^{\mathrm{e}}}{\partial t} \tag{3.97}$$

Application of the Coulomb gauge,

$$\nabla \cdot \vec{A}^{\mathrm{e}} = 0 \tag{3.98}$$

to Equation 3.97 gives

$$\nabla^2 \vec{A}^{\mathrm{e}} = \sigma\mu \frac{\partial \vec{A}^{\mathrm{e}}}{\partial t} \tag{3.99}$$

The eddy component of the vector potential vanishes when the skin depth is zero. To illustrate this fact, we calculate the circulation of the electric field over any closed contour Γ using Equation 3.92 as follows:

$$\oint_{\Gamma} \vec{E} \cdot \mathrm{d}\vec{l} = -\mu \frac{\partial}{\partial t} \oint_{\Gamma} \vec{A} \cdot \mathrm{d}\vec{l} + \oint_{\Gamma} \nabla V \cdot \mathrm{d}\vec{l} \tag{3.100}$$

Taking into account the fact that

$$\oint_{\Gamma} \nabla V \cdot d\vec{l} \equiv 0 \tag{3.101}$$

and

$$\oint_{\Gamma} \vec{A} \cdot d\vec{l} = \int_{S} \nabla \times \vec{A} \cdot d\vec{s} \quad \text{(Stokes's theorem)} \tag{3.102}$$

Equation 3.100 reduces to the following form

$$\oint_{\Gamma} \vec{E} \cdot d\vec{l} = -\mu \frac{\partial}{\partial t} \int_{S} \nabla \times \vec{A} \cdot d\vec{s} = -\mu \frac{\partial}{\partial t} \int_{S} \nabla \times \vec{A}^{e} \cdot d\vec{s} = -\mu \frac{\partial}{\partial t} \int_{S} \vec{H} \cdot d\vec{s} = -\frac{\partial \Phi}{\partial t} \tag{3.103}$$

where Φ is the magnetic flux crossing the surface S. Assuming the conductor to be stationary, the only physical process causing variation of the magnetic flux through the conductor is the diffusion of the electromagnetic field causing expansion of the penetration depth and, consequently, the surface S, which is a function of the skin layer thickness δ. Thus, it is natural to suppose that in the case of skin effect (when the skin layer remains shallow), the eddy component of the vector potential is proportional to the skin depth and, consequently, much smaller than the source component. In the PEC-limit ($\tilde{p} = 0$), the field does not penetrate inside the conductor, the magnetic flux crossing the conductor does not vary in time, and the vector potential has only a source component.

For further derivations in terms of nondimensional variables, we introduce the scale factors $\boldsymbol{A}$ and $\boldsymbol{V}$ for the magnetic vector potential and electric scalar potential, respectively. The first of these can be obtained directly from Equation 3.89 using Equation 3.14 as follows:

$$\vec{\tilde{H}}\boldsymbol{H} = D^{-1} \nabla \times \vec{\tilde{A}}\boldsymbol{A} \tag{3.104}$$

So that

$$A = \boldsymbol{H}; \quad D = \boldsymbol{I} \tag{3.105}$$

and

$$\vec{\tilde{H}} = \nabla \times \vec{\tilde{A}} \tag{3.106}$$

Substitution of Equation 3.105 into Equation 3.92 yields the scale factor for the electric scalar potential

$$\mu I T^{-1} = -\mu I T^{-1}\frac{\partial \vec{\tilde{A}}}{\partial \tilde{t}} + VD^{-1}\nabla\tilde{V} \tag{3.107}$$

so that

$$V = \mu I D T^{-1} \tag{3.108}$$

It is natural to consider how the tangential components of the magnetic vector potential at the interface can be expressed in terms of the tangential magnetic field. The shortest way is to use Equation 3.96 relating the vector potential and the electric field, and then substitute the *E–H* boundary condition obtained in Section 3.2. As a first step, we represent Equation 3.96 in terms of nondimensional variables:

$$\vec{\tilde{E}} = -\frac{\partial \vec{\tilde{A}}^{\mathrm{e}}}{\partial t} \tag{3.109}$$

Rewriting Equation 3.109 in the Laplace domain and applying the interface, we obtain

$$\left(\vec{\tilde{\tilde{A}}}^{\mathrm{e}}\right)^{\mathrm{b}} = -\tilde{s}^{-1}\vec{\tilde{\tilde{E}}}^{\mathrm{b}} \tag{3.110}$$

Substituting Equations 3.45 and 3.47 in Equation 3.110, the *A–H* relations are obtained in the various approximations:

Leontovich's approximation:

$$\left(\vec{\tilde{A}}^{\mathrm{e}}\right)^{\mathrm{b}}_{\xi_{3-k}} = \begin{cases} (-1)^{3-k}\tilde{p}\tilde{s}^{-1/2}\tilde{\tilde{H}}^{\mathrm{b}}_{\xi_k} & \mathrm{LD} \\ (-1)^{3-k}\tilde{p}\dfrac{1-j}{2}\dot{\tilde{H}}^{\mathrm{b}}_{\xi_k} & \mathrm{FD}, \quad k=1,2 \\ (-1)^{3-k}\tilde{p}\tilde{H}^{\mathrm{b}}_{\xi_k} * \tilde{T}^{\mathrm{b}}_{2} & \mathrm{TD} \end{cases} \tag{3.111}$$

Mitzner's approximation:

$$\left(\vec{\tilde{A}}^{\mathrm{e}}\right)^{\mathrm{b}}_{\xi_{3-k}} = \begin{cases} (-1)^{3-k}\tilde{p}\left(\tilde{s}^{-1/2}\tilde{\tilde{H}}^{\mathrm{b}}_{\xi_k} - \tilde{p}\tilde{s}^{-1}\tilde{\tilde{H}}^{\mathrm{b}}_{\xi_k}\dfrac{\tilde{d}_k - \tilde{d}_{3-k}}{2\tilde{d}_k\tilde{d}_{3-k}}\right) & \mathrm{LD} \\ (-1)^{3-k}\tilde{p}\left(\dfrac{1-j}{2}\dot{\tilde{H}}^{\mathrm{b}}_{\xi_k} + \dfrac{j}{2}\tilde{p}\dot{\tilde{H}}^{\mathrm{b}}_{\xi_k}\dfrac{\tilde{d}_k - \tilde{d}_{3-k}}{2\tilde{d}_k\tilde{d}_{3-k}}\right) & \mathrm{FD}, \quad k=1,2 \\ (-1)^{3-k}\tilde{p}\left(f^{\mathrm{b}}_{\xi_k} * \tilde{T}^{\mathrm{b}}_{2} - \tilde{p}\dfrac{\tilde{d}_k - \tilde{d}_{3-k}}{2\tilde{d}_k\tilde{d}_{3-k}}f^{\mathrm{b}}_{\xi_k} * \tilde{T}^{\mathrm{b}}_{3}\right) & \mathrm{TD} \end{cases} \tag{3.112}$$

Rytov's approximation:

$$\left(\vec{A}^{\mathrm{e}}\right)^{\mathrm{b}}_{\xi_{3-k}} = \begin{cases} (-1)^{3-k}\tilde{p}\left[\tilde{s}^{-1/2}\bar{\tilde{H}}^{\mathrm{b}}_{\xi_k} - \tilde{p}\tilde{s}^{-1}\bar{\tilde{H}}^{\mathrm{b}}_{\xi_k}\dfrac{\tilde{d}_k - \tilde{d}_{3-k}}{2\tilde{d}_k\tilde{d}_{3-k}}\right. \\ \left. \quad -\tilde{p}^2\tilde{s}^{-3/2}\left(\bar{\tilde{H}}^{\mathrm{b}}_{\xi_k}\dfrac{\tilde{d}_k^2 + 2\tilde{d}_k\tilde{d}_{3-k} - 3\tilde{d}_{3-k}^2}{8\tilde{d}_k^2\tilde{d}_{3-k}^2} - \dfrac{1}{2}\dfrac{\partial^2\bar{\tilde{H}}^{\mathrm{b}}_{\xi_k}}{\partial\tilde{\xi}_k^2} + \dfrac{1}{2}\dfrac{\partial^2\bar{\tilde{H}}^{\mathrm{b}}_{\xi_k}}{\partial\tilde{\xi}_{3-k}^2} - \dfrac{\partial^2\bar{\tilde{H}}^{\mathrm{b}}_{\xi_{3-k}}}{\partial\tilde{\xi}_k\partial\tilde{\xi}_{3-k}}\right)\right] & \text{LD} \\ (-1)^{3-k}\tilde{p}\left[\dfrac{1-j}{2}\dot{\tilde{H}}^{\mathrm{b}}_{\xi_k} + \dfrac{j}{2}\tilde{p}\dot{\tilde{H}}^{\mathrm{b}}_{\xi_k}\dfrac{\tilde{d}_k - \tilde{d}_{3-k}}{2\tilde{d}_k\tilde{d}_{3-k}}\right. \\ \left. \quad -\tilde{p}^2\frac{1+j}{4}\left(\dot{\tilde{H}}^{\mathrm{b}}_{\xi_k}\dfrac{\tilde{d}_k^2 + 2\tilde{d}_k\tilde{d}_{3-k} - 3\tilde{d}_{3-k}^2}{8\tilde{d}_k^2\tilde{d}_{3-k}^2} - \dfrac{1}{2}\dfrac{\partial^2\dot{\tilde{H}}^{\mathrm{b}}_{\xi_k}}{\partial\tilde{\xi}_k^2} + \dfrac{1}{2}\dfrac{\partial^2\dot{\tilde{H}}^{\mathrm{b}}_{\xi_k}}{\partial\tilde{\xi}_{3-k}^2} - \dfrac{\partial^2\dot{\tilde{H}}^{\mathrm{b}}_{\xi_{3-k}}}{\partial\tilde{\xi}_k\partial\tilde{\xi}_{3-k}}\right)\right] & \text{FD}, \quad k=1,2 \\ (-1)^{3-k}\tilde{p}\left[\tilde{H}^{\mathrm{b}}_{\xi_k} * \tilde{T}^{\mathrm{b}}_2 - \tilde{p}\dfrac{\tilde{d}_k - \tilde{d}_{3-k}}{2\tilde{d}_k\tilde{d}_{3-k}}\tilde{H}^{\mathrm{b}}_{\xi_k} * \tilde{T}^{\mathrm{b}}_3\right. \\ \left. \quad -\tilde{p}^2\left(\tilde{H}^{\mathrm{b}}_{\xi_k}\dfrac{\tilde{d}_k^2 + 2\tilde{d}_k\tilde{d}_{3-k} - 3\tilde{d}_{3-k}^2}{8\tilde{d}_k^2\tilde{d}_{3-k}^2} - \dfrac{1}{2}\dfrac{\partial^2\tilde{H}^{\mathrm{b}}_{\xi_k}}{\partial\tilde{\xi}_k^2} + \dfrac{1}{2}\dfrac{\partial^2\tilde{H}^{\mathrm{b}}_{\xi_k}}{\partial\tilde{\xi}_{3-k}^2} - \dfrac{\partial^2\tilde{H}^{\mathrm{b}}_{\xi_{3-k}}}{\partial\tilde{\xi}_k\partial\tilde{\xi}_{3-k}}\right) * \tilde{T}^{\mathrm{b}}_4\right] & \text{TD} \end{cases} \tag{3.113}$$

Switching in Equations 3.111 through 3.113 back to dimensional variables, we represent the *A*–*H* relations in the frequency- and time-domain as follows:

Leontovich's approximation:

$$\left(\vec{A}^{\mathrm{e}}\right)^{\mathrm{b}}_{\xi_{3-k}} = \begin{cases} (-1)^{3-k}(j\omega\sigma\mu)^{-1/2}\dot{H}^{\mathrm{b}}_{\xi_k} & \text{FD} \\ (-1)^{3-k}(\sigma\mu)^{-1/2}H^{\mathrm{b}}_{\xi_k} * T^{\mathrm{b}}_2 & \text{TD} \end{cases}, \quad k = 1,2 \tag{3.114}$$

Mitzner's approximation:

$$\left(\vec{A}^{\mathrm{e}}\right)^{\mathrm{b}}_{\xi_{3-k}} = \begin{cases} (-1)^{3-k}\left[(j\omega\sigma\mu)^{-1/2}\dot{H}^{\mathrm{b}}_{\xi_k} - (j\omega\sigma\mu)^{-1}\dfrac{d_k - d_{3-k}}{2d_kd_{3-k}}\dot{H}^{\mathrm{b}}_{\xi_k}\right] & \text{FD} \\ (-1)^{3-k}\left[(\sigma\mu)^{-1/2}H^{\mathrm{b}}_{\xi_k} * T^{\mathrm{b}}_2 - (\sigma\mu)^{-1}\dfrac{d_k - d_{3-k}}{2d_kd_{3-k}}H^{\mathrm{b}}_{\xi_k} * T^{\mathrm{b}}_3\right] & \text{TD} \end{cases}, \quad k=1,2 \tag{3.115}$$

Rytov's approximation:

$$\left(\vec{A}^{\mathrm{e}}\right)^{\mathrm{b}}_{\xi_{3-k}} = \begin{cases} (-1)^{3-k}\left[(j\omega\sigma\mu)^{-1/2}\dot{H}^{\mathrm{b}}_{\xi_k} - (j\omega\sigma\mu)^{-1/2}\dfrac{d_k - d_{3-k}}{2d_kd_{3-k}}\dot{H}^{\mathrm{b}}_{\xi_k}\right. \\ \left. \quad -(j\omega\sigma\mu)^{-3/2}\left(\dot{H}^{\mathrm{b}}_{\xi_k}\dfrac{d_k^2 + 2d_kd_{3-k} - 3d_{3-k}^2}{8d_k^2d_{3-k}^2} - \dfrac{1}{2}\dfrac{\partial^2\dot{H}^{\mathrm{b}}_{\xi_k}}{\partial\xi_k^2} + \dfrac{1}{2}\dfrac{\partial^2\dot{H}^{\mathrm{b}}_{\xi_k}}{\partial\xi_{3-k}^2} - \dfrac{\partial^2\dot{H}^{\mathrm{b}}_{\xi_{3-k}}}{\partial\xi_k\partial\xi_{3-k}}\right)\right] & \text{FD} \\ (-1)^{3-k}\left[(\sigma\mu)^{-1/2}H^{\mathrm{b}}_{\xi_k} * T^{\mathrm{b}}_2 - (\sigma\mu)^{-1}\dfrac{d_k - d_{3-k}}{2d_kd_{3-k}}H^{\mathrm{b}}_{\xi_k} * T^{\mathrm{b}}_3\right. \\ \left. \quad -(\sigma\mu)^{-3/2}\left(H^{\mathrm{b}}_{\xi_k}\dfrac{d_k^2 + 2d_kd_{3-k} - 3\tilde{d}_{3-k}^2}{8d_k^2d_{3-k}^2} - \dfrac{1}{2}\dfrac{\partial^2H^{\mathrm{b}}_{\xi_k}}{\partial\xi_k^2} + \dfrac{1}{2}\dfrac{\partial^2H^{\mathrm{b}}_{\xi_k}}{\partial\xi_{3-k}^2} - \dfrac{\partial^2H^{\mathrm{b}}_{\xi_{3-k}}}{\partial\xi_k\partial\xi_{3-k}}\right) * T^{\mathrm{b}}_4\right] & \text{TD} \end{cases}, \quad k=1,2 \tag{3.116}$$

The SIBCs in Equations 3.111 through 3.113 and Equations 3.114 through 3.116 can be represented in much shorter form using the surface impedance defined in Equations 3.86 through 3.88 and Equations 3.82 through 3.84 as follows:

Nondimensional form:

$$\left(\tilde{\vec{A}}^{\mathrm{e}}\right)_{\xi_{3-k}}^{\mathrm{b}} = (-1)^{3-k}\tilde{F}\left[\tilde{H}_{\xi_k}^{\mathrm{b}}\right] \tag{3.117}$$

Dimensional form:

$$\left(\vec{A}^{\mathrm{e}}\right)_{\xi_{3-k}}^{\mathrm{b}} = (-1)^{3-k}F\left[H_{\xi_k}^{\mathrm{b}}\right] \tag{3.118}$$

In some formulations [2–5], employing the vector potential formalism relationships of the following kind are required at the interface to eliminate the conducting domain from the numerical procedure:

$$\left.\frac{\partial\left(\vec{n}\times\vec{A}\right)}{\partial\vec{n}}\right|^{\mathrm{b}} = f_{\mathrm{b}}\left(\vec{n}\times\vec{A}^{\mathrm{b}}\right) \tag{3.119}$$

$$\vec{n}\times\left.\left(\nabla\times\vec{A}\right)\right|^{\mathrm{b}} = f_{\mathrm{a}}\left(\vec{n}\times\vec{A}^{\mathrm{b}}\right) \tag{3.120}$$

where f_{a} and f_{b} are functions to be derived. Since the distribution of the magnetic vector potential in the conductor is described by the diffusion equation (Equation 2.2), the surface impedance "toolbox" can again be used and application of Equations 2.199, 2.204, 2.209, 2.201, 2.206, and 2.211 directly, yields the appropriate approximations:

Leontovich's approximation:

$$\left.\frac{\partial\left(\vec{A}^{\mathrm{e}}\right)_{\xi_k}}{\partial\eta}\right|^{\mathrm{b}} = \begin{cases} -(j\omega\sigma\mu)^{1/2}\left(\dot{\vec{A}}^{\mathrm{e}}\right)_{\xi_k}^{\mathrm{b}} & \mathrm{FD} \\ -(\sigma\mu)^{1/2}\frac{\mathrm{d}}{\mathrm{d}t}\left(\left(\vec{A}^{\mathrm{e}}\right)_{\xi_k}^{\mathrm{b}} * T_2^{\mathrm{b}}\right) & \mathrm{TD} \end{cases}, \quad k=1,2 \tag{3.121}$$

$$\left(\nabla\times\vec{A}^{\mathrm{e}}\right)_{\xi_{3-k}}^{\mathrm{b}} = \begin{cases} (-1)^k(j\omega\sigma\mu)^{1/2}(\dot{\vec{A}}^{\mathrm{e}})_{\xi_k}^{\mathrm{b}} & \mathrm{FD} \\ (-1)^k(\sigma\mu)^{1/2}\frac{\mathrm{d}}{\mathrm{d}t}\left(\left(\vec{A}^{\mathrm{e}}\right)_{\xi_k}^{\mathrm{b}} * T_2^{\mathrm{b}}\right) & \mathrm{TD} \end{cases}, \quad k=1,2 \tag{3.122}$$

Mitzner's approximation:

$$\left.\frac{\partial\left(\vec{A}^{e}\right)_{\xi_k}}{\partial\eta}\right|^{b}=\begin{cases}-(j\omega\sigma\mu)^{1/2}\left(\dot{\vec{A}}^{e}\right)_{\xi_k}^{b}+\dfrac{d_k+d_{3-k}}{2d_kd_{3-k}}\left(\dot{\vec{A}}^{e}\right)_{\xi_k}^{b} & \text{FD}\\ -(\sigma\mu)^{1/2}\dfrac{\mathrm{d}}{\mathrm{d}t}\left(\left(\vec{A}^{e}\right)_{\xi_k}^{b}*T_2^{b}\right)+\dfrac{d_k+d_{3-k}}{2d_kd_{3-k}}\left(\vec{A}^{e}\right)_{\xi_k}^{b} & \text{TD}\end{cases},\quad k=1,2 \tag{3.123}$$

$$\left(\nabla\times\vec{A}^{e}\right)_{\xi_{3-k}}^{b}=\begin{cases}(-1)^{k}\left((j\omega\sigma\mu)^{1/2}\left(\dot{\vec{A}}^{e}\right)_{\xi_k}^{b}-\dfrac{d_k-d_{3-k}}{2d_kd_{3-k}}\left(\dot{\vec{A}}^{e}\right)_{\xi_k}^{b}\right) & \text{FD}\\ (-1)^{k}\left((\sigma\mu)^{1/2}\dfrac{\mathrm{d}}{\mathrm{d}t}\left(\left(\vec{A}^{e}\right)_{\xi_k}^{b}*T_2^{b}\right)-\dfrac{d_k-d_{3-k}}{2d_kd_{3-k}}\left(\vec{A}^{e}\right)_{\xi_k}^{b}\right) & \text{TD}\end{cases},\quad k=1,2 \tag{3.124}$$

Rytov's approximation:

$$\left.\frac{\partial\left(\vec{A}^{e}\right)_{\xi_k}}{\partial\eta}\right|^{b}=\begin{cases}-(j\omega\sigma\mu)^{1/2}\left(\dot{\vec{A}}^{e}\right)_{\xi_k}^{b}+\dfrac{d_k+d_{3-k}}{2d_kd_{3-k}}\left(\dot{\vec{A}}^{e}\right)_{\xi_k}^{b}\\ \quad+(j\omega\sigma\mu)^{-1/2}\left(\left(\dot{\vec{A}}^{e}\right)_{\xi_k}^{b}\dfrac{d_k^2+2d_kd_{3-k}-3d_{3-k}^2}{8d_k^2d_{3-k}^2}+\dfrac{1}{2}\displaystyle\sum_{i=1}^{2}\dfrac{\partial^2\left(\dot{\vec{A}}^{e}\right)_{\xi_k}^{b}}{\partial\xi_i^2}\right) & \text{FD}\\ -(\sigma\mu)^{1/2}\frac{\mathrm{d}}{\mathrm{d}t}\left(\left(\vec{A}^{e}\right)_{\xi_k}^{b}*T_2^{b}\right)+\dfrac{d_k+d_{3-k}}{2d_kd_{3-k}}\left(\vec{A}^{e}\right)_{\xi_k}^{b}\\ \quad+(\sigma\mu)^{-1/2}\left(\left(\vec{A}^{e}\right)_{\xi_k}^{b}\dfrac{d_k^2+2d_kd_{3-k}-3d_{3-k}^2}{8d_k^2d_{3-k}^2}+\dfrac{1}{2}\displaystyle\sum_{i=1}^{2}\dfrac{\partial^2\left(\vec{A}^{e}\right)_{\xi_k}^{b}}{\partial\xi_i^2}\right)*T_2^{b} & \text{TD}\end{cases},\quad k=1,2 \tag{3.125}$$

$$\left(\nabla\times\vec{A}^{e}\right)_{\xi_{3-k}}^{b}=\begin{cases}(-1)^{k}\left[(j\omega\sigma\mu)^{1/2}\left(\dot{\vec{A}}^{e}\right)_{\xi_k}^{b}-\dfrac{d_k-d_{3-k}}{2d_kd_{3-k}}\left(\dot{\vec{A}}^{e}\right)_{\xi_k}^{b}\right.\\ \quad-(j\omega\sigma\mu)^{-1/2}\left(\left(\dot{\vec{A}}^{e}\right)_{\xi_k}^{b}\dfrac{d_k^2+2d_kd_{3-k}-3d_{3-k}^2}{8d_k^2d_{3-k}^2}-\dfrac{1}{2}\dfrac{\partial^2\left(\dot{\vec{A}}^{e}\right)_{\xi_k}^{b}}{\partial\xi_k^2}\right.\\ \quad\left.\left.+\dfrac{1}{2}\dfrac{\partial^2\left(\dot{\vec{A}}^{e}\right)_{\xi_k}^{b}}{\partial\xi_{3-k}^2}-\dfrac{\partial^2\left(\dot{\vec{A}}^{e}\right)_{\xi_{3-k}}^{b}}{\partial\xi_k\partial\xi_{3-k}}\right)\right] & \text{FD}\\ (-1)^{k}\left[(\sigma\mu)^{1/2}\dfrac{\mathrm{d}}{\mathrm{d}t}\left((\vec{A}^{e})_{\xi_k}^{b}*T_2^{b}\right)-\dfrac{d_k-d_{3-k}}{2d_kd_{3-k}}\left(\vec{A}^{e}\right)_{\xi_k}^{b}\right.\\ \quad-(\sigma\mu)^{-1/2}\left(\left(\vec{A}^{e}\right)_{\xi_k}^{b}\dfrac{d_k^2+2d_kd_{3-k}-3d_{3-k}^2}{8d_k^2d_{3-k}^2}-\dfrac{1}{2}\dfrac{\partial^2\left(\vec{A}^{e}\right)_{\xi_k}^{b}}{\partial\xi_k^2}\right.\\ \quad\left.\left.+\dfrac{1}{2}\dfrac{\partial^2\left(\vec{A}^{e}\right)_{\xi_k}^{b}}{\partial\xi_{3-k}^2}-\dfrac{\partial^2\left(\vec{A}^{e}\right)_{\xi_{3-k}}^{b}}{\partial\xi_k\partial\xi_{3-k}}\right)*T_2^{b}\right] & \text{TD}\end{cases},\quad k=1,2 \tag{3.126}$$

Note that the right-hand side of Equation 3.121 (and, consequently, Equations 3.123 and 3.125) does not vanish when $\delta \to 0$. In other words, the product $\delta^{-1}(\vec{A}^{e})^{b}_{\xi_k}$ remains a finite quantity even when the skin layer becomes infinitely thin. Indeed, from Equation 3.114 it follows that

$$(-1)^{3-k}\frac{(1+j)}{\delta}\left(\dot{\vec{A}}^{e}\right)^{b}_{\xi_{3-k}} = \dot{H}^{b}_{\xi_k} \tag{3.127}$$

Since the tangential magnetic field in the PEC limit does not vanish and transforms at the interface to a surface current, the term on the left-hand side of Equation 3.127 has the physical meaning of a *surface current density*.

3.6 Common Representation of Various SIBCs Using a Surface Impedance Function

In the previous section, we considered approximate boundary conditions expressing the tangential electric field, the normal magnetic field/normal derivative of the magnetic scalar potential, and the tangential magnetic vector potential in terms of the tangential magnetic field. Since the derivation of all the mentioned conditions is based on the same concept of surface impedance, it is intuitively clear that the SIBCs in Equations 3.55 through 3.60, 3.70 through 3.72, 3.81, and 3.118 can be represented in a "common" form or using a "common" function to be derived in this section. The surface impedance function defined in Equations 3.82 through 3.84 is a natural candidate for this role and it has been already used to obtain Equations 3.81 and 3.118. To represent other SIBCs, we rewrite Equations 2.186 and 2.12 in the following form:

Rytov's approximation:

$$\left(\tilde{\nabla}\times\vec{f}\right)^{b}_{\xi_{3-k}} = \begin{cases} (-1)^{k}\tilde{s}\left[\tilde{s}^{-1/2}\bar{f}^{b}_{\xi_k} - \tilde{p}\tilde{s}^{-1}\bar{f}^{b}_{\xi_k}\dfrac{\tilde{d}_k-\tilde{d}_{3-k}}{2\tilde{d}_k\tilde{d}_{3-k}} \right. \\ \left. \quad -\tilde{p}^{2}\tilde{s}^{-3/2}\left(\bar{f}^{b}_{\xi_k}\dfrac{\tilde{d}_k^{2}+2\tilde{d}_k\tilde{d}_{3-k}-3\tilde{d}_{3-k}^{2}}{8\tilde{d}_k^{2}\tilde{d}_{3-k}^{2}} - \dfrac{1}{2}\dfrac{\partial^{2}\bar{f}^{b}_{\xi_k}}{\partial\tilde{\xi}_k^{2}} + \dfrac{1}{2}\dfrac{\partial^{2}\bar{f}^{b}_{\xi_k}}{\partial\tilde{\xi}_{3-k}^{2}} - \dfrac{\partial^{2}\bar{f}^{b}_{\xi_{3-k}}}{\partial\tilde{\xi}_k\partial\tilde{\xi}_{3-k}}\right)\right] & \text{LD} \\ (-1)^{k}2j\left[\dfrac{1-j}{2}\dot{f}^{b}_{\xi_k} + \tilde{p}\dfrac{j}{2}\dot{f}^{b}_{\xi_k}\dfrac{\tilde{d}_k-\tilde{d}_{3-k}}{2\tilde{d}_k\tilde{d}_{3-k}} \right. \\ \left. \quad -\tilde{p}^{2}\dfrac{1+j}{4}\left(\dot{f}^{b}_{\xi_k}\dfrac{\tilde{d}_k^{2}+2\tilde{d}_k\tilde{d}_{3-k}-3\tilde{d}_{3-k}^{2}}{8\tilde{d}_k^{2}\tilde{d}_{3-k}^{2}} - \dfrac{1}{2}\dfrac{\partial^{2}\dot{f}^{b}_{\xi_k}}{\partial\tilde{\xi}_k^{2}} + \dfrac{1}{2}\dfrac{\partial^{2}\dot{f}^{b}_{\xi_k}}{\partial\tilde{\xi}_{3-k}^{2}} - \dfrac{\partial^{2}\dot{f}^{b}_{\xi_{3-k}}}{\partial\tilde{\xi}_k\partial\tilde{\xi}_{3-k}}\right)\right] & \text{FD}, \quad k=1,2 \\ (-1)^{k}\dfrac{\mathrm{d}}{\mathrm{d}\tilde{t}}\left[f^{b}_{\xi_k}*\tilde{T}^{b}_{2} - \tilde{p}\dfrac{\tilde{d}_k-\tilde{d}_{3-k}}{2\tilde{d}_k\tilde{d}_{3-k}}f^{b}_{\xi_k}*\tilde{T}^{b}_{3} \right. \\ \left. \quad -\tilde{p}^{2}\left(f^{b}_{\xi_k}\dfrac{\tilde{d}_k^{2}+2\tilde{d}_k\tilde{d}_{3-k}-3\tilde{d}_{3-k}^{2}}{8\tilde{d}_k^{2}\tilde{d}_{3-k}^{2}} - \dfrac{1}{2}\dfrac{\partial^{2}f^{b}_{\xi_k}}{\partial\tilde{\xi}_k^{2}} + \dfrac{1}{2}\dfrac{\partial^{2}f^{b}_{\xi_k}}{\partial\tilde{\xi}_{3-k}^{2}} - \dfrac{\partial^{2}f^{b}_{\xi_{3-k}}}{\partial\tilde{\xi}_k\partial\tilde{\xi}_{3-k}}\right)*\tilde{T}^{b}_{4}\right] & \text{TD} \end{cases} \tag{3.128}$$

$$\left(\nabla\times\vec{f}\right)^{\mathrm{b}}_{\xi_{3-k}}=\begin{cases}(-1)^k j\omega\sigma\mu\left[(j\omega\sigma\mu)^{-1/2}\dot{f}^{\mathrm{b}}_{\xi_k}-(j\omega\sigma\mu)^{-1}\dfrac{d_k-d_{3-k}}{2d_kd_{3-k}}\dot{f}^{\mathrm{b}}_{\xi_k}\right.\\ \quad\left.-(j\omega\sigma\mu)^{-3/2}\left(\dot{f}^{\mathrm{b}}_{\xi_k}\dfrac{d_k^2+2d_kd_{3-k}-3d_{3-k}^2}{8d_k^2d_{3-k}^2}-\dfrac{1}{2}\dfrac{\partial^2\dot{f}^{\mathrm{b}}_{\xi_k}}{\partial\xi_k^2}+\dfrac{1}{2}\dfrac{\partial^2\dot{f}^{\mathrm{b}}_{\xi_k}}{\partial\xi_{3-k}^2}-\dfrac{\partial^2\dot{f}^{\mathrm{b}}_{\xi_{3-k}}}{\partial\xi_k\partial\xi_{3-k}}\right)\right]\quad\mathrm{FD}\\ (-1)^k\sigma\mu\dfrac{\mathrm{d}}{\mathrm{d}t}\left[(\sigma\mu)^{-1/2}f^{\mathrm{b}}_{\xi_k}*T^{\mathrm{b}}_2-(\sigma\mu)^{-1}\dfrac{d_k-d_{3-k}}{2d_kd_{3-k}}f^{\mathrm{b}}_{\xi_k}*T^{\mathrm{b}}_3\right.\\ \quad\left.-(\sigma\mu)^{-3/2}\left(f^{\mathrm{b}}_{\xi_k}\dfrac{d_k^2+2d_kd_{3-k}-3\tilde{d}_{3-k}^2}{8d_k^2d_{3-k}^2}-\dfrac{1}{2}\dfrac{\partial^2 f^{\mathrm{b}}_{\xi_k}}{\partial\xi_k^2}+\dfrac{1}{2}\dfrac{\partial^2 f^{\mathrm{b}}_{\xi_k}}{\partial\xi_{3-k}^2}-\dfrac{\partial^2 f^{\mathrm{b}}_{\xi_{3-k}}}{\partial\xi_k\partial\xi_{3-k}}\right)*T^{\mathrm{b}}_4\right]\quad\mathrm{TD}\end{cases},\quad k=1,2 \tag{3.129}$$

Taking into account Equations 3.84 and 3.88, we obtain

$$\left(\tilde{\nabla}\times\vec{f}\right)^{\mathrm{b}}_{\xi_{3-k}}=\begin{cases}(-1)^{3-k}\tilde{p}^{-1}\tilde{s}\tilde{F}^{\mathrm{Rytov}}\left[\tilde{f}^{\mathrm{b}}_{\xi_k}\right] & \mathrm{LD}\\ (-1)^{3-k}\tilde{p}^{-1}2j\tilde{F}^{\mathrm{Rytov}}\left[\dot{f}^{\mathrm{b}}_{\xi_k}\right] & \mathrm{FD},\qquad k=1,2\\ (-1)^{3-k}\tilde{p}^{-1}\frac{\partial}{\partial t}\tilde{F}^{\mathrm{Rytov}}\left[f^{\mathrm{b}}_{\xi_k}\right] & \mathrm{TD}\end{cases} \tag{3.130}$$

$$\left(\nabla\times\vec{f}\right)^{\mathrm{b}}_{\xi_{3-k}}=\begin{cases}(-1)^{3-k}j\omega\sigma\mu F^{\mathrm{Rytov}}\left[\dot{f}^{\mathrm{b}}_{\xi_k}\right] & \mathrm{FD}\\ (-1)^{3-k}\sigma\mu\frac{\partial}{\partial t}F^{\mathrm{Rytov}}\left[f^{\mathrm{b}}_{\xi_k}\right] & \mathrm{TD}\end{cases},\quad k=1,2 \tag{3.131}$$

Clearly, Equations 3.130 and 3.131 hold for the lower-order Leontovich and Mitzner approximations.

With the use of Equations 3.130 and 3.131, the SIBCs in terms of various formalism can be represented as follows:

E–H formalism:

$$E^{\mathrm{b}}_{\xi_{3-k}}=(-1)^{3-k}\mu\frac{\partial}{\partial t}F\left[H^{\mathrm{b}}_{\xi_k}\right]=(-1)^{3-k}\mu\hat{F}\left[H^{\mathrm{b}}_{\xi_k}\right],\quad k=1,2 \tag{3.132}$$

H–φ formalism:

$$\left(\frac{\partial\varphi}{\partial\eta}\right)=H^{\mathrm{s}}_{\eta}-\sum_{i=1}^{2}\frac{\partial}{\partial\xi_i}F\left[H^{\mathrm{s}}_{\xi_i}\right]+\sum_{i=1}^{2}\frac{\partial}{\partial\xi_i}F[\partial\varphi/\partial\xi_i]\quad\text{(this is Equation 3.81)}$$

$$H^{\mathrm{b}}_{\eta}=\sum_{i=1}^{2}\frac{\partial}{\partial\xi_i}F\left[H^{\mathrm{b}}_{\xi_i}\right]\quad\text{(these are Equations 3.70 through 3.72)}$$

A–V formalism:

$$\left(\vec{A}^{\mathrm{e}}\right)^{\mathrm{b}}_{\xi_{3-k}}=(-1)^{3-k}F\left[H^{\mathrm{b}}_{\xi_k}\right],\quad k=1,2\quad\text{(this is Equation 3.118)}$$

$$\left(\nabla\times\vec{A}^{\mathrm{e}}\right)^{\mathrm{b}}_{\xi_{3-k}}=(-1)^{3-k}\frac{\partial}{\partial t}F\left[\left(\vec{A}^{\mathrm{e}}\right)^{\mathrm{b}}_{\xi_k}\right],\quad k=1,2 \tag{3.133}$$

The time-derivative $\hat{F}$ of the surface impedance function can be calculated as follows:

$$\frac{\partial}{\partial t}F^{\text{Leontovich}}\left[f_{\xi_k}\right]=\hat{F}^{\text{Leontovich}}\left[f_{\xi_k}\right]=\begin{cases}(j\omega\sigma\mu)^{1/2}\dot{f}^{\text{b}}_{\xi_k} & \text{FD}\\ (\sigma\mu)^{1/2}f^{\text{b}}_{\xi_k}*\hat{T}^{\text{b}}_2 & \text{TD}\end{cases},\quad k=1,2 \tag{3.134}$$

$$\frac{\partial}{\partial t}F^{\text{Mitzner}}\left[f_{\xi_k}\right]=\hat{F}^{\text{Mitzner}}\left[f_{\xi_k}\right]=\begin{cases}(j\omega\sigma\mu)^{1/2}\dot{f}^{\text{b}}_{\xi_k}-\dfrac{d_k-d_{3-k}}{2d_kd_{3-k}}\dot{f}^{\text{b}}_{\xi_k} & \text{FD}\\ (\sigma\mu)^{1/2}f^{\text{b}}_{\xi_k}*\hat{T}^{\text{b}}_2-\dfrac{d_k-d_{3-k}}{2d_kd_{3-k}}f^{\text{b}}_{\xi_k}*\hat{T}^{\text{b}}_3 & \text{TD}\end{cases},\quad k=1,2 \tag{3.135}$$

$$\frac{\partial}{\partial t}F^{\text{Rytov}}\left[f_{\xi_k}\right]=\hat{F}^{\text{Rytov}}\left[f_{\xi_k}\right]$$
$$=\begin{cases}(j\omega\sigma\mu)^{1/2}\dot{f}^{\text{b}}_{\xi_k}-\dfrac{d_k-d_{3-k}}{2d_kd_{3-k}}\dot{f}^{\text{b}}_{\xi_k}\\ \quad-(j\omega\sigma\mu)^{-1/2}\left(\dot{f}^{\text{b}}_{\xi_k}\dfrac{d_k^2+2d_kd_{3-k}-3d_{3-k}^2}{8d_k^2d_{3-k}^2}-\dfrac{1}{2}\dfrac{\partial^2\dot{f}^{\text{b}}_{\xi_k}}{\partial\xi_k^2}+\dfrac{1}{2}\dfrac{\partial^2\dot{f}^{\text{b}}_{\xi_k}}{\partial\xi_{3-k}^2}-\dfrac{\partial^2\dot{f}^{\text{b}}_{\xi_{3-k}}}{\partial\xi_k\partial\xi_{3-k}}\right) & \text{FD}\\ (\sigma\mu)^{1/2}f^{\text{b}}_{\xi_k}*\hat{T}^{\text{b}}_2-\dfrac{d_k-d_{3-k}}{2d_kd_{3-k}}f^{\text{b}}_{\xi_k}*\hat{T}^{\text{b}}_3\\ \quad-(\sigma\mu)^{-1/2}\left(f^{\text{b}}_{\xi_k}\dfrac{d_k^2+2d_kd_{3-k}-3\tilde{d}_{3-k}^2}{8d_k^2d_{3-k}^2}-\dfrac{1}{2}\dfrac{\partial^2 f^{\text{b}}_{\xi_k}}{\partial\xi_k^2}+\dfrac{1}{2}\dfrac{\partial^2 f^{\text{b}}_{\xi_k}}{\partial\xi_{3-k}^2}-\dfrac{\partial^2 f^{\text{b}}_{\xi_{3-k}}}{\partial\xi_k\partial\xi_{3-k}}\right)*\hat{T}^{\text{b}}_4 & \text{TD}\end{cases},\quad k=1,2 \tag{3.136}$$

The functions $\hat{T}_m$ are obtained using the Duhamel theorem and written in the form:

$$\begin{aligned}\frac{\text{d}}{\text{d}t}(T_m(t)*f(\vec{r},t))&=\frac{\text{d}T_m(t)}{\text{d}t}*f(\vec{r},t)+T_m(0)f(\vec{r},t)=\frac{\text{d}T_m(t)}{\text{d}t}*f(\vec{r},t)\\&=\hat{T}_m(t)*f(\vec{r},t),\quad m=2,\,3,\,4\end{aligned} \tag{3.137}$$

$$\hat{T}^{\text{b}}_2=\frac{\text{d}T^{\text{b}}_2}{\text{d}t}=-(4\pi t^3)^{-1/2}\quad\text{(this is Equation 2.217)}$$

$$\hat{T}^{\text{b}}_3=\frac{\text{d}T^{\text{b}}_3}{\text{d}t}=U'=T^{\text{b}}_1 \tag{3.138}$$

$$\hat{T}^{\text{b}}_4=\frac{\text{d}T^{\text{b}}_4}{\text{d}t}=(\pi t)^{-1/2}=T^{\text{b}}_2 \tag{3.139}$$

The operator-function F may be called a *surface impedance function* since it describes the perturbation of the external field surrounding the body due to the field diffusion into the body and dissipation of its energy by the body. The first term on the right-hand side of Equation 3.84 gives the contribution from the field diffusion in the direction normal to the planar surface. The second and third terms describe the field diffusion in the direction normal to the curvilinear surface. The fourth term takes into account the field diffusion in the directions tangential to the planar surface. The functions T_m or $\hat{T}_m$ demonstrate the evolution of these processes in time.

3.7 Surface Impedance near Corners and Edges

As was mentioned in the Introduction chapter, the surface impedance concept has been originally developed under the assumption that the electromagnetic field distribution in the conductor may be described by a one-dimensional equation in the direction normal to the conductor surface. Clearly, this approximation is only valid for smooth bodies. To modify the SIBC near a 90°-conducting wedge, infinitely long in the z-direction (two-dimensional problem), Deeley [6] postulated that the transverse magnetic field can be described in terms of the superposition of two modes:

$$H_x = H_{xo}\exp(-\zeta y); \quad H_y = H_{yo}\exp(-\zeta x), \quad \zeta^2 = -j\omega\sigma\mu \tag{3.140}$$

where H_{xo} and H_{yo} are assumed to be constant. Jingguo and Lavers [7] generalized Deeley's approach for an arbitrary corner angle.

However, rigorous consideration of this problem requires the field equations being solved in both the conducting and nonconducting regions near the corner. It is well known that in the classical limiting case of a perfect conductor, the distribution of the magnetic field along the surface of the edge is singular and can be described by the following function [8–11]:

$$H_\rho = C\rho^\gamma \tag{3.141}$$

where
- C is a known coefficient
- ρ and α are polar coordinates (see Figure 3.1)
- $\gamma = -1/3$ for a 90° edge

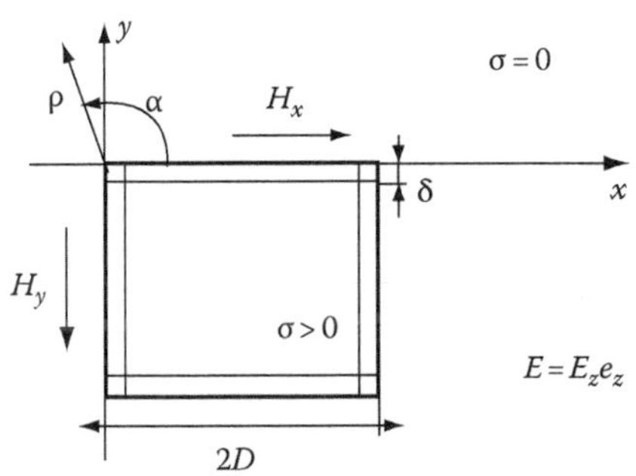

FIGURE 3.1
Derivation of the SIBCs near corners: geometry of the problem.

It is natural to suppose that discontinuities also occur near imperfectly conducting wedges. Clearly then, the assumption in [6] is not correct because it ignores the singularity.

In the present work, we apply the approach previously used to obtain SIBCs of high order of approximation, namely, the use of the perturbation technique in the small parameter equal to the ratio of the skin depth δ and characteristic size D of the conducting domain. The solution of the problem for the perfect conductor is naturally considered as the zero-order approximation. In the first approximation, the distribution of the magnetic field along the perfectly conducting edge is used as a boundary condition for the field diffusion problem inside the conductor. Thus the solution of this problem gives the correction to the zero-order approximation by taking into account the finite conductivity of the body. In other words, we seek a generalization of the solution for the perfectly conducting edge.

We consider an eddy current problem consisting of a rectangular conductor in the external electric field directed along the z-axis so that the magnetic field has only x- and y-components (Figure 3.1). We also assume that the dimension D of the conductor remains small compared to the wave length λ. In this case, the distributions of the electric and magnetic fields in the conducting and nonconducting regions can be described by the Maxwell equations in the following form:

Conductor:

$$\frac{\partial H_y}{\partial x} - \frac{\partial H_x}{\partial y} = \sigma E \tag{3.142}$$

$$j\omega\mu H_x = \frac{\partial E}{\partial y} \tag{3.143}$$

$$j\omega\mu H_y = -\frac{\partial E}{\partial x} \tag{3.144}$$

Dielectric:

$$\frac{\partial H_y}{\partial x} - \frac{\partial H_x}{\partial y} = 0 \tag{3.145}$$

$$j\omega\mu H_x = \frac{\partial E}{\partial y} \tag{3.146}$$

$$j\omega\mu H_y = -\frac{\partial E}{\partial x} \tag{3.147}$$

Boundary conditions:

$$\vec{H}^{\text{cond}}\Big|_{\text{surf}} = \vec{H}^{\text{diel}}\Big|_{\text{surf}}; \quad \vec{E}^{\text{cond}}\Big|_{\text{surf}} = \vec{E}^{\text{diel}}\Big|_{\text{surf}} \tag{3.148}$$

Assuming that the condition in Equation 2.1 is satisfied, we seek the surface impedance in the region near an edge, i.e., we need to calculate the following functions:

$$Z_x = \left[\frac{E_z}{H_x}\right]_{\alpha=2\pi}; \quad Z_y = \left[\frac{E_z}{H_y}\right]_{\alpha=3\pi/2} \quad \text{for} \quad \rho \ll D \tag{3.149}$$

where ρ is the distance from the edge.

Following the perturbation theory, we now switch to dimensionless variables by choosing scale factors defined in Equations 3.13 and 3.14 for the magnetic and electric fields, respectively. The skin depth δ is a natural choice as scale factor for the polar coordinate ρ (as well as both Cartesian coordinates x and y). With nondimensional variables, equations in Equations 3.142 through 3.144 take the form:

$$\tilde{E} = \tilde{p}\left(\frac{\partial \tilde{H}_y}{\partial \tilde{x}} - \frac{\partial \tilde{H}_x}{\partial \tilde{y}}\right) \tag{3.150}$$

$$\tilde{p}\tilde{k}^2\tilde{H}_x = \frac{\partial \tilde{E}}{\partial \tilde{y}} \tag{3.151}$$

$$\tilde{p}\tilde{k}^2\tilde{H}_y = -\frac{\partial \tilde{E}}{\partial \tilde{x}} \tag{3.152}$$

$$\tilde{p} = \delta/D = \sqrt{2/(\omega\sigma\mu D^2)} \ll 1 \tag{3.153}$$

where $\tilde{k}^2 = 2j$. Substituting the scale factors into Equation 3.141 gives

$$\tilde{H}_\rho I D^{-1} = C\tilde{\rho}^\gamma(\tilde{p}D)^\gamma \quad \Leftrightarrow \quad \tilde{H}_\rho = \tilde{p}^\gamma\tilde{\rho}^\gamma C I^{-1} D^{\gamma+1} \tag{3.154}$$

Introducing the following scale factor for C

$$C = ID^{-\gamma-1} \tag{3.155}$$

we represent Equation 3.141 in the form:

$$\tilde{H}_\rho = \tilde{p}\tilde{\rho}^\gamma\tilde{C} \tag{3.156}$$

As a result of transfer to nondimensional variables, the small parameter $\tilde{p}$ appears in the governing equations. We will seek the magnetic and electric fields in the conducting region in the form of power series in the small parameter $\tilde{p}$:

$$\vec{\tilde{H}} = \tilde{p}^\gamma\vec{\tilde{H}}_0 + \tilde{p}^{\gamma+1}\vec{\tilde{H}}_1 + \cdots; \quad \vec{\tilde{E}} = \tilde{p}^\gamma\vec{\tilde{E}}_0 + \tilde{p}^{\gamma+1}\vec{\tilde{E}}_1 + \cdots \tag{3.157}$$

Since $\gamma+1>0$ (recall that $\gamma>0$) the terms in Equation 3.157, beginning from the second vanish when $\tilde{p}=0$. Thus the first terms $\tilde{H}_0$ and $\tilde{E}_0$ in Equation 3.157 describe the field distribution in the PEC limit whereas the others take into account the field diffusion into the conductor. Substituting Equation 3.157 into Equations 3.150 through 3.152, and 3.156 and equating the coefficients of equal powers of $\tilde{p}$, we obtain the following equations for the coefficients in Equation 3.157:

$\tilde{p}^{\gamma}$:

$$\tilde{E}_0 = 0 \tag{3.158}$$

$$\left(\vec{\tilde{H}}_0\right)_\rho\Big|_{\text{surface}} = \tilde{\rho}^{\gamma}\tilde{C} \tag{3.159}$$

$\tilde{p}^{\gamma+1}$:

$$\tilde{E}_1 = \frac{\partial(\tilde{H}_0)_y}{\partial\tilde{x}} - \frac{\partial(\tilde{H}_0)_x}{\partial\tilde{y}} \tag{3.160}$$

$$\tilde{k}\left(\vec{\tilde{H}}_0\right)_x = \partial\tilde{E}_1/\partial\tilde{y} \tag{3.161}$$

$$\tilde{k}\left(\vec{\tilde{H}}_0\right)_y = -\partial\tilde{E}_1/\partial\tilde{x} \tag{3.162}$$

Equation 3.158 indicates that the electric field is zero everywhere inside the perfect conductor including on the surface. Therefore, the expansions in Equation 3.157 have the same physical meaning as in the classical surface impedance theory for smooth surfaces in spite of the fact that in the latter case $\gamma=0$ in the power series. To obtain $\tilde{E}_1$ using Equation 3.160, we must know the normal derivatives of the functions $(\vec{\tilde{H}}_0)_x$ and $(\vec{\tilde{H}}_0)_y$ along the conductor's surface.

For this purpose we consider the boundary value problem of distribution of the magnetic field inside the conductor. The governing equations are

$$\nabla^2 H_x = j\omega\sigma\mu H_x \tag{3.163}$$

$$\nabla^2 H_y = j\omega\sigma\mu H_y \tag{3.164}$$

By switching in Equations 3.163 and 3.164 to nondimensional variables and substituting Equation 3.157, the following equations are obtained:

$$\Delta\left(\vec{\tilde{H}}_0\right)_x = \tilde{k}^2\left(\vec{\tilde{H}}_0\right)_x \tag{3.165}$$

$$\Delta\left(\vec{\tilde{H}}_0\right)_y = \tilde{k}^2\left(\vec{\tilde{H}}_0\right)_y \tag{3.166}$$

The problems in Equations 3.165 and 3.166 can be supplemented by the following boundary conditions based on Equation 3.159:

$$\left.\begin{array}{l} \alpha = 2\pi\colon \left(\vec{\tilde{H}}_0\right)_x = \tilde{\rho}^{\gamma}\tilde{C};\quad \alpha = 3\pi/2\colon \left(\vec{\tilde{H}}_0\right)_x = 0 \\ \tilde{\rho} = 0\colon \left(\vec{\tilde{H}}_0\right)_x \to \infty;\quad \tilde{\rho} \to \infty\colon \left(\vec{\tilde{H}}_0\right)_x \to 0 \end{array}\right\} \tag{3.167}$$

$$\left.\begin{array}{l} \alpha = 2\pi\colon \left(\vec{\tilde{H}}_0\right)_y = 0;\quad \alpha = 3\pi/2\colon \left(\vec{\tilde{H}}_0\right)_y = \tilde{\rho}^{\gamma}\tilde{C} \\ \tilde{\rho} = 0\colon \left(\vec{\tilde{H}}_0\right)_y \to \infty;\quad \tilde{\rho} \to \infty\colon \left(\vec{\tilde{H}}_0\right)_y \to 0 \end{array}\right\} \tag{3.168}$$

In Equations 3.167 and 3.168 we used the fact that the normal magnetic field vanishes on the surface of a perfect conductor. Note that the formulations in Equations 3.165 and 3.168 can provide only approximate solutions of the problem of the magnetic field diffusion inside a *real* conductor near the edge because the boundary conditions for the *ideal* conductor are used. This means that terms containing $\tilde{H}_1$ have been neglected, but it is a suitable approximation for calculation of $\vec{\tilde{E}}_1$ as follows from Equations 3.158 through 3.162.

Equations 3.165 and/or 3.167 and Equations 3.166 and/or 3.168 are similar so we can restrict ourselves by consideration of the problem for $(\tilde{H}_0)_x$. Let $f(\alpha)$ be an integrable function. Then it can be proven that Equation 3.165 or Equation 3.167 is equivalent to the following boundary value problem in the domain *ABcDE* (Figure 3.2):

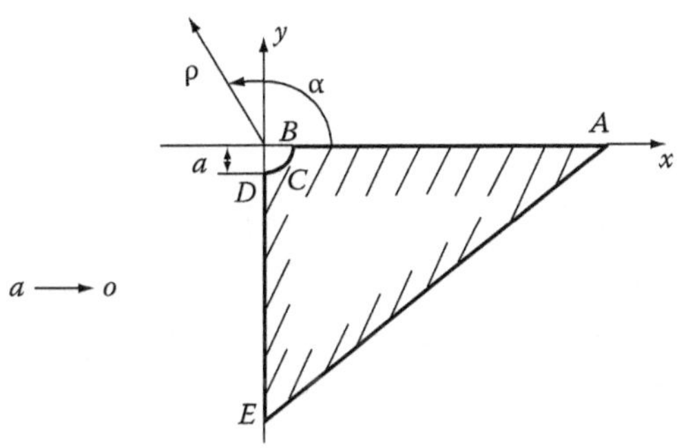

FIGURE 3.2
The boundary value problem under consideration.

$$\frac{\partial^2\left(\vec{\tilde{H}}_0\right)_x}{\partial\tilde{\rho}^2}+\frac{1}{\tilde{\rho}}\frac{\partial\left(\vec{\tilde{H}}_0\right)_x}{\partial\tilde{\rho}}+\frac{1}{\tilde{\rho}^2}\frac{\partial^2\left(\vec{\tilde{H}}_0\right)_x}{\partial\alpha^2}-k^2\left(\vec{\tilde{H}}_0\right)_x=0$$
$$\left(\vec{\tilde{H}}_0\right)_x\Big|_{\alpha=2\pi}=\tilde{C}\tilde{\rho}^{\gamma};\quad \left(\vec{\tilde{H}}_0\right)_x\Big|_{\alpha=3/2\pi}=0 \tag{3.169}$$
$$\left(\vec{\tilde{H}}_0\right)_x\Big|_{\rho=a}=f(\alpha);\quad \left(\vec{\tilde{H}}_0\right)_x\Big|_{\rho=\infty}=0$$

provided that the radius a of the arc BcD tends to zero. The problem in Equation 3.169 has been solved using the Green function technique and the solution can be represented in the form:

$$\left(\vec{\tilde{H}}_0\right)_x=\Phi_1+\Phi_2 \tag{3.170}$$

where

$$\Phi_1=\kappa\sum_{n=1}^{\infty}\frac{K_{\lambda_n}(\kappa\tilde{\chi})}{K_{\lambda_n}(\kappa)}F_n\sin(2\alpha n/3) \tag{3.171}$$

$$\Phi_2=\tilde{C}\kappa e^{-i\pi/2}\sum_{n=1}^{\infty}\frac{K_{\lambda_n}(\kappa\tilde{\chi})}{K_{\lambda_n}(\kappa)}\lambda_n\sin(2\alpha n/3)\int_1^{\chi}\chi'\gamma(I_{\lambda_n}(\kappa\tilde{\chi}')K_{\lambda_n}(\kappa)-I_{\lambda_n}(\kappa)K_{\lambda_n}(\kappa\tilde{\chi}'))d\chi'$$
$$+\tilde{C}\kappa i\sum_{n=1}^{\infty}\lambda_n\sin(2\alpha n/3)\left(I_{\lambda_n}(\kappa\tilde{\chi})-\frac{I_{\lambda_n}(\kappa)}{K_{\lambda_n}(\kappa)}K_{\lambda_n}(\kappa\chi)\right)\int_{\chi}^{\infty}\tilde{\chi}'\gamma K_{\lambda_n}(\kappa\tilde{\chi}')d\tilde{\chi}' \tag{3.172}$$

$$\lambda_n=2n/3;\quad \tilde{\chi}=\rho/a;\quad \kappa=ka \tag{3.173}$$

F_n is a Fourier transform of the function $f(\alpha)$

I_m and K_m are the m-order modified Bessel and Hankel functions, respectively

It can be shown that Γ_1 vanishes when $a\to 0$ and Equation 3.170 takes the form:

$$\left(\vec{\tilde{H}}_0\right)_x\Big|_{a\to 0}=\sum_{n=1}^{\infty}\frac{\pi n}{\alpha}e^{-i\pi/2}\sin(2n\alpha/3)\tilde{C}\left\{K_{\lambda_n}(k\tilde{\rho})\int_0^{\rho}\tilde{\rho}'\gamma I_{\lambda_n}(k\tilde{\rho}')d\tilde{\rho}'+I_{\lambda_n}(k\tilde{\rho})\int_{\rho}^{\infty}\tilde{\rho}'\gamma K_{\lambda_n}(k\tilde{\rho}')d\tilde{\rho}'\right\} \tag{3.174}$$

The problem described in Equation 3.166/3.168 has been solved in the same way and the result can be written in the form:

$$\left(\vec{\tilde{H}}_0\right)_y\Big|_{a\to 0} = \sum_{n=1}^{\infty}\frac{2n}{3}e^{-i\pi/2}(-1)^n\sin(2n\alpha/3)\tilde{C}\left\{K_{\lambda_n}(k\tilde{\rho})\int_0^{\rho}\tilde{\rho}'\gamma I_{\lambda_n}(k\tilde{\rho}')d\tilde{\rho}' + I_{\lambda_n}(k\tilde{\rho})\int_{\rho}^{\infty}\tilde{\rho}'\gamma K_{\lambda_n}(k\tilde{\rho}')d\tilde{\rho}'\right\} \tag{3.175}$$

Asymptotic analysis of Equations 3.174 and 3.175 yields distributions of the functions $(\vec{\tilde{H}}_0)_x$ and $(\vec{\tilde{H}}_0)_y$ in the conductor, near the corner:

$$\left(\vec{\tilde{H}}_0\right)_x = \tilde{C}\left(\frac{2}{k\tilde{\rho}}\right)^{1/3}\frac{\cos(\alpha/3)}{\cos(2\pi/3)} \tag{3.176}$$

$$\left(\vec{\tilde{H}}_0\right)_y = \tilde{C}\left(\frac{2}{k\tilde{\rho}}\right)^{1/3}\frac{\sin((2\pi-\alpha)/3)}{\cos(2\pi/3)} \tag{3.177}$$

Substituting Equations 3.176 and 3.177 into Equation 3.160, the electric field is obtained as

$$\tilde{E}_1 = \frac{2}{3}\tilde{C}\tilde{\rho}^{-4/3}\left(\frac{2}{3}\sin\alpha\cos\frac{\alpha}{3} + \frac{2}{3}\sin\frac{2\pi-\alpha}{3}\cos\alpha + \frac{1}{3}\sin\frac{4\alpha}{3} + \frac{1}{3}\sin\frac{2\pi-4\alpha}{3}\right) \tag{3.178}$$

To illustrate the preceding theory, consider the problem of diffraction of plane waves by an imperfectly conducting edge as shown in Figure 3.3. Solution of the problem for a perfect conductor is known [8] and can be represented in the form:

$$H_x\big|_{\rho\ll 1} \approx \frac{\sqrt{3\pi k}}{\Gamma(2/3)}H^{\text{inc}}e^{i\pi/3}\left(\frac{2}{k\rho}\right)^{1/3}\sin\frac{2\psi^{\text{inc}}}{2}\cos\frac{\alpha}{3} \tag{3.179}$$

$$H_y\big|_{\rho\ll 1} \approx \frac{\sqrt{3\pi k}}{\Gamma(2/3)}H^{\text{inc}}e^{i\pi/3}\left(\frac{2}{k\rho}\right)^{1/3}\sin\frac{2\psi^{\text{inc}}}{2}\sin\frac{\alpha}{3} \tag{3.180}$$

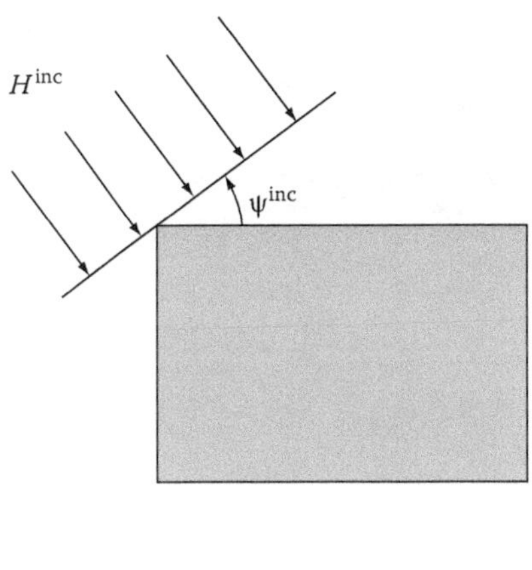

FIGURE 3.3
Diffraction of a plane wave by a conducting edge.

Thus, in this case the coefficient C takes the form:

$$C = H^{\text{inc}} e^{i\pi/3} \frac{\sqrt{3\pi k}}{\Gamma(2/3)} \sin\left(2\psi^{\text{inc}}/3\right) \tag{3.181}$$

where Γ denotes the gamma-function. The distribution of the magnetic field intensity in the conducting and dielectric regions near the edge was calculated using Equations 3.176 and 3.177 and shown in Figures 3.4 and 3.5.

To calculate the surface impedance, we switch to nondimensional variables and substitute the series (Equation 3.157), resulting in the following formula:

$$\tilde{Z}_i = \frac{\tilde{E}}{H_i} = \frac{\tilde{p}\tilde{E}_1 + O(\tilde{p}^2)}{\left(\vec{\tilde{H}}_0\right)_i + \tilde{p}\left(\vec{\tilde{H}}_1\right)_i + O(\tilde{p}^2)} = \frac{\tilde{p}\tilde{E}_1}{\left(\vec{\tilde{H}}_0\right)_i} + O(\tilde{p}^2), \quad i = x, y \tag{3.182}$$

Substitution of Equations 3.176 through 3.178 into Equation 3.182 leads to the following expressions:

$$\tilde{Z}_x = -\tilde{p}\frac{\tan(\pi/6)}{\tilde{\rho}}; \quad \tilde{Z}_y = \tilde{p}\frac{\tan(\pi/6)}{\tilde{\rho}} \tag{3.183}$$

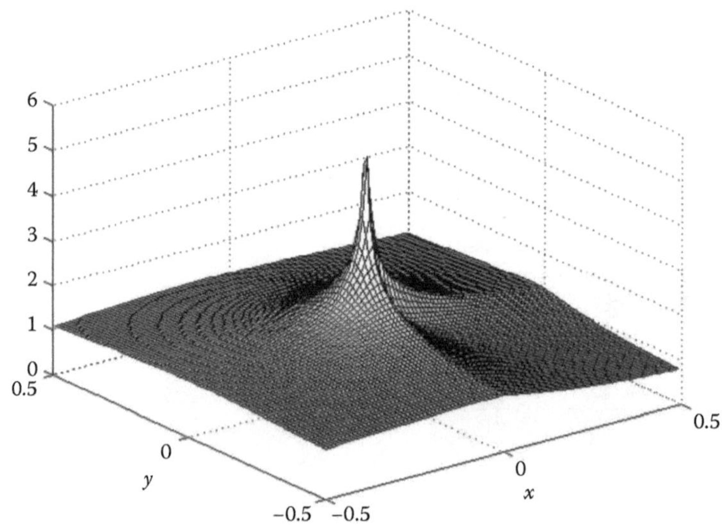

FIGURE 3.4
The magnetic field intensity near the corner.

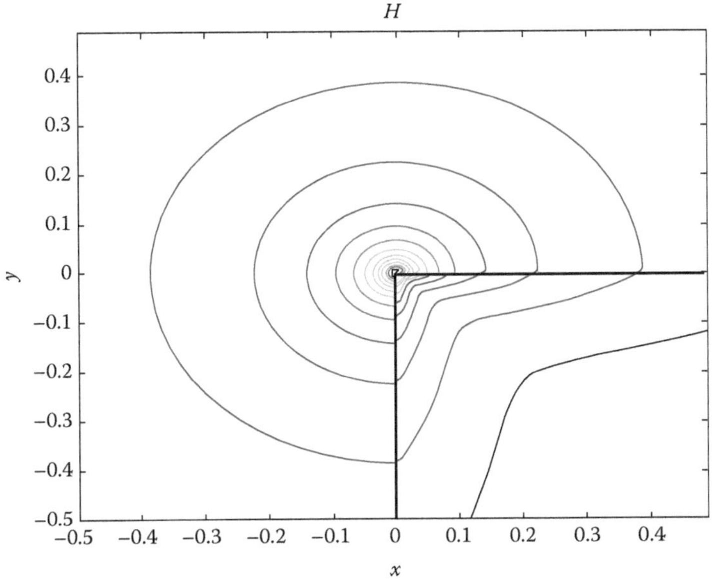

FIGURE 3.5
The magnetic field intensity near the corner—contour plot.

Returning in Equation 3.183 to dimensional variables, we finally obtain

$$Z_x = -\frac{\tan(\pi/6)}{\sigma\rho}; \quad Z_y = \frac{\tan(\pi/6)}{\sigma\rho} \tag{3.184}$$

References

1. C.R.I. Emson and J. Simkin, An optimal method for 3-D eddy currents, *IEEE Transactions on Magnetics*, MAG-21(6), November 1985, 2231–2234.
2. W.M. Rucker and K.R. Richter, A BEM code for 3-D eddy current calculations, *IEEE Transactions on Magnetics*, 26(2), March 1990, 462–465.
3. H. Igarashi, A. Kost, and T. Honma, A three dimensional analysis of magnetic field around thin magnetic conductive layer using vector potential, *IEEE Transactions on Magnetics*, 34(5), September 1998, 2539–2542.
4. M. Cao and P.P. Biringer, BIE formulations for skin and proximity effect problems of parallel conductors, *IEEE Transactions on Magnetics*, 26(5), September 1990, 2768–2770.
5. D. Rodger and H.C. Lai, A surface impedance method for 3D time transient problems, *IEEE Transactions on Magnetics*, 35(3), May 1999, 1369–1371.

6. E.M. Deeley, Surface impedance near edges and corners in three dimensional media, *IEEE Transactions on Magnetics*, 26(2), March 1990, 712–714.
7. W. Jingguo and J.D. Lavers, Modified surface impedance boundary conditions for 3D eddy current problems, *IEEE Transactions on Magnetics*, 29(2), March 1993, 1826–1829.
8. G.A. Greenberg, On a method of solving the fundamental problem of electrostatics and allied problems, *Russian Journal of Experimental and Theoretical Physics*, Part I, 8(3), March 1938, 221–252; Part II, 9(6), June 1939, 725–728 (in Russian).
9. J. Meixner, The behavior of electromagnetic fields at edges, *IEEE Transactions on Antennas and Propagation*, 20(4), July 1972, 442–446.
10. J. Van Bladel, *Electromagnetic Fields*, Chapters 5 and 12, *A SUMMA Book*, Hemisphere, Washington, DC, 1985.
11. K. Nikoskinen and I. Lindell, Image solution for Poisson's equation in wedge geometry, *IEEE Transactions on Antennas and Propagation*, 43(2), February 1995, 179–187.

4

Calculation of the Electromagnetic Field Characteristics in the Conductor's Skin Layer

4.1 Introduction

In Chapters 2 and 3, we derived the distributions of the magnetic field, electric field, and vector potential "inside" the conductor as a function of the tangential magnetic field at "the conductor's surface." This information is sufficient for calculation of such important practical quantities as current density, energy stored in the electromagnetic field, inductance, and power losses. In the case of constant material properties of the conductor, calculation of the quantities mentioned is possible without numerical consideration of the conducting domain. In other words, the surface impedance concept should provide not only local boundary relations between fields or potentials, but also analytical formulae for calculation of "integral" or "volume" quantities using only the distributions of the EM field at the surface of the conductor.

Once the distribution of the electric and magnetic fields inside a domain V is obtained, energy-related quantities such as ohmic power loss (Joule loss) P_W and the magnetic field energy W_m associated with V can be calculated as follows:

$$P_W = \int_V \sigma^{-1} J^2 \, dv \tag{4.1}$$

$$W_m = \frac{1}{2} \int_V \vec{H} \cdot \vec{B} \, dv \tag{4.2}$$

where $\vec{J}$ and $\vec{B}$ are the current density and the magnetic flux density, respectively. In circuit theory, these quantities are defined as

$$P_W = I^2 R \tag{4.3}$$

$$W_m = \frac{1}{2} L I^2 \tag{4.4}$$

where R and L are the resistance and inductance of the conductors, respectively.

From a computational point of view, it is very attractive to solve the problem of electromagnetic field distribution in the domain under consideration, use the solution to calculate "extract" lumped parameters and then apply them in circuit analysis. However, any combination of field and circuit representations must be done with great care because the procedure is applicable only in particular cases. For example, it is impossible to guarantee in the general case that resistance defined using Equations 4.3 and 4.1 can be then used to calculate voltage drop between two selected points. Here, we will restrict ourselves to the practically important case of long parallel conductors where description of the electromagnetic processes using lumped parameters is possible.

In this chapter, we demonstrate how to calculate the following quantities based on known distributions of the tangential magnetic field along the conductor's surface in the practically important case of time-harmonic fields:

- Distribution of the electric and magnetic fields across the skin layer
- Power losses, energy in the magnetic field, and force acting on the conductor
- Lumped parameters such as resistance and internal inductance per unit length of the conductor

4.2 Distributions across the Skin Layer

In the previous chapters, we focused on derivation of relations between electromagnetic quantities at the interface between the dielectric and the conductor. Since the goals of this chapter require performing integrations over the conductor's volume, we start from the distributions of magnetic and electric fields, and the magnetic vector potential throughout the skin layer of a good conductor (where displacement current may be neglected).

Consider a straight long current-carrying conductor of constant cross-section as shown in Figure 4.1. In this case, the electric current flowing in

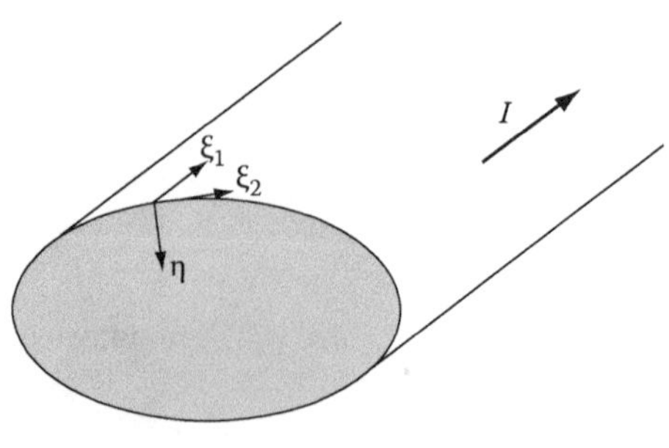

FIGURE 4.1
The two-dimensional problem of electromagnetic field diffusion into a long straight cylindrical conductor of constant cross-section.

the conductor has only a longitudinal component and the problem of the electromagnetic field distribution inside the conductor can be considered as two dimensional in its cross section. Let coordinate ξ_1 of our surface-related coordinate system (ξ_1, ξ_2, η) be directed along the conductor so that

$$\frac{\partial}{\partial \xi_1} = 0; \quad d_1 \to \infty \tag{4.5}$$

Then, the electric and magnetic field vectors can be represented in the form:

$$\vec{E} = \vec{e}_{\xi_1} E; \quad \vec{H} = \vec{e}_{\xi_2} H_{\xi_2} + \vec{e}_{\eta} H_{\eta} \tag{4.6}$$

Substituting Equation 4.5 into the "toolbox" formulae (Equations 2.92 through 2.94, 2.158, 2.162, and 2.166), we reduce them to the following form:

First-order approximation:

$$\dot{f}_{\xi_k} = \dot{f}^b_{\xi_k} \exp(-(1+j)\tilde{\eta}), \quad k = 1,2 \tag{4.7}$$

$$\dot{g}_{\xi_{3-k}} = (-1)^k (1+j) \dot{f}^b_{\xi_k} \exp(-(1+j)\tilde{\eta}), \quad k = 1,2 \tag{4.8}$$

Second-order approximation:

$$\dot{f}_{\xi_k} = \left(\dot{f}^b_{\xi_k} + \tilde{p}\tilde{\eta} \dot{f}^b_{\xi_k} \frac{1}{2\tilde{d}_2} \right) \exp(-(1+j)\tilde{\eta}), \quad k = 1,2 \tag{4.9}$$

$$\dot{g}_{\xi_1} = (1+j) \left[1 + \tilde{p}\tilde{\eta} \frac{1}{2\tilde{d}_2} + \tilde{p} \frac{1-j}{2} \frac{1}{2\tilde{d}_2} \right] \dot{f}^b_{\xi_2} \exp(-(1+j)\tilde{\eta}) \tag{4.10}$$

$$\dot{g}_{\xi_2} = -(1+j) \left[1 + \tilde{p}\tilde{\eta} \frac{1}{2\tilde{d}_2} - \tilde{p} \frac{1-j}{2} \frac{1}{2\tilde{d}_2} \right] \dot{f}^b_{\xi_1} \exp(-(1+j)\tilde{\eta}) \tag{4.11}$$

Third-order approximation:

$$\dot{f}_{\xi_1} = \left[\dot{f}^b_{\xi_1} + \tilde{p}\tilde{\eta} \dot{f}^b_{\xi_1} \frac{1}{2\tilde{d}_2} + \tilde{p}^2 \tilde{\eta}^2 \dot{f}^b_{\xi_1} \frac{3}{8\tilde{d}_2^2} + \tilde{p}^2 \frac{1-j}{2} \tilde{\eta} \left(\dot{f}^b_{\xi_1} \frac{1}{8\tilde{d}_2^2} + \frac{1}{2} \frac{\partial^2 \dot{f}^b_{\xi_1}}{\partial \tilde{\xi}_2^2} \right) \right] \exp(-(1+j)\tilde{\eta}) \tag{4.12}$$

$$\dot{f}_{\xi_2} = \left[\dot{f}^b_{\xi_2} + \tilde{p}\tilde{\eta} \dot{f}^b_{\xi_2} \frac{1}{2\tilde{d}_2} + \tilde{p}^2 \tilde{\eta}^2 \dot{f}^b_{\xi_2} \frac{3}{8\tilde{d}_2^2} + \tilde{p}^2 \frac{1-j}{2} \tilde{\eta} \left(-\dot{f}^b_{\xi_2} \frac{3}{8\tilde{d}_2^2} + \frac{1}{2} \frac{\partial^2 \dot{f}^b_{\xi_2}}{\partial \tilde{\xi}_2^2} \right) \right] \exp(-(1+j)\tilde{\eta}) \tag{4.13}$$

$$\dot{g}_{\xi_1} = (1+j)\left[\dot{f}^{\mathrm{b}}_{\xi_2}\left(1+\tilde{p}\tilde{\eta}\frac{1}{2\tilde{d}_2}+\tilde{p}^2\tilde{\eta}^2\frac{3}{8\tilde{d}_2^2}\right)+\tilde{p}\frac{1-j}{2}\left(\dot{f}^{\mathrm{b}}_{\xi_2}\frac{1}{2\tilde{d}_2}+\tilde{p}\tilde{\eta}\dot{f}^{\mathrm{b}}_{\xi_2}\frac{3}{2\tilde{d}_2^2}+\tilde{p}\tilde{\eta}\frac{1}{2}\frac{\partial^2\dot{f}^{\mathrm{b}}_{\xi_2}}{\partial\tilde{\xi}_2^2}\right)\right.$$
$$\left.-\tilde{p}^2\frac{j}{2}\left(\dot{f}^{\mathrm{b}}_{\xi_2}\frac{3}{4\tilde{d}_2^2}+\frac{\partial^2\dot{f}^{\mathrm{b}}_{\xi_2}}{\partial\tilde{\xi}_2^2}\right)\right]\exp(-(1+j)\tilde{\eta}) \tag{4.14}$$

$$\dot{g}_{\xi_2} = -(1+j)\left[\dot{f}^{\mathrm{b}}_{\xi_1}\left(1+\tilde{p}\tilde{\eta}\frac{1}{2\tilde{d}_2}+\tilde{p}^2\tilde{\eta}^2\frac{3}{8\tilde{d}_2^2}\right)-\tilde{p}\frac{1-j}{2}\left(\dot{f}^{\mathrm{b}}_{\xi_1}\frac{1}{2\tilde{d}_2}+\tilde{p}\tilde{\eta}\dot{f}^{\mathrm{b}}_{\xi_1}\frac{5}{2\tilde{d}_2^2}-\tilde{p}\tilde{\eta}\frac{1}{2}\frac{\partial^2\dot{f}^{\mathrm{b}}_{\xi_1}}{\partial\tilde{\xi}_2^2}\right)\right.$$
$$\left.+\tilde{p}^2\frac{j}{2}\left(\dot{f}^{\mathrm{b}}_{\xi_1}\frac{1}{4\tilde{d}_2^2}+\frac{\partial^2\dot{f}^{\mathrm{b}}_{\xi_1}}{\partial\tilde{\xi}_2^2}\right)\right]\exp(-(1+j)\tilde{\eta}) \tag{4.15}$$

Letting $\dot{E}=\dot{f}_{\xi_1}$ and $\dot{H}_{\xi_2}=\dot{f}_{\xi_2}$ in Equations 4.7, 4.9, 4.12, and 4.13 and switching to dimensional variables, we obtain distributions of the tangential electric and magnetic fields as a function of η (across the skin layer) in the various approximations:

Leontovich's approximation:

$$\dot{E} = \dot{E}^{\mathrm{b}}\exp(-\beta\eta) \tag{4.16}$$

$$\dot{H}_{\xi} = \dot{H}^{\mathrm{b}}_{\xi}\exp(-\beta\eta) \tag{4.17}$$

Mitzner's approximation:

$$\dot{E} = \dot{E}^{\mathrm{b}}\left(1+\frac{\eta}{2d}\right)\exp(-\beta\eta) \tag{4.18}$$

$$\dot{H}_{\xi} = \dot{H}^{\mathrm{b}}_{\xi}\left(1+\frac{\eta}{2d}\right)\exp(-\beta\eta) \tag{4.19}$$

Rytov's approximation:

$$\dot{E} = \left[\dot{E}^{\mathrm{b}}+\dot{E}^{\mathrm{b}}\frac{\eta}{2d}+\dot{E}^{\mathrm{b}}\frac{3\eta^2}{8d^2}+\delta\frac{1-j}{2}\eta\left(\dot{E}^{\mathrm{b}}\frac{1}{8d^2}+\frac{1}{2}\frac{\partial^2\dot{E}^{\mathrm{b}}}{\partial\xi^2}\right)\right]\exp(-\beta\eta) \tag{4.20}$$

$$\dot{H}_{\xi} = \left[\dot{H}^{\mathrm{b}}_{\xi}+\dot{H}^{\mathrm{b}}_{\xi}\frac{\eta}{2d}+\dot{H}^{\mathrm{b}}_{\xi}\frac{3\eta^2}{8d^2}+\delta\frac{1-j}{2}\eta\left(-\dot{H}^{\mathrm{b}}_{\xi}\frac{3}{8d^2}+\frac{1}{2}\frac{\partial^2\dot{H}^{\mathrm{b}}_{\xi}}{\partial\xi^2}\right)\right]\exp(-\beta\eta) \tag{4.21}$$

where

$$\beta = \sqrt{j\omega\sigma\mu} = \frac{1+j}{\delta} \tag{4.22}$$

and the following notations have been introduced:

$$\xi = \xi_2; \quad d = d_2 \tag{4.23}$$

By denoting $\dot{E} = \dot{g}_{\xi_1}$ and $\dot{H}_{\xi_2} = \dot{f}_{\xi_2}$ in Equations 4.8, 4.10, 4.11, 4.14, and 4.15, we obtain the tangential electric field distribution in terms of the tangential magnetic field at the interface:

Leontovich's approximation:

$$\dot{E} = \frac{1+j}{\sigma\delta}\dot{H}_\xi^{\mathrm{b}} \exp(-\beta\eta) \tag{4.24}$$

Mitzner's approximation:

$$\dot{E} = \frac{1+j}{\sigma\delta}\left[1 + \frac{\eta}{2d} + \frac{1-j}{4}\frac{\delta}{d}\right]\dot{H}_\xi^{\mathrm{b}} \exp(-\beta\eta) \tag{4.25}$$

Rytov's approximation:

$$\dot{E} = \frac{1+j}{\sigma\delta}\left[\dot{H}_\xi^{\mathrm{b}}\left(1 + \frac{\eta}{2d} + \frac{3\eta^2}{8d^2}\right) + \delta\frac{1-j}{4}\left(\dot{H}_\xi^{\mathrm{b}}\frac{1}{d} + \dot{H}_\xi^{\mathrm{b}}\frac{3\eta}{d^2} + \eta\frac{\partial^2 \dot{H}_\xi^{\mathrm{b}}}{\partial\xi^2}\right)\right.$$
$$\left. - \delta^2\frac{j}{2}\left(\dot{H}_\xi^{\mathrm{b}}\frac{3}{4d^2} + \frac{\partial^2 \dot{H}_\xi^{\mathrm{b}}}{\partial\xi^2}\right)\right]\exp(-\beta\eta) \tag{4.26}$$

Improvement in accuracy of results by taking into account high-order terms at large values of $\tilde{p} = \delta/d$ can be demonstrated through comparison of Equations 4.16 through 4.21 and Equations 4.24 through 4.26 with the exact solution, which is not limited by the skin layer approximation. An exact solution is available for the single current-carrying round conductor, shown in Figure 4.2, where the fields vary only radially and there is no variation in

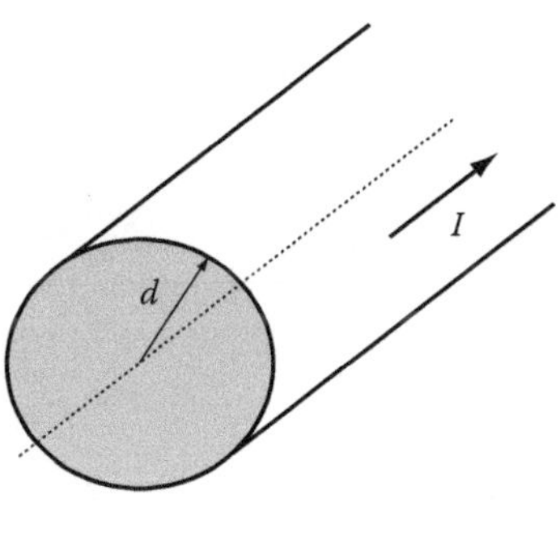

FIGURE 4.2
A solid, round, current-carrying conductor.

the tangential direction (second derivatives in Equations 4.16 through 4.21 and Equations 4.24 through 4.26 vanish).

Due to symmetry, the magnetic field has only one component:

$$\vec{H} = \vec{e}_\xi H_\xi = \vec{e}_\xi H \tag{4.27}$$

The diffusion equations describing the electric and magnetic fields in the common cylindrical coordinates are the following:

$$\frac{\partial^2 \dot{E}}{\partial r^2} + \frac{1}{r}\frac{\partial \dot{E}}{\partial r} = \beta^2 E \tag{4.28}$$

$$\frac{\partial^2 \dot{H}}{\partial r^2} + \frac{1}{r}\frac{\partial \dot{H}}{\partial r} = \left(\beta^2 + \frac{1}{r^2}\right)\dot{H} \tag{4.29}$$

The radial coordinate r is directed from the center of the cylinder to its surface. The coordinates r and η are related as

$$r = d - \eta \tag{4.30}$$

where d is the radius of the cylinder.

Taking into account the fact that the electromagnetic field at the center of the cylinder ($r = 0$) has finite value, the solutions of these equations can be represented in the form:

$$\dot{E}(r) = \dot{E}^{\mathrm{b}} \frac{I_0(\beta r)}{I_0(\beta d)} \tag{4.31}$$

$$\dot{H}(r) = \dot{H}^{\mathrm{b}} \frac{I_1(\beta r)}{I_1(\beta d)} \tag{4.32}$$

Here I_ν is the modified Bessel function of the first kind and order, ν. The distributions of the functions I_ν are shown in Figure 4.3.

The function I_1 can be calculated as the derivative of I_0 with respect to βr:

$$I_1(\beta r) = I_0'(\beta r) \tag{4.33}$$

Note that the solution in Equation 4.32 is equivalent to Equation 2.223. Indeed, comparing Equations 2.222 and 4.22, we get

$$\beta = j\vartheta; \quad I_1(\beta r) = I_1(j\vartheta r) = jJ_1(\vartheta r) \tag{4.34}$$

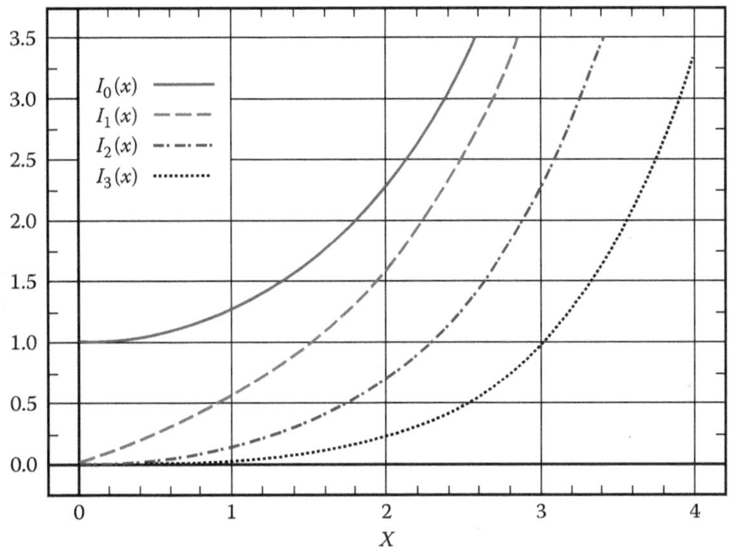

FIGURE 4.3
Distributions of modified Bessel functions of the first kind I_ν, $\nu = 0, 1, 2$, and 3.

where J_1 is the Bessel function of the first kind and first order. With the use of Equation 4.34, we can represent Equation 4.32 in the form:

$$\dot{H}(r) = \dot{H}^{\text{b}} \frac{J_1(\vartheta r)}{J_1(\vartheta d)} \tag{4.35}$$

In the particular case of a single circular cylinder, Equation 3.1 takes the form

$$j\omega\mu\dot{H} = \frac{\mathrm{d}\dot{E}}{\mathrm{d}r} \tag{4.36}$$

Substituting Equation 4.31 into Equation 4.36, we obtain

$$\dot{H} = (j\omega\mu)^{-1} \frac{\dot{E}^{\text{b}}}{I_0(\beta d)} \frac{\mathrm{d}I_0(\beta r)}{\mathrm{d}r} = \frac{\beta}{j\omega\mu} \dot{E}^{\text{b}} \frac{I_1(\beta r)}{I_0(\beta d)} \tag{4.37}$$

Comparison of Equations 4.32 and 4.37 gives

$$\dot{H}^{\text{b}} = \frac{\beta}{j\omega\mu} \dot{E}^{\text{b}} \frac{I_1(\beta d)}{I_0(\beta d)} \quad \text{or} \quad \dot{E}^{\text{b}} = \frac{j\omega\mu}{\beta} \dot{H}^{\text{b}} \frac{I_0(\beta d)}{I_1(\beta d)} \tag{4.38}$$

Therefore

$$\dot{E}(r) = \frac{j\omega\mu}{\beta}\dot{H}^{\mathrm{b}}\frac{I_0(\beta r)}{I_1(\beta d)} = \frac{\beta}{\sigma}\dot{H}^{\mathrm{b}}\frac{I_0(\beta r)}{I_1(\beta d)} = \frac{1+j}{\sigma\delta}\dot{H}^{\mathrm{b}}\frac{I_0(\beta r)}{I_1(\beta d)} \tag{4.39}$$

Figures 4.4 through 4.9 show distributions of the real parts, imaginary parts, and amplitudes of the functions $\dot{E} = f(\dot{E}^{\mathrm{b}}, \tilde{\eta})$ and $\dot{E} = g(\dot{H}^{\mathrm{b}}, \tilde{\eta})$ for three different values of the small parameter $\tilde{p} = \delta/d$, namely 0.1, 0.3, and 0.5. E^{b} and $\delta^{-1}\dot{H}^{\mathrm{b}}$ are assumed to be equal to 1 and the nondimensional variable $\tilde{\eta}$ varies from 0 (surface) to 0.5 (at half the distance from the surface to the center of the cylinder). Each figure gives four curves: approximate solutions in the Leontovich, Mitzner, and Rytov approximation and the exact solution. The disagreement between the curves gives the approximation error in our example. From Figures 4.4 and 4.7, it can be seen that for $\tilde{p} = 0.1$ (very thin skin layer) all curves are very close to each other. This means that the formula (Equation 4.16) of the order of the Leontovich approximation is sufficiently accurate and there is no need to take into account even the radius of curvature of the conductor's surface. The situation naturally changes as the parameter $\tilde{p}$ increases. According to the results shown in Figures 4.5 and 4.8, application of Equation 4.16 in the case $\tilde{p} = 0.3$ leads to significant errors whereas Equation 4.18 obtained in the Mitzner approximation works

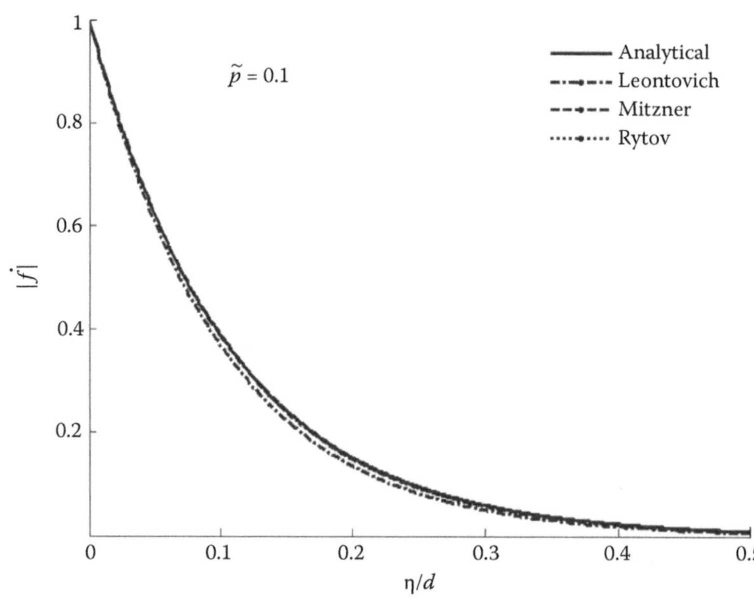

FIGURE 4.4
Amplitude of $\dot{E} = f(\dot{E}^{\mathrm{b}}, \tilde{\eta})$ for $\tilde{p} = 0.1$ and $\dot{E}^{\mathrm{b}} = 1$.

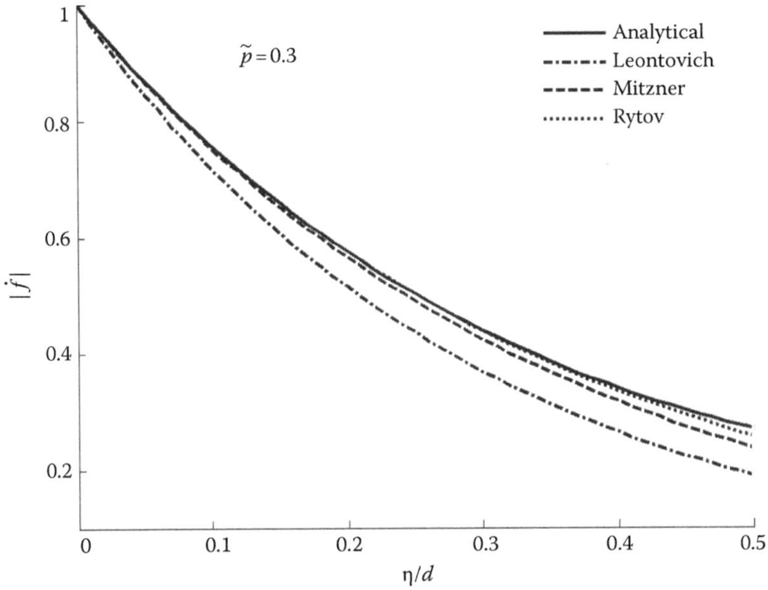

FIGURE 4.5
Amplitude of $\dot{E} = f(\dot{E}^b, \tilde{\eta})$ for $\tilde{p} = 0.3$ and $\dot{E}^b = 1$.

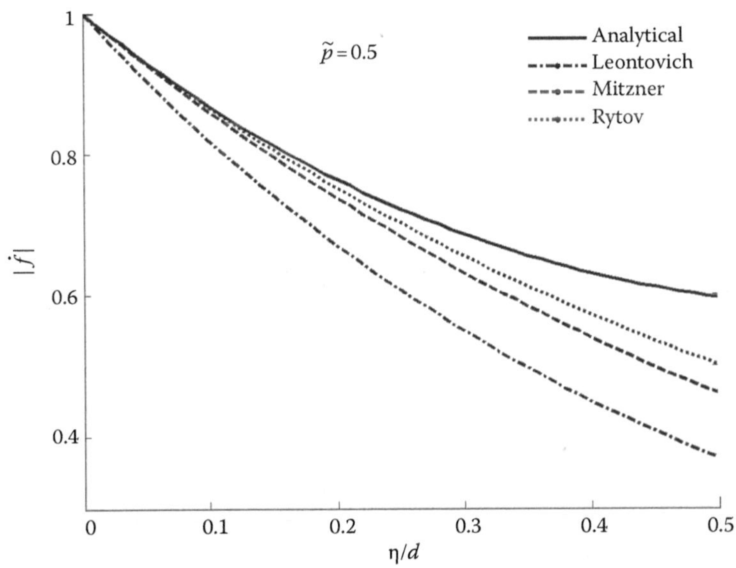

FIGURE 4.6
Amplitude of $\dot{E} = f(\dot{E}^b, \tilde{\eta})$ for $\tilde{p} = 0.5$ and $\dot{E}^b = 1$.

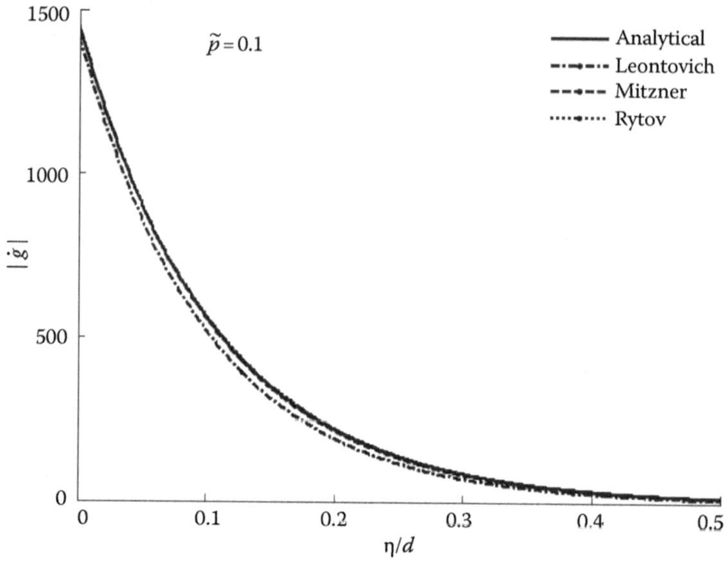

FIGURE 4.7
Amplitude of $\dot{E} = g(\dot{H}^b, \tilde{\eta})$ for $\tilde{p} = 0.1$ and $\dot{H}^b\delta^{-1} = 1$.

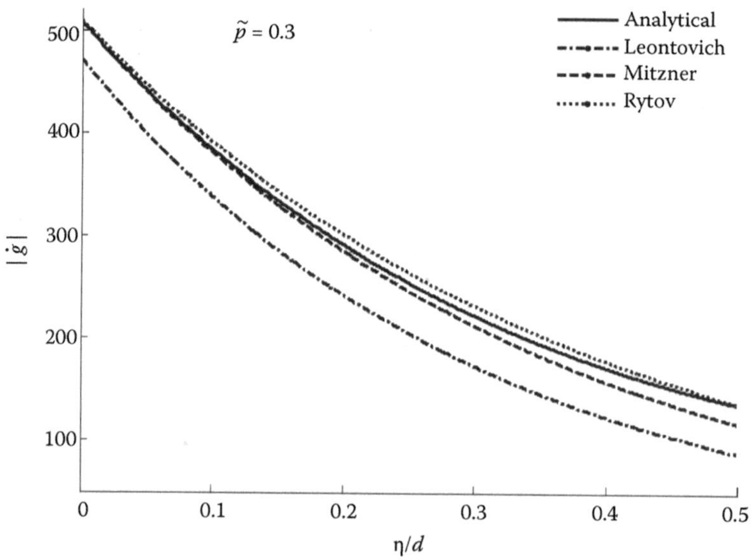

FIGURE 4.8
Amplitude of $\dot{E} = g(\dot{H}^b, \tilde{\eta})$ for $\tilde{p} = 0.3$ and $\dot{H}^b\delta^{-1} = 1$.

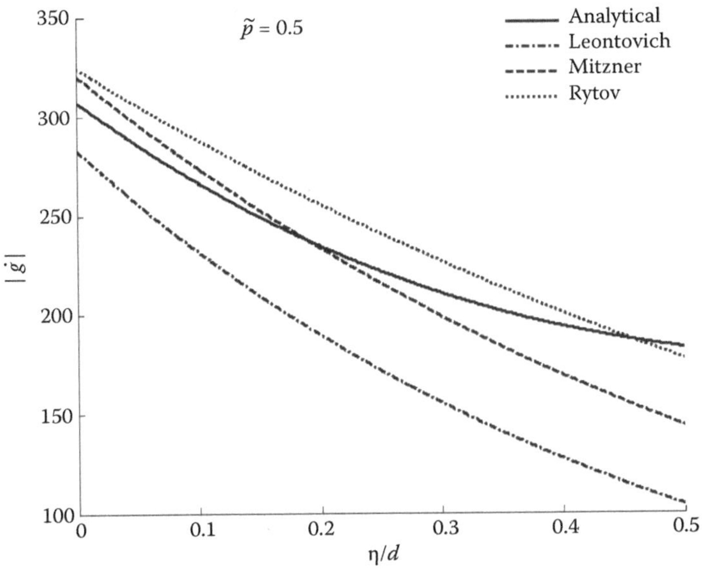

FIGURE 4.9
Amplitude of $\dot{E} = g(\dot{H}^{\mathrm{b}}, \tilde{\eta})$ for $\tilde{p} = 0.3$ and $\dot{H}^{\mathrm{b}}\delta^{-1} = 1$.

satisfactorily, and Equation 4.20 (Rytov's approximation) provides high accuracy. When $\tilde{p} = 0.5$ even the Rytov approximation cannot provide agreement with the exact solution (Figures 4.6 and 4.9). Note that in the latter case one can expect more significant errors because the skin depth is so large that the surface impedance concept should not be generally applied (see Chapter 9 for range of applicability of the surface impedance concept). However, the existence of symmetry in the circular conductor reduces the error. Analysis of the functions $H = f_2(\dot{H}^{\mathrm{b}}, \tilde{\eta})$ and $\dot{E} = g(\dot{H}^{\mathrm{b}}, \tilde{\eta})$ leads to the same conclusions.

It is readily demonstrated that under the conditions of skin effect ($|\beta r| \gg 1$) Equations 4.31, 4.32, and 4.39 lead to Equations 4.16, 4.17, and 4.24, respectively. Taking into account the asymptotic behavior of the Bessel functions at large arguments:

$$I_\nu(z) \approx \frac{\exp(z)}{\sqrt{2\pi z}}, \quad \text{when } |z| \to \infty \tag{4.40}$$

we represent Equation 4.31 in the form:

$$\dot{E}(r) = \dot{E}^{\mathrm{b}}\sqrt{\frac{d}{r}}\exp(\beta(r-d)) = \dot{E}^{\mathrm{b}}\sqrt{\frac{d}{d-\eta}}\exp(-\beta\eta) \approx \dot{E}^{\mathrm{b}}\exp(-\beta\eta) \tag{4.41}$$

Equations 4.32 and 4.39 are transformed in the same way. Note that application of Equation 4.39 for $|\beta r| \gg 1$ yields the classical Leontovich SIBC:

$$\dot{E}^{\mathrm{b}} = \frac{1+j}{\sigma\delta}\dot{H}^{\mathrm{b}}\frac{I_0(\beta r)}{I_1(\beta d)} \approx \frac{1+j}{\sigma\delta}\dot{H}^{\mathrm{b}} \tag{4.42}$$

Since under the conditions of skin effect problems the current is concentrated near the surface, it is natural to introduce the "surface current" density $\vec{K}$ as

$$\vec{K}(\xi_1,\xi_2) = \int_0^\infty \vec{J}(\xi_1,\xi_2,\eta)d\eta = \int_0^\infty \nabla \times \vec{H}(\xi_1,\xi_2,\eta)d\eta \tag{4.43}$$

In the two-dimensional case under consideration here, the surface current density is scalar. Substituting Equations 4.16 through 4.21 into Equation 4.43, we obtain

Leontovich's approximation:

$$\dot{K} = \frac{1+j}{\delta\beta}\dot{H}_\xi^{\mathrm{b}} = \dot{H}_\xi^{\mathrm{b}} \tag{4.44}$$

Mitzner's approximation:

$$\dot{K} = \beta\left(\frac{1}{\beta} + \frac{1}{\beta^2 2d} + \frac{1-j}{4}\frac{\delta}{\beta d}\right)\dot{H}_\xi^{\mathrm{b}} = \left(1 + \frac{1-j}{2}\frac{\delta}{d}\right)\dot{H}_\xi^{\mathrm{b}} \tag{4.45}$$

Rytov's approximation:

$$\begin{aligned}
\dot{K} &= \beta\left[\dot{H}_\xi^{\mathrm{b}}\left(\frac{1}{\beta} + \frac{1}{\beta^2 2d} + \frac{3}{2\beta^3 8d^2}\right) + \delta\frac{1-j}{4}\left(\dot{H}_\xi^{\mathrm{b}}\frac{1}{\beta d} + \dot{H}_\xi^{\mathrm{b}}\frac{3}{\beta^2 d^2} + \frac{1}{\beta^2}\frac{\partial^2 \dot{H}_\xi^{\mathrm{b}}}{\partial \xi^2}\right) - \frac{\delta^2}{\beta}\frac{j}{2}\left(\dot{H}_\xi^{\mathrm{b}}\frac{3}{4d^2} + \frac{\partial^2 \dot{H}_\xi^{\mathrm{b}}}{\partial \xi^2}\right)\right] \\
&= \dot{H}_\xi^{\mathrm{b}}\left(1 + \frac{1}{\beta 2d} + \frac{3}{2\beta^2 8d^2}\right) + \delta\frac{1-j}{4}\left(\dot{H}_\xi^{\mathrm{b}}\frac{1}{d} + \dot{H}_\xi^{\mathrm{b}}\frac{3}{\beta d^2} + \frac{1}{\beta}\frac{\partial^2 \dot{H}_\xi^{\mathrm{b}}}{\partial \xi^2}\right) - j\delta^2\left(\dot{H}_\xi^{\mathrm{b}}\frac{3}{8d^2} + \frac{1}{2}\frac{\partial^2 \dot{H}_\xi^{\mathrm{b}}}{\partial \xi^2}\right) \\
&\approx \dot{H}_\xi^{\mathrm{b}} + \frac{1-j}{2}\frac{\delta}{d}\dot{H}_\xi^{\mathrm{b}} - j\delta^2\left(\frac{\dot{H}_\xi^{\mathrm{b}}}{d^2} + \frac{3}{2}\frac{\partial^2 \dot{H}_\xi^{\mathrm{b}}}{\partial \xi^2}\right)
\end{aligned} \tag{4.46}$$

In terms of the electric field at the interface:

Leontovich's approximation:

$$\dot{K} = \dot{E}^{\mathrm{b}}\sigma\int_0^\infty \exp(-\beta\eta)d\eta = \dot{E}^{\mathrm{b}}\sigma\beta^{-1} = \dot{E}^{\mathrm{b}}\frac{\sigma\delta}{1+j} = \dot{E}^{\mathrm{b}}\frac{\sigma\delta}{2}(1-j) \tag{4.47}$$

Mitzner's approximation:

$$\dot{K} = \dot{E}^{\text{b}}\sigma \int_0^\infty \left(1 + \frac{\eta}{2d}\right) \exp(-\beta\eta) d\eta = \dot{E}^{\text{b}} \frac{\sigma\delta}{1+j}\left(1 + \frac{1}{\beta 2d}\right) = \dot{E}^{\text{b}} \frac{\sigma\delta}{2}\left[1 - j\left(1 + \frac{\delta}{2d}\right)\right] \tag{4.48}$$

Rytov's approximation:

$$\begin{aligned}
\dot{K} &= \sigma \int_0^\infty \left[\dot{E}^{\text{b}} + \dot{E}^{\text{b}} \frac{\eta}{2d} + \dot{E}^{\text{b}} \frac{3\eta^2}{8d^2} + \delta \frac{1-j}{2} \eta \left(\dot{E}^{\text{b}} \frac{1}{8d^2} + \frac{1}{2} \frac{\partial^2 \dot{E}^{\text{b}}}{\partial \xi^2} \right) \right] \exp(-\beta\eta) \\
&= \frac{\sigma\delta}{1+j} \left[\dot{E}^{\text{b}} + \dot{E}^{\text{b}} \frac{1}{\beta 2d} + \dot{E}^{\text{b}} \frac{3}{\beta^2 16 d^2} + \frac{1}{\beta^2} \left(\dot{E}^{\text{b}} \frac{1}{8d^2} + \frac{1}{2} \frac{\partial^2 \dot{E}^{\text{b}}}{\partial \xi^2} \right) \right] \\
&\approx \frac{\sigma\delta}{2} \left\{ \dot{E}^{\text{b}} \left[1 - j \left(1 + \frac{\delta}{2d} + \frac{\delta^2}{8d^2} \right) \right] - j \frac{\delta^2}{4} \frac{\partial^2 \dot{E}^{\text{b}}}{\partial \xi^2} \right\}
\end{aligned} \tag{4.49}$$

The surface current density is an important quantity in the surface impedance concept and is frequently used in practical calculations.

4.3 Resistance and Internal Inductance

We first consider an idealized problem in which a current-carrying conductor fills the region $x \geq 0$. The conductor is assumed to be of finite length ℓ in the ξ-direction as shown in Figure 4.10.

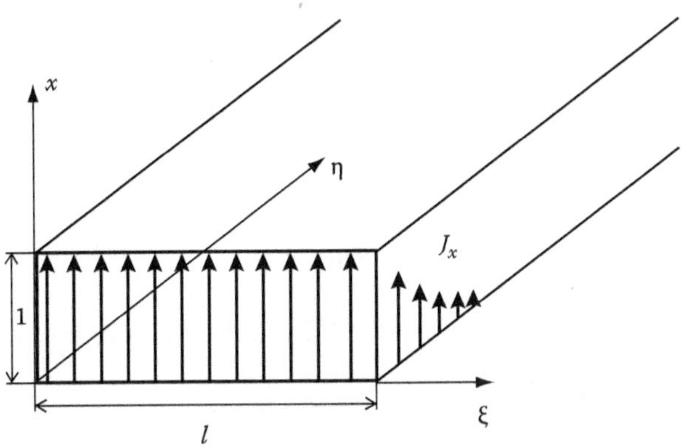

FIGURE 4.10
A current-carrying conducting slab with the current density distribution indicated.

The ratio of the electric field $\dot{E}^{\mathrm{b}}$ at the conductor's surface to the total current $\dot{I}_\ell$ flowing through the conductor gives the internal impedance. The latter can be represented in the form:

$$\frac{\dot{E}^{\mathrm{b}}}{\dot{I}_\ell} = R'_\ell + j\omega L'_{i\ell} \tag{4.50}$$

In this case, the total current is

$$\dot{I}_\ell = \int_0^\ell \dot{K} \mathrm{d}\xi = \ell \dot{K} \tag{4.51}$$

In the Leontovich approximation, Equation 4.51 takes the form:

$$\dot{I}_\ell = \ell \dot{E}^{\mathrm{b}} \frac{\sigma\delta}{1+j} \tag{4.52}$$

Substituting Equation 4.52 into Equation 4.50, we obtain the resistance and internal inductance per unit length and per unit width in the Leontovich approximation:

$$R'_\ell = \frac{1}{\ell\sigma\delta} \tag{4.53}$$

$$L'_{i\ell} = \frac{1}{\ell\omega\sigma\delta} \tag{4.54}$$

In the case of a single circular cylindrical conductor (see Figure 4.2) $\ell = 2\pi d$ and the resistance and internal inductance per unit length of the conductor can be calculated as

$$R' = \frac{1}{2\pi d\sigma\delta} \tag{4.55}$$

$$L'_i = \frac{1}{2\pi d\omega\sigma\delta} \tag{4.56}$$

More accurate results can be obtained by substituting Equation 4.48 into Equation 4.51:

$$\dot{I} = 2\pi \mathrm{d} \dot{E}^{\mathrm{b}} \frac{\sigma\delta}{2} \left[1 - j\left(1 + \frac{\delta}{2d} \right) \right] \tag{4.57}$$

Thus

$$\frac{\dot{E}^{\mathrm{b}}}{\dot{I}_{\ell}} = \frac{1}{2\pi d\sigma\delta} 2\left[1 - j\left(1 + \frac{\delta}{2d}\right)\right]^{-1} = \frac{1}{2\pi d\sigma\delta}\left[1 - \frac{\delta}{2d} + j + O\left(\frac{\delta^2}{d^2}\right)\right] \tag{4.58}$$

and

$$R' = \frac{1}{2\pi d\sigma\delta}\left(1 - \frac{\delta}{2d}\right) \tag{4.59}$$

$$L_i' = \frac{1}{2\pi d\omega\sigma\delta} \tag{4.60}$$

These formulae have been obtained assuming no tangential variation of the field. In the case of homogeneous distribution of the field along the interface, one can introduce the following quantities defined at each point of the cross section's contour:

$$\frac{\dot{E}^{\mathrm{b}}}{\dot{K}} = R'' + j\omega L_i'' \tag{4.61}$$

$$R'' = \frac{1}{\sigma\delta}\left(1 - \frac{\delta}{2d}\right) \tag{4.62}$$

$$L_i'' = \frac{1}{\omega\sigma\delta} \tag{4.63}$$

where the quantities $R'' = R''(\xi)$ and $L_i'' = L_i''(\xi)$ are functions of the contour coordinate ξ.

Resistance and inductance can also be calculated using the Poynting vector:

$$-\oint_S \left(\dot{\vec{E}} \times \dot{\vec{H}}^*\right) \cdot \vec{n}\mathrm{d}s = |\dot{I}|^2 (R + j\omega L_i) \tag{4.64}$$

where $\dot{\vec{H}}^*$ represents the conjugate of the magnetic field phasor and the normal unit vector $\vec{n}$ is directed outward to volume V bounded by surface S. In our two-dimensional case, Equation 4.64 can be represented in the following form using surface-related coordinates (x, ξ, η):

$$\oint_L \dot{E}^{\mathrm{b}}\left(\dot{H}_{\xi}^{\mathrm{b}}\right)^* \mathrm{d}\xi = |\dot{I}|^2 (R' + j\omega L_i') \tag{4.65}$$

where L is the contour of the cross section of the conductor. Using Equation 3.55, we can calculate the Poynting vector in the Leontovich approximation as follows:

$$\oint_L \dot{E}^{\mathrm{b}} \left(\dot{H}_\xi^{\mathrm{b}}\right)^* \mathrm{d}\xi = \frac{1+j}{\sigma\delta} \oint_L \left|\dot{H}_\xi^{\mathrm{b}}\right|^2 \mathrm{d}\xi \tag{4.66}$$

Substituting this result in Equation 4.64, we obtain

$$R' = \frac{1}{\sigma\delta|\dot{I}|^2} \oint_L \left|\dot{H}_\xi^{\mathrm{b}}\right|^2 \mathrm{d}\xi \tag{4.67}$$

$$L_i' = \frac{1}{\omega\sigma\delta|\dot{I}|^2} \oint_L \left|\dot{H}_\xi^{\mathrm{b}}\right|^2 \mathrm{d}\xi \tag{4.68}$$

or, taking into account Equation 4.44,

$$R' = \frac{1}{\sigma\delta|\dot{I}|^2} \oint_L |\dot{K}|^2 \mathrm{d}\xi \tag{4.69}$$

$$L_i' = \frac{1}{\omega\sigma\delta|\dot{I}|^2} \oint_L |\dot{K}|^2 \mathrm{d}\xi \tag{4.70}$$

It is easily demonstrated that Equations 4.69 and 4.70 reduce to Equations 4.55 and 4.56 in the case of the single circular conductor considered above. Indeed, in that case,

$$\dot{K} = \frac{\dot{I}}{2\pi d} = \mathrm{const} \tag{4.71}$$

and substitution of Equation 4.71 into Equations 4.69 and 4.70 results in Equations 4.55 and 4.56.

In the next, higher approximation

$$\oint_L \dot{E}^{\mathrm{b}} \left(\dot{H}_\xi^{\mathrm{b}}\right)^* \mathrm{d}\xi = \frac{1+\delta/(2d)+j}{\sigma\delta} \oint_L \left|\dot{H}_\xi^{\mathrm{b}}\right|^2 \mathrm{d}\xi \tag{4.72}$$

and the improved R' is

$$R' = \frac{1+\delta/(2d)}{\sigma\delta} \frac{\oint\limits_L \left|\dot{H}_\xi^{\mathrm{b}}\right|^2 \mathrm{d}\xi}{\left|\dot{I}\right|^2} \quad (4.73)$$

Another way to obtain R' is to use the formula for ohmic power losses per unit length given by the Poynting theorem:

$$P'_{\mathrm{W}} = R'\left|\dot{I}\right|^2 = \oint\limits_L \int\limits_0^\infty \sigma \dot{E}\dot{J}^* \mathrm{d}\xi \quad (4.74)$$

Substituting Equations 4.24 or 4.25 into Equation 4.74, performing the required transformations and neglecting the terms of the order δ^2/d^2 and higher, we obtain Equations 4.67 or 4.73, respectively. Note that Equations 4.67 and 4.68 allowing for nonhomogeneous distribution of the field along the surface have broader applicability than Equations 4.53 and 4.54 or Equations 4.55 and 4.56.

4.4 Forces Acting on the Conductor

The force on a charged particle moving through a steady magnetic field may be written as the differential force $\mathrm{d}\vec{F}$ exerted on a differential element of charge $\mathrm{d}Q$

$$\mathrm{d}\vec{F} = \mathrm{d}Q\vec{v} \times \vec{B} \quad (4.75)$$

where $\vec{u}$ is the velocity of the charge. The differential element of charge may also be expressed in terms of the volume charge density ρ_v and, consequently, current density $\vec{J}$:

$$\mathrm{d}Q = \rho_v \mathrm{d}v \quad (4.76)$$

Substituting Equation 4.76 into Equation 4.75, we obtain

$$\mathrm{d}\vec{F} = \vec{u} \times \vec{B}\rho_v \mathrm{d}v \quad (4.77)$$

Taking into account that

$$\vec{J} = \rho_v \vec{u} \tag{4.78}$$

we bring Equation 4.77 to the form:

$$d\vec{F} = \vec{J} \times \vec{B} dv \tag{4.79}$$

In long parallel conductors, the force is directed normal to the surface. Representing Equation 4.79 using surface-related coordinates, we obtain the differential force per unit length for time-harmonic fields:

$$d\dot{F}'_\eta = \dot{J}\dot{B}^*_\xi dv \tag{4.80}$$

Integrating Equation 4.80 over the cross section of the conductor results in the formula for force per unit length:

$$\dot{F}'_\eta = \oint_L d\xi \int_0^\infty \dot{J}\dot{B}^*_\xi d\eta = \oint_L d\xi \int_0^\infty \sigma\mu\dot{E}\dot{H}^*_\xi d\eta \tag{4.81}$$

Substituting Equations 4.17 and 4.24 into Equation 4.81 and performing the integration, we obtain the force per unit length in the Leontovich approximation:

$$\dot{F}'_\eta = \frac{1+j}{\delta}\mu \oint_L \left|\dot{H}^b_\xi\right|^2 d\xi \int_0^\infty \exp(-2\eta/\delta) = (1+j)\frac{\mu}{2}\oint_L \left|\dot{H}^b_\xi\right|^2 d\xi \tag{4.82}$$

Using Equation 4.44, we can represent this formula in terms of the surface current density:

$$\dot{F}'_\eta = (1+j)\frac{\mu}{2}\oint_L \dot{H}^b_\xi \dot{K}^* d\xi = (1+j)\frac{\mu}{2}\oint_L \left|\dot{K}\right|^2 d\xi \tag{4.83}$$

Correction of the order of the Mitzner approximation can be performed by substituting Equations 4.19 and 4.25 into Equation 4.81:

$$\begin{aligned}\dot{F}'_\eta &= \frac{1}{\delta}\mu\oint_L \left|\dot{H}^b_\xi\right|^2 d\xi \int_0^\infty \left[1+j+\frac{\delta}{2d}+\eta\left(\frac{1+j}{d}+\frac{\delta}{4d^2}\right)+\eta^2\frac{1+j}{4d^2}\right]\exp(-2\eta/\delta)\\ &= \frac{1}{2}\mu\oint_L \left|\dot{H}^b_\xi\right|^2\left[1+j+\frac{\delta}{2d}+\frac{1+j}{2d}\delta+O\left(\frac{\delta^2}{d^2}\right)\right]d\xi \approx \left[1+\frac{\delta}{d}+j\left(1+\frac{\delta}{2d}\right)\right]\frac{\mu}{2}\oint_L \left|\dot{H}^b_\xi\right|^2 d\xi\end{aligned} \tag{4.84}$$

Finally, substituting Equation 4.45 into Equation 4.84 and neglecting the terms of the order δ^2/d^2 and higher, we express the force in terms of the surface current density:

$$\dot{F}'_{\eta} \approx \left[1 + \frac{\delta}{2d} + j\left(1 + \frac{\delta}{d}\right)\right] \frac{\mu}{2} \oint_L |\dot{K}|^2 \mathrm{d}\xi \tag{4.85}$$

In this chapter, we demonstrated that for calculation of resistance, internal inductance, energy, power losses, and forces in the conductor under the condition of skin effect it is sufficient to know the distribution of the tangential component of the magnetic field along the surface of the conductor.

5

Derivation of SIBCs for Nonlinear and Nonhomogeneous Problems

5.1 Introduction

In previous chapters, we assumed the properties of the conducting medium were constant. In practice, however, there are problems in which these assumptions are not valid and the governing Maxwell's equations for the electromagnetic field distribution in conductors must be supplemented by additional equation(s) describing material properties. For example, the magnetic flux and magnetic induction in ferromagnetic materials are related through a nonlinear function. When currents flowing through the conductors are high enough, Joule heating starts to play a role and the dependence between conductivity and temperature cannot be neglected. In the latter case the heat transfer equation must be considered together with Maxwell's equations. Objects consisting of several conductive layers or conductors coated with dielectrics belong to yet another class of problems where assumption of constant material properties is invalid.

Intuitively it is clear that the SIBCs considered in Chapters 2 through 4 cannot be used directly in modeling nonlinear problems. However, the general application area of the surface impedance concept is not restricted to homogeneous conductors with constant material properties. The perturbation approach developed in Chapter 2 for derivation of SIBCs consists of two major stages:

1. Approximation of the original three-dimensional equations of the electromagnetic field distribution inside the conductor by one-dimensional equations in the direction normal to the conductor's surface
2. Derivation of the boundary conditions from analytical solution(s) of the one-dimensional equation(s) of diffusion of the magnetic field inside the conductor

Stage 1 can be performed if the electromagnetic field in the skin layer varies in the normal direction much faster than in the tangential directions so that tangential derivatives are much smaller than normal derivatives. But this is exactly the main condition of applicability of the surface impedance concept to the problem under consideration. This means that the conditions of applicability of the surface impedance concept do not restrict the transformations performed at stage 2. In other words, even if the equations considered at stage 2 do not provide SIBCs in the classical single frequency form or do not allow exact analytical solution at all, it does not mean that the surface impedance concept cannot be applied.

The objective of this chapter is to demonstrate how the perturbation approach developed in the previous chapters can be applied to derive boundary conditions for nonlinear and nonhomogeneous problems. It is expected that these SIBCs will be more complex than the classical conditions for homogeneous conductors. Thus we will restrict ourselves to low-order approximation to reduce the number of transformations. High-order terms can be derived following the same approach.

5.2 Coupled Electromagnetic–Thermal Problems

When dependencies between the characteristics of the conductor's material and temperature, T cannot be neglected, the equations governing electromagnetic field distribution in the conductor must be supplemented by the heat transfer equation in the form:

$$c\rho \frac{\partial T}{\partial t} = \nabla \cdot (\lambda \nabla T) + \frac{J^2}{\sigma} \tag{5.1}$$

where J, c, ρ, and λ are the current density, specific heat, mass density, and thermal conductivity, respectively. We assume σ and λ to be functions of temperature and denote the electrical and thermal conductivities at $T = 293$ K as σ_∞ and λ_∞, respectively. With the use of the following Maxwell equation:

$$\vec{J} = \nabla \times \vec{H} \tag{5.2}$$

we rewrite Equation 5.1 in the form:

$$c\rho \frac{\partial T}{\partial t} = \nabla \cdot (\lambda \nabla T) + \sigma^{-1} |\nabla \times \vec{H}|^2 \tag{5.3}$$

Following the perturbation approach, we rewrite Equation 5.3 in nondimensional variables using the scale factors for basic quantities given in Table 3.1:

$$\boldsymbol{T} \frac{\partial \tilde{T}}{\partial \tilde{t}} = \boldsymbol{T} \frac{\lambda^* \tau}{c\rho\delta^2} \tilde{\nabla} \cdot (\tilde{\lambda} \tilde{\nabla} \tilde{T}) + |\tilde{\nabla} \times \tilde{\vec{H}}|^2 \frac{\tau}{c\rho\tilde{\sigma}\sigma^*\delta^2} \frac{\boldsymbol{I}^2}{\boldsymbol{D}^2} \tag{5.4}$$

where

$$\tilde{\sigma} = \frac{\sigma}{\sigma^*}; \quad \sigma^* = \sigma(\boldsymbol{T}) \tag{5.5}$$

$$\tilde{\lambda} = \frac{\lambda}{\lambda^*}; \quad \lambda^* = \lambda(\boldsymbol{T}) \tag{5.6}$$

Here σ^* and λ^* are the electrical conductivity and the thermal conductivity respectively, at the characteristic temperature T used as a scale factor.

Representing the characteristic skin depth δ and small parameter $\tilde{p}$ in the forms:

$$\delta = \sqrt{\frac{\tau}{\sigma^*\mu}} \tag{5.7}$$

$$\tilde{p} = \frac{\delta}{D} = \sqrt{\frac{\tau}{\sigma^*\mu D^2}} \ll 1 \tag{5.8}$$

transforms Equation 5.4 into the following form:

$$\boldsymbol{T}\frac{\partial \tilde{T}}{\partial \tilde{t}} = \boldsymbol{T}\tilde{\varepsilon}^2\tilde{\nabla} \cdot (\tilde{\lambda}\tilde{\nabla}\tilde{T}) + \tilde{\sigma}^{-1}\left(\tilde{\nabla} \times \vec{\tilde{H}}\right)^2 \frac{\mu}{c\rho}\frac{\boldsymbol{I}^2}{\boldsymbol{D}^2} \tag{5.9}$$

where

$$\tilde{\varepsilon} = (\lambda^*\sigma^*\mu/(c\rho))^{1/2} \tag{5.10}$$

Introducing a scale factor for temperature as

$$\boldsymbol{T} = \frac{\mu}{c\rho}\frac{\boldsymbol{I}^2}{\boldsymbol{D}^2} \tag{5.11}$$

and substituting Equation 5.11 into Equation 5.9, we finally write the heat transfer equation in nondimensional form:

$$\frac{\partial \tilde{T}}{\partial \tilde{t}} = \tilde{\varepsilon}^2\tilde{\nabla} \cdot (\tilde{\lambda}\tilde{\nabla}\tilde{T}) + \tilde{\sigma}^{-1}\left(\tilde{\nabla} \times \vec{\tilde{H}}\right)^2 \tag{5.12}$$

Equation 5.12 must be considered together with Equations 3.16 and 3.18 in describing the distribution of the tangential components of the electromagnetic field in the conductor. Substituting Equations 2.12 through 2.15, which give the vector operators in local coordinates $\tilde{\xi}_1$, $\tilde{\xi}_2$, $\tilde{\eta}$, into Equation 5.12 and neglecting terms containing p^2 (Leontovich's approximation) in Equations 5.12, 3.16, and 3.18, we obtain

$$\frac{\partial \tilde{T}}{\partial \tilde{t}} = \tilde{\varepsilon}^2\frac{\partial}{\partial \tilde{\eta}}\left(\tilde{\lambda}\frac{\partial \tilde{T}}{\partial \tilde{\eta}}\right) + \tilde{\sigma}^{-1}\sum_{i=1}^{2}\left(\frac{\partial \tilde{H}_{\xi_i}}{\partial \tilde{\eta}}\right)^2 \tag{5.13}$$

$$\tilde{p}\frac{\partial \tilde{H}_{\xi_k}}{\partial \tilde{\eta}} = (-1)^{3-k}\tilde{\sigma}\tilde{E}_{\xi_{3-k}}, \quad k = 1,2 \tag{5.14}$$

$$\frac{\partial \tilde{E}_{\xi_{3-k}}}{\partial \tilde{\eta}} = (-1)^{3-k}\tilde{p}\frac{\partial \tilde{H}_{\xi_k}}{\partial \tilde{t}}, \quad k = 1,2 \tag{5.15}$$

As follows from Equation 5.13, the distribution of temperature in the Leontovich approximation is described by a one-dimensional equation in the direction normal to the conductor's surface. The tangential electric field at the interface can be calculated from Equation 5.14:

$$\tilde{E}^{\mathrm{b}}_{\xi_{3-k}} = (-1)^k\tilde{p}\frac{\partial}{\partial \tilde{t}}\int_0^\infty \tilde{H}_{\xi_k}\mathrm{d}\tilde{\eta} \tag{5.16}$$

The use of Equation 5.16 is impossible without solving the equation of diffusion of the magnetic field. The latter can be obtained, from Equations 5.14 and 5.15, and written in the form:

$$\frac{\partial}{\partial \tilde{\eta}}\left(\tilde{\sigma}^{-1}\frac{\partial \tilde{H}_{\xi_k}}{\partial \tilde{\eta}}\right) = \frac{\partial \tilde{H}_{\xi_k}}{\partial \tilde{t}}, \quad k = 1,2 \tag{5.17}$$

Equations 5.13 and 5.17 must be solved together with the following boundary and initial conditions:

$$\tilde{\eta} = 0: \tilde{H} = \tilde{H}^{\mathrm{b}}; \quad \tilde{\eta} \to \infty: \tilde{H} \to 0; \quad \tilde{t} = 0: \tilde{H} = 0 \tag{5.18}$$

$$\tilde{\eta} = 0: \quad \frac{\partial \tilde{T}}{\partial \tilde{\eta}} = \gamma(\tilde{T} - T_\infty/\boldsymbol{T}); \quad \eta \to \infty: \quad \tilde{T} \to T_\infty/\boldsymbol{T}; \quad \tilde{t} = 0: \quad \tilde{T} = \frac{T_\infty}{\boldsymbol{T}} \tag{5.19}$$

Solving the one-dimensional coupled problem (Equations 5.13 and 5.17 through 5.19), and substituting the results in Equation 5.16, we obtain the relationships between the electric field and the magnetic field at the conductor's surface. Thus Equations 5.13 and 5.16 through Equation 5.19 can be treated as SIBCs for coupled electromagnetic–thermal problems in implicit form.

Returning to dimensional variables in Equations 5.13 and 5.16 through Equation 5.19 gives

$$\frac{\partial}{\partial \eta}\left((\sigma\mu)^{-1}\frac{\partial H_{\xi_k}}{\partial \eta}\right) = \frac{\partial H_{\xi_k}}{\partial t}, \quad k = 1,2 \tag{5.20}$$

$$c\rho\frac{\partial T}{\partial t} = \frac{\partial}{\partial \eta}\left(\lambda\frac{\partial T}{\partial \tilde{\eta}}\right) + \sigma^{-1}\sum_{i=1}^{2}\left(\frac{\partial H_{\xi_i}}{\partial \eta}\right)^2, \quad k = 1,2 \tag{5.21}$$

$$E^{\mathrm{b}}_{\xi_{3-k}} = (-1)^k\mu\frac{\partial}{\partial t}\int_0^\infty H_{\xi_k}\mathrm{d}\eta \tag{5.22}$$

As an example, consider the following temperature dependencies of conductivity and thermal conductivity:

$$\sigma = \frac{\sigma_\infty}{1+\alpha T} \tag{5.23}$$

$$\lambda = \chi\sigma T \tag{5.24}$$

where the coefficients α and χ can be assumed to be constant for pure materials (for example, in copper $\chi = 2.23 \times 10^{-8}$). Substitution of Equations 5.23 and 5.24 into Equations 5.20 and 5.21 leads to the following equations:

$$\begin{cases} \dfrac{\sigma^*}{\sigma_\infty}\dfrac{\partial}{\partial\tilde{\eta}}\left((1+\alpha T\tilde{T})\dfrac{\partial\tilde{H}_{\xi_k}}{\partial\tilde{\eta}}\right) = \dfrac{\partial\tilde{H}_{\xi_k}}{\partial\tilde{t}} \\ \dfrac{\partial\tilde{T}}{\partial\tilde{t}} = \tilde{\varepsilon}^2\dfrac{\lambda_\infty}{\lambda^*}\dfrac{\partial}{\partial\tilde{\eta}}\left(\dfrac{\tilde{T}}{1+\alpha T\tilde{T}}\dfrac{\partial\tilde{T}}{\partial\tilde{\eta}}\right) + \tilde{\sigma}^{-1}\displaystyle\sum_{i=1}^{2}\left(\dfrac{\partial\tilde{H}_{\xi_i}}{\partial\tilde{\eta}}\right)^2 \end{cases} \tag{5.25}$$

where

$$\lambda_\infty = \chi\sigma_\infty T_\infty \tag{5.26}$$

With dimensional variables, Equation 5.25 can be written in the form

$$\begin{cases} \dfrac{\partial}{\partial\tilde{\eta}}\left((1+\alpha T\tilde{T})\dfrac{\partial\tilde{H}_{\xi_k}}{\partial\tilde{\eta}}\right) = \sigma_\infty\mu\dfrac{\partial\tilde{H}_{\xi_k}}{\partial\tilde{t}} \\ c\rho\dfrac{\partial T}{\partial t} = \chi\sigma_\infty\dfrac{\partial}{\partial\eta}\left(\dfrac{T}{1+\alpha T}\dfrac{\partial T}{\partial\eta}\right) + \sigma^{-1}\displaystyle\sum_{i=1}^{2}\left(\dfrac{\partial H_{\xi_i}}{\partial\eta}\right)^2 \end{cases} \tag{5.27}$$

Under conditions of skin effect, the temperature is maximum at the surface of current-carrying conductors and decays monotonically inside the conductor with the increase in distance from the surface. The distributions of the magnetic field inside the conductor in the linear and nonlinear cases are similar, but the behavior of the current density may change significantly due to the decrease in conductivity at high temperatures (as follows from Equation 5.23). This results in a shift of the maximum in the current density distribution from the surface to the areas where Joule heating is not as significant (and consequently, the conductivity is constant). This phenomenon can be visualized by solving the one-dimensional problem of electromagnetic field diffusion (Equation 5.27) into the conducting half-space shown in Figure 5.1.

Typical distributions of the current density inside copper at high currents (~1 MA) are shown in Figure 5.2.

To demonstrate the importance of taking into account the temperature-dependent properties in Equations 5.23 and 5.24 for practical calculation of the surface current density along the conductor's surface, consider a pair of

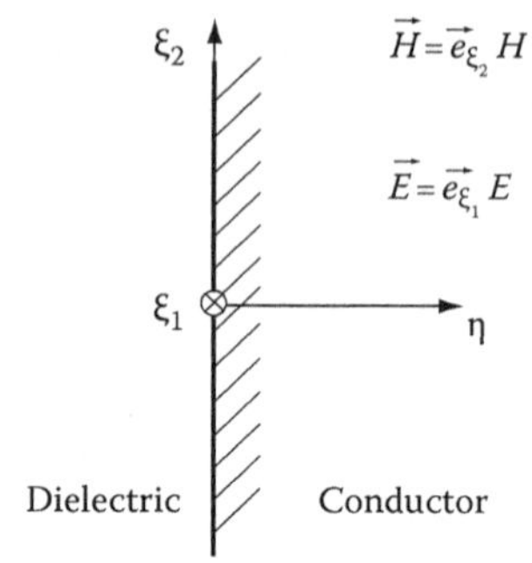

FIGURE 5.1
One-dimensional problem of electromagnetic field diffusion into a conducting half-space.

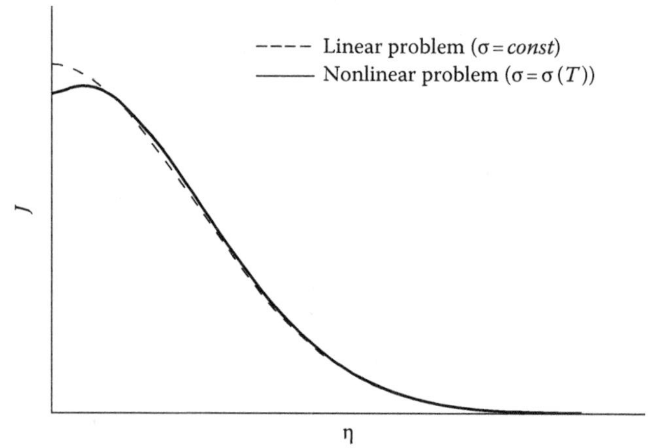

FIGURE 5.2
Distribution of the current density inside the conducting half-space at high currents (~1 MA).

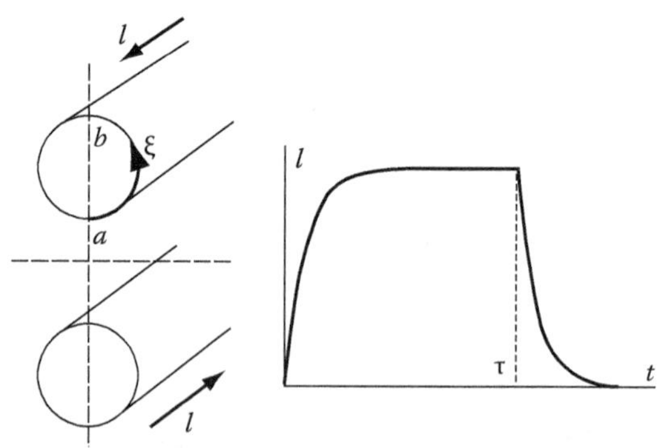

FIGURE 5.3
Statement of problem: Electromagnetic field diffusion due to passage of a current pulse in a system of two parallel conductors.

infinitely long identical parallel conductors with circular cross sections with equal and oppositely directed current pulses flowing from an external source as shown in Figure 5.3. Under these conditions the current density has only one component directed along the conductors. The radius of each conductor

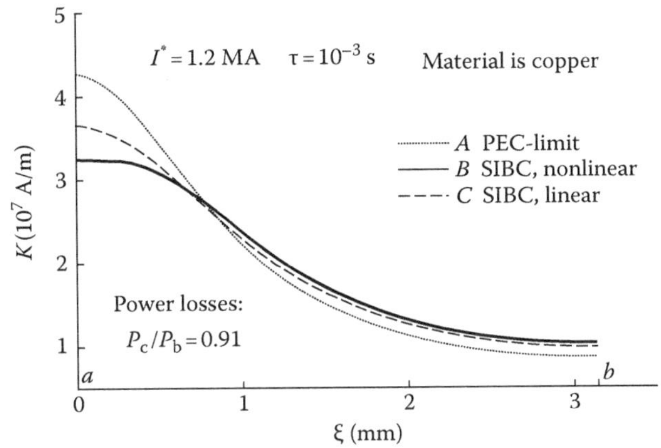

FIGURE 5.4
Distributions of the surface current density along half the contour around the cross section of one conductor.

and the distance between the conductors are equal to 0.01 m (this is the characteristic scale D). The magnitude and duration of the current pulse are equal to 1 MA and 0.001 s, respectively.

Figure 5.4 shows distributions of the surface current density K, obtained using different approximations of the electromagnetic field diffusion into the conductors, as a function of the surface coordinate ξ. The origin is at point a and proceeds around the contour, ending at point b (due to double symmetry only half the contour around the cross section of one conductor is shown). Curve A was obtained using the perfect electrical conductor (PEC) limit (the electromagnetic field diffusion is neglected). Curve C was obtained using the classical Leontovich SIBC in which the conductors' properties are assumed to be constant and curve B was obtained using the nonlinear SIBC in Equation 5.27.

Obviously, the solution of the problem obtained in the PEC-limit does not depend on the properties of the conductor because the field diffusion into the body is not taken into account. The diffusion induces a redistribution of the current along the conductor surface with time. The effect of this process is to smooth out nonuniformities in the surface current density distribution. Increase of the resistivity with temperature leads to deeper penetration of the field into the conductor compared to penetration in the linear material. Thus the maximum surface current density is significantly lower in the case of the nonlinear SIBC than the PEC-limit or the SIBC in which thermal properties of the conductors are neglected.

Figure 5.5 shows the increase in surface temperature with time at point a of the conductor cross section as a result of Joule heating. It is easy to see that the $\sigma(T)$ dependence becomes a very important factor limiting the maximum surface temperature of the conductor (the difference in the surface

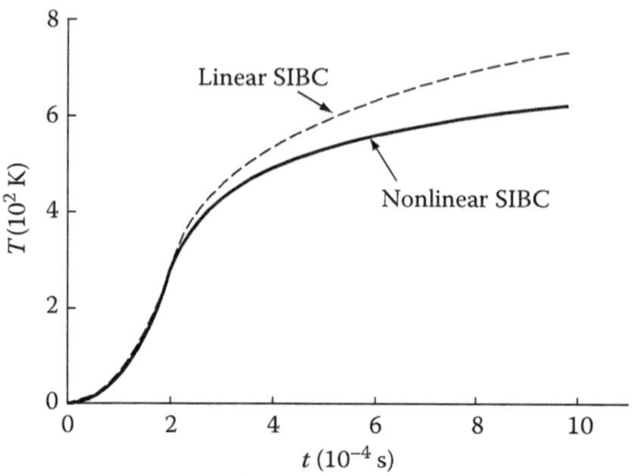

FIGURE 5.5
Increase in temperature of the conductor's surface due to passage of the current pulse.

temperature between the curves obtained using the linear and nonlinear SIBCs is about 15%).

As a conclusion of this section we now demonstrate how SIBCs for coupled problems can be derived in terms of the magnetic vector potential formalism. Substituting Equation 3.91, (which relates the magnetic vector potential and current density), into Equation 5.1, we obtain

$$c\rho\frac{\partial T}{\partial t} = \nabla\cdot(\lambda\nabla T) + \sigma^{-1}\left|\nabla^2\vec{A}\right|^2 \tag{5.28}$$

As we have seen in Chapter 3, the distribution of the vector potential in the conductor is described in Equation 3.99. The scale factor for the magnetic vector potential is given in Equation 3.105. Turning to nondimensional variables in Equations 3.99 and 5.28 and neglecting terms of the order $O(p^2)$ yields the following coupled one-dimensional equations:

$$\tilde{\sigma}^{-1}\frac{\partial^2\tilde{A}^{\mathrm{e}}_{\xi_k}}{\partial\tilde{\eta}^2} = \frac{\partial\tilde{A}^{\mathrm{e}}_{\xi_k}}{\partial\tilde{t}},\quad k = 1, 2 \tag{5.29}$$

$$\frac{\partial\tilde{T}}{\partial\tilde{t}} = \tilde{\varepsilon}^2\frac{\partial}{\partial\tilde{\eta}}\left(\tilde{\lambda}\frac{\partial\tilde{T}}{\partial\tilde{\eta}}\right) + \tilde{\sigma}^{-1}\sum_{i=1}^{2}\left(\frac{\partial^2\tilde{A}^{\mathrm{e}}_{\xi_k}}{\partial\tilde{\eta}^2}\right)^2 \tag{5.30}$$

These must be supplemented by the boundary and initial conditions:

$$\tilde{\eta} = 0\colon \tilde{A}^{\mathrm{e}}_{\xi_k} = \left(\vec{\tilde{A}}^{\mathrm{e}}_{\xi_k}\right)^{\mathrm{b}};\quad \tilde{\eta}\to\infty\colon \tilde{A}^{\mathrm{e}}_{\xi_k}\to 0;\quad \tilde{t} = 0\colon \tilde{A}^{\mathrm{e}}_{\xi_k} = 0 \tag{5.31}$$

Returning to dimensional variables in Equations 5.29 and 5.30 gives the required SIBC:

$$\begin{cases} \dfrac{\partial^2 A^{\mathrm{e}}_{\xi_k}}{\partial \eta^2} = \sigma\mu \dfrac{\partial A^{\mathrm{e}}_{\xi_k}}{\partial t} \\ c\rho \dfrac{\partial T}{\partial t} = \dfrac{\partial}{\partial \eta}\left(\lambda \dfrac{\partial T}{\partial \eta}\right) + \sigma^{-1} \displaystyle\sum_{i=1}^{2} \left(\dfrac{\partial^2 A^{\mathrm{e}}_{\xi_k}}{\partial \eta^2}\right)^2 \end{cases} \tag{5.32}$$

5.3 Magnetic Materials

The presence of magnetic materials complicates the calculation of the surface impedance significantly, due to the following main reasons:

- The relationship between the magnetic field intensity (H) and magnetic flux density (B) in magnetic materials is nonlinear. If the relationship between the two is plotted for increasing levels of field intensity, it will follow a curve up to a point where further increases in magnetic field strength will result in no further change in flux density. This condition is called magnetic saturation.
- The hysteresis loop due to the "history-dependent" nature of magnetization of the ferromagnetic material. Once the material has been driven to saturation, the magnetizing filed can be dropped to zero and the material will retain most of its magnetization.

The equation of the magnetic field diffusion taking into account both factors does not admit analytical solution in the general case. Therefore, the classical approach in practical electrical engineering calculations was replacement of the real BH-dependence by idealized functions such as equivalent rectangular or ellipsoidal functions.

The first of these was introduced in the so-called limit theory where

$$B(H) = B_0 = \text{const} \tag{5.33}$$

where B_0 is the saturation magnetic flux density corresponding to the amplitude of the magnetic field H_{m} at the conductor's surface. Consider now the one-dimensional problem of the electromagnetic field diffusion into a ferromagnetic half-space and assume time-harmonic variation of the magnetic field along the conductor's surface:

$$H = H_{\mathrm{m}} \sin(\omega t + \psi_{\mathrm{c}}) \tag{5.34}$$

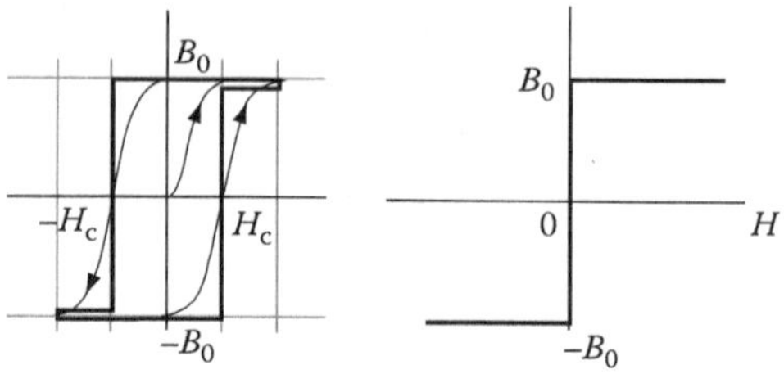

FIGURE 5.6
Idealized magnetization curves in the limit theory: with hysteresis (left) and without hysteresis (right).

where ψ_c is given by the initial condition (Figure 5.6, left):

$$H|_{t=0} = H_m \sin \psi_c = H_c \tag{5.35}$$

If the hysteresis loop can be neglected then ψ_c is assumed to be zero and the BH-curve takes the form shown in Figure 5.6 (right). According to Equation 5.33, the magnetic flux density B is assumed to have a constant value from the surface to a certain depth. Beyond this depth, the flux density has an opposite value. The reversal of the flux defines the wave front which sweeps into the conductor with velocity $v = d\eta/dt$. When this wave front reaches its maximum depth, a new wave front starts from the surface. The wave front moving inside the conductor induces an electric field that has constant value from the surface up to the wave front:

$$E = -2B_0 v \tag{5.36}$$

where the multiplier 2 was introduced because the magnetic flux changes values from $-B_0$ to $+B_0$. With these conditions, Equation 3.2 reduces to the form:

$$\frac{dH}{d\eta} = \sigma E = -2\sigma B_0 v \tag{5.37}$$

Solution of Equation 5.37, which shows a linear decrease of the magnetic field inside the conductor, can be written in the form:

$$H = -2\sigma B_0 v \eta + \text{const}, \quad 0 < \eta < \eta_m \tag{5.38}$$

where η_m is the distance that the wave front travels during time t_m. The integration constant in Equation 5.37 is obtained from the condition in Equation 5.35:

$$\text{const} = H|_{\eta=0;\, t=0} = H_c \tag{5.39}$$

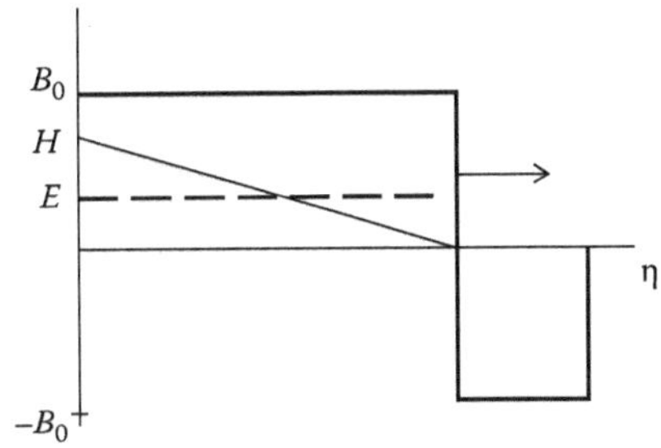

FIGURE 5.7
Distribution of electromagnetic field quantities inside a ferromagnetic half-space with a rectangular *BH*-curve.

The distributions of the functions *B*, *H*, and *E* inside the conductor are shown in Figure 5.7.

Substituting Equation 5.39 into Equation 5.38 and performing simple transformations, we obtain

$$\frac{H - H_c}{\sigma B_0} = 2v = 2\eta\frac{d\eta}{dt} = \frac{d(\eta^2)}{dt} \tag{5.40}$$

Dividing both sides of Equation 5.40 by ω:

$$\frac{H - H_c}{\omega\sigma B_0} = \frac{d(\eta^2)}{d(\omega t)} \tag{5.41}$$

Substituting Equation 5.34 into Equation 5.41 and integrating with respect to the variable ωt, we obtain

$$\eta_m^2 = \frac{H_m}{\omega\sigma B_0}\int_0^{\omega t}[\sin(\omega t + \psi_c) - \sin\psi_c]d(\omega t) \tag{5.42}$$

Simplifying Equation 5.42 gives the depth the wave front reaches at the end of each half-cycle. The wave front penetration as a function of time is therefore:

$$\eta_m = \frac{\delta_{nl}}{\sqrt{2}}[\cos\psi_c - \cos(\omega t + \psi_c) - \omega t\sin\psi_c]^{1/2} \tag{5.43}$$

where

$$\delta_{nl} = \sqrt{\frac{2H_m}{\omega\sigma B_0}} \tag{5.44}$$

The subscript "nl" in Equations 5.43 and 5.44 means nonlinear. By neglecting hysteresis ($\psi_c = 0$) and setting $\omega t = \pi$, Equation 5.43 reduces to the following form:

$$\eta_m|_{\omega t=\pi} = \frac{\delta_{nl}}{\sqrt{2}}[1 - \cos\omega t]^{1/2}\Big|_{\omega t=\pi} = \delta_{nl} \tag{5.45}$$

Therefore, δ_{nl} is the maximum electromagnetic penetration depth in ferromagnetic materials and occurs when hysteresis is neglected.

According to the limit theory in [1] and [2] the surface impedance for rectangular *BH*-curves and sinusoidal magnetic fields can be calculated using the following formula:

$$Z_{nl}^{s} = \frac{8}{3\pi\sigma\delta_{nl}}(2 + j) \tag{5.46}$$

The sinusoidal magnetic field formula is valid only for a small number of eddy current problems where the magnetic field reaches the surface of the ferromagnetic (steel) body mostly in a tangential direction, since the tangential magnetic field remains unchanged through the interface. In all other cases the magnetic field reaches the surface of the conductor in a mostly normal direction due to its high permeability. In such cases the electric field, instead of the magnetic field, is assumed to be sinusoidal and another formula of surface impedance has been proposed in [3]:

$$Z_{nl}^{s} = \frac{27\pi^3}{2\sqrt{5}\sigma\delta_{nl}}\left(1 + \frac{4}{3\pi}j\right) \tag{5.47}$$

The main weakness of the limit theory is that it is only accurate for high fields with strongly saturated materials but not for low fields where the magnetic characteristics of the material can be assumed to be linear. Deeley [4] and Guerin et al. [5] presented a balanced method in which the linear and nonlinear models due to Agarwal [1] are combined and weighted by a function $f(H^s)$ taking into account the degree of saturation:

$$Z_{magnetic}^{s} = f(H_m)Z^s + (1 - f(H_m))Z_{nl}^{s} \tag{5.48}$$

where Z^s is the Leontovich surface impedance for linear problems. The choice of function f is empirical. In [4] and [5], the following function was used:

$$f(H_m) = \frac{1}{1 + k\frac{H_m}{H_{knee}}} \tag{5.49}$$

where

H_{knee} corresponds to the value of the magnetic field at the knee of the *BH*-curve

k is a coefficient to be chosen

Another weighting function has been proposed in [6] in the form:

$$f(H_{\text{m}}) = 2\left(1 - \frac{\int_0^{H_{\text{m}}} B(H)\text{d}H}{B(H_{\text{m}})H_{\text{m}}}\right) \tag{5.50}$$

In the models considered above the surface impedance in the case of nonlinear *BH*-dependence was represented in a form similar to that of single frequency SIBCs for linear materials. However, in the general case neither magnetic nor electric fields in the conductor with nonlinear properties are actually sinusoidal. To overcome this difficulty we consider below the derivation of SIBCs for magnetic materials in the time domain.

In the general case of nonlinear *BH*-dependence one should use Maxwell's equations in the following form instead of Equations 3.1 and 3.2:

$$\nabla \times \vec{H} = \sigma \vec{E} \tag{5.51}$$

$$\nabla \times \vec{E} = -\frac{\partial \vec{B}}{\partial t} \tag{5.52}$$

$$\nabla \cdot \vec{B} = 0 \tag{5.53}$$

$$\|\vec{H}\| / \|\vec{B}\| = \nu \tag{5.54}$$

We assume σ to be constant and the *BH*-curve to be known.

Equations 5.51 through 5.53 in the local coordinates ξ_1, ξ_2, η take the following forms:

$$\frac{\partial H_{\xi_k}}{\partial \eta} - \frac{d_k}{d_k - \eta}\frac{\partial H_\eta}{\partial \xi_k} - \frac{H_{\xi_k}}{d_k - \eta} = (-1)^{3-k}\sigma E_{\xi_{3-k}}, \quad k = 1, 2 \tag{5.55}$$

$$\sum_{i=1}^{2} (-1)^{3-i} \frac{d_i}{d_i - \eta}\frac{\partial H_{\xi_{3-i}}}{\partial \xi_i} = \sigma E_\eta \tag{5.56}$$

$$\frac{\partial E_{\xi_k}}{\partial \eta} - \frac{d_k}{d_k - \eta}\frac{\partial E_\eta}{\partial \xi_k} - \frac{E_{\xi_k}}{d_k - \eta} = (-1)^k \frac{\partial B_{\xi_{3-k}}}{\partial t}, \quad k = 1, 2 \tag{5.57}$$

$$\sum_{i=1}^{2}(-1)^i \frac{d_i}{d_i - \eta} \frac{\partial E_{\xi_i}}{\partial \xi_{3-i}} = \frac{\partial B_\eta}{\partial t} \tag{5.58}$$

$$\frac{\partial B_\eta}{\partial \eta} + \sum_{i=1}^{2} \frac{d_i}{d_i - \eta} \frac{\partial B_{\xi_i}}{\partial \xi_i} = B_\eta \sum_{i=1}^{2} \frac{1}{d_i - \eta} \tag{5.59}$$

Following the perturbation approach, we now switch to dimensionless variables by introducing basic scale factors. The basic scale factors ***I***, ***D***, and τ are introduced in the same way as in Chapter 3. Equation 3.14 giving ***H*** also remains valid. The reluctivity corresponding to the characteristic magnetic field ***H*** is called characteristic reluctivity and denoted as ν^*. With these, the scale factors ***B*** and ***E*** for the magnetic flux and electric field, respectively, can be introduced as

$$\boldsymbol{B} = \frac{\boldsymbol{H}}{\nu^*} = \frac{\boldsymbol{I}}{\nu^* \boldsymbol{D}} \tag{5.60}$$

$$\boldsymbol{E} = \frac{\boldsymbol{D}\boldsymbol{H}}{\tau \nu^*} = \frac{\boldsymbol{I}}{\nu^* \tau} \tag{5.61}$$

The scale factor δ (characteristic penetration depth) for the normal coordinate η can be written in the form:

$$\delta = \sqrt{\tau \nu^* / \sigma} \tag{5.62}$$

With dimensionless variables, Equations 5.55 through 5.59 are written in the form:

$$\tilde{p} \frac{\partial \tilde{H}_{\xi_k}}{\partial \tilde{\eta}} - \tilde{p}^2 \frac{\tilde{d}_k}{\tilde{d}_k - \tilde{p}\tilde{\eta}} \frac{\partial \tilde{H}_\eta}{\partial \tilde{\xi}_k} - \tilde{p}^2 \frac{\tilde{H}_{\xi_k}}{\tilde{d}_k - \tilde{p}\tilde{\eta}} = (-1)^{3-k} \tilde{E}_{\xi_{3-k}}, \quad k = 1, 2 \tag{5.63}$$

$$p^2 \sum_{i=1}^{2} (-1)^{3-i} \frac{\tilde{d}_i}{\tilde{d}_i - p\tilde{\eta}} \frac{\partial \tilde{H}_{\xi_{3-i}}}{\partial \tilde{\xi}_i} = \tilde{E}_\eta \tag{5.64}$$

$$\frac{\partial \tilde{E}_{\xi_k}}{\partial \tilde{\eta}} - \tilde{p} \frac{\tilde{d}_k}{\tilde{d}_k - \tilde{p}\tilde{\eta}} \frac{\partial \tilde{E}_\eta}{\partial \tilde{\xi}_k} - \tilde{p} \frac{\tilde{E}_{\xi_k}}{\tilde{d}_k - \tilde{p}\tilde{\eta}} = (-1)^k \tilde{p} \frac{\partial \tilde{B}_{\xi_{3-k}}}{\partial \tilde{t}}, \quad k = 1, 2 \tag{5.65}$$

$$\sum_{i=1}^{2} (-1)^i \frac{\tilde{d}_i}{\tilde{d}_i - \tilde{p}\tilde{\eta}} \frac{\partial \tilde{E}_i}{\partial \tilde{\xi}_{3-i}} = -\frac{\partial \tilde{B}_\eta}{\partial \tilde{t}} \tag{5.66}$$

$$\frac{\partial \tilde{B}_\eta}{\partial \tilde{\eta}} + \tilde{p} \sum_{i=1}^{2} \frac{\tilde{d}_i}{\tilde{d}_i - p\tilde{\eta}} \frac{\partial \tilde{B}_{\xi i}}{\partial \tilde{\xi}_i} = \tilde{p}\tilde{B}_\eta \sum_{i=1}^{2} \frac{1}{\tilde{d}_i - \tilde{p}\tilde{\eta}} \tag{5.67}$$

$$\tilde{H}/\tilde{B} = \nu/\nu^* = \tilde{\nu} \tag{5.68}$$

Neglecting terms containing $\tilde{p}^2$ transforms (Equations 5.63 through 5.67) into the following form:

$$\tilde{p}\frac{\partial \tilde{H}_{\xi_k}}{\partial \tilde{\eta}} = (-1)^{3-k}\tilde{E}_{\xi_{3-k}}, \quad k = 1,2 \tag{5.69}$$

$$\tilde{E}_{\eta} = 0 \tag{5.70}$$

$$\frac{\partial \tilde{E}_{\xi_k}}{\partial \tilde{\eta}} = (-1)^k \tilde{p}\frac{\partial \tilde{B}_{\xi_{3-k}}}{\partial \tilde{t}}, \quad k = 1,2 \tag{5.71}$$

$$\sum_{i=1}^{2}(-1)^i \frac{\partial \tilde{E}_i}{\partial \tilde{\xi}_{3-i}} = -\frac{\partial \tilde{B}_{\eta}}{\partial \tilde{t}} \tag{5.72}$$

$$\frac{\partial \tilde{B}_{\eta}}{\partial \tilde{\eta}} = -\tilde{p}\sum_{i=1}^{2}\frac{\partial \tilde{B}_{\xi i}}{\partial \tilde{\xi}_i} \tag{5.73}$$

The tangential electric field can be expressed in terms of the magnetic field or magnetic flux using Equations 5.69, 5.71, and 5.68 as follows:

$$\tilde{E}^{\mathrm{b}}_{\xi_k} = (-1)^k \tilde{p}\frac{\partial \tilde{H}_{\xi_{3-k}}}{\partial \tilde{\eta}}\bigg|^{\mathrm{b}} \tag{5.74}$$

$$\tilde{E}^{\mathrm{b}}_{\xi_k} = (-1)^{3-k}\tilde{p}\frac{\partial}{\partial \tilde{t}}\int_0^{\infty} \tilde{B}_{\xi_{3-k}}\mathrm{d}\tilde{\eta} = (-1)^{3-k}\tilde{p}\frac{\partial}{\partial \tilde{t}}\int_0^{\infty} \frac{\tilde{H}_{\xi_{3-k}}}{\tilde{\nu}}\mathrm{d}\tilde{\eta} \tag{5.75}$$

$$\tilde{E}^{\mathrm{b}}_{\xi_k} = (-1)^k \tilde{p}\frac{\partial \tilde{H}_{\xi_{3-k}}}{\partial \tilde{\eta}}\bigg|^{\mathrm{b}} = (-1)^k \tilde{p}\frac{\partial (\tilde{\nu}\tilde{B})_{\xi_{3-k}}}{\partial \tilde{\eta}}\bigg|^{\mathrm{b}} \tag{5.76}$$

$$\tilde{E}^{\mathrm{b}}_{\xi_k} = (-1)^{3-k}\tilde{p}\frac{\partial}{\partial \tilde{t}}\int_0^{\infty} \tilde{B}_{\xi_{3-k}}\mathrm{d}\tilde{\eta} \tag{5.77}$$

As was shown in Chapter 2, by neglecting terms of the order of magnitude of $\tilde{p}^2$ and higher, the distribution of the electromagnetic field inside the conductor is approximated by a one-dimensional equations of diffusion in the direction normal to the conductor's surface, which is assumed to be planar (the Leontovich approximation). Substituting Equation 5.69 into Equation 5.71, we obtain

$$\frac{\partial^2 \tilde{H}_{\xi_k}}{\partial \tilde{\eta}^2} - \frac{\partial \tilde{B}_{\xi_k}}{\partial \tilde{t}} = 0, \quad k = 1,2 \tag{5.78}$$

Boundary and initial conditions for Equation 5.78 can be formulated as follows:

$$\tilde{\eta} = 0 : \tilde{B} = \tilde{B}^{\mathrm{b}}; \quad \tilde{\eta} \to \infty : \tilde{B} \to 0; \quad \tilde{t} = 0 : \tilde{B} = 0 \tag{5.79}$$

or

$$\tilde{\eta} = 0 : \tilde{H} = \tilde{H}^{\mathrm{b}}; \quad \tilde{\eta} \to \infty : \tilde{H} \to 0; \quad \tilde{t} = 0 : \tilde{H} = 0 \tag{5.80}$$

The question of which function (H or B) should be used as a state variable to solve the problem in Equation 5.78 is defined by the specific BH-curve used in a given problem. In most practical cases the magnetization curve is obtained by approximation of experimental data. Appropriate approximating functions should represent the whole range from the origin to saturation with a minimum error over the whole range. This subject has been widely considered in the literature using power series, transcendental functions, Fourier transforms, or exponential functions. Exponentials have been used in [7] to generate the following curve fit:

$$B(H) = a_0 + \sum_{i=1}^{4} a_i \exp(-\alpha_i H) \tag{5.81}$$

Table 5.1 gives the coefficients a_i and α_i for several magnetic materials.

Figures 5.8 through 5.11 show very good agreement between measured and computed magnetization curves for the materials given in Table 5.1.

Another approximation for the BH-curve for various materials has been proposed in [8]:

$$H = \nu(B^2)B = \left[\tilde{k}_1 \exp\left(\tilde{k}_2 B^2\right) + \tilde{k}_3\right] B \tag{5.82}$$

TABLE 5.1

Values for the Coefficients a_i and α_i (Equation 5.81) for Different Magnetic Materials

	Cold Rolled Steel	Cast Steel	Annealed Sheet Steel	Cast Iron
a_0	4.128	3.69	4.295	2.338
a_1	−1.191	−0.797	−0.358	−0.266
a_2	−2.638	−2.173	−2.887	−1.724
a_3	−1.24	−1.045	3.809	1.177
a_4	0.852	0.325	−4.859	−1.525
α_1	8.89×10^{-3}	3.81×10^{-3}	1.78×10^{-3}	1.07×10^{-3}
α_2	1.4×10^{-5}	1.27×10^{-5}	1.73×10^{-5}	3.61×10^{-5}
α_3	1.91×10^{-3}	1.01×10^{-3}	22.9×10^{-3}	4.57×10^{-3}
α_4	15.2×10^{-3}	12.7×10^{-3}	17.9×10^{-3}	3.3×10^{-3}

Source: El-Sherbiny, M., *IEEE Trans. Mag*, 9(1), 60, March 1973. With permission.

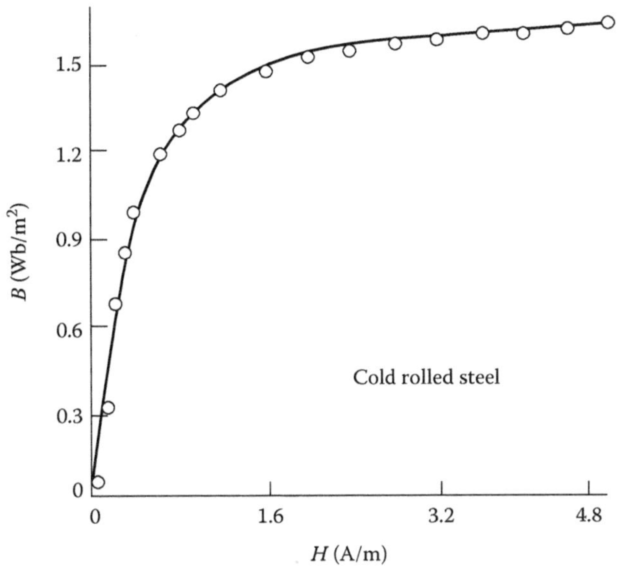

FIGURE 5.8
Comparison between measured *BH*-curve for cold rolled steel (solid points) and simulation using Equation 5.81 (solid curve).

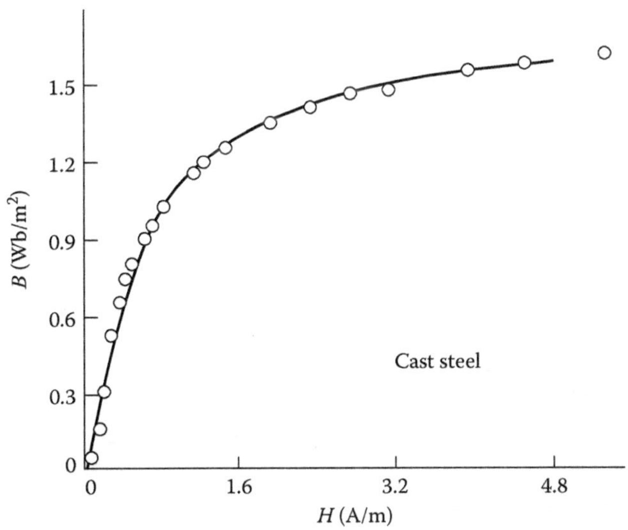

FIGURE 5.9
Comparison between measured *BH*-curve for cast steel (solid points) and simulation using Equation 5.81 (solid curve).

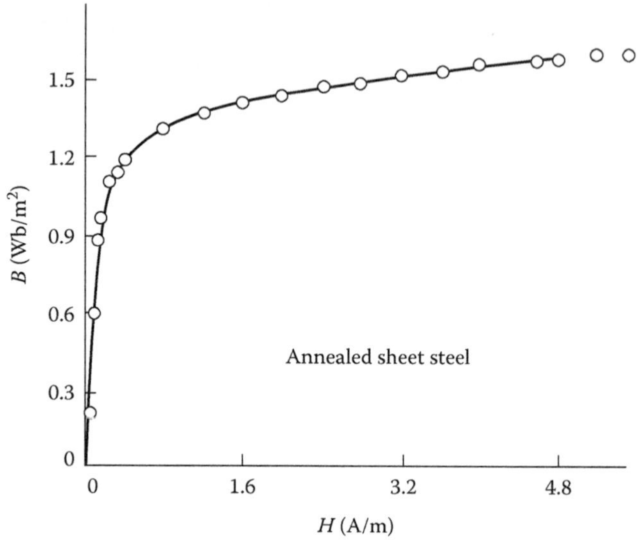

FIGURE 5.10
Comparison between measured *BH*-curve for annealed sheet steel (solid points) and simulation using Equation 5.81 (solid curve).

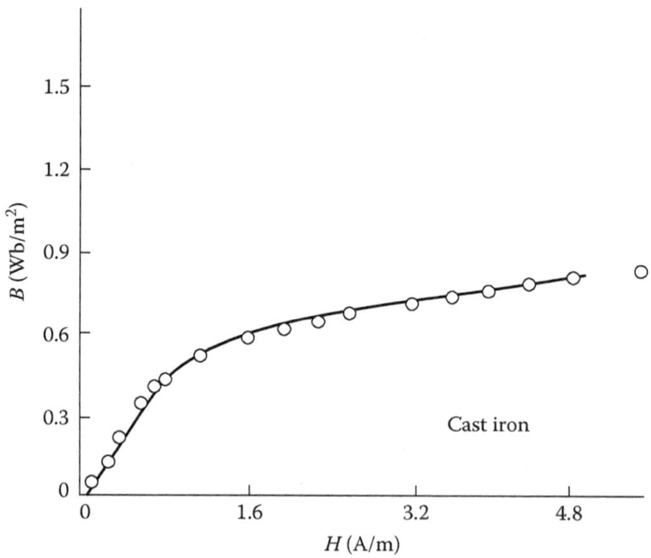

FIGURE 5.11
Comparison between measured *BH*-curve for cast iron (solid points) and simulation using Equation 5.81 (solid curve).

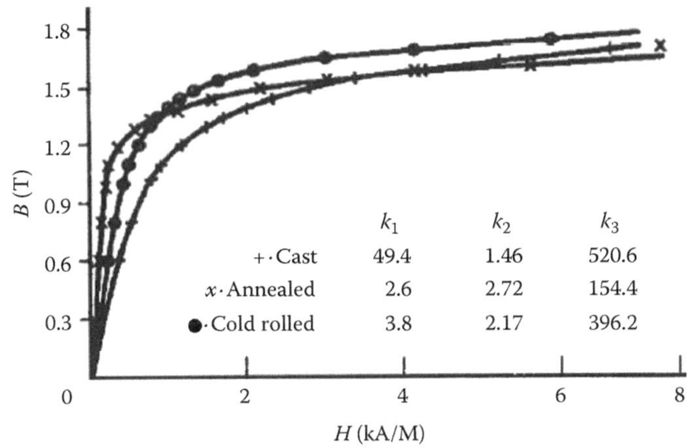

FIGURE 5.12
Comparison between measured and computed *BH*-curves using Equation 5.82 for three types of steel. (From Brauer, J., *IEEE Trans. Magnet.*, 11(1), 81, January 1975. With permission.)

where k_j are values determined from experimental data for any type of steel. In particular, for annealed steel $k_1 = 2.6$, $k_2 = 2.72$, $k_3 = 154.4$. Equation 5.82 also gives very good agreement between measured and computed data (see Figure 5.12).

In dimensionless variables, Equation 5.82 is written in the following form by removing all terms higher than $O(\tilde{p}^2)$:

$$\tilde{H} = \tilde{\nu}\tilde{B}; \quad \tilde{\nu} = (\nu^*)^{-1}\left[\tilde{k}_1 \exp\left(\tilde{k}_2 (B^*)^2 (\tilde{B}_0)^2\right) + \tilde{k}_3\right] \tag{5.83}$$

With the aid of Equation 5.83, the problem in Equations 5.78 through 5.80 can be expressed in the form:

$$\begin{cases} \partial^2(\tilde{\nu}\tilde{B})/\partial\tilde{\eta}^2 - \partial\tilde{B}/\partial\tilde{t} = 0; \\ \tilde{\eta} = 0: \quad \tilde{\nu}\tilde{B}^{\mathrm{b}} = \tilde{H}^{\mathrm{b}} \\ \tilde{\eta} \to \infty : \tilde{B} \to 0; \quad \tilde{t} = 0 : \tilde{B} = 0 \end{cases} \tag{5.84}$$

If we can specify the function $\tilde{H}_{\xi_k}$ on the surface of the conductor, the distributions of the function $\tilde{B}_{\xi_k}$ over the boundary layer can be obtained by solving the one-dimensional nonlinear problem in Equation 5.61. Substituting the results in Equation 5.76 or 5.77, we obtain the desired approximate boundary relation. Therefore, the relation in Equations 5.76 through 5.80 can be treated as the SIBC in *implicit* form. Note that the one-dimensional relation in Equation 5.78 is much more appealing for numerical implementation as compared with the initial equations in Equations 5.51 through 5.53.

The results obtained by solving the one-dimensional problem in Equations 5.78 through 5.80 can also be used to calculate the normal components of the magnetic field or magnetic flux density at the conductor' surface, using Equation 5.73:

$$\tilde{B}_{\eta}^{\mathrm{b}} = \tilde{p}\sum_{i=1}^{2}\frac{\partial}{\partial\tilde{\xi}_i}\int_0^{\infty}\tilde{B}_{\xi_i}\mathrm{d}\tilde{\eta} \tag{5.85}$$

$$\tilde{H}_{\eta}^{\mathrm{b}} = \tilde{p}\tilde{\nu}\sum_{i=1}^{2}\frac{\partial}{\partial\tilde{\xi}_i}\int_0^{\infty}\frac{\tilde{B}_{\xi_i}}{\tilde{\nu}}\mathrm{d}\tilde{\eta} \tag{5.86}$$

To illustrate these derivations consider again the passage of a current pulse in the system of conductors shown in Figure 5.3. The conductor is now cast steel (with the *BH*-curve given in [8]), $I = 100$ A and $\tau = 0.1$ s. The calculations are performed using four alternatives and the numerical results are shown in Figure 5.13: the PEC boundary conditions (curve *A*), the proposed nonlinear SIBC (Equations 5.76 through 5.80) (curve *B*), and two linear SIBCs (curves C and *D*). Curve *D* requires a preliminary nonlinear run. From this nonlinear run the point of largest field is used to determine the constant ν. This max-field method of choosing ν leads to a 40% error in the calculated power losses. The accuracy of the power losses computed using a linear SIBC can be

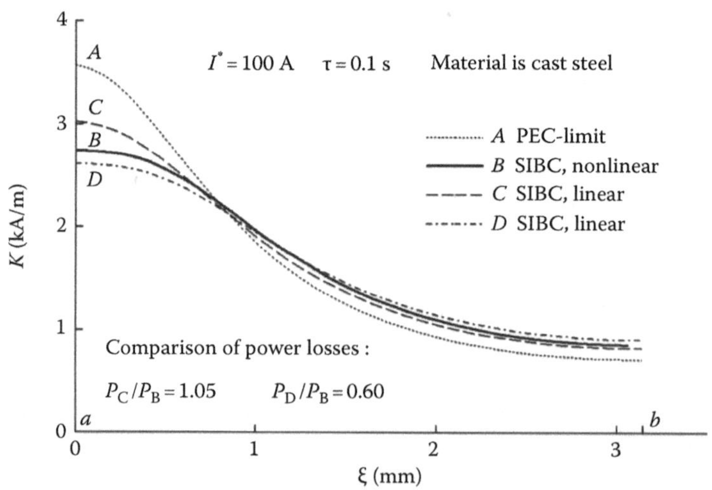

FIGURE 5.13
Distributions of the surface current density along half the contour around the cross section of one conductor shown in Figure 5.3.

improved if the characteristic reluctivity ν^* is used (curve *C*) instead of the max-field ν. However, the difference in the surface current density between curves *B* and *C* is larger than that between curves *B* and *D*.

In some practical problems (for example, in induction heating) several nonlinear effects ($\sigma(T)$, $\lambda(T)$, and $\nu\ (B)$) play a combined role and have to be taken into account [9,10]. In such cases, the characteristic skin depth δ and small parameter $\tilde{p}$ take the form:

$$\delta = \sqrt{\frac{\tau\nu^*}{\sigma^*}} \tag{5.87}$$

$$\tilde{p} = \frac{\delta}{D} = \sqrt{\frac{\tau\nu^*}{\sigma^* D^2}} \ll 1 \tag{5.88}$$

and the time-domain SIBC in implicit form can be represented as follows:

$$\begin{cases} \dfrac{\partial}{\partial\tilde{\eta}}\left(\tilde{\sigma}^{-1}\dfrac{\partial\tilde{H}_{\xi_k}}{\partial\tilde{\eta}}\right) = \dfrac{\partial\tilde{B}_{\xi_k}}{\partial\tilde{t}} \\ \dfrac{\partial\tilde{T}}{\partial\tilde{t}} = \tilde{\varepsilon}^2\dfrac{\partial}{\partial\tilde{\eta}}\left(\tilde{\lambda}\dfrac{\partial\tilde{T}}{\partial\tilde{\eta}}\right) + \tilde{\sigma}^{-1}\displaystyle\sum_{i=1}^{2}\left(\dfrac{\partial\tilde{H}_{\xi_i}}{\partial\tilde{\eta}}\right)^2 \end{cases} \tag{5.89}$$

where the scale factor for temperature and the small parameter $\tilde{\varepsilon}$ have been used in the following form:

$$\boldsymbol{T} = \frac{1}{c\rho\nu^*}\frac{\boldsymbol{I}^2}{\boldsymbol{D}^2}; \quad \tilde{\varepsilon} = \left(\lambda^*\sigma^*/(\nu^* c\rho)\right)^{1/2} \tag{5.90}$$

The one-dimensional equations in Equation 5.89 must be solved with the boundary and initial conditions given in Equations 5.79 and 5.80, and Equations 5.19 and 5.20. The results are then be substituted into Equations 5.74 and 5.75, or Equations 5.76 and 5.77 to obtain the tangential electric field at the interface.

Finally, we represent Equation 5.89 in dimensional variables as follows:

$$\begin{cases} \dfrac{\partial}{\partial\eta}\left(\sigma^{-1}\dfrac{\partial\tilde{H}_{\xi_k}}{\partial\eta}\right) = \dfrac{\partial B_{\xi_k}}{\partial t} \\ c\rho\dfrac{\partial T}{\partial t} = \dfrac{\partial}{\partial\eta}\left(\lambda\dfrac{\partial T}{\partial\eta}\right) + \sigma^{-1}\displaystyle\sum_{i=1}^{2}\left(\dfrac{\partial H_{\xi_i}}{\partial\eta}\right)^2 \end{cases} \tag{5.91}$$

Implicit time-domain SIBCs require more effort in the numerical implementation than frequency domain conditions based on limit theory.

However, time-domain SIBCs are not related to a single frequency and are therefore suitable for any *BH*-curve. In that sense, they have a wider application area.

5.4 Nonhomogeneous Conductors

5.4.1 PEC-Backed Lossy Dielectric Layer

In this section, we consider the SIBC formulation for a thin lossy dielectric layer (Figure 5.14). The distribution of the electric and magnetic fields in the layer is described by the following equations:

$$\nabla \times \vec{H} = \sigma\vec{E} + \varepsilon\frac{\partial\vec{E}}{\partial t} \tag{5.92}$$

$$\nabla \times \vec{E} = -\mu\frac{\partial\vec{H}}{\partial t} \tag{5.93}$$

Let the thickness h of the layer be much thinner than the characteristic size D of the conductors surface so the following conditions are satisfied:

$$h \ll D \tag{5.94}$$

We introduce the scale factor $\boldsymbol{\eta}$ for coordinate η in the form

$$\boldsymbol{\eta} = h \tag{5.95}$$

and represent the small parameter $\tilde{p}$ in the form:

$$\tilde{p} = \boldsymbol{\eta}/D \ll 1 \tag{5.96}$$

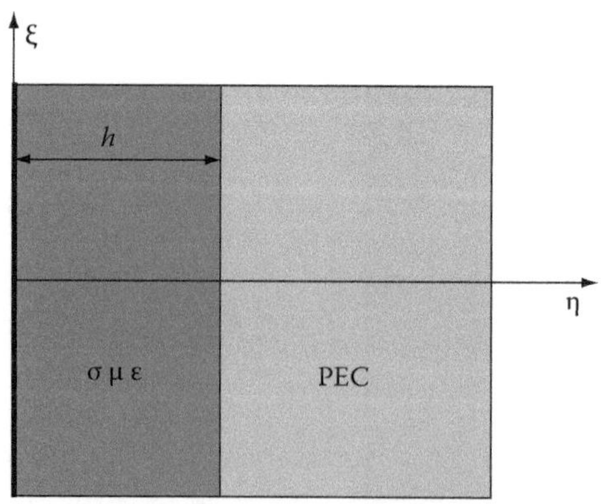

FIGURE 5.14
A PEC-backed thin lossy dielectric layer.

Scale factors for the electric and magnetic fields in both regions are obtained from Equations 3.14 and 3.15:

$$\boldsymbol{H} = \boldsymbol{I}/D; \quad \boldsymbol{E} = \frac{\mu_1}{\tau}\boldsymbol{I} \tag{5.97}$$

Now, Equations 5.92 and 5.93 can be written with nondimensional variables neglecting the terms of order $O(\tilde{p}^2)$ (Leontovich's approximation) as follows:

$$\tilde{p}\frac{\partial \tilde{H}_{\xi_k}}{\partial \tilde{\eta}} = (-1)^{3-k}\left(\tilde{E}_{\xi_{3-k}} + \tilde{a}\frac{\partial \tilde{E}_{\xi_{3-k}}}{\partial \tilde{t}}\right), \quad k = 1,2 \tag{5.98}$$

$$\frac{\partial \tilde{E}_{\xi_{3-k}}}{\partial \tilde{\eta}} = (-1)^{3-k}\tilde{p}\frac{\partial \tilde{H}_{\xi_k}}{\partial \tilde{t}}, \quad k = 1,2 \tag{5.99}$$

where $\tilde{a} = \varepsilon/(\sigma\tau)$. Equations 5.98 and 5.99 in the Laplace domain take the form:

$$\tilde{p}\frac{\partial \bar{\tilde{H}}_{\xi_k}}{\partial \tilde{\eta}} = (-1)^{3-k}\bar{\tilde{E}}_{\xi_{3-k}}(1 + \tilde{a}\tilde{s}), \quad k = 1,2 \tag{5.100}$$

$$\frac{\partial \bar{\tilde{E}}_{\xi_{3-k}}}{\partial \tilde{\eta}} = (-1)^{3-k}\tilde{p}\tilde{s}\bar{\tilde{H}}_{\xi_k}, \quad k = 1,2 \tag{5.101}$$

The electric field at the surface of the body ($\tilde{\eta} = 0$) is

$$\bar{\tilde{E}}^{\mathrm{b}}_{\xi_{3-k}} = \tilde{p}\frac{(-1)^{3-k}}{1 + \tilde{s}\tilde{a}}\frac{\partial \bar{\tilde{H}}_{\xi_k}}{\partial \tilde{\eta}}\Bigg|^{\mathrm{b}}, \quad k = 1,2 \tag{5.102}$$

At the opposite side of the layer ($\tilde{\eta} = \tilde{h} = h/\eta = 1$) the tangential electric field is zero

$$\tilde{\eta} = \tilde{h}\colon \bar{\tilde{E}}_{\xi_{3-k}} = \frac{\partial \bar{\tilde{H}}_{\xi_k}}{\partial \tilde{\eta}} = 0, \quad k = 1,2 \tag{5.103}$$

The distribution of the magnetic field is described by the following one-dimensional equations of diffusion:

$$\frac{\partial^2 \bar{\tilde{H}}_{\xi_k}}{\partial \tilde{\eta}^2} - (\tilde{s} + \tilde{s}^2\tilde{a})\bar{\tilde{H}}_{\xi_k} = 0, \quad k = 1,2 \tag{5.104}$$

The boundary conditions for the equations in Equation 5.107 can be written in the form ($k=1,2$):

$$\tilde{\eta} = 0: \bar{\tilde{H}}_{\xi_k} = \bar{\tilde{H}}^{\mathrm{b}}_{\xi_k} \tag{5.105}$$

$$\tilde{\eta} = \tilde{h}: \frac{\partial \bar{\tilde{H}}_{\xi_k}}{\partial \tilde{\eta}} = 0 \tag{5.106}$$

The solution of the problem in Equations 5.104 through 5.106 can be written in the form:

$$\bar{\tilde{H}}_{\xi_k} = \bar{\tilde{H}}^{\mathrm{b}}_{\xi_k} \frac{\exp\left(-\left(\tilde{\eta} - \tilde{h}\right)\sqrt{\tilde{s} + \tilde{s}^2\tilde{a}}\right) + \exp\left(\left(\tilde{\eta} - \tilde{h}\right)\sqrt{\tilde{s} + \tilde{s}^2\tilde{a}}\right)}{\exp\left(-\tilde{h}\sqrt{\tilde{s} + \tilde{s}^2\tilde{a}}\right) + \exp\left(\tilde{h}\sqrt{\tilde{s} + \tilde{s}^2\tilde{a}}\right)} \tag{5.107}$$

Taking the derivative with respect to $\tilde{\eta}$, we obtain

$$\frac{\partial \bar{\tilde{H}}_{\xi_k}}{\partial \tilde{\eta}} = \bar{\tilde{H}}^{\mathrm{b}}_{\xi_k} \sqrt{\tilde{s} + \tilde{s}^2\tilde{a}} \frac{-\exp\left(-\left(\tilde{\eta} - \tilde{h}\right)\sqrt{\tilde{s} + \tilde{s}^2\tilde{a}}\right) + \exp\left(\left(\tilde{\eta} - \tilde{h}\right)\sqrt{\tilde{s} + \tilde{s}^2\tilde{a}}\right)}{\exp\left(-\tilde{h}\sqrt{\tilde{s} + \tilde{s}^2\tilde{a}}\right) + \exp\left(\tilde{h}\sqrt{\tilde{s} + \tilde{s}^2\tilde{a}}\right)} \tag{5.108}$$

Substituting Equation 5.108 into Equation 5.102, we obtain

$$\begin{aligned}\bar{\tilde{E}}^{\mathrm{b}}_{\xi_{3-k}} &= (-1)^{3-k}\tilde{p}\frac{\tilde{s}}{\sqrt{\tilde{s} + \tilde{s}^2\tilde{a}}}\bar{\tilde{H}}^{\mathrm{b}}_{\xi_k} \frac{\exp\left(-\tilde{h}\sqrt{\tilde{s} + \tilde{s}^2\tilde{a}}\right) - \exp\left(\tilde{h}\sqrt{\tilde{s} + \tilde{s}^2\tilde{a}}\right)}{\exp\left(-\tilde{h}\sqrt{\tilde{s} + \tilde{s}^2\tilde{a}}\right) + \exp\left(\tilde{h}\sqrt{\tilde{s} + \tilde{s}^2\tilde{a}}\right)} \\ &= (-1)^k\tilde{p}\frac{\tilde{s}}{\sqrt{\tilde{s} + \tilde{s}^2\tilde{a}}}\bar{\tilde{H}}^{\mathrm{b}}_{\xi_k} \tan h(\tilde{h}\sqrt{\tilde{s} + \tilde{s}^2\tilde{a}})\end{aligned} \tag{5.109}$$

In the case of a thin layer

$$\tan h\left(\tilde{h}\sqrt{\tilde{s} + \tilde{s}^2\tilde{a}}\right) \approx \tilde{h}\sqrt{\tilde{s} + \tilde{s}^2\tilde{a}} \tag{5.110}$$

and Equation 5.109 becomes

$$\bar{\tilde{E}}^{\mathrm{b}}_{\xi_{3-k}} = (-1)^k\tilde{p}\tilde{s}\tilde{h}\bar{\tilde{H}}^{\mathrm{b}}_{\xi_k}, \quad k = 1,2 \tag{5.111}$$

Frequency- and time-domain analogs of Equation 5.111 can be written in the forms:

$$\dot{\tilde{E}}^{\mathrm{b}}_{\xi_{3-k}} = (-1)^k 2j\tilde{p}\tilde{h}\dot{\tilde{H}}^{\mathrm{b}}_{\xi_k}, \quad k = 1, 2 \tag{5.112}$$

$$\tilde{E}^{\mathrm{b}}_{\xi_{3-k}} = (-1)^k \tilde{p}\tilde{h}\frac{\mathrm{d}\tilde{H}^{\mathrm{b}}_{\xi_k}}{\mathrm{d}\tilde{t}}, \quad k = 1, 2 \tag{5.113}$$

These formulae are represented with dimensional variables in the following forms:

$$\dot{E}^{\mathrm{b}}_{\xi_{3-k}} = (-1)^k j\omega\mu h\dot{H}^{\mathrm{b}}_{\xi_k}, \quad k = 1, 2 \tag{5.114}$$

$$E^{\mathrm{b}}_{\xi_{3-k}} = (-1)^k \mu h\frac{\mathrm{d}H^{\mathrm{b}}_{\xi_k}}{\mathrm{d}t}, \quad k = 1, 2 \tag{5.115}$$

5.4.2 Two-Layer Conducting Structure

We now consider derivation of the impedance boundary conditions at the surface of a body consisting of two layers of different materials shown in Figure 5.15. Material properties are assumed to be constant. The electromagnetic field distribution in each layer can be described by the following equations:

$$\nabla \times \vec{H}^{(i)} = \sigma_i \vec{E}^{(i)} \tag{5.116}$$

$$\nabla \times \vec{E}^{(i)} = -\mu_i \frac{\partial \vec{H}^{(i)}}{\partial t} \tag{5.117}$$

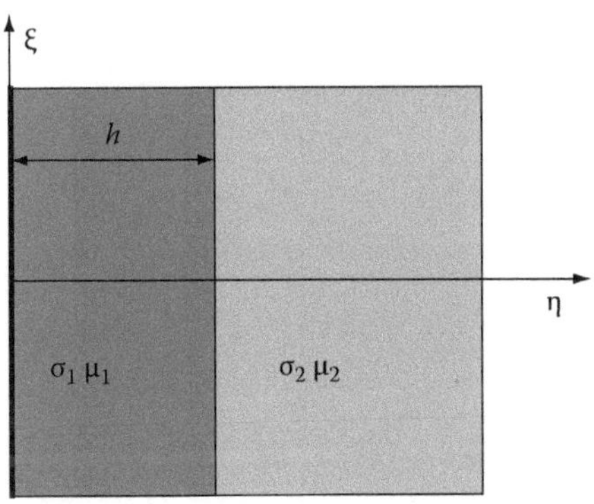

FIGURE 5.15
A two-layer conducting half-space.

Here the index i denotes the region (1 or 2). The electromagnetic depth of penetration in each region is

$$\delta_i = \sqrt{\tau/(\sigma_i \mu_i)} \tag{5.118}$$

Let the thickness h of the first layer and electromagnetic penetration depth $\delta^{(2)}$ in the second layer be much thinner than the characteristic size D of the conductors surface so that the following conditions are met:

$$h \ll D; \quad \delta_1 \ll D; \quad \delta_2 \ll D; \quad h + \delta_2 \ll D \tag{5.119}$$

It is natural to suppose that

$$h < \delta_1 \tag{5.120}$$

The scale factor $\boldsymbol{\eta}$ for coordinate η is now introduced in the form

$$\eta = h + \delta_2 \tag{5.121}$$

and the small parameter $\tilde{p}$ in the form

$$\tilde{p} = \boldsymbol{\eta}/D \tag{5.122}$$

The scale factors for the electric and magnetic fields in both regions are

$$\boldsymbol{H} = \boldsymbol{I}/D; \quad \boldsymbol{E} = \frac{\mu_1}{\tau}\boldsymbol{I} \tag{5.123}$$

Under these conditions Equations 5.116 and 5.117, which provide the distributions of the tangential electric and magnetic fields in the Leontovich approximation in each region can be written in the forms:

$$\tilde{p}\frac{\partial \tilde{H}^{(i)}_{\xi_k}}{\partial \tilde{\eta}} = (-1)^{3-k}\tilde{\gamma}_i^2 \tilde{E}^{(i)}_{\xi_{3-k}}, \quad k = 1,2, \quad i = 1,2 \tag{5.124}$$

$$\frac{\partial \tilde{E}^{(i)}_{\xi_{3-k}}}{\partial \tilde{\eta}} = (-1)^{3-k}\tilde{p}\frac{\partial \tilde{H}^{(i)}_{\xi_k}}{\partial \tilde{t}}, \quad k = 1,2, \quad i = 1,2 \tag{5.125}$$

where the nondimensional parameters $\tilde{\gamma}_i$ are

$$\tilde{\gamma}_1 = \boldsymbol{\eta}/\delta_1; \quad \tilde{\gamma}_2 = \boldsymbol{\eta}\mu_2/(\delta_2\mu_1) \tag{5.126}$$

Equations 5.123 and 5.124 in the Laplace domain take the forms:

$$\tilde{p}\frac{\partial \bar{\tilde{H}}^{(i)}_{\xi_k}}{\partial \tilde{\eta}} = (-1)^{3-k}\tilde{\gamma}_i^2 \bar{\tilde{E}}^{(i)}_{\xi_{3-k}}, \quad k = 1,2, \quad i = 1,2 \tag{5.127}$$

$$\frac{\partial \bar{\tilde{E}}^{(i)}_{\xi_{3-k}}}{\partial \tilde{\eta}} = (-1)^{3-k}\tilde{p}\tilde{s}\bar{\tilde{H}}^{(i)}_{\xi_k}, \quad k = 1,2, \quad i = 1,2 \tag{5.128}$$

The electric field at the surface of the body ($\tilde{\eta} = 0$) is

$$\bar{\tilde{E}}^{(1)}_{\xi_{3-k}} = (-1)^{3-k}\tilde{p}\tilde{\gamma}_1^{-2}\frac{\partial \bar{\tilde{H}}^{(1)}_{\xi_k}}{\partial \tilde{\eta}} \tag{5.129}$$

The distribution of the magnetic field in both regions is described by the following one-dimensional equations of diffusion:

$$\frac{\partial^2 \tilde{H}^{(i)}_{\xi_k}}{\partial \tilde{\eta}^2} - \tilde{\gamma}_i^2\frac{\partial \tilde{H}^{(i)}_{\xi_k}}{\partial \tilde{t}} = 0, \quad k = 1,2, \quad i = 1,2 \tag{5.130}$$

Equation 5.130 is represented in the Laplace domain as

$$\frac{\partial^2 \bar{\tilde{H}}^{(i)}_{\xi_k}}{\partial \tilde{\eta}^2} - \tilde{s}\tilde{\gamma}_i^2 \bar{\tilde{H}}^{(i)}_{\xi_k} = 0, \quad k = 1,2, \quad i = 1,2 \tag{5.131}$$

Boundary conditions for the equations in Equation 5.131 can be written in the form ($k = 1,2$):

$$\tilde{\eta} = 0: \bar{\tilde{H}}^{(1)}_{\xi_k} = \bar{\tilde{H}}^{(1)}_{\xi_k}\Big|^{b} \tag{5.132}$$

$$\tilde{\eta} = \tilde{h}: \bar{\tilde{H}}^{(1)}_{\xi_k} = \bar{\tilde{H}}^{(2)}_{\xi_k}; \quad \frac{\partial \bar{\tilde{H}}^{(1)}_{\xi_k}}{\partial \tilde{\eta}} = \frac{\partial \bar{\tilde{H}}^{(2)}_{\xi_k}}{\partial \tilde{\eta}}; \quad \tilde{h} = h/\boldsymbol{\eta} \tag{5.133}$$

$$\tilde{\eta} \to \infty: \bar{\tilde{H}}^{(2)}_{\xi_k} = 0 \tag{5.134}$$

Solutions of the problems in Equation 5.131 can be written in the forms:

$$\bar{\tilde{H}}^{(1)}_{\xi_k} = \bar{\tilde{X}}^{(1)}_k \exp\left(-\tilde{\gamma}_1\tilde{\eta}\sqrt{\tilde{s}}\right) + \bar{\tilde{Y}}^{(1)}_k \exp\left(\tilde{\gamma}_1\tilde{\eta}\sqrt{\tilde{s}}\right) \tag{5.135}$$

$$\bar{\tilde{H}}^{(2)}_{\xi_k} = \bar{\tilde{X}}^{(2)}_k \exp\left(-\tilde{\gamma}_2(\tilde{\eta} - \tilde{h})\sqrt{\tilde{s}}\right) + \bar{\tilde{Y}}^{(2)}_k \exp\left(\tilde{\gamma}_2(\tilde{\eta} - \tilde{h})\sqrt{\tilde{s}}\right) \tag{5.136}$$

Taking into account the boundary conditions in Equations 5.132 through 5.134, the coefficients $\bar{\tilde{X}}_k^{(i)}$ and $\bar{\tilde{Y}}_k^{(i)}$ can be obtained and written in the forms:

$$\bar{\tilde{X}}_k^{(1)} = \bar{\tilde{H}}_{\xi_k}^{(1)}\Big|^b \frac{(\tilde{\gamma}_1 + \tilde{\gamma}_2)\exp\left(\tilde{\gamma}_1\tilde{h}\sqrt{\bar{s}}\right)}{(\tilde{\gamma}_1 + \tilde{\gamma}_2)\exp\left(\tilde{\gamma}_1\tilde{h}\sqrt{\bar{s}}\right) + (\tilde{\gamma}_1 - \tilde{\gamma}_2)\exp\left(-\tilde{\gamma}_1\tilde{h}\sqrt{\bar{s}}\right)}$$

$$= \bar{\tilde{H}}_{\xi_k}^{(1)}\Big|^b \frac{(\tilde{\gamma}_1 + \tilde{\gamma}_2)\exp\left(\tilde{\gamma}_1\tilde{h}\sqrt{\bar{s}}\right)}{2\tilde{\gamma}_1 ch\left(\tilde{\gamma}_1\tilde{h}\sqrt{\bar{s}}\right) + 2\tilde{\gamma}_2 sh\left(\tilde{\gamma}_1\tilde{h}\sqrt{\bar{s}}\right)} \tag{5.137}$$

$$\bar{\tilde{Y}}_k^{(1)} = \bar{\tilde{H}}_{\xi_k}^{(1)}\Big|^b \frac{(\tilde{\gamma}_1 - \tilde{\gamma}_2)\exp\left(-\tilde{\gamma}_1\tilde{h}\sqrt{\bar{s}}\right)}{(\tilde{\gamma}_1 + \tilde{\gamma}_2)\exp\left(\tilde{\gamma}_1\tilde{h}\sqrt{\bar{s}}\right) + (\tilde{\gamma}_1 - \tilde{\gamma}_2)\exp\left(-\tilde{\gamma}_1\tilde{h}\sqrt{\bar{s}}\right)}$$

$$= \bar{\tilde{H}}_{\xi_k}^{(1)}\Big|^b \frac{(\tilde{\gamma}_1 - \tilde{\gamma}_2)\exp\left(-\tilde{\gamma}_1\tilde{h}\sqrt{\bar{s}}\right)}{2\tilde{\gamma}_1 ch\left(\tilde{\gamma}_1\tilde{h}\sqrt{\bar{s}}\right) + 2\tilde{\gamma}_2 sh\left(\tilde{\gamma}_1\tilde{h}\sqrt{\bar{s}}\right)} \tag{5.138}$$

$$\bar{\tilde{X}}_k^{(2)} = \bar{\tilde{H}}_{\xi_k}^{(1)}\Big|^b \frac{2\tilde{\gamma}_1}{(\tilde{\gamma}_1 + \tilde{\gamma}_2)\exp\left(\tilde{\gamma}_1\tilde{h}\sqrt{\bar{s}}\right) + (\tilde{\gamma}_1 - \tilde{\gamma}_2)\exp\left(-\tilde{\gamma}_1\tilde{h}\sqrt{\bar{s}}\right)}$$

$$= \bar{\tilde{H}}_{\xi_k}^{(1)}\Big|^b \frac{2\tilde{\gamma}_1}{2\tilde{\gamma}_1 ch\left(\tilde{\gamma}_1\tilde{h}\sqrt{\bar{s}}\right) + 2\tilde{\gamma}_2 sh\left(\tilde{\gamma}_1\tilde{h}\sqrt{\bar{s}}\right)} \tag{5.139}$$

$$\bar{\tilde{Y}}_k^{(2)} = 0 \tag{5.140}$$

where

$$ch(x) = \frac{\exp(-x) + \exp(x)}{2}; \quad sh(x) = \frac{\exp(x) - \exp(-x)}{2} \tag{5.141}$$

The normal derivative of the tangential magnetic field in the first region is obtained directly from Equation 5.135:

$$\frac{\partial \bar{\tilde{H}}_{\xi_k}^{(1)}}{\partial \eta} = \tilde{\gamma}_1\sqrt{\bar{s}}\left(-\bar{\tilde{X}}_k^{(1)}\exp\left(-\tilde{\gamma}_1\tilde{\eta}\sqrt{\bar{s}}\right) + \bar{\tilde{Y}}_k^{(1)}\exp\left(\tilde{\gamma}_1\tilde{\eta}\sqrt{\bar{s}}\right)\right) \tag{5.142}$$

Application of Equation 5.142 on the surface of the conductor ($\tilde{\eta}=0$) gives

$$\left.\frac{\partial \bar{\tilde{H}}_{\xi_k}^{(1)}}{\partial \eta}\right|^{\mathrm{b}} = \sqrt{\tilde{s}}\tilde{\gamma}_1\left(-\tilde{X}_k^{(1)} + \tilde{Y}_k^{(1)}\right)\sqrt{\tilde{s}}\tilde{\gamma}_1 \left.\bar{\tilde{H}}_{\xi_k}^{(1)}\right|^{\mathrm{b}}\left[-\frac{(\tilde{\gamma}_1+\tilde{\gamma}_2)\exp\left(\tilde{\gamma}_1\tilde{h}\sqrt{\tilde{s}}\right)}{2\tilde{\gamma}_1 ch\left(\tilde{\gamma}_1\tilde{h}\sqrt{\tilde{s}}\right)+2\tilde{\gamma}_2 sh\left(\tilde{\gamma}_1\tilde{h}\sqrt{\tilde{s}}\right)}\right.$$

$$\left.+\frac{(\tilde{\gamma}_1-\tilde{\gamma}_2)\exp\left(-\tilde{\gamma}_1\tilde{h}\sqrt{\tilde{s}}\right)}{2\tilde{\gamma}_1 ch\left(\tilde{\gamma}_1\tilde{h}\sqrt{\tilde{s}}\right)+2\tilde{\gamma}_2 sh\left(\tilde{\gamma}_1\tilde{h}\sqrt{\tilde{s}}\right)}\right]$$

$$= -\sqrt{\tilde{s}}\tilde{\gamma}_1 \left.\bar{\tilde{H}}_{\xi_k}^{(1)}\right|^{\mathrm{b}} \frac{\tilde{\gamma}_1 sh\left(\tilde{\gamma}_1\tilde{h}\sqrt{\tilde{s}}\right)+\tilde{\gamma}_2 ch\left(\tilde{\gamma}_1\tilde{h}\sqrt{\tilde{s}}\right)}{\tilde{\gamma}_1 ch\left(\tilde{\gamma}_1\tilde{h}\sqrt{\tilde{s}}\right)+\tilde{\gamma}_2 sh\left(\tilde{\gamma}_1\tilde{h}\sqrt{\tilde{s}}\right)} \tag{5.143}$$

Substituting Equation 5.143 into Equation 5.129, we obtain the desired SIBC between the tangential electric and magnetic fields at the conductor's surface:

$$\left.\bar{\tilde{E}}_{\xi_{3-k}}^{(1)}\right|^{\mathrm{b}} = (-1)^k\tilde{p}\tilde{\gamma}_1^{-1}\sqrt{\tilde{s}}\frac{\tilde{\gamma}_1 sh\left(\tilde{\gamma}_1\tilde{h}\sqrt{\tilde{s}}\right)+\tilde{\gamma}_2 ch\left(\tilde{\gamma}_1\tilde{h}\sqrt{\tilde{s}}\right)}{\tilde{\gamma}_1 ch\left(\tilde{\gamma}_1\tilde{h}\sqrt{\tilde{s}}\right)+\tilde{\gamma}_2 sh\left(\tilde{\gamma}_1\tilde{h}\sqrt{\tilde{s}}\right)}\left.\bar{\tilde{H}}_{\xi_k}^{(1)}\right|^{\mathrm{b}}$$

$$= (-1)^k\tilde{p}\sqrt{\tilde{s}}\frac{sh\left(\tilde{\gamma}_1\tilde{h}\sqrt{\tilde{s}}\right)+\frac{\tilde{\gamma}_2}{\tilde{\gamma}_1}ch\left(\tilde{\gamma}_1\tilde{h}\sqrt{\tilde{s}}\right)}{ch\left(\tilde{\gamma}_1\tilde{h}\sqrt{\tilde{s}}\right)+\frac{\tilde{\gamma}_2}{\tilde{\gamma}_1}sh\left(\tilde{\gamma}_1\tilde{h}\sqrt{\tilde{s}}\right)}\left.\bar{\tilde{H}}_{\xi_k}^{(1)}\right|^{\mathrm{b}} \tag{5.144}$$

From Equation 5.126 it follows that

$$\frac{\tilde{\gamma}_2}{\tilde{\gamma}_1} = \frac{\mu_2\delta_1}{\mu_1\delta_2} = \frac{\sigma_1\delta_1}{\sigma_2\delta_2} \tag{5.145}$$

Substitution of Equation 5.145 into Equation 5.144 gives

$$\left.\bar{\tilde{E}}_{\xi_{3-k}}^{(1)}\right|^{\mathrm{b}} = (-1)^k\tilde{p}\sqrt{\tilde{s}}\frac{sh\left(\tilde{\gamma}_1\tilde{h}\sqrt{\tilde{s}}\right)+\frac{\sigma_1\delta_1}{\sigma_2\delta_2}ch\left(\tilde{\gamma}_1\tilde{h}\sqrt{\tilde{s}}\right)}{ch\left(\tilde{\gamma}_1\tilde{h}\sqrt{\tilde{s}}\right)+\frac{\sigma_1\delta_1}{\sigma_2\delta_2}sh\left(\tilde{\gamma}_1\tilde{h}\sqrt{\tilde{s}}\right)}\left.\bar{\tilde{H}}_{\xi_k}^{(1)}\right|^{\mathrm{b}} \tag{5.146}$$

Equation 5.146 is too complex for the inverse Laplace transform. We therefore restrict ourselves to its representation in the frequency domain:

$$\left.\dot{\tilde{E}}_{\xi_{3-k}}^{(1)}\right|^{\mathrm{b}} = (-1)^k\tilde{p}\sqrt{2j}\frac{sh\left(\tilde{\gamma}_1\tilde{h}\sqrt{2j}\right)+\frac{\sigma_1\delta_1}{\sigma_2\delta_2}ch\left(\tilde{\gamma}_1\tilde{h}\sqrt{2j}\right)}{ch\left(\tilde{\gamma}_1\tilde{h}\sqrt{2j}\right)+\frac{\sigma_1\delta_1}{\sigma_2\delta_2}sh\left(\tilde{\gamma}_1\tilde{h}\sqrt{2j}\right)}\left.\dot{\tilde{H}}_{\xi_k}^{(1)}\right|^{\mathrm{b}} \tag{5.147}$$

Return to dimensional variables yields:

$$\dot{E}^{(1)}_{\xi_{3-k}}\Big|^{b} = (-1)^k \frac{1+j}{\delta_1\sigma_1} \frac{sh((1+j)h/\delta_1) + \dfrac{\sigma_1\delta_1}{\sigma_2\delta_2} ch((1+j)h/\delta_1)}{ch((1+j)h/\delta_1) + \dfrac{\sigma_1\delta_1}{\sigma_2\delta_2} sh((1+j)h/\delta_1)} \dot{H}^{(1)}_{\xi_k}\Big|^{b} \tag{5.148}$$

Therefore the surface impedance is

$$Z = \frac{1+j}{\delta_1\sigma_1} \frac{sh((1+j)h/\delta_1) + \dfrac{\sigma_1\delta_1}{\sigma_2\delta_2} ch((1+j)h/\delta_1)}{ch((1+j)h/\delta_1) + \dfrac{\sigma_1\delta_1}{\sigma_2\delta_2} sh((1+j)h/\delta_1)} \tag{5.149}$$

Note that the surface impedance of layer "1" is

$$Z^{(1)} = \frac{1+j}{\delta_1\sigma_1} \tag{5.150}$$

Equation 5.149 can be represented in another form:

$$Z = Z^{(1)} \frac{sh((1+j)h/\delta_1) + \dfrac{\sigma_1\delta_1}{\sigma_2\delta_2} ch((1+j)h/\delta_1)}{ch((1+j)h/\delta_1) + \dfrac{\sigma_1\delta_1}{\sigma_2\delta_2} sh((1+j)h/\delta_1)} \tag{5.151}$$

It is easy to see that

$$Z \to Z^{(1)} \quad \text{when } h/\delta_1 \to \infty \tag{5.152}$$

If layer "1" is so thin that

$$sh\left(\tilde{\theta}^{(1)}\tilde{h}\right) \approx \tilde{\theta}^{(1)}\tilde{h}; \quad ch\left(\tilde{\theta}^{(1)}\tilde{h}\right) \approx 1 \tag{5.153}$$

Equation 5.146 can be reduced to the following form:

$$\begin{aligned}
\bar{\bar{E}}^{(1)}_{\xi_{3-k}}\Big|^{b} &= (-1)^k \tilde{p}\sqrt{\bar{s}} \frac{\tilde{\gamma}_1\tilde{h}\sqrt{\bar{s}} + \dfrac{\sigma_1\delta_1}{\sigma_2\delta_2}}{1 + \dfrac{\sigma_1\delta_1}{\sigma_2\delta_2}\tilde{\gamma}_1\tilde{h}\sqrt{\bar{s}}} \bar{\bar{H}}^{(1)}_{\xi_k}\Big|^{b} \approx (-1)^k \tilde{p}\sqrt{\bar{s}}\bar{\bar{H}}^{(1)}_{\xi_k}\Big|^{b} \\
&\quad \times \left(\tilde{\gamma}_1\tilde{h}\sqrt{\bar{s}} + \frac{\sigma_1\delta_1}{\sigma_2\delta_2}\right)\left(1 - \frac{\sigma_1\delta_1}{\sigma_2\delta_2}\tilde{\gamma}_1\tilde{h}\sqrt{\bar{s}}\right) \\
&= (-1)^k \tilde{p}\sqrt{\bar{s}}\bar{\bar{H}}^{(1)}_{\xi_k}\Big|^{b} \frac{\sigma_1\delta_1}{\sigma_2\delta_2}\left(1 + \frac{\sigma_2\delta_2}{\sigma_1\delta_1}\tilde{\gamma}_1\tilde{h}\sqrt{\bar{s}}\right)\left(1 - \frac{\sigma_1\delta_1}{\sigma_2\delta_2}\tilde{\gamma}_1\tilde{h}\sqrt{\bar{s}}\right) \\
&\approx (-1)^k \tilde{p}\sqrt{\bar{s}}\bar{\bar{H}}^{(1)}_{\xi_k}\Big|^{b} \frac{\sigma_1\delta_1}{\sigma_2\delta_2}\left[1 + \tilde{\gamma}_1\tilde{h}\sqrt{\bar{s}}\frac{(\sigma_2\delta_2)^2 - (\sigma_1\delta_1)^2}{\sigma_1\sigma_2\delta_1\delta_2}\right]
\end{aligned} \tag{5.154}$$

Representing the SIBC, in Equation 5.154, in the frequency and time domains, we obtain

$$\dot{\tilde{E}}^{(1)}_{\xi_{3-k}}\Big|^{\mathrm{b}} \approx (-1)^k \tilde{p}\sqrt{2j}\dot{\tilde{H}}^{(1)}_{\xi_k}\Big|^{\mathrm{b}} \frac{\sigma_1\delta_1}{\sigma_2\delta_2}\left[1+\tilde{\gamma}_1\tilde{h}\sqrt{2j}\frac{(\sigma_2\delta_2)^2-(\sigma_1\delta_1)^2}{\sigma_1\sigma_2\delta_1\delta_2}\right] \tag{5.155}$$

$$\tilde{E}^{(1)}_{\xi_{3-k}}\Big|^{\mathrm{b}} \approx (-1)^k \tilde{p}\frac{\sigma_1\delta_1}{\sigma_2\delta_2}\left[\tilde{H}^{(1)}_{\xi_k}\Big|^{\mathrm{b}} * \tilde{\tilde{T}}^{\mathrm{b}}_2 + \tilde{\gamma}_1\tilde{h}\frac{(\sigma_2\delta_2)^2-(\sigma_1\delta_1)^2}{\sigma_1\sigma_2\delta_1\delta_2}\frac{\mathrm{d}}{\mathrm{d}\tilde{t}}\left(\tilde{H}^{(1)}_{\xi_k}\Big|^{\mathrm{b}}\right)\right] \tag{5.156}$$

Switching to dimensional variables in Equations 5.155 and 5.156, we obtain

$$\begin{aligned}\dot{E}^{(1)}_{\xi_{3-k}}\Big|^{\mathrm{b}} &\approx (-1)^k \frac{1+j}{\sigma_1\delta_1}\dot{\tilde{H}}^{(1)}_{\xi_k}\Big|^{\mathrm{b}} \frac{\sigma_1\delta_1}{\sigma_2\delta_2}\left[1+(1+j)\frac{h}{\delta_1}\frac{(\sigma_2\delta_2)^2-(\sigma_1\delta_1)^2}{\sigma_1\sigma_2\delta_1\delta_2}\right]\\ &= (-1)^k \frac{1+j}{\sigma_2\delta_2}\dot{\tilde{H}}^{(1)}_{\xi_k}\Big|^{\mathrm{b}}\left[1+(1+j)\frac{h}{\delta_1}\frac{(\sigma_2\delta_2)^2-(\sigma_1\delta_1)^2}{\sigma_1\sigma_2\delta_1\delta_2}\right]\end{aligned} \tag{5.157}$$

$$E^{(1)}_{\xi_{3-k}}\Big|^{\mathrm{b}} \approx (-1)^k \sqrt{\frac{\mu_1}{\sigma_1}\frac{\sigma_1\delta_1}{\sigma_2\delta_2}}\left[H^{(1)}_{\xi_k}\Big|^{\mathrm{b}} * \hat{T}^{\mathrm{b}}_2 + h\frac{(\sigma_2\delta_2)^2-(\sigma_1\delta_1)^2}{\sigma_1\sigma_2\delta_1\delta_2}\frac{\mathrm{d}}{\mathrm{d}t}\left(H^{(1)}_{\xi_k}\Big|^{\mathrm{b}}\right)\right] \tag{5.158}$$

Naturally Equations 5.157 and 5.158 transform to Equations 3.55 and 3.56, respectively, by setting $\sigma_1 = \sigma_2$ and $\mu_1 = \mu_2$.

References

1. P.G. Agarwal, Eddy current losses in solid and laminated iron, *AIEE Transactions*, 78, Part 1, May 1959, 169–181.
2. W. MacLean, Theory of strong electromagnetic waves in massive iron, *Journal of Applied Physics*, 25(10), October 1954, 1267–1270.
3. D. Lowther and E. Wyatt, The computation of eddy current losses in solid iron under various surface conditions, COMPUMAG Conference, Oxford, S.W. 7, 1976.
4. E.M. Deeley, Flux penetration in two dimensions into saturating iron and the use of surface equations, *Proceedings of the IEE*, 126(2), February 1979, 204–208.
5. C. Guerin, G. Meunier, and G. Tanneau, Surface impedance for 3D nonlinear eddy current problems—Application to loss computation in transformers, *IEEE Transactions Magnetics*, 32(3), May 1996, 808–811.
6. W. Mai and C. Henneberger, Field and temperature calculations in transverse flux inductive heating devices heating nonparamagnetic materials using surface impedance formulations for nonlinear eddy-current problems, *IEEE Transactions on Magnetics*, 25(3), Part 1, May 1999, 1590–1593.

7. M. El-Sherbiny, Representation of the magnetization characteristic by a sum of exponentials, *IEEE Transactions on Magnetics*, 9(1), March 1973, 60–61.
8. J. Brauer, Simple equations for the magnetization and reluctivity curves of steel, *IEEE Transactions on Magnetics*, 11(1), January 1975, 81.
9. F. Azzouz and M. Feliachi, Non-linear surface impedance taking account of thermal effect induction heating, *IEEE Transactions on Magnetics*, 37(5), Part 1, September 2001, 3175–3177.
10. J. Nerg and J. Partanen, A simplified FEM based calculation model for 3-D induction heating problems using surface impedance formulations, *IEEE Transactions on Magnetics*, 37(5), Part 1, September 2001, 3719–3722.

6

Implementation of SIBCs for the Boundary Integral Equation Method: Low-Frequency Problems

6.1 Introduction

The numerical method best suited for use with SIBCs is the boundary element method (BEM) because in both the BEM and the SIBCs the functions are approximated at the same points on the interface between the media. When applied to an eddy current problem consisting of conducting and nonconducting regions, the BEM yields a system of two integral equations over the conductor's surface with respect to two unknowns: the required function and its normal derivative at the conductor/dielectric interface [1]. Under the conditions of skin effect, the electromagnetic field behavior in the conductor is known, the surface impedance concept may be applied, and the formulation can be reduced to a single integral equation employing the fundamental solution of the Laplace equation. The extra unknown is eliminated using the SIBCs relating the function and its normal derivative at the conductor's surface. This approach is almost ideal for solving time-harmonic problems. Because of this, BEM–SIBC formulations have been widely used for the analysis of skin and proximity effect problems of multiconductor systems [2–7].

In the general case of transient excitation of the surface integral equation, including the time-domain SIBCs in the form Equation 3.81, 3.118, or 3.132, must be solved at every time step due to the time convolution terms. This leads to a significant increase in the computational cost required for the solution and, for all practical purposes, renders its numerical implementation impractical. The best way to avoid this difficulty is the separation of variables in the formulation into space and time components. In this case, the integral equation for the space component needs to be solved only once for a given system of conductors, and the result multiplied by the time component to obtain the solution of the problem for any time dependence of the source.

It is easy to see that the variables in the surface impedance function given in Equations 3.82 through 3.84 cannot be separated because the functions T_m are different. However, in the low-frequency case (the so-called quasistatic approximation, in which the displacement current can be neglected) each term in Equations 3.82 through 3.84, when considered independently, admits separations of variables. This circumstance and the fact that the SIBC in Equations 3.82 through 3.84 can be represented as power series in the small parameter are keys to the development of the formulation desired. Indeed, it is natural to suppose that use of the perturbation technique as described in Chapter 2 will lead to a set of integral equations so that every equation includes only one term of Equations 3.82 through 3.84 and, consequently, admits separation of variables. Such a formulation has a clear physical meaning: the zero-order integral equation gives the solution in the well-known perfect electrical conductor limit and the other equations contribute corrections of the order of Leontovich, Mitzner, and Rytov approximations. Thus the total number of integral equations in the formulation will not exceed four. In fact, it may be even less depending on the problem as will be shown in Chapter 9. Derivation of this type of formulation is the aim of this chapter.

Consider a system of N cylindrical conductors of arbitrary cross sections surrounded by a homogeneous nonconducting space as shown in Figure 6.1. The parameters of the conducting and nonconducting media are assumed to be constants. To simplify derivations and without loss of generality, let us assume that the magnetic permeabilities of the conducting and dielectric regions are the same. Let an external source produce quasisteady current pulses flowing through the conductors so that the condition of applicability of the surface impedance concept in Equation 2.1 is satisfied. In low-frequency problems, the electromagnetic field distribution in both the dielectric and conducting regions can be described by the following equations:

Conducting domain:

$$\nabla \times \vec{E} = -\mu \frac{\partial \vec{H}}{\partial t} \tag{6.1}$$

$$\nabla \times \vec{H} = \sigma \vec{E} \tag{6.2}$$

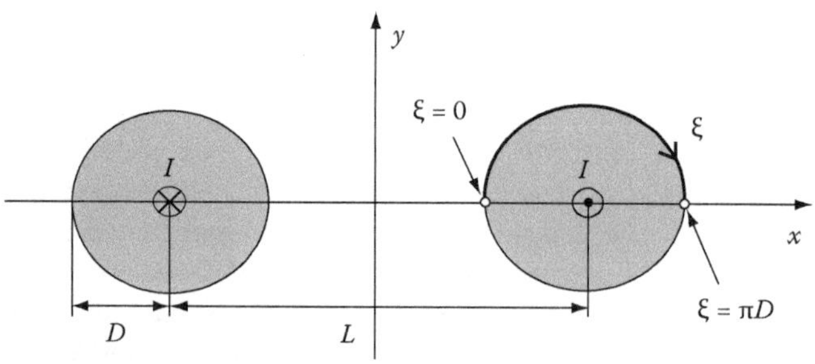

FIGURE 6.1
Simulation setup.

$$\nabla \cdot \vec{H} = 0 \tag{6.3}$$

$$\nabla \cdot \vec{E} = 0 \tag{6.4}$$

Nonconducting domain:

$$\nabla \times \vec{E} = -\mu \frac{\partial \vec{H}}{\partial t} \tag{6.5}$$

$$\nabla \times \vec{H} = 0 \tag{6.6}$$

$$\nabla \cdot \vec{H} = 0 \tag{6.7}$$

$$\nabla \cdot \vec{E} = 0 \tag{6.8}$$

6.2 Two-Dimensional Problems

Consider the particular case of long parallel conductors of constant cross section as shown in Figure 6.1. We direct the tangential coordinate, ξ_1, along the conductors; set $d_1 \to \infty$ and assume that no variation of the electromagnetic field takes place in this direction

$$\frac{\partial f}{\partial \xi_1} = 0 \tag{6.9}$$

where f denotes any function. Under these conditions, the problem of the electromagnetic field distribution can be considered as two-dimensional in the plane of cross sections of the conductors, and the vectors $\vec{I}_i$, $\vec{E}$, $\vec{H}$, and $\vec{A}_1$ can be represented in the following form:

$$\vec{I}_i = (I_i)_{\xi_1}\vec{e}_1; \quad \vec{E} = E_{\xi_1}\vec{e}_1; \quad \vec{H} = H_{\xi_2}\vec{e}_2 + H_{\eta}\vec{e}_3; \quad \vec{A} = A_{\xi_1}\vec{e}_1 \tag{6.10}$$

Further derivations for the two-dimensional case will be performed in terms of electric–magnetic fields and vector potential formalisms.

6.2.1 *E–H* Formalism

In the two-dimensional case, Equations 6.1 through 6.8 are written in the following form with local variables (ξ_2, η):

Conducting domain:

$$\frac{\partial E}{\partial \eta} = -\mu \frac{\partial H_{\xi_2}}{\partial t} \tag{6.11}$$

$$\nabla \times \vec{H} = \sigma \vec{E} \tag{6.12}$$

$$\nabla \cdot \vec{H} = 0 \tag{6.13}$$

$$\nabla \cdot \vec{E} = 0 \tag{6.14}$$

Nonconducting domain:

$$\nabla^2 E = 0 \tag{6.15}$$

Since the distribution of the electric field in the dielectric region obeys the Laplace equation, the boundary integral equation method [1] yields the following surface integral equation

$$\tilde{c}E + \sum_{i=1}^{N} \oint_{L_i} E \frac{\partial G^{(2D)}}{\partial \vec{n}} \mathrm{d}\xi_2 = \sum_{i=1}^{N} \oint_{L_i} G^{(2D)} \frac{\partial E}{\partial \vec{n}} \mathrm{d}\xi_2 \tag{6.16}$$

where

- L_i is the contour of the cross section of the ith conductor
- $\tilde{c} = \tilde{c}(\xi_2)$ is a nondimensional coefficient depending on the contour (for a smooth contour $\tilde{c} = 0.5$)
- $\vec{n}$ is the normal unit vector directed inside the conductor
- $G^{(2D)}$ is the fundamental solution of the two-dimensional Laplace equation in free space

$$G^{(2D)}\left(\xi_2', \xi_2\right) = -(2\pi)^{-1} \ln\left(R\left(\xi_2', \xi_2\right)\right) \tag{6.17}$$

where R is the distance between the observation and integration points

$$R\left(\xi_2', \xi_2\right) = \left|\vec{r}(\xi_2) - \vec{r}\left(\xi_2'\right)\right| = \left|\vec{r} - \vec{r}'\right| \tag{6.18}$$

In the two-dimensional case, the SIBCs of the order of Rytov's approximation given in Equations 3.59 and 3.60 reduce to the following form:

Frequency domain:

$$\begin{aligned} \dot{E}^{\mathrm{b}} &= \frac{1}{\sigma}\left[\sqrt{j\omega\sigma\mu}\dot{H}^{\mathrm{b}}_{\xi_2} + \frac{1}{2d_2}\dot{H}^{\mathrm{b}}_{\xi_2} + \frac{1}{\sqrt{j\omega\sigma\mu}}\left(\frac{3}{8d_2^2}\dot{H}^{\mathrm{b}}_{\xi_2} + \frac{1}{2}\frac{\partial^2 \dot{H}^{\mathrm{b}}_{\xi_2}}{\partial \xi_2^2}\right)\right] \\ &= \frac{j\omega\mu}{\sqrt{j\omega\sigma\mu}}\left[\dot{H}^{\mathrm{b}}_{\xi_2} + \frac{1}{2d_2}\frac{\dot{H}^{\mathrm{b}}_{\xi_2}}{\sqrt{j\omega\sigma\mu}} + \frac{1}{j\omega\sigma\mu}\left(\frac{3}{8d_2^2}\dot{H}^{\mathrm{b}}_{\xi_2} + \frac{1}{2}\frac{\partial^2 \dot{H}^{\mathrm{b}}_{\xi_2}}{\partial \xi_2^2}\right)\right] \end{aligned} \tag{6.19}$$

Time domain:

$$E^{\mathrm{b}} = \frac{1}{\sigma}\left[\sqrt{\sigma\mu}\frac{\partial}{\partial t}\left(H^{\mathrm{b}}_{\xi_2} * T^{\mathrm{b}}_2\right) + \frac{1}{2d_2}H^{\mathrm{b}}_{\xi_2} + \frac{1}{\sqrt{\sigma\mu}}\left(\frac{3}{8d_2^2}H^{\mathrm{b}}_{\xi_2} + \frac{1}{2}\frac{\partial^2 H^{\mathrm{b}}_{\xi_2}}{\partial \xi_2^2}\right) * T^{\mathrm{b}}_2\right]$$
$$= \frac{1}{\sigma}\frac{\partial}{\partial t}\left[\sqrt{\sigma\mu}H^{\mathrm{b}}_{\xi_2} * T^{\mathrm{b}}_2 + \frac{1}{2d_2}H^{\mathrm{b}}_{\xi_2} * T^{\mathrm{b}}_3 + \frac{1}{\sqrt{\sigma\mu}}\left(\frac{3}{8d_2^2}H^{\mathrm{b}}_{\xi_2} + \frac{1}{2}\frac{\partial^2 H^{\mathrm{b}}_{\xi_2}}{\partial \xi_2^2}\right) * T^{\mathrm{b}}_4\right] \tag{6.20}$$

where the functions T^{b}_i, $i = 2, 3, 4$, are given in Equations 2.214 through 2.216, respectively. The index "b" will be omitted in this chapter since all quantities are related to the conductor's surface.

Let us represent the electric field as a combination of the source component, E^{s}, and eddy component, E^{e}.

$$E(\xi_2, \eta, t) = E^{\mathrm{s}}(t) + E^{\mathrm{e}}(\xi_2, \eta, t) \tag{6.21}$$

Note that in the two-dimensional case the source field is a function of time only. Substituting Equation 6.21 in Equation 6.16, we obtain

$$E^{\mathrm{s}} + \tilde{c}E^{\mathrm{e}} + \sum_{i=1}^{N} \oint_{L_i} E^{\mathrm{e}} \frac{\partial G^{(2\mathrm{D})}}{\partial \vec{n}} \mathrm{d}\xi_2 = \sum_{i=1}^{N} \oint_{L_i} G^{(2\mathrm{D})} \frac{\partial E^{\mathrm{e}}}{\partial \vec{n}} \mathrm{d}\xi_2 \tag{6.22}$$

where the following identity was used:

$$\oint_{L_i} E^{\mathrm{s}} \frac{\partial G^{(2\mathrm{D})}}{\partial \vec{n}} \mathrm{d}\xi_2 = E^{\mathrm{s}} \oint_{L_i} \frac{\partial G^{(2\mathrm{D})}}{\partial \vec{n}} \mathrm{d}\xi_2 = (1 - \tilde{c})E^{\mathrm{s}} \tag{6.23}$$

The normal derivative of the electric field in Equation 6.22 can be replaced by the time derivative of the magnetic field using Equation 6.11 so that Equation 6.22 can be rewritten in the form:

$$E^{\mathrm{s}} + \tilde{c}E^{\mathrm{e}} + \sum_{i=1}^{N} \oint_{L_i} E^{\mathrm{e}} \frac{\partial G^{(2\mathrm{D})}}{\partial \vec{n}} \mathrm{d}\xi_2 = \mu \sum_{i=1}^{N} \oint_{L_i} G^{(2\mathrm{D})} \frac{\partial H_{\xi_2}}{\partial t} \mathrm{d}\xi_2 \quad \text{Time domain} \tag{6.24}$$

$$\dot{E}^{\mathrm{s}} + \tilde{c}\dot{E}^{\mathrm{e}} + \sum_{i=1}^{N} \oint_{L_i} \dot{E}^{\mathrm{e}} \frac{\partial G^{(2\mathrm{D})}}{\partial \vec{n}} \mathrm{d}\xi_2 = j\omega\mu \sum_{i=1}^{N} \oint_{L_i} G^{(2\mathrm{D})} \dot{H}_{\xi_2} \mathrm{d}\xi_2 \quad \text{Frequency domain} \tag{6.25}$$

The boundary integral equation (Equation 6.24) contains three unknowns, namely E^{s}, E^{e}, and $\partial H_{\xi_2}/\partial t$. Suppose the contours L_i, $i = 1, 2, \ldots, N$, are discretized using M nodes alltogether, so that the total number of unknowns is $(2M + N)$ because E^{s} has separate values for each conductor. Discretization of Equations 6.24 and 6.25 can provide only M algebraic equations. Another set of M equations can be obtained using the SIBCs in Equations 6.19 and 6.20 relating the functions E^{e} and $\partial H_{\xi_2}/\partial t$ at every node. And finally we have to add N equations by setting the total current flowing through each conductor:

$$\oint_{L_i} H_{\xi_2} \mathrm{d}\xi_2 = I_i, \quad i = 1, 2, \ldots, N \tag{6.26}$$

In principle, the system of equations (Equations 6.24 through 6.26 and Equations 6.19 and 6.20) is now ready to be solved using an appropriate numerical technique. However, its direct implementation is impractical because substitution of Equations 6.19 and 6.20 into Equations 6.24 and 6.25 allocates terms containing time convolution products on the left-hand side of the integral equation. Obviously, numerical implementation would require considerably fewer computer resources if those terms moved to the right-hand side. In the following, we demonstrate how this can be achieved using the perturbation approach developed in Chapter 2.

As a first step, we represent the relations in Equations 6.24 through 6.26 and Equations 6.19 and 6.20 in the Laplace domain with nondimensional variables using the scale factors given in Table 3.1:

$$\bar{\bar{\tilde{E}}}^{\mathrm{s}} + \tilde{c}\bar{\bar{\tilde{E}}}^{\mathrm{e}} + \sum_{i=1}^{N} \oint_{L_i} \bar{\bar{\tilde{E}}}^{\mathrm{e}} \frac{\partial \tilde{G}^{(2\mathrm{D})}}{\partial \vec{n}} \mathrm{d}\tilde{\xi}_2 = \tilde{s} \sum_{i=1}^{N} \oint_{L_i} \tilde{G}^{(2\mathrm{D})} \bar{\bar{\tilde{H}}}_{\xi_2} \mathrm{d}\tilde{\xi}_2 \tag{6.27}$$

$$\bar{\bar{\tilde{E}}}^{\mathrm{e}} = \tilde{s}\tilde{p}\left(\tilde{s}^{-1/2} + \tilde{p}\tilde{s}^{-1}\frac{1}{2\tilde{d}_2} + \tilde{p}^2\tilde{s}^{-3/2}\frac{3}{8\tilde{d}_2^2}\bar{\bar{\tilde{H}}}_{\xi_{3-k}} + \tilde{p}^2\tilde{s}^{-3/2}\frac{1}{2}\frac{\partial^2 \bar{\bar{\tilde{H}}}_{\xi_2}}{\partial \tilde{\xi}_2^2} \right) \tag{6.28}$$

$$\oint_{L_i} \bar{\bar{\tilde{H}}}_{\xi_2} \mathrm{d}\xi_2 = \bar{\bar{\tilde{I}}}_i, \quad i = 1, 2 \ldots, N \tag{6.29}$$

Note that Equation 6.28 can be obtained by reducing Equation 3.47 to the two-dimensional case.

We now expand the functions for which the solutions are sought in power series in the small parameter $\tilde{p}$:

$$\bar{\bar{\tilde{E}}}^{\mathrm{e}} = \sum_{m=1}^{\infty} \tilde{p}\bar{\bar{\tilde{E}}}_m^{\mathrm{e}}; \quad \bar{\bar{\tilde{E}}}^{\mathrm{s}} = \sum_{m=0}^{\infty} \tilde{p}\bar{\bar{\tilde{E}}}_m^{\mathrm{s}}; \quad \vec{\bar{\bar{\tilde{H}}}} = \sum_{m=0}^{\infty} \tilde{p}\vec{\bar{\bar{\tilde{H}}}}_m \tag{6.30}$$

Substituting Equation 6.30 into Equations 6.27 through 6.29 and equating coefficients of equal powers of $\tilde{p}$, the following equations for the coefficients of expansions are obtained:

PEC limit ($m = 0$):

$$\bar{\tilde{E}}_0^{\mathrm{s}} - \tilde{s} \sum_{i=1}^{N} \oint_{L_i} \tilde{G}^{(2\mathrm{D})} \left(\bar{\tilde{\vec{H}}}_0 \right)_{\xi_2} \mathrm{d}\tilde{\xi}_2 = 0 \tag{6.31}$$

$$\oint_{L_i} \left(\bar{\tilde{\vec{H}}}_0 \right)_{\xi_2} \mathrm{d}\tilde{\xi}_2 = \bar{\tilde{I}}_i \tag{6.32}$$

Leontovich's approximation ($m = 1$):

$$\bar{\tilde{E}}_1^{\mathrm{e}} = \tilde{s} \left[\tilde{s}^{-1/2} \left(\bar{\tilde{\vec{H}}}_0 \right)_{\xi_2} \right] \tag{6.33}$$

$$\bar{\tilde{E}}_1^{\mathrm{s}} - \tilde{s} \sum_{i=1}^{N} \oint_{L_i} \tilde{G}^{(2\mathrm{D})} \left(\bar{\tilde{\vec{H}}}_1 \right)_{\xi_2} \mathrm{d}\tilde{\xi}_2 = -\tilde{c} \bar{\tilde{E}}_1^{\mathrm{e}} - \sum_{i=1}^{N} \oint_{L_i} \bar{\tilde{E}}_1^{\mathrm{e}} \frac{\partial \tilde{G}^{(2\mathrm{D})}}{\partial \vec{n}} \mathrm{d}\tilde{\xi}_2 \tag{6.34}$$

$$\oint_{L_i} \left(\bar{\tilde{\vec{H}}}_1 \right)_{\xi_2} \mathrm{d}\tilde{\xi}_2 = 0 \tag{6.35}$$

Mitzner's approximation ($m = 2$):

$$\bar{\tilde{E}}_2^{\mathrm{e}} = \tilde{s} \left[\tilde{s}^{-1/2} \left(\bar{\tilde{\vec{H}}}_1 \right)_{\xi_2} + \tilde{s}^{-1} \frac{1}{2\tilde{d}_2} \left(\bar{\tilde{\vec{H}}}_0 \right)_{\xi_2} \right] \tag{6.36}$$

$$\bar{\tilde{E}}_2^{\mathrm{s}} - \tilde{s} \sum_{i=1}^{N} \oint_{L_i} \tilde{G}^{(2\mathrm{D})} \left(\bar{\tilde{\vec{H}}}_2 \right)_{\xi_2} \mathrm{d}\tilde{\xi}_2 = -\tilde{c} \bar{\tilde{E}}_2^{\mathrm{e}} - \sum_{i=1}^{N} \oint_{L_i} \bar{\tilde{E}}_2^{\mathrm{e}} \frac{\partial \tilde{G}^{(2\mathrm{D})}}{\partial \vec{n}} \mathrm{d}\tilde{\xi}_2 \tag{6.37}$$

$$\oint_{L_i} \left(\bar{\tilde{\vec{H}}}_2 \right)_{\xi_2} \mathrm{d}\tilde{\xi}_2 = 0 \tag{6.38}$$

Rytov's approximation ($m = 3$):

$$\bar{\tilde{E}}_3^{\mathrm{e}} = \tilde{s} \left[\tilde{s}^{-1/2} \left(\bar{\tilde{\vec{H}}}_2 \right)_{\xi_2} + \tilde{s}^{-1} \frac{1}{2\tilde{d}_2} \left(\bar{\tilde{\vec{H}}}_1 \right)_{\xi_2}^{\mathrm{b}} + \tilde{s}^{-3/2} \frac{3}{8\tilde{d}_2^2} \left(\bar{\tilde{\vec{H}}}_0 \right)_{\xi_2} + \tilde{s}^{-3/2} \frac{1}{2} \frac{\partial^2 \left(\bar{\tilde{\vec{H}}}_0 \right)_{\xi_2}}{\partial \tilde{\xi}_2^2} \right] \tag{6.39}$$

$$\bar{\tilde{E}}_3^{\mathrm{s}} - \tilde{s}\sum_{i=1}^{N}\oint_{L_i}\tilde{G}^{(2D)}\left(\vec{\bar{\tilde{H}}}_3\right)_{\xi_2}\mathrm{d}\tilde{\xi}_2 = -\tilde{c}\bar{\tilde{E}}_3^{\mathrm{e}} - \sum_{i=1}^{N}\oint_{L_i}\bar{\tilde{E}}_3^{\mathrm{e}}\frac{\partial\tilde{G}^{(2D)}}{\partial\vec{n}}\mathrm{d}\tilde{\xi}_2 \tag{6.40}$$

$$\oint_{L_i}\left(\vec{\bar{\tilde{H}}}_3\right)_{\xi_2}\mathrm{d}\tilde{\xi}_2 = 0 \tag{6.41}$$

Returning in Equations 6.31 through 6.41 to the frequency and time domains gives

Frequency domain:

$$\dot{\tilde{E}}_0^{\mathrm{s}} - 2j\sum_{i=1}^{N}\oint_{L_i}\tilde{G}^{(2D)}\left(\vec{\dot{\tilde{H}}}_0\right)_{\xi_2}\mathrm{d}\tilde{\xi}_2 = 0 \tag{6.42}$$

$$\oint_{L_i}\left(\vec{\dot{\tilde{H}}}_0\right)_{\xi_2}\mathrm{d}\tilde{\xi}_2 = \dot{\tilde{I}}_i \tag{6.43}$$

$$\dot{\tilde{E}}_1^{\mathrm{e}} = \sqrt{2j}\left(\vec{\dot{\tilde{H}}}_0\right)_{\xi_2} \tag{6.44}$$

$$\dot{\tilde{E}}_1^{\mathrm{s}} - 2j\sum_{i=1}^{N}\oint_{L_i}\tilde{G}^{(2D)}\left(\vec{\dot{\tilde{H}}}_1\right)_{\xi_2}\mathrm{d}\tilde{\xi}_2 = -\tilde{c}\dot{\tilde{E}}_1^{\mathrm{e}} - \sum_{i=1}^{N}\oint_{L_i}\dot{\tilde{E}}_1^{\mathrm{e}}\frac{\partial\tilde{G}^{(2D)}}{\partial\vec{n}}\mathrm{d}\tilde{\xi}_2 \tag{6.45}$$

$$\oint_{L_i}\left(\vec{\dot{\tilde{H}}}_1\right)_{\xi_2}\mathrm{d}\tilde{\xi}_2 = 0 \tag{6.46}$$

$$\dot{\tilde{E}}_2^{\mathrm{e}} = \sqrt{2j}\left(\vec{\dot{\tilde{H}}}_1\right)_{\xi_2} + \frac{1}{2\tilde{d}_2}\left(\vec{\dot{\tilde{H}}}_0\right)_{\xi_2} \tag{6.47}$$

$$\dot{\tilde{E}}_2^{\mathrm{s}} - 2j\sum_{i=1}^{N}\oint_{L_i}\tilde{G}^{(2D)}\left(\vec{\dot{\tilde{H}}}_2\right)_{\xi_2}\mathrm{d}\tilde{\xi}_2 = -\tilde{c}\dot{\tilde{E}}_2^{\mathrm{e}} - \sum_{i=1}^{N}\oint_{L_i}\dot{\tilde{E}}_2^{\mathrm{e}}\frac{\partial\tilde{G}^{(2D)}}{\partial\vec{n}}\mathrm{d}\tilde{\xi}_2 \tag{6.48}$$

$$\oint_{L_i}\left(\vec{\dot{\tilde{H}}}_2\right)_{\xi_2}\mathrm{d}\tilde{\xi}_2 = 0 \tag{6.49}$$

$$\dot{\tilde{E}}_3^{\mathrm{e}} = \sqrt{2j}\left(\vec{\dot{\tilde{H}}}_2\right)_{\xi_2} + \frac{1}{2\tilde{d}_2}\left(\vec{\dot{\tilde{H}}}_1\right)_{\xi_2}^{\mathrm{b}} + \frac{3}{8\tilde{d}_2^2\sqrt{2j}}\left(\vec{\dot{\tilde{H}}}_0\right)_{\xi_2} + \frac{1}{2\sqrt{2j}}\frac{\partial^2\left(\vec{\dot{\tilde{H}}}_0\right)_{\xi_2}}{\partial\tilde{\xi}_2^2} \tag{6.50}$$

$$\dot{\tilde{E}}_3^{\mathrm{s}} - 2j\sum_{i=1}^{N} \oint\limits_{L_i} \tilde{G}^{(2D)} \left(\dot{\vec{\tilde{H}}}_3\right)_{\xi_2} \mathrm{d}\tilde{\xi}_2 = -\tilde{c}\dot{\tilde{E}}_3^{\mathrm{e}} - \sum_{i=1}^{N} \oint\limits_{L_i} \dot{\tilde{E}}_3^{\mathrm{e}} \frac{\partial \tilde{G}^{(2D)}}{\partial \vec{n}} \mathrm{d}\tilde{\xi}_2 \tag{6.51}$$

$$\oint\limits_{L_i} \left(\dot{\vec{\tilde{H}}}_3\right)_{\xi_2} \mathrm{d}\tilde{\xi}_2 = 0 \tag{6.52}$$

Time domain:

$$\tilde{E}_0^{\mathrm{s}} - \frac{\partial}{\partial \tilde{t}} \sum_{i=1}^{N} \oint\limits_{L_i} \tilde{G}^{(2D)} \left(\vec{\tilde{H}}_0\right)_{\xi_2} \mathrm{d}\tilde{\xi}_2 = 0 \tag{6.53}$$

$$\oint\limits_{L_i} \left(\vec{\tilde{H}}_0\right)_{\xi_2} \mathrm{d}\tilde{\xi}_2 = \tilde{I}_i \tag{6.54}$$

$$\tilde{E}_1^{\mathrm{e}} = \frac{\partial}{\partial \tilde{t}} \left[\left(\vec{\tilde{H}}_0\right)_{\xi_2} * \tilde{T}_2^{\mathrm{b}}\right] = \left(\vec{\tilde{H}}_0\right)_{\xi_2} * \dot{\tilde{T}}_2^{\mathrm{b}} \tag{6.55}$$

$$\tilde{E}_1^{\mathrm{s}} - \frac{\partial}{\partial \tilde{t}} \sum_{i=1}^{N} \oint\limits_{L_i} \tilde{G}^{(2D)} \left(\vec{\tilde{H}}_1\right)_{\xi_2} \mathrm{d}\tilde{\xi}_2 = -\tilde{c}\tilde{E}_1^{\mathrm{e}} - \sum_{i=1}^{N} \oint\limits_{L_i} \tilde{E}_1^{\mathrm{e}} \frac{\partial \tilde{G}^{(2D)}}{\partial \vec{n}} \mathrm{d}\tilde{\xi}_2 \tag{6.56}$$

$$\oint\limits_{L_i} \left(\vec{\tilde{H}}_1\right)_{\xi_2} \mathrm{d}\tilde{\xi}_2 = 0 \tag{6.57}$$

$$\tilde{E}_2^{\mathrm{e}} = \frac{\partial}{\partial \tilde{t}} \left[\left(\vec{\tilde{H}}_1\right)_{\xi_2} * \tilde{T}_2^{\mathrm{b}} + \frac{1}{2\tilde{d}_2}\left(\vec{\tilde{H}}_0\right)_{\xi_2} * \tilde{T}_3^{\mathrm{b}}\right] = \left(\vec{\tilde{H}}_1\right)_{\xi_2} * \dot{\tilde{T}}_2^{\mathrm{b}} + \frac{1}{2\tilde{d}_2}\left(\vec{\tilde{H}}_0\right)_{\xi_2} * \dot{\tilde{T}}_2^{\mathrm{b}} \tag{6.58}$$

$$\tilde{E}_2^{\mathrm{s}} - \frac{\partial}{\partial \tilde{t}} \sum_{i=1}^{N} \oint\limits_{L_i} \tilde{G}^{(2D)} \left(\vec{\tilde{H}}_2\right)_{\xi_2} \mathrm{d}\tilde{\xi}_2 = -\tilde{c}\tilde{E}_2^{\mathrm{e}} - \sum_{i=1}^{N} \oint\limits_{L_i} \tilde{E}_2^{\mathrm{e}} \frac{\partial \tilde{G}^{(2D)}}{\partial \vec{n}} \mathrm{d}\tilde{\xi}_2 \tag{6.59}$$

$$\oint\limits_{L_i} \left(\vec{\tilde{H}}_2\right)_{\xi_2} \mathrm{d}\tilde{\xi}_2 = 0 \tag{6.60}$$

$$\begin{aligned} \tilde{E}_3^{\mathrm{e}} &= \frac{\partial}{\partial \tilde{t}} \left[\left(\vec{\tilde{H}}_2\right)_{\xi_2} * \tilde{T}_2^{\mathrm{b}} + \frac{1}{2\tilde{d}_2}\left(\vec{\tilde{H}}_1\right)_{\xi_2} * \tilde{T}_3^{\mathrm{b}} + \frac{3}{8\tilde{d}_2^2}\left(\vec{\tilde{H}}_0\right)_{\xi_2} * \tilde{T}_4^{\mathrm{b}} + \frac{1}{2}\frac{\partial^2\left(\vec{\tilde{H}}_0\right)_{\xi_2}}{\partial \tilde{\xi}_2^2} * \tilde{T}_4^{\mathrm{b}}\right] \\ &= \left(\vec{\tilde{H}}_2\right)_{\xi_2} * \dot{\tilde{T}}_2^{\mathrm{b}} + \frac{1}{2\tilde{d}_2}\left(\vec{\tilde{H}}_1\right)_{\xi_2} * \dot{\tilde{T}}_3^{\mathrm{b}} + \frac{3}{8\tilde{d}_2^2}\left(\vec{\tilde{H}}_0\right)_{\xi_2} * \dot{\tilde{T}}_4^{\mathrm{b}} + \frac{1}{2}\frac{\partial^2\left(\vec{\tilde{H}}_0\right)_{\xi_2}}{\partial \tilde{\xi}_2^2} * \dot{\tilde{T}}_4^{\mathrm{b}} \end{aligned} \tag{6.61}$$

$$\tilde{E}_3^{\mathrm{s}} - \frac{\partial}{\partial \tilde{t}} \sum_{i=1}^{N} \oint_{L_i} \tilde{G}^{(2\mathrm{D})} \left(\vec{\tilde{H}}_3 \right)_{\xi_2} \mathrm{d}\tilde{\xi}_2 = -\tilde{c}\tilde{E}_3^{\mathrm{e}} - \sum_{i=1}^{N} \oint_{L_i} \tilde{E}_3^{\mathrm{e}} \frac{\partial \tilde{G}^{(2\mathrm{D})}}{\partial \vec{n}} \mathrm{d}\tilde{\xi}_2 \tag{6.62}$$

$$\oint_{L_i} \left(\vec{\tilde{H}}_3 \right)_{\xi_2} \mathrm{d}\tilde{\xi}_2 = 0 \tag{6.63}$$

The functions $\tilde{T}_i^{\mathrm{b}}$, $i = 2, 3, 4$, are given in Equations 2.99, 2.115, and 2.116, respectively. The functions $\tilde{\tilde{T}}_i^{\mathrm{b}}$, $i = 2, 3, 4$, are obtained from Equations 2.217, 3.138, and 3.139 and written in the form:

$$\tilde{\tilde{T}}_2^{\mathrm{b}} = -\left(4\pi\tilde{t}^3\right)^{-1/2} \tag{6.64}$$

$$\tilde{\tilde{T}}_3^{\mathrm{b}} = \tilde{U}' \tag{6.65}$$

$$\tilde{\tilde{T}}_4^{\mathrm{b}} = (\pi\tilde{t})^{-1/2} \tag{6.66}$$

We emphasize the following basic advantages of the formulations in Equations 6.42 through 6.63:

1. The integral equations for the terms of the expansions differ only in the form of the right-hand side and can be solved by the same solution procedures. Therefore, new computational complications do not arise as compared with solving the problem using the well-known perfect conductor limit ($m = 0$).
2. The convolution integrals in the integral equations are on the right-hand sides only and can be calculated and tabulated before numerically solving the appropriate integral equation so that the computer resources required for the computations are greatly reduced as compared with the "original" equations in Equations 6.19, 6.20, 6.24 through 6.26.

6.2.2 *A–K* Formalism

Derivations in terms of *A–K* are very similar to the *E–H* formulation we discussed in Section 6.2.1. In the low-frequency case, the distribution of the magnetic vector potential in nonconducting domains is described by the Laplace equation:

$$\nabla^2 A = 0 \tag{6.67}$$

Therefore, the application of the boundary integral equation method gives the following integral equation over contours of the cross sections of the conductors:

$$\tilde{c}A + \sum_{i=1}^{N} \oint_{L_i} A \frac{\partial G^{(2D)}}{\partial \vec{n}} \mathrm{d}\xi_2 = \sum_{i=1}^{N} \oint_{L_i} G^{(2D)} \frac{\partial A}{\partial \vec{n}} \mathrm{d}\xi_2 \tag{6.68}$$

Applying the representation in Equation 3.94 to the two-dimensional case, we obtain

$$A(\xi_2, \eta, t) = A^{\mathrm{s}}(t) + A^{\mathrm{e}}(\xi_2, \eta, t) \tag{6.69}$$

Similar to E^{s}, A^{s} is a function of time alone, but it has separate values for each conductor. Substitution of Equation 6.69 into Equation 6.68 yields

$$A^{\mathrm{s}} + \tilde{c}A^{\mathrm{e}} + \sum_{i=1}^{N} \oint_{L_i} A^{\mathrm{e}} \frac{\partial G^{(2D)}}{\partial \vec{n}} \mathrm{d}\xi_2 = \sum_{i=1}^{N} \oint_{L_i} G^{(2D)} \frac{\partial A}{\partial \vec{n}} \mathrm{d}\xi_2 \tag{6.70}$$

The additional relation between the magnetic vector potential and its normal derivative is given by Equation 3.125 which can be reduced to two dimensions and written in the following form:

Frequency domain:

$$\begin{aligned}\frac{\partial \dot{A}^{\mathrm{e}}}{\partial \eta} &= -\sqrt{j\omega\sigma\mu}\dot{A}^{\mathrm{e}} + \frac{1}{2d_2}\dot{A}^{\mathrm{e}} + \frac{1}{\sqrt{j\omega\sigma\mu}}\left(\frac{3}{8d_2^2}\dot{A}^{\mathrm{e}} + \frac{1}{2}\frac{\partial^2 \dot{A}^{\mathrm{e}}}{\partial \xi_2^2}\right) \\ &= \sqrt{j\omega\sigma\mu}\left[-\dot{A}^{\mathrm{e}} + \frac{1}{2d_2}\frac{\dot{A}^{\mathrm{e}}}{\sqrt{j\omega\sigma\mu}} + \frac{1}{j\omega\sigma\mu}\left(\frac{3}{8d_2^2}\dot{A}^{\mathrm{e}} + \frac{1}{2}\frac{\partial^2 \dot{A}^{\mathrm{e}}}{\partial \xi_2^2}\right)\right]\end{aligned} \tag{6.71}$$

Time domain:

$$\begin{aligned}\frac{\partial A^{\mathrm{e}}}{\partial \eta} &= -\sqrt{\sigma\mu}\frac{\partial}{\partial t}(A^{\mathrm{e}} * T_2^{\mathrm{b}}) + \frac{1}{2d_2}H_{\xi_2}^{\mathrm{b}} + \frac{1}{\sqrt{\sigma\mu}}\left(\frac{3}{8d_2^2}H_{\xi_2}^{\mathrm{b}} + \frac{1}{2}\frac{\partial^2 H_{\xi_2}^{\mathrm{b}}}{\partial \xi_2^2}\right) * T_2^{\mathrm{b}} \\ &= \frac{\partial}{\partial t}\left[-\sqrt{\sigma\mu}A^{\mathrm{e}} * T_2^{\mathrm{b}} + \frac{1}{2d_2}A^{\mathrm{e}} * T_3^{\mathrm{b}} + \frac{1}{\sqrt{\sigma\mu}}\left(\frac{3}{8d_2^2}A^{\mathrm{e}} + \frac{1}{2}\frac{\partial^2 A^{\mathrm{e}}}{\partial \xi_2^2}\right) * T_4^{\mathrm{b}}\right] \\ &= -\sqrt{\sigma\mu}A^{\mathrm{e}} * \hat{T}_2^{\mathrm{b}} + \frac{1}{2d_2}A^{\mathrm{e}} * \hat{T}_3^{\mathrm{b}} + \frac{1}{\sqrt{\sigma\mu}}\left(\frac{3}{8d_2^2}A^{\mathrm{e}} + \frac{1}{2}\frac{\partial^2 A^{\mathrm{e}}}{\partial \xi_2^2}\right) * \hat{T}_4^{\mathrm{b}}\end{aligned} \tag{6.72}$$

The remaining equation is the condition in Equation 6.29 imposed on the total current flowing in the conductor. To represent it in terms of the vector potential formalism, we first represent Equation 3.89 using Equation 2.5 in the form

$$H_{\xi_2} = \frac{\partial A^{\mathrm{e}}}{\partial \eta} = K \tag{6.73}$$

where K is the surface current density. Thus, the condition in Equation 6.29 in terms of the magnetic vector potential for two-dimensional cases takes the form:

$$\oint_{L_i} \frac{\partial A^{\mathrm{e}}}{\partial \eta} \mathrm{d}\xi_2 = \oint_{L_i} K \mathrm{d}\xi_2 = I_i \tag{6.74}$$

We emphasize that Equation 6.73 is valid for all approximations (Leontovich, Mitzner, and Rytov) unlike the three-dimensional case, in which the normal derivative of the vector potential is equal to the surface magnetic field/surface current density only in the Leontovich approximation.

Equations 6.70 through 6.72 and Equation 6.74 can be solved to obtain the functions $A^{\mathrm{s}}(t)$, $A^{\mathrm{e}}(\xi_2, t)$, and $K(\xi_2, t)$. To improve the efficiency of the numerical implementation, we transform Equations 6.70 and 6.74 to the Laplace domain in terms of nondimensional variables:

$$\bar{\tilde{A}}^{\mathrm{s}} + \tilde{c}\bar{\tilde{A}}^{\mathrm{e}} + \sum_{i=1}^{N} \oint_{L_i} \bar{\tilde{A}}^{\mathrm{e}} \frac{\partial \tilde{G}^{(2\mathrm{D})}}{\partial \vec{n}} \mathrm{d}\tilde{\xi}_2 = \tilde{p}^{-1} \sum_{i=1}^{N} \oint_{L_i} \tilde{G}^{(2\mathrm{D})} \bar{\tilde{K}} \mathrm{d}\tilde{\xi}_2 \tag{6.75}$$

$$\oint_{L_i} \bar{\tilde{K}} \mathrm{d}\tilde{\xi}_2 = \bar{\tilde{I}}_i \tag{6.76}$$

Representation of the functions $A^{\mathrm{s}}(t)$, $A^{\mathrm{e}}(\xi_2, t)$, and $K(\xi_2, t)$ in asymptotic expansions in the small parameter $\tilde{p}$ is as follows:

$$\tilde{A}^{\mathrm{e}} = \sum_{m=1}^{\infty} \tilde{p}^m \tilde{A}^{\mathrm{e}}_m; \quad \tilde{A}^{\mathrm{s}} = \sum_{m=0}^{\infty} \tilde{p}^m \tilde{A}^{\mathrm{s}}_m; \quad \tilde{K} = \sum_{m=0}^{\infty} \tilde{p}^m \tilde{K}_m \tag{6.77}$$

Taking into account that

$$\tilde{K}_m = \frac{\partial \tilde{A}^{\mathrm{e}}_{m+1}}{\partial \tilde{\eta}}, \quad m = 1, 2, 3 \tag{6.78}$$

we obtain the following integral equations:

PEC limit ($m = 0$):

$$\bar{\tilde{A}}^{\mathrm{s}}_0 - \sum_{i=1}^{N} \oint_{L_i} \tilde{G}^{(2\mathrm{D})} \bar{\tilde{K}}_0 \mathrm{d}\tilde{\xi}_2 = 0 \tag{6.79}$$

$$\oint\limits_{L_i} \bar{\bar{K}}_0 \mathrm{d}\tilde{\xi}_2 = \bar{\bar{I}}_i \tag{6.80}$$

Leontovich's approximation ($m = 1$):

$$\bar{\bar{A}}_1^{\mathrm{s}} - \sum_{i=1}^{N} \oint\limits_{L_i} \tilde{G}^{(2D)} \bar{\bar{K}}_1 \mathrm{d}\tilde{\xi}_2 = -\tilde{c}\bar{\bar{A}}_1^{\mathrm{e}} - \sum_{i=1}^{N} \oint\limits_{L_i} \bar{\bar{A}}_1^{\mathrm{e}} \frac{\partial \tilde{G}^{(2D)}}{\partial \vec{n}} \mathrm{d}\tilde{\xi}_2 \tag{6.81}$$

$$\oint\limits_{L_i} \bar{\bar{K}}_1 \mathrm{d}\tilde{\xi}_2 = 0 \tag{6.82}$$

Mitzner's approximation ($m = 2$):

$$\bar{\bar{A}}_2^{\mathrm{s}} - \sum_{i=1}^{N} \oint\limits_{L_i} \tilde{G}^{(2D)} \bar{\bar{K}}_2 \mathrm{d}\tilde{\xi}_2 = -\tilde{c}\bar{\bar{A}}_2^{\mathrm{e}} - \sum_{i=1}^{N} \oint\limits_{L_i} \bar{\bar{A}}_2^{\mathrm{e}} \frac{\partial \tilde{G}^{(2D)}}{\partial \vec{n}} \mathrm{d}\tilde{\xi}_2 \tag{6.83}$$

$$\oint\limits_{L_i} \bar{\bar{K}}_2 \mathrm{d}\tilde{\xi}_2 = 0 \tag{6.84}$$

Rytov's approximation ($m = 4$):

$$\bar{\bar{A}}_3^{\mathrm{s}} - \sum_{i=1}^{N} \oint\limits_{L_i} \tilde{G}^{(2D)} \bar{\bar{K}}_3 \mathrm{d}\tilde{\xi}_2 = -\tilde{c}\bar{\bar{A}}_3^{\mathrm{e}} - \sum_{i=1}^{N} \oint\limits_{L_i} \bar{\bar{A}}_3^{\mathrm{e}} \frac{\partial \tilde{G}^{(2D)}}{\partial \vec{n}} \mathrm{d}\tilde{\xi}_2 \tag{6.85}$$

$$\oint\limits_{L_i} \bar{\bar{K}}_3 \mathrm{d}\tilde{\xi}_2 = 0 \tag{6.86}$$

Additional equations relating $\bar{\bar{A}}_{m+1}^{\mathrm{e}}$ and $\bar{\bar{K}}_m$, $m = 1, 2, 3$, can be obtained from the SIBCs (Equations 6.71 and 6.72). Switching to nondimensional variables in Equations 6.71 and 6.72, and substituting the expansions in Equation 6.77, we obtain

$$\frac{\partial \bar{\bar{A}}_1^{\mathrm{e}}}{\partial \tilde{\eta}} = -\tilde{s}^{1/2} \bar{\bar{A}}_1^{\mathrm{e}} \tag{6.87}$$

$$\frac{\partial \bar{\bar{A}}_2^{\mathrm{e}}}{\partial \tilde{\eta}} = -\tilde{s}^{1/2} \bar{\bar{A}}_2^{\mathrm{e}} + \frac{1}{2\tilde{d}_2} \bar{\bar{A}}_1^{\mathrm{e}} \tag{6.88}$$

$$\frac{\partial \bar{\tilde{A}}_3^{\mathrm{e}}}{\partial \tilde{\eta}} = -\tilde{s}^{1/2}\bar{\tilde{A}}_3^{\mathrm{e}} + \frac{1}{2\tilde{d}_2}\bar{\tilde{A}}_2^{\mathrm{e}} + \tilde{s}^{-1/2}\left(\frac{3}{8\tilde{d}_2^2}\bar{\tilde{A}}_1^{\mathrm{e}} + \frac{1}{2}\frac{\partial^2 \bar{\tilde{A}}_1^{\mathrm{e}}}{\partial \tilde{\xi}_2^2}\right) \tag{6.89}$$

Since Equations 6.83 and 6.85 require the coefficients $\bar{\tilde{A}}_{m+1}^{\mathrm{e}}$ being expressed in terms of $\bar{\tilde{K}}_m$, the relations in Equations 6.87 through 6.89 must be represented in the following form:

$$\bar{\tilde{A}}_1^{\mathrm{e}} = -\tilde{s}^{-1/2}\frac{\partial \bar{\tilde{A}}_1^{\mathrm{e}}}{\partial \tilde{\eta}} = -\tilde{s}^{-1/2}\bar{\tilde{K}}_0 \tag{6.90}$$

$$\bar{\tilde{A}}_2^{\mathrm{e}} = -\tilde{s}^{-1/2}\frac{\partial \bar{\tilde{A}}_2^{\mathrm{e}}}{\partial \tilde{\eta}} + \tilde{s}^{-1/2}\frac{1}{2\tilde{d}_2}\bar{\tilde{A}}_1^{\mathrm{e}} = -\left[\tilde{s}^{-1/2}\frac{\partial \bar{\tilde{A}}_2^{\mathrm{e}}}{\partial \tilde{\eta}} + \tilde{s}^{-1}\frac{1}{2\tilde{d}_2}\frac{\partial \bar{\tilde{A}}_1^{\mathrm{e}}}{\partial \tilde{\eta}}\right] = -\left[\tilde{s}^{-1/2}\bar{\tilde{K}}_1 + \tilde{s}^{-1}\frac{1}{2\tilde{d}_2}\bar{\tilde{K}}_0\right] \tag{6.91}$$

$$\begin{aligned}
\bar{\tilde{A}}_3^{\mathrm{e}} &= -\tilde{s}^{-1/2}\frac{\partial \bar{\tilde{A}}_3^{\mathrm{e}}}{\partial \tilde{\eta}} + \tilde{s}^{-1/2}\frac{1}{2\tilde{d}_2}\bar{\tilde{A}}_2^{\mathrm{e}} + \tilde{s}^{-1}\frac{3}{8\tilde{d}_2^2}\bar{\tilde{A}}_1^{\mathrm{e}} + \frac{1}{2}\tilde{s}^{-1}\frac{\partial^2 \bar{\tilde{A}}_1^{\mathrm{e}}}{\partial \tilde{\xi}_2^2} \\
&= -\left[\tilde{s}^{-1/2}\frac{\partial \bar{\tilde{A}}_3^{\mathrm{e}}}{\partial \tilde{\eta}} + \frac{1}{2\tilde{d}_2}\tilde{s}^{-1}\frac{\partial \bar{\tilde{A}}_2^{\mathrm{e}}}{\partial \tilde{\eta}} + \frac{3}{8\tilde{d}_2^2}\tilde{s}^{-3/2}\frac{\partial \bar{\tilde{A}}_1^{\mathrm{e}}}{\partial \tilde{\eta}} + \frac{1}{2}\tilde{s}^{-3/2}\frac{\partial^2}{\partial \tilde{\xi}_2^2}\frac{\partial \bar{\tilde{A}}_1^{\mathrm{e}}}{\partial \tilde{\eta}}\right] \\
&= -\left[\tilde{s}^{-1/2}\bar{\tilde{K}}_2 + \frac{1}{2\tilde{d}_2}\tilde{s}^{-1}\bar{\tilde{K}}_1 + \frac{3}{8\tilde{d}_2^2}\tilde{s}^{-3/2}\bar{\tilde{K}}_0 + \frac{1}{2}\tilde{s}^{-3/2}\frac{\partial^2 \bar{\tilde{K}}_0}{\partial \tilde{\xi}_2^2}\right]
\end{aligned} \tag{6.92}$$

Finally, Equations 6.79 through 6.86 and Equations 6.90 through 6.92 must be returned to the frequency and time domains:

Frequency domain:

$$\dot{\tilde{A}}_0^{\mathrm{s}} - \sum_{i=1}^{N}\oint_{L_i}\tilde{G}^{(2\mathrm{D})}\dot{\tilde{K}}_0 \mathrm{d}\tilde{\xi}_2 = 0 \tag{6.93}$$

$$\oint_{L_i}\dot{\tilde{K}}_0 \mathrm{d}\tilde{\xi}_2 = \dot{\tilde{I}}_i \tag{6.94}$$

$$\dot{\tilde{A}}_1^{\mathrm{e}} = -(2j)^{-1/2}\dot{\tilde{K}}_0 = -\frac{1-j}{2}\dot{\tilde{K}}_0 \tag{6.95}$$

$$\dot{\tilde{A}}_1^{\mathrm{s}} - \sum_{i=1}^{N}\oint_{L_i}\tilde{G}^{(2\mathrm{D})}\dot{\tilde{K}}_1 \mathrm{d}\tilde{\xi}_2 = -\tilde{c}\dot{\tilde{A}}_1^{\mathrm{e}} - \sum_{i=1}^{N}\oint_{L_i}\dot{\tilde{A}}_1^{\mathrm{e}}\frac{\partial \tilde{G}^{(2\mathrm{D})}}{\partial \tilde{n}}\mathrm{d}\tilde{\xi}_2 \tag{6.96}$$

$$\oint_{L_i}\dot{\tilde{K}}_1 \mathrm{d}\tilde{\xi}_2 = 0 \tag{6.97}$$

$$\dot{\tilde{A}}_2^{\rm e} = -\left[(2j)^{-1/2}\dot{\tilde{K}}_1 + (2j)^{-1}\frac{1}{2\tilde{d}_2}\dot{\tilde{K}}_0\right] = -\left[\frac{1-j}{2}\dot{\tilde{K}}_1 - \frac{j}{4\tilde{d}_2}\dot{\tilde{K}}_0\right] \tag{6.98}$$

$$\dot{\tilde{A}}_2^{\rm s} - \sum_{i=1}^{N}\oint_{L_i}\tilde{G}^{\rm (2D)}\dot{\tilde{K}}_2{\rm d}\tilde{\xi}_2 = -\tilde{c}\dot{\tilde{A}}_2^{\rm e} - \sum_{i=1}^{N}\oint_{L_i}\dot{\tilde{A}}_2^{\rm e}\frac{\partial\tilde{G}^{\rm (2D)}}{\partial\vec{n}}{\rm d}\tilde{\xi}_2 \tag{6.99}$$

$$\oint_{L_i}\dot{\tilde{K}}_2{\rm d}\tilde{\xi}_2 = 0 \tag{6.100}$$

$$\begin{aligned}\dot{\tilde{A}}_3^{\rm e} &= -\left[(2j)^{-1/2}\dot{\tilde{K}}_2 + \frac{1}{2\tilde{d}_2}(2j)^{-1}\dot{\tilde{K}}_1 + \frac{3}{8\tilde{d}_2^2}(2j)^{-3/2}\dot{\tilde{K}}_0 + \frac{1}{2}(2j)^{-3/2}\frac{\partial^2\dot{\tilde{K}}_0}{\partial\tilde{\xi}_2^2}\right]\\ &= -\left[\frac{1-j}{2}\dot{\tilde{K}}_2 - \frac{j}{4\tilde{d}_2}\dot{\tilde{K}}_1 - \frac{3(1+j)}{32\tilde{d}_2^2}\dot{\tilde{K}}_0 - \frac{1+j}{8}\frac{\partial^2\dot{\tilde{K}}_0}{\partial\tilde{\xi}_2^2}\right]\end{aligned} \tag{6.101}$$

$$\dot{\tilde{A}}_3^{\rm s} - \sum_{i=1}^{N}\oint_{L_i}\tilde{G}^{\rm (2D)}\dot{\tilde{K}}_3{\rm d}\tilde{\xi}_2 = -\tilde{c}\dot{\tilde{A}}_3^{\rm e} - \sum_{i=1}^{N}\oint_{L_i}\dot{\tilde{A}}_3^{\rm e}\frac{\partial\tilde{G}^{\rm (2D)}}{\partial\vec{n}}{\rm d}\tilde{\xi}_2 \tag{6.102}$$

$$\oint_{L_i}\dot{\tilde{K}}_3{\rm d}\tilde{\xi}_2 = 0 \tag{6.103}$$

Time domain:

$$\tilde{A}_0^{\rm s} - \sum_{i=1}^{N}\oint_{L_i}\tilde{G}^{\rm (2D)}\tilde{K}_0{\rm d}\tilde{\xi}_2 = 0 \tag{6.104}$$

$$\oint_{L_i}\tilde{K}_0{\rm d}\tilde{\xi}_2 = \tilde{I}_i \tag{6.105}$$

$$\tilde{A}_1^{\rm e} = -\frac{\partial}{\partial\tilde{t}}\left[\tilde{K}_0 * \tilde{T}_2^{\rm b}\right] = -\tilde{K}_0 * \tilde{\tilde{T}}_2^{\rm b} \tag{6.106}$$

$$\tilde{A}_1^{\rm s} - \sum_{i=1}^{N}\oint_{L_i}\tilde{G}^{\rm (2D)}\tilde{K}_1{\rm d}\tilde{\xi}_2 = -\tilde{c}\tilde{A}_1^{\rm e} - \sum_{i=1}^{N}\oint_{L_i}\tilde{A}_1^{\rm e}\frac{\partial\tilde{G}^{\rm (2D)}}{\partial\vec{n}}{\rm d}\tilde{\xi}_2 \tag{6.107}$$

$$\oint_{L_i}\tilde{K}_1{\rm d}\tilde{\xi}_2 = 0 \tag{6.108}$$

$$\tilde{A}_2^{\rm e} = -\frac{\partial}{\partial\tilde{t}}\left[\tilde{K}_1 * \tilde{T}_2^{\rm b} + \frac{1}{2\tilde{d}_2}\tilde{K}_0 * \tilde{T}_3^{\rm b}\right] = -\left[\tilde{K}_1 * \tilde{\tilde{T}}_2^{\rm b} + \frac{1}{2\tilde{d}_2}\tilde{K}_0 * \tilde{\tilde{T}}_3^{\rm b}\right] \tag{6.109}$$

$$\tilde{A}_2^{\mathrm{s}} - \sum_{i=1}^{N} \oint_{L_i} \tilde{G}^{(2\mathrm{D})} \tilde{K}_2 \mathrm{d}\tilde{\xi}_2 = -\tilde{c}\tilde{A}_2^{\mathrm{e}} - \sum_{i=1}^{N} \oint_{L_i} \tilde{A}_2^{\mathrm{e}} \frac{\partial \tilde{G}^{(2\mathrm{D})}}{\partial \vec{n}} \mathrm{d}\tilde{\xi}_2 \tag{6.110}$$

$$\oint_{L_i} \tilde{K}_2 \mathrm{d}\tilde{\xi}_2 = 0 \tag{6.111}$$

$$\begin{aligned} \tilde{A}_3^{\mathrm{e}} &= -\frac{\partial}{\partial \tilde{t}} \left[\tilde{K}_2 * \tilde{T}_2^{\mathrm{b}} + \frac{1}{2\tilde{d}_2} \tilde{K}_1 * \tilde{T}_3^{\mathrm{b}} + \frac{3}{8\tilde{d}_2^2} \tilde{K}_0 * \tilde{T}_4^{\mathrm{b}} + \frac{1}{2} \frac{\partial^2 \tilde{K}_0}{\partial \tilde{\xi}_2^2} * \tilde{T}_4^{\mathrm{b}} \right] \\ &= -\left[\tilde{K}_2 * \tilde{\dot{T}}_2^{\mathrm{b}} + \frac{1}{2\tilde{d}_2} \tilde{K}_1 * \tilde{\dot{T}}_3^{\mathrm{b}} + \frac{3}{8\tilde{d}_2^2} \tilde{K}_0 * \tilde{\dot{T}}_4^{\mathrm{b}} + \frac{1}{2} \frac{\partial^2 \tilde{K}_0}{\partial \tilde{\xi}_2^2} * \tilde{\dot{T}}_4^{\mathrm{b}} \right] \end{aligned} \tag{6.112}$$

$$\tilde{A}_3^{\mathrm{s}} - \sum_{i=1}^{N} \oint_{L_i} \tilde{G}^{(2\mathrm{D})} \tilde{K}_3 \mathrm{d}\tilde{\xi}_2 = -\tilde{c}\tilde{A}_3^{\mathrm{e}} - \sum_{i=1}^{N} \oint_{L_i} \tilde{A}_3^{\mathrm{e}} \frac{\partial \tilde{G}^{(2\mathrm{D})}}{\partial \vec{n}} \mathrm{d}\tilde{\xi}_2 \tag{6.113}$$

$$\oint_{L_i} \tilde{K}_3 \mathrm{d}\tilde{\xi}_2 = 0 \tag{6.114}$$

Equations 6.93 through 6.114 have the same structure of the left-hand sides and, therefore, they incorporate all advantages listed at the end of Section 6.2.1 for Equations 6.42 through 6.63.

6.2.3 Common Representation

As has been shown in Sections 6.2.1 and 6.2.2, integral equation formulations in terms of *A–K* and *E–H* formalisms are so similar that it is natural to find a common representation for them. Thus the purpose of this section is derivation of an integral equation formulation that can be used in terms of both formalisms. First, we will represent the equations relating $\tilde{E}_m^{\mathrm{e}}$ and $\partial\big(\tilde{\vec{H}}_{m-1}\big)_{\xi_2} \big/ \partial\tilde{t}$, $m = 1, 2, 3$, and the equations relating $\tilde{A}_m^{\mathrm{e}}$ and $\tilde{K}_{m-1}$ using the surface impedance operator-function F defined in Equations 3.84 through 3.88. In the particular two-dimensional low-frequency case, the operator-function F of the order of Rytov's approximation reduces to the following form:

Dimensional variables:

$$F^{(2\mathrm{D})}[f] = \begin{cases} (j\omega\sigma\mu)^{-1/2}\dot{f} + (j\omega\sigma\mu)^{-1}\dfrac{1}{2d_2}\dot{f} + (j\omega\sigma\mu)^{-3/2}\dfrac{3}{8d_2^2}\dot{f} + (j\omega\sigma\mu)^{-3/2}\dfrac{1}{2}\dfrac{\partial^2 \dot{f}}{\partial \xi_2^2} & \text{FD} \\ f * T_2^{\mathrm{b}} + \dfrac{1}{2\tilde{d}_2} f * T_3^{\mathrm{b}} + \dfrac{3}{8d_2^2} f * T_4^{\mathrm{b}} + \dfrac{1}{2}\dfrac{\partial^2 f}{\partial \xi_2^2} * T_4^{\mathrm{b}} & \text{TD} \end{cases} \tag{6.115}$$

Nondimensional variables:

$$\tilde{F}^{(2D)}\left[\tilde{f}\right] = \begin{cases} (2j)^{-1/2}\dot{\tilde{f}} + (2j)^{-1}\dfrac{1}{2\tilde{d}_2}\dot{\tilde{f}} + (2j)^{-3/2}\dfrac{3}{8\tilde{d}_2^2}\dot{\tilde{f}} + (2j)^{-3/2}\dfrac{1}{2}\dfrac{\partial^2\dot{\tilde{f}}}{\partial\tilde{\xi}_2^2} & \text{FD} \\ \tilde{f} * \tilde{T}_2^{\mathrm{b}} + \dfrac{1}{2\tilde{d}_2}\tilde{f} * \tilde{T}_3^{\mathrm{b}} + \dfrac{3}{8\tilde{d}_2^2}\tilde{f} * \tilde{T}_4^{\mathrm{b}} + \dfrac{1}{2}\dfrac{\partial^2\tilde{f}}{\partial\tilde{\xi}_2^2} * \tilde{T}_4^{\mathrm{b}} & \text{TD} \end{cases} \tag{6.116}$$

We introduce the operators $\Lambda_m^{(2)}$, $m = 1, 2, 3$, as follows:

Dimensional variables:

$$\Lambda_1^{(2D)}[f] = \begin{cases} (j\omega\sigma\mu)^{-1/2}\dot{f} & \text{FD} \\ f * T_2^{\mathrm{b}} & \text{TD} \end{cases} \tag{6.117}$$

Here and below the index (2D) denotes quantities related to the two-dimensional case.

$$\Lambda_2^{(2D)}[f] = \begin{cases} (j\omega\sigma\mu)^{-1}\dfrac{1}{2d_2}\dot{f} & \text{FD} \\ \dfrac{1}{2d_2}f * T_3^{\mathrm{b}} & \text{TD} \end{cases} \tag{6.118}$$

$$\Lambda_3^{(2D)}[f] = \begin{cases} (j\omega\sigma\mu)^{-3/2}\dfrac{3}{8d_2^2}\dot{f} + (j\omega\sigma\mu)^{-3/2}\dfrac{1}{2}\dfrac{\partial^2\dot{f}}{\partial\xi_2^2} & \text{FD} \\ \dfrac{3}{8d_2^2}f * T_4^{\mathrm{b}} + \dfrac{1}{2}\dfrac{\partial^2 f}{\partial\xi_2^2} * T_4^{\mathrm{b}} & \text{TD} \end{cases} \tag{6.119}$$

Nondimensional variables:

$$\tilde{\Lambda}_1^{(2D)}\left[\tilde{f}\right] = \begin{cases} (2j)^{-1/2}\dot{\tilde{f}} & \text{FD} \\ \tilde{f} * \tilde{T}_2^{\mathrm{b}} & \text{TD} \end{cases} \tag{6.120}$$

$$\tilde{\Lambda}_2^{(2D)}\left[\tilde{f}\right] = \begin{cases} (2j)^{-1}\dfrac{1}{2\tilde{d}_2}\dot{\tilde{f}} & \text{FD} \\ \dfrac{1}{2\tilde{d}_2}\tilde{f} * \tilde{T}_3^{\mathrm{b}} & \text{TD} \end{cases} \tag{6.121}$$

$$\tilde{\Lambda}_3^{(2D)}\left[\tilde{f}\right] = \begin{cases} \dfrac{3}{8\tilde{d}_2^2}(2j)^{-3/2}\dot{\tilde{f}} + (2j)^{-3/2}\dfrac{1}{2}\dfrac{\partial^2\dot{\tilde{f}}}{\partial\tilde{\xi}_2^2} & \text{FD} \\ \dfrac{3}{8\tilde{d}_2^2}\tilde{f} * \tilde{T}_4^{\mathrm{b}} + \dfrac{1}{2}\dfrac{\partial^2\tilde{f}}{\partial\tilde{\xi}_2^2} * \tilde{T}_3^{\mathrm{b}} & \text{TD} \end{cases} \tag{6.122}$$

The operators $\Lambda_m^{(2)}$ are related with the surface impedance operator-function F of various approximation orders as follows:

Dimensional variables:

$$F^{(2D)}[f]\Big|^{\text{Leontovich}} = \Lambda_1^{(2D)}[f] \tag{6.123}$$

$$F^{(2D)}[f]\Big|^{\text{Mitzner}} = \Lambda_1^{(2D)}[f] + \Lambda_2^{(2D)}[f] \tag{6.124}$$

$$F^{(2D)}[f]\Big|^{\text{Rytov}} = \Lambda_1^{(2D)}[f] + \Lambda_2^{(2D)}[f] + \Lambda_3^{(2D)}[f] \tag{6.125}$$

Nondimensional variables:

$$\tilde{F}^{(2D)}\left[\tilde{f}\right]\Big|^{\text{Leontovich}} = \tilde{\Lambda}_1^{(2D)}\left[\tilde{f}\right] \tag{6.126}$$

$$\tilde{F}^{(2D)}\left[\tilde{f}\right]\Big|^{\text{Mitzner}} = \tilde{\Lambda}_1^{(2D)}\left[\tilde{f}\right] + \tilde{p}\tilde{\Lambda}_2^{(2D)}\left[\tilde{f}\right] \tag{6.127}$$

$$\tilde{F}^{(2D)}\left[\tilde{f}\right]\Big|^{\text{Rytov}} = \tilde{\Lambda}_1^{(2D)}\left[\tilde{f}\right] + \tilde{p}\tilde{\Lambda}_2^{(2D)}\left[\tilde{f}\right] + \tilde{p}^2\tilde{\Lambda}_3^{(2D)}\left[\tilde{f}\right] \tag{6.128}$$

Substituting Equations 6.120 through 6.122 into Equations 6.55, 6.58, and 6.61 for the *E–H* formalism and Equations 6.106, 6.109, and 6.112 for the *A–K* formalism, we obtain the following equations:

E–H formalism:

$$\tilde{E}_1^{\text{e}} = \frac{\partial}{\partial t}\left(\tilde{\Lambda}_1^{(2D)}\left[\left(\vec{\tilde{H}}_0\right)_{\xi_2}\right]\right) = \tilde{\Lambda}_1^{(2D)}\left[\frac{\partial\left(\vec{\tilde{H}}_0\right)_{\xi_2}}{\partial\tilde{t}}\right] \tag{6.129}$$

$$\tilde{E}_2^{\text{e}} = \tilde{\Lambda}_2^{(2D)}\left[\frac{\partial\left(\vec{\tilde{H}}_0\right)_{\xi_2}}{\partial\tilde{t}}\right] + \tilde{\Lambda}_1^{(2D)}\left[\frac{\partial\left(\vec{\tilde{H}}_1\right)_{\xi_2}}{\partial\tilde{t}}\right] \tag{6.130}$$

$$\tilde{E}_3^{\text{e}} = \tilde{\Lambda}_3^{(2D)}\left[\frac{\partial\left(\vec{\tilde{H}}_0\right)_{\xi_2}}{\partial\tilde{t}}\right] + \tilde{\Lambda}_2^{(2D)}\left[\frac{\partial\left(\vec{\tilde{H}}_1\right)_{\xi_2}}{\partial\tilde{t}}\right] + \tilde{\Lambda}_1^{(2D)}\left[\frac{\partial\left(\vec{\tilde{H}}_2\right)_{\xi_2}}{\partial\tilde{t}}\right] \tag{6.131}$$

A–K formalism:

$$\tilde{A}_1^{\text{e}} = -\tilde{\Lambda}_1^{(2D)}\left[\tilde{K}_0\right] \tag{6.132}$$

$$\tilde{A}_2^{\text{e}} = -\tilde{\Lambda}_2^{(2D)}\left[\tilde{K}_0\right] - \tilde{\Lambda}_1^{(2D)}\left[\tilde{K}_1\right] \tag{6.133}$$

$$\tilde{A}_3^{\text{e}} = -\tilde{\Lambda}_3^{(2D)}\left[\tilde{K}_0\right] - \tilde{\Lambda}_2^{(2D)}\left[\tilde{K}_1\right] - \tilde{\Lambda}_1^{(2D)}\left[\tilde{K}_2\right] \tag{6.134}$$

Introducing nondimensional functions $\tilde{u}$ and $\tilde{v}$ which can take the following forms depending on the formalism (*E–H* or *A–K*) used, we get

$$\tilde{u} = \begin{cases} \partial \tilde{H}_{\xi_2}/\partial \tilde{t} & E\text{–}H \text{ formalism} \\ -\tilde{K} & A\text{–}K \text{ formalism} \end{cases} \tag{6.135}$$

$$\tilde{v} = \begin{cases} \tilde{E}^{\mathrm{s}} & E\text{–}H \text{ formalism} \\ \tilde{A}^{\mathrm{s}} & A\text{–}K \text{ formalism} \end{cases} \tag{6.136}$$

The functions $\tilde{u}$ and $\tilde{v}$ can be expanded into power series in the small parameter $\tilde{p}$:

$$\tilde{u} = \sum_{m=0}^{\infty} \tilde{p}^m \tilde{u}_m; \quad \tilde{v} = \sum_{m=0}^{\infty} \tilde{p}^m \tilde{v}_m \tag{6.137}$$

The coefficients of expansions in Equation 6.137 are related with $\tilde{E}^{\mathrm{s}}_m$, $\left(\vec{\tilde{H}}_m\right)_{\xi_2}$, $\tilde{A}^{\mathrm{s}}_m$, and $\tilde{K}_m$ as follows:

$$\tilde{u}_m = \begin{cases} \partial \left(\vec{\tilde{H}}_m\right)_{\xi_2} \Big/ \partial \tilde{t} & E\text{–}H \text{ formalism} \\ -\tilde{K}_m & A\text{–}K \text{ formalism} \end{cases} \tag{6.138}$$

$$\tilde{v}_m = \begin{cases} \tilde{E}^{\mathrm{s}}_m & E\text{–}H \text{ formalism} \\ \tilde{A}^{\mathrm{s}}_m & A\text{–}K \text{ formalism} \end{cases} \tag{6.139}$$

Finally, we represent the formulations in Equations 6.42 through 6.63 and Equations 6.93 through 6.114 with the variables $\tilde{u}_m$ and $\tilde{v}_m$ using Equations 6.129 through 6.132 in the frequency and time domains:

Frequency domain:

$$\dot{\tilde{v}}_0 - \sum_{i=1}^{N} \oint_{L_i} \tilde{G}^{(2D)} \dot{\tilde{u}}_0 \mathrm{d}\tilde{\xi}_2 = 0 \tag{6.140}$$

$$\oint_{L_i} \dot{\tilde{u}}_0 \mathrm{d}\tilde{\xi}_2 = \begin{cases} 2j\dot{\tilde{I}}_i & E\text{–}H \text{ formalism} \\ \dot{\tilde{I}}_i & A\text{–}K \text{ formalism} \end{cases} \tag{6.141}$$

$$\dot{\tilde{v}}_1 - \sum_{i=1}^{N} \oint_{L_i} \tilde{G}^{(2D)} \dot{\tilde{u}}_1 \mathrm{d}\tilde{\xi}_2 = -\tilde{c}\tilde{\Lambda}_1^{(2D)}\left[\dot{\tilde{u}}_0\right] - \sum_{i=1}^{N} \oint_{L_i} \frac{\partial \tilde{G}^{(2D)}}{\partial \vec{n}} \tilde{\Lambda}_1^{(2D)}\left[\dot{\tilde{u}}_0\right] \mathrm{d}\tilde{\xi}_2 \tag{6.142}$$

$$\oint_{L_i} \dot{\tilde{u}}_1 \mathrm{d}\tilde{\xi}_2 = 0 \tag{6.143}$$

$$\dot{\tilde{v}}_2 - \sum_{i=1}^{N} \oint_{L_i} \tilde{G}^{(2)} \dot{\tilde{u}}_2 d\tilde{\xi}_2 = -\tilde{c}\left(\tilde{\Lambda}_1^{(2D)}\left[\dot{\tilde{u}}_1\right] + \tilde{\Lambda}_2^{(2D)}\left[\dot{\tilde{u}}_0\right]\right) - \sum_{i=1}^{N} \oint_{L_i} \frac{\partial \tilde{G}^{(2)}}{\partial \vec{n}} \left(\tilde{\Lambda}_1^{(2D)}\left[\dot{\tilde{u}}_1\right] + \tilde{\Lambda}_2^{(2D)}\left[\dot{\tilde{u}}_0\right]\right) d\tilde{\xi}_2 \tag{6.144}$$

$$\oint_{L_i} \dot{\tilde{u}}_2 d\tilde{\xi}_2 = 0 \tag{6.145}$$

$$\dot{\tilde{v}}_3 - \sum_{i=1}^{N} \oint_{L_i} \tilde{G}^{(2D)} \dot{\tilde{u}}_3 d\tilde{\xi}_2 = -\tilde{c}\left(\tilde{\Lambda}_1^{(2D)}\left[\dot{\tilde{u}}_2\right] + \tilde{\Lambda}_2^{(2D)}\left[\dot{\tilde{u}}_1\right] + \tilde{\Lambda}_3^{(2D)}\left[\dot{\tilde{u}}_0\right]\right)$$
$$- \sum_{i=1}^{N} \oint_{L_i} \frac{\partial \tilde{G}^{(2D)}}{\partial \vec{n}} \left(\tilde{\Lambda}_1^{(2D)}\left[\dot{\tilde{u}}_2\right] + \tilde{\Lambda}_2^{(2D)}\left[\dot{\tilde{u}}_1\right] + \tilde{\Lambda}_3^{(2D)}\left[\dot{\tilde{u}}_0\right]\right) d\tilde{\xi}_2 \tag{6.146}$$

$$\oint_{L_i} \dot{\tilde{u}}_3 d\tilde{\xi}_2 = 0 \tag{6.147}$$

Time domain:

$$\tilde{v}_0 - \sum_{i=1}^{N} \oint_{L_i} \int_{L_i} \tilde{G}^{(2D)} \tilde{u}_0 d\tilde{\xi}_2 = 0 \tag{6.148}$$

$$\oint_{L_i} \tilde{u}_0 d\tilde{\xi}_2 = \begin{cases} \partial \tilde{I}_i / \partial \tilde{t} & E\text{–}H \text{ formalism} \\ \tilde{I}_i & A\text{–}K \text{ formalism} \end{cases} \tag{6.149}$$

$$\tilde{v}_1 - \sum_{i=1}^{N} \oint_{L_i} \tilde{G}^{(2D)} \tilde{u}_1 d\tilde{\xi}_2 = -\tilde{c}\tilde{\Lambda}_1^{(2D)}[\tilde{u}_0] - \sum_{i=1}^{N} \oint_{L_i} \frac{\partial \tilde{G}^{(2D)}}{\partial \vec{n}} \tilde{\Lambda}_1^{(2D)}[\tilde{u}_0] d\tilde{\xi}_2 \tag{6.150}$$

$$\oint_{L_i} \tilde{u}_1 d\tilde{\xi}_2 = 0 \tag{6.151}$$

$$\tilde{v}_2 - \sum_{i=1}^{N} \oint_{L_i} \tilde{G}^{(2D)} \tilde{u}_2 d\tilde{\xi}_2 = -\tilde{c}\left(\tilde{\Lambda}_1^{(2D)}[\tilde{u}_1] + \tilde{\Lambda}_2^{(2D)}[\tilde{u}_0]\right) - \sum_{i=1}^{N} \oint_{L_i} \frac{\partial \tilde{G}^{(2D)}}{\partial \vec{n}} \left(\tilde{\Lambda}_1^{(2D)}[\tilde{u}_1] + \tilde{\Lambda}_2^{(2D)}[\tilde{u}_0]\right) d\tilde{\xi}_2 \tag{6.152}$$

$$\oint_{L_i} \tilde{u}_2 d\tilde{\xi}_2 = 0 \tag{6.153}$$

$$\tilde{v}_3 - \sum_{i=1}^{N} \oint_{L_i} \tilde{G}^{(2D)} \tilde{u}_3 d\tilde{\xi}_2 = -\tilde{c}\left(\tilde{\Lambda}_1^{(2D)}[\tilde{u}_2] + \tilde{\Lambda}_2^{(2D)}[\tilde{u}_1] + \tilde{\Lambda}_3^{(2D)}[\tilde{u}_0]\right)$$
$$- \sum_{i=1}^{N} \oint_{L_i} \frac{\partial \tilde{G}^{(2D)}}{\partial \vec{n}} \left(\tilde{\Lambda}_1^{(2D)}[\tilde{u}_2] + \tilde{\Lambda}_2^{(2D)}[\tilde{u}_1] + \tilde{\Lambda}_3^{(2D)}[\tilde{u}_0]\right) d\tilde{\xi}_2 \tag{6.154}$$

$$\oint_{L_i} \tilde{u}_3 d\tilde{\xi}_2 = 0 \tag{6.155}$$

The boundary integral equation formulations (Equations 6.140 through 6.155) can be solved with respect to $\tilde{u}_m$ and $\tilde{v}_m$ and then the output variables of *E–H* or *A–K* formalisms can be obtained using Equations 6.135 and 6.136. The only equation depending on the formalism selected is Equation 6.141 or 6.149: the circulation of $\tilde{u}_0$ along the contour of the cross section of each conductor is set to equal the total current flowing in the conductor when *A–K* formalism is used or the time derivative of the total current in the case of *E–H* formalism. Therefore, one can expect that separation of variables into space and time components may transform Equation 6.141 or 6.149 (and, consequently, all equations in Equations 6.140 through 6.155) to a form invariant on the formalism selected. Derivation of such an invariant formulation for two-dimensional problems will be considered in Section 6.5.

6.3 Three-Dimensional Problems

Solution of three-dimensional problems is most readily done in terms of the scalar potential formalism. To ensure uniqueness of the solution, the magnetic scalar potential ϕ in nonconducting domains is introduced following the approach described in [8]:

$$\vec{H} = \vec{H}^s - \nabla\phi \tag{6.156}$$

where

$\vec{H}^s$ is the "source" magnetic field, which only meets the following requirement:

$$\oint_{L_i} \vec{H}^s \cdot d\vec{l} = I_i \tag{6.157}$$

$$\vec{H}^s = \sum_{i=1}^{N} \vec{H}_i^s \tag{6.158}$$

L_i is an arbitrary path enclosing the ith conductor
$\vec{H}_i^s$ is the magnetic field generated by an imagined filamentary conductor carrying the current $\vec{I}_i$

The field $\vec{H}_i^s$ can be calculated using the Biot–Savart law:

$$\vec{H}_i^s = \frac{1}{4\pi} \int_{L_i} \vec{I}_i(\vec{r}', t) \times \frac{\vec{r} - \vec{r}'}{|\vec{r} - \vec{r}'|^3} \, dl \tag{6.159}$$

Substituting Equation 6.156 into Equation 6.6 provides the governing equation for the scalar potential:

$$\nabla^2 \phi = 0 \tag{6.160}$$

The representation in Equations 6.159 and 6.160 not only ensures uniqueness of the magnetic scalar potential, but also provides explicit introduction of the source term containing the total currents in the scalar potential formulation. The presence of the source term means that the trivial solution does not satisfy the problem.

Since the scalar potential in free space obeys the Laplace equation, the boundary integral equation method yields the following surface integral equation:

$$\tilde{c}\phi + \sum_{i=1}^{N} \int_{S_i} \phi \frac{\partial G^{(3D)}}{\partial \vec{n}} ds = \sum_{i=1}^{N} \int_{S_i} G^{(3D)} \frac{\partial \phi}{\partial \vec{n}} ds; \quad G^{(3D)} = (4\pi|\vec{r} - \vec{r}'|)^{-1} \tag{6.161}$$

where
$G^{(3D)}$ is the fundamental solution of Laplace's equation in free space
S_i is the ith conductor's surface
$\vec{n}$ is the unit normal vector is chosen to point inwards

Equation 6.161 includes two unknowns and must be supplemented by Equation 3.81 relating the magnetic scalar potential and its normal derivative:

$$\left(\frac{\partial \phi}{\partial \eta}\right) = H_\eta^s - \sum_{i=1}^{2} \frac{\partial}{\partial \xi_i} F\left[H_{\xi_i}^s\right] + \sum_{i=1}^{2} \frac{\partial}{\partial \xi_i} F[\partial \phi / \partial \xi_i] = H_\eta^s - \sum_{i=1}^{2} \frac{\partial}{\partial \xi_i} F\left[H_{\xi_i}^s\right] + \sum_{i=1}^{2} \frac{\partial}{\partial \xi_i} F\left[(\nabla \phi)_{\xi_i}\right] \tag{6.162}$$

where the surface impedance operator-function F for three-dimensional problems is given by Equations 3.82 through 3.84 and Equations 3.86 through 3.88 in terms of dimensional and nondimensional variables, respectively. Substitution of Equation 6.161 into Equation 6.161 gives

$$\tilde{c}\phi+\sum_{i=1}^{N}\int_{S_i}\left(\frac{\partial G^{(3D)}}{\partial\vec{n}}\phi-G^{(3)}\sum_{i=1}^{2}\frac{\partial}{\partial\xi_i}F\left[(\nabla\phi)_{\xi_i}\right]\right)\mathrm{d}s=\sum_{i=1}^{N}\int_{S_i}G^{(3D)}\left(H_\eta^{\mathrm{s}}-\sum_{i=1}^{2}\frac{\partial}{\partial\xi_i}F\left[H_{\xi_i}^{\mathrm{s}}\right]\right)\mathrm{d}s \tag{6.163}$$

Although the surface integral equation (Equation 6.163) can be solved with respect to ϕ, it has to be done at every time step, a process that renders this numerical procedure impractical. However, Equation 6.163 can be transformed into a form suitable for efficient numerical implementation using the perturbation technique as was done in the previous section in terms of E–H or A–K formalisms. Transfer to nondimensional variables in Equation 6.162 yields

$$\tilde{c}\tilde{\phi}+\sum_{i=1}^{N}\int_{S_i}\left(\frac{\partial\tilde{G}^{(3D)}}{\partial\vec{n}}\tilde{\phi}-\tilde{G}^{(3D)}\tilde{p}\sum_{i=1}^{2}\frac{\partial}{\partial\tilde{\xi}_i}\tilde{F}\left[(\nabla\tilde{\phi})_{\xi_i}\right]\right)\mathrm{d}\tilde{s}=\sum_{i=1}^{N}\int_{S_i}\tilde{G}^{(3D)}\left(\tilde{H}_\eta^{\mathrm{s}}-\tilde{p}\sum_{i=1}^{2}\frac{\partial}{\partial\tilde{\xi}_i}\tilde{F}\left[\tilde{H}_{\xi_i}^{\mathrm{s}}\right]\right)\mathrm{d}\tilde{s} \tag{6.164}$$

where

$$\vec{\tilde{H}}^{\mathrm{s}}=\sum_{i=1}^{N}\int_{L_i}\vec{\tilde{I}}_i\times\frac{\vec{\tilde{R}}}{\tilde{R}^3}\mathrm{d}\tilde{l} \tag{6.165}$$

Following the approach developed above for the two-dimensional case, we introduce the operators $\Lambda_m^{(3D)}$, $m=1,2,3$, in the following form:

$$\Lambda_1^{(3D)}\left[\vec{f}\right]=\begin{cases}(j\omega\sigma\mu)^{-1/2}\displaystyle\sum_{i=1}^{2}\frac{\partial\dot{f}_{\xi_i}}{\partial\xi_i} & \text{FD}\\ T_2^{\mathrm{b}}*\displaystyle\sum_{i=1}^{2}\frac{\partial f_{\xi_i}}{\partial\xi_i} & \text{TD}\end{cases} \tag{6.166}$$

$$\Lambda_2^{(3D)}\left[\vec{f}\right]=\begin{cases}(j\omega\sigma\mu)^{-1}\displaystyle\sum_{i=1}^{2}\frac{d_{3-i}-d_i}{2d_id_{3-i}}\frac{\partial\dot{f}_{\xi_i}}{\partial\xi_i} & \text{FD}\\ T_3^{\mathrm{b}}*\displaystyle\sum_{i=1}^{2}\frac{d_{3-i}-d_i}{2d_id_{3-i}}\frac{\partial f_{\xi_i}}{\partial\xi_i} & \text{TD}\end{cases} \tag{6.167}$$

$$\Lambda_3^{(3D)}\left[\vec{f}\right] = \begin{cases} (j\omega\sigma\mu)^{-3/2} \sum_{i=1}^{2} \left[\frac{3d_{3-i}^2 - d_i^2 - 2d_i d_{3-i}}{8d_i^2 d_{3-i}^2} \frac{\dot{f}_{\xi_i}}{\partial \xi_i} + \frac{1}{2} \frac{\partial}{\partial \xi_i} \left(\frac{\partial^2 \dot{f}_{\xi_i}}{\partial \xi_{3-i}^2} + \frac{\partial^2 \dot{f}_{\xi_i}}{\partial \xi_i^2} \right) \right] & \text{FD} \\ T_4^{\text{b}} * \sum_{i=1}^{2} \left[\frac{3d_{3-i}^2 - d_i^2 - 2d_i d_{3-i}}{8d_i^2 d_{3-i}^2} \frac{\partial f_{\xi_i}}{\partial \xi_i} + \frac{1}{2} \frac{\partial}{\partial \xi_i} \left(\frac{\partial^2 f_{\xi_i}}{\partial \xi_{3-i}^2} + \frac{\partial^2 f_{\xi_i}}{\partial \xi_i^2} \right) \right] & \text{TD} \end{cases} \tag{6.168}$$

Here the following transformation was used:

$$\sum_{i=1}^{2} \left(-\frac{\partial^3 f_{\xi_i}}{\partial \xi_i \partial \xi_{3-i}^2} + \frac{\partial^3 f_{\xi_i}}{\partial \xi_i^3} + 2\frac{\partial^3 f_{\xi_{3-i}}}{\partial \xi_i^2 \partial \xi_{3-i}} \right) = \sum_{i=1}^{2} \frac{\partial}{\xi_i} \left(\frac{\partial^2 f_{\xi_i}}{\partial \xi_{3-i}^2} + \frac{\partial^2 f_{\xi_i}}{\partial \xi_i^2} \right) \tag{6.169}$$

The "nondimensional" versions of Equations 6.166 and 6.167 are the following:

$$\tilde{\Lambda}_1^{(3D)}\left[\vec{\tilde{f}}\right] = \begin{cases} (2j)^{-1/2} \sum_{i=1}^{2} \frac{\partial \dot{\tilde{f}}_{\xi_i}}{\partial \tilde{\xi}_i} & \text{FD} \\ \tilde{T}_2^{\text{b}} * \sum_{i=1}^{2} \frac{\partial \tilde{f}_{\xi_i}}{\partial \tilde{\xi}_i} & \text{TD} \end{cases} \tag{6.170}$$

$$\tilde{\Lambda}_2^{(3D)}\left[\vec{\tilde{f}}\right] = \begin{cases} (2j)^{-1} \sum_{i=1}^{2} \frac{\tilde{d}_{3-i} - \tilde{d}_i}{2\tilde{d}_i \tilde{d}_{3-i}} \frac{\partial \dot{\tilde{f}}_{\xi_i}}{\partial \tilde{\xi}_i} & \text{FD} \\ \tilde{T}_3^{\text{b}} * \sum_{i=1}^{2} \frac{\tilde{d}_{3-i} - \tilde{d}_i}{2\tilde{d}_i \tilde{d}_{3-i}} \frac{\partial \tilde{f}_{\xi_i}}{\partial \tilde{\xi}_i} & \text{TD} \end{cases} \tag{6.171}$$

$$\tilde{\Lambda}_3^{(3D)}\left[\vec{\tilde{f}}\right] = \begin{cases} (2j)^{-3/2} \sum_{i=1}^{2} \left[\frac{3\tilde{d}_{3-i}^2 - \tilde{d}_i^2 - 2\tilde{d}_i \tilde{d}_{3-i}}{8\tilde{d}_i^2 \tilde{d}_{3-i}^2} \frac{\partial \tilde{f}_{\xi_i}}{\partial \tilde{\xi}_i} + \frac{1}{2} \frac{\partial}{\partial \tilde{\xi}_i} \left(\frac{\partial^2 \tilde{f}_{\xi_i}}{\partial \tilde{\xi}_{3-i}^2} + \frac{\partial^2 \tilde{f}_{\xi_i}}{\partial \tilde{\xi}_i^2} \right) \right] & \text{FD} \\ \tilde{T}_4^{\text{b}} * \sum_{i=1}^{2} \left[\frac{3\tilde{d}_{3-i}^2 - \tilde{d}_i^2 - 2\tilde{d}_i \tilde{d}_{3-i}}{8\tilde{d}_i^2 \tilde{d}_{3-i}^2} \frac{\partial \tilde{f}_{\xi_i}}{\partial \tilde{\xi}_i} + \frac{1}{2} \frac{\partial}{\partial \tilde{\xi}_i} \left(\frac{\partial^2 \tilde{f}_{\xi_i}}{\partial \tilde{\xi}_{3-i}^2} + \frac{\partial^2 \tilde{f}_{\xi_i}}{\partial \tilde{\xi}_i^2} \right) \right] & \text{TD} \end{cases} \tag{6.172}$$

The operators $\Lambda_m^{(3D)}$ are related to the functions F as follows:

Dimensional variables:

$$\Lambda_1^{(3D)}\left[\vec{f}\right] = \sum_{i=1}^{2} \frac{\partial}{\partial \xi_i} F\left[f_{\xi_i}\right] \Big|^{\text{Leontovich}} \tag{6.173}$$

$$\Lambda_1^{(3D)}\left[\vec{f}\right] + \Lambda_2^{(3D)}\left[\vec{f}\right] = \sum_{i=1}^{2} \frac{\partial}{\partial \xi_i} F\left[f_{\xi_i}\right] \Big|^{\text{Mitzner}} \tag{6.174}$$

$$\Lambda_1^{(3D)}\left[\vec{f}\right]+\Lambda_2^{(3D)}\left[\vec{f}\right]+\Lambda_3^{(3D)}\left[\vec{f}\right]=\sum_{i=1}^{2}\frac{\partial}{\partial\xi_i}F\left[f_{\xi_i}\right]\Big|^{\text{Rytov}} \tag{6.175}$$

Nondimensional variables:

$$\tilde{\Lambda}_1^{(3D)}\left[\vec{\tilde{f}}\right]=\sum_{i=1}^{2}\frac{\partial}{\partial\tilde{\xi}_i}\tilde{F}\left[\tilde{f}_{\xi_i}\right]\Big|^{\text{Leontovich}} \tag{6.176}$$

$$\tilde{\Lambda}_1^{(3D)}\left[\vec{\tilde{f}}\right]+\tilde{p}\tilde{\Lambda}_2^{(3D)}\left[\vec{\tilde{f}}\right]=\sum_{i=1}^{2}\frac{\partial}{\partial\tilde{\xi}_i}\tilde{F}\left[\tilde{f}_{\xi_i}\right]\Big|^{\text{Mitzner}}_{\tilde{a}\ll 1} \tag{6.177}$$

$$\tilde{\Lambda}_1^{(3D)}\left[\vec{\tilde{f}}\right]+\tilde{p}\tilde{\Lambda}_2^{(3D)}\left[\vec{\tilde{f}}\right]+\tilde{p}^2\tilde{\Lambda}_3^{(3D)}\left[\vec{\tilde{f}}\right]=\sum_{i=1}^{2}\frac{\partial}{\partial\tilde{\xi}_i}\tilde{F}\left[\tilde{f}_{\xi_i}\right]\Big|^{\text{Rytov}}_{\tilde{a}\ll 1} \tag{6.178}$$

Substituting Equations 6.176 through 6.178 in the integral equation (Equation 6.163), we obtain

$$\frac{\tilde{\phi}}{2}+\sum_{i=1}^{N}\int_{S_i}\left(\tilde{\phi}\frac{\partial\tilde{G}^{(3D)}}{\partial\vec{n}}-\tilde{G}^{(3D)}\sum_{m=1}^{3}\tilde{p}^m\tilde{\Lambda}_m^{(3D)}[\nabla\phi]\right)\mathrm{d}\tilde{s}=\sum_{i=1}^{N}\int_{S_i}\tilde{G}^{(3D)}\left(\vec{n}\cdot\vec{\tilde{H}}_{fil}-\sum_{m=1}^{3}\tilde{p}^m\tilde{\Lambda}_m^{(3D)}\left[\vec{\tilde{H}}_{fil}\right]\right)\mathrm{d}\tilde{s} \tag{6.179}$$

We represent the function $\tilde{\phi}$ in the form of expansions in the small parameter p:

$$\tilde{\phi}=\sum_{m=1}^{\infty}p^m\tilde{\phi}_m \tag{6.180}$$

Substituting the expansions in Equation 6.180 into Equation 6.179 and equating the coefficients of equal powers of p, we obtain the following equations for the expansion coefficients:

PEC-limit ($m=0$):

$$\tilde{c}\tilde{\phi}_0+\sum_{i=1}^{N}\int_{S_i}\tilde{\phi}_0\frac{\partial\tilde{G}^{(3D)}}{\partial\vec{n}}\mathrm{d}\tilde{s}=\sum_{i=1}^{N}\int_{S_i}\tilde{G}^{(3)}\left(\vec{n}\cdot\vec{\tilde{H}}^{\mathrm{s}}\right)\mathrm{d}\tilde{s} \tag{6.181}$$

Leontovich's approximation ($m=1$):

$$\tilde{c}\tilde{\phi}_1+\sum_{i=1}^{N}\int_{S_i}\tilde{\phi}_1\frac{\partial\tilde{G}^{(3D)}}{\partial\vec{n}}\mathrm{d}\tilde{s}=-\sum_{i=1}^{N}\int_{S_i}\tilde{G}^{(3D)}\tilde{\Lambda}_1^{(3D)}\left[\vec{\tilde{H}}^{\mathrm{s}}-\nabla\tilde{\phi}_0\right]\mathrm{d}\tilde{s} \tag{6.182}$$

Mitzner's approximation ($m=2$):

$$\tilde{c}\tilde{\phi}_2 + \sum_{i=1}^{N} \int\limits_{S_i} \tilde{\phi}_2 \frac{\partial \tilde{G}^{(3D)}}{\partial \vec{n}} d\tilde{s} = -\sum_{i=1}^{N} \int\limits_{S_i} \tilde{G}^{(3D)} \left(\tilde{\Lambda}_2^{(3D)} \left[\tilde{\vec{H}}^s - \nabla \tilde{\phi}_0 \right] + \tilde{\Lambda}_1^{(3D)} \left[-\nabla \tilde{\phi}_1 \right] \right) d\tilde{s} \tag{6.183}$$

Rytov's approximation ($m=3$):

$$\tilde{c}\tilde{\phi}_3 + \sum_{i=1}^{N} \int\limits_{S_i} \tilde{\phi}_3 \frac{\partial \tilde{G}^{(3D)}}{\partial \vec{n}} d\tilde{s} = -\sum_{i=1}^{N} \int\limits_{S_i} \tilde{G}^{(3D)} \left(\tilde{\Lambda}_3^{(3D)} \left[\tilde{\vec{H}}^s - \nabla \tilde{\phi}_0 \right] + \tilde{\Lambda}_2^{(3D)} \left[-\nabla \tilde{\phi}_1 \right] + \tilde{\Lambda}_1^{(3D)} \left[-\nabla \tilde{\phi}_2 \right] \right) d\tilde{s} \tag{6.184}$$

Obviously, the integral equations (Equations 6.181 through 6.184) can be reduced to the two-dimensional case of long parallel conductors. In this case, the operators $\Lambda_m^{(3D)}$ are related to $\Lambda_m^{(2D)}$ as follows:

$$\Lambda_1^{(3D)}[\nabla \phi]\Big|_{\substack{d_1 \to \infty \\ \partial/\partial \xi_1 = 0}} = \begin{cases} (j\omega\sigma\mu)^{-1/2} \dfrac{\partial}{\partial \xi_2} \left(\dfrac{\partial \dot{\phi}}{\partial \xi_2} \right) & \text{FD} \\ \hat{T}_2^b * \dfrac{\partial}{\partial \xi_2} \left(\dfrac{\partial \phi}{\partial \xi_2} \right) & \text{TD} \end{cases} = \Lambda_1^{(2D)} \left[\frac{\partial^2 \phi}{\partial \xi_2^2} \right] \tag{6.185}$$

$$\Lambda_2^{(3D)}[\nabla \phi]\Big|_{\substack{d_1 \to \infty \\ \partial/\partial \xi_1 = 0}} = \Lambda_2^{(2D)} \left[\frac{\partial^2 \phi}{\partial \xi_2^2} \right] \tag{6.186}$$

$$\Lambda_3^{(3D)}[\nabla \phi]\Big|_{\substack{d_1 \to \infty \\ \partial/\partial \xi_1 = 0}} = \Lambda_3^{(2D)} \left[\frac{\partial^2 \phi}{\partial \xi_2^2} \right] \tag{6.187}$$

Taking into account Equations 6.185 through 6.187, we adjust the integral equations (Equations 6.181 through 6.184) to the two-dimensional case in the following form:

$$\tilde{c}\tilde{\phi}_0 + \sum_{i=1}^{N} \oint\limits_{L_i} \tilde{\phi}_0 \frac{\partial \tilde{G}^{(2D)}}{\partial \vec{n}} d\tilde{\xi}_2 = \sum_{i=1}^{N} \oint\limits_{L_i} \tilde{G}^{(2D)} \left(\vec{n} \cdot \tilde{\vec{H}}^s \right) d\tilde{\xi}_2 \tag{6.188}$$

$$\tilde{c}\tilde{\phi}_1 + \sum_{i=1}^{N} \oint\limits_{L_i} \tilde{\phi}_1 \frac{\partial \tilde{G}^{(2D)}}{\partial \vec{n}} d\tilde{\xi}_2 = -\sum_{i=1}^{N} \oint\limits_{L_i} \tilde{G}^{(2D)} \tilde{\Lambda}_1^{(2D)} \left[\frac{\partial \tilde{H}_{\xi_2}^s}{\partial \tilde{\xi}_2} - \frac{\partial^2 \tilde{\phi}_0}{\partial \tilde{\xi}_2^2} \right] d\tilde{\xi}_2 \tag{6.189}$$

$$\tilde{c}\tilde{\phi}_2 + \sum_{i=1}^{N} \oint_{L_i} \tilde{\phi}_2 \frac{\partial \tilde{G}^{(2D)}}{\partial \vec{n}} d\tilde{\xi}_2 = -\sum_{i=1}^{N} \oint_{L_i} \tilde{G}^{(2D)} \left(\tilde{\Lambda}_2^{(2D)} \left[\frac{\partial \tilde{H}^s_{\xi_2}}{\partial \tilde{\xi}_2} - \frac{\partial^2 \tilde{\phi}_0}{\partial \tilde{\xi}_2^2} \right] + \tilde{\Lambda}_1^{(2D)} \left[-\frac{\partial^2 \tilde{\phi}_1}{\partial \tilde{\xi}_2^2} \right] \right) d\tilde{\xi}_2 \tag{6.190}$$

$$\tilde{c}\tilde{\phi}_3 + \sum_{i=1}^{N} \oint_{L_i} \tilde{\phi}_3 \frac{\partial \tilde{G}^{(2D)}}{\partial \vec{n}} d\tilde{\xi}_2$$
$$= -\sum_{i=1}^{N} \oint_{L_i} \tilde{G}^{(2D)} \left(\tilde{\Lambda}_3^{(2D)} \left[\frac{\partial \tilde{H}^s_{\xi_2}}{\partial \tilde{\xi}_2} - \frac{\partial^2 \tilde{\phi}_0}{\partial \tilde{\xi}_2^2} \right] + \tilde{\Lambda}_2^{(2D)} \left[-\frac{\partial^2 \tilde{\phi}_1}{\partial \tilde{\xi}_2^2} \right] + \tilde{\Lambda}_1^{(2D)} \left[-\frac{\partial^2 \tilde{\phi}_2}{\partial \tilde{\xi}_2^2} \right] \right) d\tilde{\xi}_2 \tag{6.191}$$

6.4 Properties of the Surface Impedance Function

1. *The operators $\Lambda_m^{(2D)}$ and $\Lambda_m^{(3D)}$ admit separation of variables into space and time components.*

We assume that a function can be split into space and time components as follows:

$$f(\vec{r}, t) = g(\vec{r})h(t) \quad \text{2D case} \tag{6.192}$$

$$\vec{f}(\vec{r}, t) = \vec{g}(\vec{r})h(t) \quad \text{3D case} \tag{6.193}$$

Then the operators $\Lambda_m^{(2D)}[f]$ and $\Lambda_m^{(3D)}\left[\vec{f}\right]$, $m = 1, 2, 3$, can be represented in the form of superposition of the space operators $\Psi_m^{(2D)}[u]$ and $\Psi_m^{(3D)}[\vec{u}]$, and the time operators $\Omega_m\,[v]$:

$$\Lambda_m^{(2D)}[f] = \Omega_m[v]\Psi_m^{(2D)}[u], \quad m = 1, 2, 3 \tag{6.194}$$

$$\Lambda_m^{(3D)}\left[\vec{f}\right] = \Omega_m[v]\Psi_m^{(3D)}[\vec{u}], \quad m = 1, 2, 3 \tag{6.195}$$

where

$$\Psi_1^{(2D)}[u] = u \tag{6.196}$$

$$\Psi_2^{(2D)}[u] = \frac{1}{2d_2} u \tag{6.197}$$

$$\Psi_3^{(2D)}[u] = \frac{3}{8d_2^2} u + \frac{1}{2} \frac{\partial^2 u}{\partial \xi_2^2} \tag{6.198}$$

$$\Psi_1^{(3D)}[\vec{u}] = \sum_{i=1}^{2} \frac{\partial u_{\xi_i}}{\partial \xi_i} \tag{6.199}$$

$$\Psi_2^{(3D)}[\vec{u}] = \sum_{i=1}^{2} \frac{d_{3-i} - d_i}{2d_i d_{3-i}} \frac{\partial u_{\xi_i}}{\partial \xi_i} \tag{6.200}$$

$$\Psi_3^{(3D)}[\vec{u}] = \sum_{i=1}^{2} \left\{ \frac{3d_{3-i}^2 - d_i^2 - 2d_i d_{3-i}}{8d_i^2 d_{3-i}^2} \frac{\partial u_{\xi_i}}{\partial \xi_i} + \frac{1}{2} \frac{\partial}{\partial \xi_i} \left(\frac{\partial^2 u_{\xi_i}}{\partial \xi_{3-i}^2} + \frac{\partial^2 u_{\xi_i}}{\partial \xi_i^2} \right) \right\} \tag{6.201}$$

and

$$\Omega_1[h] = \begin{cases} (j\omega\sigma\mu)^{-1/2} \dot{h} & \text{FD} \\ (\sigma\mu)^{-1/2}\, T_2^{\mathrm{b}} * h & \text{TD} \end{cases} \tag{6.202}$$

$$\Omega_2[h] = \begin{cases} (j\omega\sigma\mu)^{-1} \dot{h} & \text{FD} \\ (\sigma\mu)^{-1}\, T_3^{\mathrm{b}} * h & \text{TD} \end{cases} \tag{6.203}$$

$$\Omega_3[h] = \begin{cases} (j\omega\sigma\mu)^{-3/2} \dot{h} & \text{FD} \\ (\sigma\mu)^{-3/2}\, T_4^{\mathrm{b}} * h & \text{TD} \end{cases} \tag{6.204}$$

With nondimensional variables, the operators $\Psi_m^{(2D)}[u]$ and $\Psi_m^{(3D)}[\vec{u}]$, and the time operators $\Omega_m[\nu]$ take the form:

$$\tilde{\Psi}_1^{(2D)}[\tilde{u}] = \tilde{u} \tag{6.205}$$

$$\tilde{\Psi}_2^{(2D)}[\tilde{u}] = \frac{1}{2\tilde{d}_2} \tilde{u} \tag{6.206}$$

$$\tilde{\Psi}_3^{(2D)}[\tilde{u}] = \frac{3}{8\tilde{d}_2^2} \tilde{u} + \frac{1}{2} \frac{\partial^2 \tilde{u}}{\partial \tilde{\xi}_2^2} \tag{6.207}$$

$$\tilde{\Psi}_1^{(3D)}\left[\vec{\tilde{u}}\right] = \sum_{i=1}^{2} \frac{\partial \tilde{u}_{\xi_i}}{\partial \tilde{\xi}_i} \tag{6.208}$$

$$\tilde{\Psi}_2^{(3D)}\left[\vec{\tilde{u}}\right] = \sum_{i=1}^{2} \frac{\tilde{d}_{3-i} - \tilde{d}_i}{2\tilde{d}_i \tilde{d}_{3-i}} \frac{\partial \tilde{u}_{\xi_i}}{\partial \tilde{\xi}_i} \tag{6.209}$$

$$\tilde{\Psi}_3^{(3D)}\left[\vec{\tilde{u}}\right] = \sum_{i=1}^{2} \left\{ \frac{3\tilde{d}_{3-i}^2 - \tilde{d}_i^2 - 2\tilde{d}_i \tilde{d}_{3-i}}{8\tilde{d}_i^2 \tilde{d}_{3-i}^2} \frac{\partial \tilde{u}_{\xi_i}}{\partial \tilde{\xi}_i} + \frac{1}{2} \frac{\partial}{\partial \tilde{\xi}_i} \left(\frac{\partial^2 \tilde{u}_{\xi_i}}{\partial \tilde{\xi}_{3-i}^2} + \frac{\partial^2 \tilde{u}_{\xi_i}}{\partial \tilde{\xi}_i^2} \right) \right\} \tag{6.210}$$

$$\tilde{\Omega}_1\left[\tilde{h}\right] = \begin{cases} (2j)^{-1/2} \dot{\tilde{h}} & \text{FD} \\ \tilde{T}_2^{b} * \tilde{h} & \text{TD} \end{cases} \tag{6.211}$$

$$\tilde{\Omega}_2\left[\tilde{h}\right] = \begin{cases} (2j)^{-1}\dot{\tilde{h}} & \text{FD} \\ \tilde{T}_3^b * \tilde{h} & \text{TD} \end{cases} \tag{6.212}$$

$$\tilde{\Omega}_3\left[\tilde{h}\right] = \begin{cases} (2j)^{-3/2}\dot{\tilde{h}} & \text{FD} \\ \tilde{T}_4^b * \tilde{h} & \text{TD} \end{cases} \tag{6.213}$$

This property allows separation of variables into space and time components in the surface impedance function, F, using the perturbation technique.

2. *The time operators* $\Omega_m\,[\nu]$ *have the following property*:

$$\Omega_2[\nu] = \Omega_1[\Omega_1[\nu]] \quad \text{and} \quad \Omega_3[\nu] = \Omega_2[\Omega_1[\nu]] = \Omega_1[\Omega_2[\nu]] \tag{6.214}$$

This property is easily proven in the Laplace domain where the operators in Equations 6.202 through 6.204 take the form:

$$\bar{\Omega}_m[\bar{\nu}] = \bar{T}_m\bar{\nu}, \quad m = 1, 2, 3 \tag{6.215}$$

where

$$\bar{T}_1(s) = (\sigma\mu)^{-1/2}s^{-1/2}; \quad \bar{T}_2(s) = (\sigma\mu)^{-1}s^{-1}; \quad \bar{T}_3(s) = (\sigma\mu)^{-3/2}s^{-3/2} \tag{6.216}$$

From Equation 6.216, we obtain

$$\bar{T}_2 = \bar{T}_1\bar{T}_1; \quad \bar{T}_3 = \bar{T}_1\bar{T}_1\bar{T}_1 = \bar{T}_1\bar{T}_2 \tag{6.217}$$

Thus

$$\bar{\Omega}_2[\bar{\nu}] = \bar{T}_2\nu = \bar{T}_1\bar{T}_1\nu = \bar{T}_1\bar{\Omega}_1[\bar{\nu}] = \bar{\Omega}_1\left[\bar{\Omega}_1[\bar{\nu}]\right] \tag{6.218}$$

$$\bar{\Omega}_3[\bar{\nu}] = \bar{T}_3\nu = \bar{T}_2\bar{T}_1\nu = \bar{T}_2\bar{\Omega}_1[\bar{\nu}] = \bar{\Omega}_2\left[\bar{\Omega}_1[\bar{\nu}]\right] \tag{6.219}$$

$$\bar{\Omega}_3[\bar{\nu}] = \bar{T}_3\nu = \bar{T}_1\bar{T}_2\nu = \bar{T}_1\bar{\Omega}_2[\bar{\nu}] = \bar{\Omega}_1\left[\bar{\Omega}_2[\bar{\nu}]\right] \tag{6.220}$$

Application of the inverse Laplace transform identities to Equations 6.218 through 6.220 leads to Equation 6.214.

6.5 Boundary Element Formulations for Two- and Three-Dimensional Problems in Invariant Form

Let an external source produce quasisteady currents flowing through the conductors so that

Two-dimensional case:

$$I_i(t) = \begin{cases} I_i^{\mathrm{sp}} \exp(j\omega t) & \text{Time-harmonic regime} \\ I_i^{\mathrm{sp}} \tilde{\theta}_0(t) & \text{Pulsed regime} \end{cases}, \quad i = 1, 2, \ldots, N \tag{6.221}$$

Three-dimensional case:

$$\vec{I}_i(\vec{r}, t) = \begin{cases} \vec{I}_i^{\,\mathrm{sp}} \exp(j\omega t) & \text{Time-harmonic regime} \\ \vec{I}_i^{\,\mathrm{sp}} \tilde{\theta}_0(t) & \text{Pulsed regime} \end{cases}, \quad i = 1, 2, \ldots, N \tag{6.222}$$

where $\tilde{\theta}_0$ is a known nondimensional time-dependent function that is the same for all conductors and I_i^{sp} or $\vec{I}_i^{\mathrm{sp}}$, $i = 1, 2, \ldots, N$, are the magnitude (in two-dimensional case) or magnitude and direction (in three-dimensional case) of the total current, respectively. The index "sp" stands for spatial. From Equations 6.221 and 6.222 it follows that

Two-dimensional case:

$$\frac{\mathrm{d}I_i(t)}{\mathrm{d}t} = \begin{cases} j\omega I_i^{\mathrm{sp}} \exp(j\omega t) & \text{Time-harmonic regime} \\ I_i^{\mathrm{sp}} \dfrac{\mathrm{d}\tilde{\theta}_0(t)}{\mathrm{d}t} & \text{Pulsed regime} \end{cases} \tag{6.223}$$

Three-dimensional case:

$$\frac{\partial \vec{I}_i(\vec{r}, t)}{\partial t} = \begin{cases} j\omega \vec{I}_i^{\,\mathrm{sp}} \exp(j\omega t) & \text{Time-harmonic regime} \\ \vec{I}_i^{\,\mathrm{sp}} \dfrac{\mathrm{d}\tilde{\theta}_0(t)}{\mathrm{d}t} & \text{Pulsed regime} \end{cases} \tag{6.224}$$

Taking into account Equations 6.221 and 6.222, the filamentary magnetic field in Equations 6.158 and 6.159 can be represented as

$$\vec{\tilde{H}}^{\mathrm{s}}\left(\vec{\tilde{r}}, \tilde{t}\right) = \begin{cases} \vec{\tilde{H}}^{\mathrm{s}}_{\mathrm{sp}}\left(\vec{\tilde{r}}\right) \exp(2j\tilde{t}) & \text{Time-harmonic regime} \\ \vec{\tilde{H}}^{\mathrm{s}}_{\mathrm{sp}}\left(\vec{\tilde{r}}\right) \tilde{\theta}_0\left(\tilde{t}\right) & \text{Pulsed regime} \end{cases} \tag{6.225}$$

$$\vec{\tilde{H}}^{\mathrm{s}}_{\mathrm{sp}} = \sum_{i=1}^{N} \int_{L_i} \vec{\tilde{I}}_i^{\mathrm{sp}}\left(\vec{\tilde{r}}'\right) \times \frac{\vec{\tilde{r}} - \vec{\tilde{r}}'}{\left|\vec{\tilde{r}} - \vec{\tilde{r}}'\right|^3} \mathrm{d}\tilde{l} \tag{6.226}$$

Introducing the following nondimensional time-dependent functions $\tilde{\theta}_m$, $m = 1, 2, 3$:

$$\tilde{\theta}_m(\tilde{t}) = \begin{cases} \tilde{\Omega}_m[\exp(2j\tilde{t})] = (2j)^{-m/2}\exp(2j\tilde{t}) & \text{Time-harmonic regime} \\ \tilde{\Omega}_m[\tilde{\theta}_0(\tilde{t})] = \tilde{T}_m(\tilde{t}) \tilde{*} \tilde{\theta}_0(\tilde{t}) & \text{Pulsed regime} \end{cases}, \quad m = 1,2,3 \tag{6.227}$$

and using Equation 6.214 gives

$$\tilde{\theta}_2 = \tilde{\Omega}_2[\tilde{\theta}_0] = \tilde{\Omega}_1\left[\tilde{\Omega}_1[\tilde{\theta}_0]\right] = \tilde{\Omega}_1[\tilde{\theta}_1] \tag{6.228}$$

$$\tilde{\theta}_3 = \tilde{\Omega}_3[\tilde{\theta}_0] = \tilde{\Omega}_1\left[\tilde{\Omega}_2[\tilde{\theta}_0]\right] = \tilde{\Omega}_1[\tilde{\theta}_2] \quad \text{and} \quad \tilde{\theta}_3 = \tilde{\Omega}_3[\tilde{\theta}_0] = \tilde{\Omega}_2\left[\tilde{\Omega}_1[\tilde{\theta}_0]\right] = \tilde{\Omega}_2[\tilde{\theta}_1] \tag{6.229}$$

We seek functions $\tilde{u}_m$, $\tilde{v}_m$, and $\tilde{\phi}_m$, for which the integral equations in Equations 6.140 through 6.155 and Equations 6.188 through 6.191 in the two-dimensional case and Equations 6.181 through 6.184 in the three-dimensional case have to be solved, in the following form:

Two-dimensional case:

$$\tilde{u}_m\left(\vec{\tilde{r}}, t\right) = \tilde{u}_m^{\text{sp}}\left(\vec{\tilde{r}}\right)\tilde{\theta}_m(t); \quad \tilde{v}_m\left(\vec{\tilde{r}}, t\right) = \tilde{v}_m^{\text{sp}}\left(\vec{\tilde{r}}\right)\tilde{\theta}_m(t), \quad m = 0, 1, 2, 3 \tag{6.230}$$

Three-dimensional case:

$$\tilde{\phi}_m\left(\vec{\tilde{r}}, t\right) = \tilde{\phi}_m^{\text{sp}}\left(\vec{\tilde{r}}\right)\tilde{\theta}_m(t), \quad m = 0, 1, 2, 3 \tag{6.231}$$

Then the operators $\tilde{\Lambda}_m^{(2D)}$ and $\tilde{\Lambda}_m^{(3D)}$ can be represented in the following form according to properties proven in the previous section:

Two-dimensional case:

$$\tilde{\Lambda}_1^{(2D)}[\tilde{u}_0] = \tilde{\Omega}_1[\tilde{\theta}_0]\tilde{\Psi}_1^{(2D)}[\tilde{u}_0^{\text{sp}}] = \tilde{\theta}_1\tilde{\Psi}_1^{(2D)}[\tilde{u}_0^{\text{sp}}] \tag{6.232}$$

$$\tilde{\Lambda}_2^{(2D)}[\tilde{u}_0] = \tilde{\Omega}_2[\tilde{\theta}_0]\tilde{\Psi}_2^{(2D)}[\tilde{u}_0^{\text{sp}}] = \tilde{\theta}_2\tilde{\Psi}_2^{(2D)}[\tilde{u}_0^{\text{sp}}] \tag{6.233}$$

$$\tilde{\Lambda}_1^{(2D)}[\tilde{u}_1] = \tilde{\Omega}_1[\tilde{\theta}_1]\tilde{\Psi}_1^{(2D)}[\tilde{u}_1^{\text{sp}}] = \tilde{\theta}_2\tilde{\Psi}_1^{(2D)}[\tilde{u}_1^{\text{sp}}] \tag{6.234}$$

$$\tilde{\Lambda}_3^{(2D)}[\tilde{u}_0] = \tilde{\Omega}_3[\tilde{\theta}_0]\tilde{\Psi}_3^{(2D)}[\tilde{u}_0^{\text{sp}}] = \tilde{\theta}_3\tilde{\Psi}_3^{(2D)}[\tilde{u}_0^{\text{sp}}] \tag{6.235}$$

$$\tilde{\Lambda}_2^{(2D)}[\tilde{u}_1] = \tilde{\Omega}_2[\tilde{\theta}_1]\tilde{\Psi}_2^{(2D)}[\tilde{u}_1^{sp}] = \tilde{\theta}_3\tilde{\Psi}_2^{(2D)}[\tilde{u}_1^{sp}] \tag{6.236}$$

$$\tilde{\Lambda}_1^{(2D)}[\tilde{u}_2] = \tilde{\Omega}_1[\tilde{\theta}_2]\tilde{\Psi}_1^{(2D)}[\tilde{u}_2^{sp}] = \tilde{\theta}_3\tilde{\Psi}_1^{(2D)}[\tilde{u}_2^{sp}] \tag{6.237}$$

Three-dimensional case:

$$\tilde{\Lambda}_1^{(3D)}\left[\vec{\tilde{H}}^s - \nabla\tilde{\phi}_0\right] = \tilde{\Omega}_1[\tilde{\theta}_0]\tilde{\Psi}_1^{(3D)}\left[\vec{\tilde{H}}_{sp}^s - \nabla\tilde{\phi}_0^{sp}\right] = \tilde{\theta}_1\tilde{\Psi}_1^{(3D)}\left[\vec{\tilde{H}}_{sp}^s - \nabla\tilde{\phi}_0^{sp}\right] \tag{6.238}$$

$$\tilde{\Lambda}_2^{(3D)}\left[\vec{\tilde{H}}^s - \nabla\tilde{\phi}_0\right] = \tilde{\Omega}_2[\tilde{\theta}_0]\tilde{\Psi}_2^{(3D)}\left[\vec{\tilde{H}}_{sp}^s - \nabla\tilde{\phi}_0^{sp}\right] = \tilde{\theta}_2\tilde{\Psi}_2^{(3D)}\left[\vec{\tilde{H}}_{sp}^s - \nabla\tilde{\phi}_0^{sp}\right] \tag{6.239}$$

$$\tilde{\Lambda}_1^{(3D)}[-\nabla\tilde{\phi}_1] = \tilde{\Omega}_1[\tilde{\theta}_1]\tilde{\Psi}_1^{(3D)}[-\nabla\tilde{\phi}_1^{sp}] = \tilde{\theta}_2\tilde{\Psi}_1^{(3D)}[-\nabla\tilde{\phi}_1^{sp}] \tag{6.240}$$

$$\tilde{\Lambda}_3^{(3D)}\left[\vec{\tilde{H}}^s - \nabla\tilde{\phi}_0\right] = \tilde{\Omega}_3[\tilde{\theta}_0]\tilde{\Psi}_3^{(3D)}\left[\vec{\tilde{H}}_{sp}^s - \nabla\tilde{\phi}_0^{sp}\right] = \tilde{\theta}_3\tilde{\Psi}_3^{(3D)}\left[\vec{\tilde{H}}_{sp}^s - \nabla\tilde{\phi}_0^{sp}\right] \tag{6.241}$$

$$\tilde{\Lambda}_2^{(3D)}[-\nabla\tilde{\phi}_1] = \tilde{\Omega}_2[\tilde{\theta}_1]\tilde{\Psi}_2^{(3D)}[-\nabla\tilde{\phi}_1^{sp}] = \tilde{\theta}_3\tilde{\Psi}_2^{(3D)}[-\nabla\tilde{\phi}_1^{sp}] \tag{6.242}$$

$$\tilde{\Lambda}_1^{(3D)}[-\nabla\tilde{\phi}_2] = \tilde{\Omega}_1[\tilde{\theta}_2]\tilde{\Psi}_1^{(3D)}[-\nabla\tilde{\phi}_2^{sp}] = \tilde{\theta}_3\tilde{\Psi}_1^{(3D)}[-\nabla\tilde{\phi}_2^{sp}] \tag{6.243}$$

Substituting Equations 6.225 through 6.227 and 6.232 through 6.243 into Equations 6.140 through 6.155, 6.181 through 6.184, and 6.188 through 6.191, we obtain the following equations for the spatial components $\tilde{v}_m^{sp}$, $\tilde{u}_m^{sp}$ (two-dimensional case), and $\tilde{\varphi}_m^{sp}$ (two- and three-dimensional cases):

Two-dimensional case:

E–H and *A–K* formalisms:

$$\tilde{v}_0^{sp} - \sum_{i=1}^{N}\oint_{L_i}\tilde{G}^{(2D)}\tilde{u}_0^{sp}\,d\tilde{\xi}_2 = 0 \tag{6.244}$$

$$\oint_{L_i}\tilde{u}_0^{sp}\,d\tilde{\xi}_2 = \tilde{I}_{sp}^{(i)} \tag{6.245}$$

$$\tilde{v}_1^{sp} - \sum_{i=1}^{N}\oint_{L_i}\tilde{G}^{(2D)}\tilde{u}_1^{sp}\,d\tilde{\xi}_2 = -\tilde{c}\tilde{\Psi}_1^{(2D)}[\tilde{u}_0^{sp}] - \sum_{i=1}^{N}\oint_{L_i}\frac{\partial\tilde{G}^{(2D)}}{\partial\vec{n}}\tilde{\Psi}_1^{(2D)}[\tilde{u}_1^{sp}]\,d\tilde{\xi}_2 \tag{6.246}$$

$$\oint_{L_i}\tilde{u}_1^{sp}\,d\tilde{\xi}_2 = 0 \tag{6.247}$$

$$\tilde{v}_2^{sp} - \sum_{i=1}^{N} \oint_{L_i} \tilde{G}^{(2D)} \tilde{u}_2^{sp} d\tilde{\xi}_2 = -\tilde{c}\left(\tilde{\Psi}_1^{(2D)}\left[\tilde{u}_1^{sp}\right] + \tilde{\Psi}_2^{(2D)}\left[\tilde{u}_0^{sp}\right]\right) - \sum_{i=1}^{N} \oint_{L_i} \frac{\partial \tilde{G}^{(2D)}}{\partial \vec{n}} \left(\tilde{\Psi}_1^{(2D)}\left[\tilde{u}_1^{sp}\right] + \tilde{\Psi}_2^{(2D)}\left[\tilde{u}_0^{sp}\right]\right) d\tilde{\xi}_2 \quad (6.248)$$

$$\oint_{L_i} \tilde{u}_2^{sp} d\tilde{\xi}_2 = 0 \quad (6.249)$$

$$\tilde{v}_3^{sp} - \sum_{i=1}^{N} \oint_{L_i} \tilde{G}^{(2D)} \tilde{u}_3^{sp} d\tilde{\xi}_2 = -\tilde{c}\left(\tilde{\Psi}_1^{(2D)}\left[\tilde{u}_2^{sp}\right] + \tilde{\Psi}_2^{(2D)}\left[\tilde{u}_1^{sp}\right] + \tilde{\Psi}_3^{(2D)}\left[\tilde{u}_0^{sp}\right]\right) - \sum_{i=1}^{N} \oint_{L_i} \frac{\partial \tilde{G}^{(2D)}}{\partial \vec{n}} \left(\tilde{\Psi}_1^{(2D)}\left[\tilde{u}_2^{sp}\right] + \tilde{\Psi}_2^{(2D)}\left[\tilde{u}_1^{sp}\right] + \tilde{\Psi}_3^{(2D)}\left[\tilde{u}_0^{sp}\right]\right) d\tilde{\xi} \quad (6.250)$$

$$\oint_{L_i} \tilde{u}_3^{sp} d\tilde{\xi}_2 = 0 \quad (6.251)$$

H–ϕ formalism:

$$\tilde{c}\tilde{\phi}_0^{sp} + \sum_{i=1}^{N} \oint_{L_i} \tilde{\phi}_0^{sp} \frac{\partial \tilde{G}^{(2D)}}{\partial \vec{n}} d\tilde{\xi}_2 = \sum_{i=1}^{N} \oint_{L_i} \tilde{G}^{(2D)} \left(\vec{n} \cdot \vec{\tilde{H}}_{sp}^{s}\right) d\tilde{\xi}_2 \quad (6.252)$$

$$\tilde{c}\tilde{\phi}_1^{sp} + \sum_{i=1}^{N} \oint_{L_i} \tilde{\phi}_1^{sp} \frac{\partial \tilde{G}^{(2D)}}{\partial \vec{n}} d\tilde{\xi}_2 = -\sum_{i=1}^{N} \oint_{L_i} \tilde{G}^{(2D)} \tilde{\Psi}_1^{(2D)} \left[\frac{\partial \left(\vec{\tilde{H}}_{sp}^{s}\right)_{\xi_2}}{\partial \tilde{\xi}_2} - \frac{\partial^2 \tilde{\phi}_0^{sp}}{\partial \tilde{\xi}_2^2}\right] d\tilde{\xi}_2 \quad (6.253)$$

$$\tilde{c}\tilde{\phi}_2^{sp} + \sum_{i=1}^{N} \oint_{L_i} \tilde{\phi}_2^{sp} \frac{\partial \tilde{G}^{(2D)}}{\partial \vec{n}} d\tilde{\xi}_2 = -\sum_{i=1}^{N} \oint_{L_i} \tilde{G}^{(2D)} \left(\tilde{\Psi}_2^{(2D)} \left[\frac{\partial \left(\vec{\tilde{H}}_{sp}^{s}\right)_{\xi_2}}{\partial \tilde{\xi}_2} - \frac{\partial^2 \tilde{\phi}_0^{sp}}{\partial \tilde{\xi}_2^2}\right] + \tilde{\Psi}_1^{(2D)} \left[-\frac{\partial^2 \tilde{\phi}_1^{sp}}{\partial \tilde{\xi}_2^2}\right]\right) d\tilde{\xi}_2 \quad (6.254)$$

$$\tilde{c}\tilde{\phi}_3^{sp} + \sum_{i=1}^{N} \oint_{L_i} \tilde{\phi}_3^{sp} \frac{\partial \tilde{G}^{(2D)}}{\partial \vec{n}} d\tilde{\xi}_2 = -\sum_{i=1}^{N} \oint_{L_i} \tilde{G}^{(2D)} \left(\tilde{\Psi}_3^{(2D)} \left[\frac{\partial \left(\vec{\tilde{H}}_{sp}^{s}\right)_{\xi_2}}{\partial \tilde{\xi}_2} - \frac{\partial^2 \tilde{\phi}_0^{sp}}{\partial \tilde{\xi}_2^2}\right] + \tilde{\Psi}_2^{(2D)} \left[-\frac{\partial^2 \tilde{\phi}_1^{sp}}{\partial \tilde{\xi}_2^2}\right] + \tilde{\Psi}_1^{(2D)} \left[-\frac{\partial^2 \tilde{\phi}_2^{sp}}{\partial \tilde{\xi}_2^2}\right]\right) d\tilde{\xi}_2 \quad (6.255)$$

Three-dimensional case:

H–ϕ formalism:

$$\tilde{c}\tilde{\phi}_0^{sp} + \sum_{i=1}^{N} \int_{S_i} \tilde{\phi}_0^{sp} \frac{\partial \tilde{G}^{(3D)}}{\partial \vec{n}} d\tilde{s} = \sum_{i=1}^{N} \int_{S_i} \tilde{G}^{(3)} \left(\vec{n} \cdot \tilde{\vec{H}}_{sp}^{s} \right) d\tilde{s} \quad (6.256)$$

$$\tilde{c}\tilde{\phi}_1^{sp} + \sum_{i=1}^{N} \int_{S_i} \tilde{\phi}_1^{sp} \frac{\partial \tilde{G}^{(3D)}}{\partial \vec{n}} d\tilde{s} = -\sum_{i=1}^{N} \int_{S_i} \tilde{G}^{(3D)} \tilde{\Psi}_1^{(3D)} \left[\tilde{\vec{H}}_{sp}^{s} - \nabla \tilde{\phi}_0^{sp} \right] d\tilde{s} \quad (6.257)$$

$$\tilde{c}\tilde{\phi}_2^{sp} + \sum_{i=1}^{N} \int_{S_i} \tilde{\phi}_2^{sp} \frac{\partial \tilde{G}^{(3D)}}{\partial \vec{n}} d\tilde{s} = -\sum_{i=1}^{N} \int_{S_i} \tilde{G}^{(3D)} \left(\tilde{\Psi}_2^{(3D)} \left[\tilde{\vec{H}}_{sp}^{s} - \nabla \tilde{\phi}_0^{sp} \right] + \tilde{\Psi}_1^{(3D)} \left[-\nabla \tilde{\phi}_1^{sp} \right] \right) d\tilde{s} \quad (6.258)$$

$$\tilde{c}\tilde{\phi}_3^{sp} + \sum_{i=1}^{N} \int_{S_i} \tilde{\phi}_3^{sp} \frac{\partial \tilde{G}^{(3D)}}{\partial \vec{n}} d\tilde{s} = -\sum_{i=1}^{N} \int_{S_i} \tilde{G}^{(3D)} \left(\tilde{\Psi}_3^{(3D)} \left[\tilde{\vec{H}}_{sp}^{s} - \nabla \tilde{\phi}_0^{sp} \right] + \tilde{\Psi}_2^{(3D)} \left[-\nabla \tilde{\phi}_1^{sp} \right] + \tilde{\Psi}_1^{(3D)} \left[-\nabla \tilde{\phi}_2^{sp} \right] \right) d\tilde{s} \quad (6.259)$$

We have obtained three sets of integral equations: Equations 6.244 through 6.251 for two-dimensional problems in terms of *E*–*H* and *A*–*K* formalisms; Equations 6.252 through 6.255 for two-dimensional problems in terms of the *H*–ϕ formalism, and Equations 6.256 through 6.259 for three-dimensional problems in terms of the *H*–ϕ formalism. The first equation in each set provides the solution in the perfect electrical conductor limit. The other integral equations give the corrections of the order of Leontovich, Mitzner, and Rytov's approximations.

We emphasize here that the formulation in Equations 6.244 through 6.251 does not depend on the formalism. In other words, the *E*–*H* and *A*–*K* formalisms in the two-dimensional produce the same integral equations for the spatial components.

Note that the *E*–*H* and *A*–*K* approaches give the Fredholm surface integral equations of the first kind whereas the *H*–ϕ approach leads to the Fredholm surface integral equations of the second kind. The latter is preferable from the numerical implementation point of view since the Fredholm surface integral equation of the first kind is the so-called ill-posed problem and requires special regularization algorithms [3].

Returning to dimensional variables using the scale factors given in Table 6.1 enables us to write Equations 6.244 through 6.259 in the following final form:

TABLE 6.1

Nondimensional Quantities and Their Scale Factors

Nondimensional Quantity	Scale Factor	Unit
$\tilde{u}_k^{sp}$, $k=0,1,2,3$	$I^*D^{-(k+1)}$	A $\mathrm{m}^{-(k+1)}$
$\tilde{v}_k^{sp}$, $k=0,1,2,3$, *E–H* formalism	$\mu I^* \tau^{-1} D^{-k}$	V $\mathrm{m}^{-(k+1)}$
$\tilde{v}_k^{sp}$, $k=0,1,2,3$, *A–K* formalism	I^*D^{-k}	A m^{-k}
$\tilde{\phi}_k^{sp}$, $k=0,1,2,3$	I^*D^{-k}	A m^{-k}
$\tilde{H}_{sp}^{s}$	I^*D^{-1}	A m^{-1}
$\tilde{G}^{(2D)}$	Nondimensional	Nondimensional
$\partial\tilde{G}^{(2D)}/\partial\vec{n}$	D^{-1}	m^{-1}
$\tilde{G}^{(3D)}$	D^{-1}	m^{-1}
$\partial\tilde{G}^{(3D)}/\partial\vec{n}$	D^{-2}	m^{-2}
$\mathrm{d}\tilde{\xi}_2$	D	m
$\mathrm{d}\tilde{s} = \mathrm{d}\tilde{\xi}_1\mathrm{d}\tilde{\xi}_2$	D^2	m^2
$\partial/\partial\tilde{\xi}_1$, $\partial/\partial\tilde{\xi}_2$	D^{-1}	m^{-1}
$\tilde{\Psi}_k^{(2D)}[\,]$, $k=1,2,3$	D^{-k+1}	$\mathrm{m}^{-(k+1)}$
$\tilde{\Psi}_k^{(3D)}[\,]$, $k=1,2,3$	D^{-k}	m^{-k}
$\tilde{T}_k$, $k=2,3,4$	$\tau^{(k-3)/2}$	$\mathrm{s}^{(k-3)/2}$
$\mathrm{d}\tilde{t}$ (in time convolution product)	τ	s

Two-dimensional case:

E–H and *A–K* formalisms:

$$v_0^{sp} - \sum_{i=1}^{N} \oint_{L_i} G^{(2D)} u_0^{sp} \mathrm{d}\xi_2 = 0 \tag{6.260}$$

$$\oint_{L_i} u_0^{sp} \mathrm{d}\xi_2 = I_{sp}^{(i)} \tag{6.261}$$

$$v_1^{sp} - \sum_{i=1}^{N} \oint_{L_i} G^{(2D)} u_1^{sp} \mathrm{d}\xi_2 = -\tilde{c}\Psi_1^{(2D)}\left[u_0^{sp}\right] - \sum_{i=1}^{N} \oint_{L_i} \frac{\partial G^{(2D)}}{\partial \vec{n}} \Psi_1^{(2D)}\left[u_1^{sp}\right] \mathrm{d}\xi_2 \tag{6.262}$$

$$\oint_{L_i} u_1^{sp} \mathrm{d}\xi_2 = 0 \tag{6.263}$$

$$v_2^{sp} - \sum_{i=1}^{N} \oint_{L_i} G^{(2D)} u_2^{sp} \mathrm{d}\xi_2 = -\tilde{c}\left(\Psi_1^{(2D)}\left[u_1^{sp}\right] + \Psi_2^{(2D)}\left[u_0^{sp}\right]\right) - \sum_{i=1}^{N} \oint_{L_i} \frac{\partial G^{(2D)}}{\partial \vec{n}} \left(\Psi_1^{(2D)}\left[u_1^{sp}\right] + \Psi_2^{(2D)}\left[u_0^{sp}\right]\right) \mathrm{d}\xi_2 \tag{6.264}$$

$$\oint_{L_i} u_2^{sp} d\xi_2 = 0 \tag{6.265}$$

$$v_3^{sp} - \sum_{i=1}^{N} \oint_{L_i} G^{(2D)} u_3^{sp} d\xi_2 = -\tilde{c}\left(\Psi_1^{(2D)}\left[u_2^{sp}\right] + \Psi_2^{(2D)}\left[u_1^{sp}\right] + \Psi_3^{(2D)}\left[u_0^{sp}\right]\right)$$
$$- \sum_{i=1}^{N} \oint_{L_i} \frac{\partial G^{(2D)}}{\partial \vec{n}} \left(\Psi_1^{(2D)}\left[u_2^{sp}\right] + \Psi_2^{(2D)}\left[u_1^{sp}\right] + \Psi_3^{(2D)}\left[u_0^{sp}\right]\right) d\xi_2 \tag{6.266}$$

$$\oint_{L_i} u_3^{sp} d\xi_2 = 0 \tag{6.267}$$

H–ϕ formalism:

$$\tilde{c}\phi_0^{sp} + \sum_{i=1}^{N} \oint_{L_i} \phi_0^{sp} \frac{\partial G^{(2D)}}{\partial \vec{n}} d\xi_2 = \sum_{i=1}^{N} \oint_{L_i} G^{(2D)} \left(\vec{n} \cdot \vec{H}_{sp}^{s}\right) d\xi_2 \tag{6.268}$$

$$\tilde{c}\phi_1^{sp} + \sum_{i=1}^{N} \oint_{L_i} \phi_1^{sp} \frac{\partial G^{(2D)}}{\partial \vec{n}} d\xi_2 = -\sum_{i=1}^{N} \oint_{L_i} G^{(2D)} \Psi_1^{(2D)} \left[\frac{\partial \left(\vec{H}_{sp}^{s}\right)_{\xi_2}}{\partial \xi_2} - \frac{\partial^2 \phi_0^{sp}}{\partial \xi_2^2}\right] d\xi_2 \tag{6.269}$$

$$\tilde{c}\phi_2^{sp} + \sum_{i=1}^{N} \oint_{L_i} \phi_2^{sp} \frac{\partial G^{(2D)}}{\partial \vec{n}} d\xi_2 = -\sum_{i=1}^{N} \oint_{L_i} G^{(2D)} \left(\Psi_2^{(2D)} \left[\frac{\partial \left(\vec{H}_{sp}^{s}\right)_{\xi_2}}{\partial \xi_2} - \frac{\partial^2 \phi_0^{sp}}{\partial \xi_2^2}\right] + \Psi_1^{(2D)} \left[-\frac{\partial^2 \phi_1^{sp}}{\partial \xi_2^2}\right]\right) d\xi_2 \tag{6.270}$$

$$\tilde{c}\phi_3^{sp} + \sum_{i=1}^{N} \oint_{L_i} \phi_3^{sp} \frac{\partial G^{(2D)}}{\partial \vec{n}} d\xi_2 = -\sum_{i=1}^{N} \oint_{L_i} G^{(2D)}$$
$$\left(\Psi_3^{(2D)} \left[\frac{\partial \left(\vec{H}_{sp}^{s}\right)_{\xi_2}}{\partial \xi_2} - \frac{\partial^2 \phi_0^{sp}}{\partial \xi_2^2}\right] + \Psi_2^{(2D)} \left[-\frac{\partial^2 \phi_1^{sp}}{\partial \xi_2^2}\right] + \Psi_1^{(2D)} \left[-\frac{\partial^2 \phi_2^{sp}}{\partial \xi_2^2}\right]\right) d\xi_2 \tag{6.271}$$

Three-dimensional case:

H–ϕ formalism:

$$\tilde{c}\phi_0^{sp} + \sum_{i=1}^{N} \int_{S_i} \phi_0^{sp} \frac{\partial G^{(3D)}}{\partial \vec{n}} ds = \sum_{i=1}^{N} \int_{S_i} G^{(3D)} \left(\vec{n} \cdot \vec{H}_{sp}^{s}\right) ds \tag{6.272}$$

$$\tilde{c}\phi_1^{\mathrm{sp}} + \sum_{i=1}^{N} \int_{S_i} \phi_1^{\mathrm{sp}} \frac{\partial G^{(3D)}}{\partial \vec{n}} \mathrm{d}\tilde{s} = -\sum_{i=1}^{N} \int_{S_i} G^{(3D)} \Psi_1^{(3D)} \left[\vec{H}_{\mathrm{sp}}^{\mathrm{s}} - \nabla \phi_0^{\mathrm{sp}} \right] \mathrm{d}s \tag{6.273}$$

$$\tilde{c}\phi_2^{\mathrm{sp}} + \sum_{i=1}^{N} \int_{S_i} \phi_2^{\mathrm{sp}} \frac{\partial G^{(3D)}}{\partial \vec{n}} \mathrm{d}\tilde{s} = -\sum_{i=1}^{N} \int_{S_i} G^{(3D)} \left(\Psi_2^{(3D)} \left[\vec{H}_{\mathrm{sp}}^{\mathrm{s}} - \nabla \phi_0^{\mathrm{sp}} \right] + \Psi_1^{(3D)} \left[-\nabla \phi_1^{\mathrm{sp}} \right] \right) \mathrm{d}s \tag{6.274}$$

$$\tilde{c}\phi_3^{\mathrm{sp}} + \sum_{i=1}^{N} \int_{S_i} \phi_3^{\mathrm{sp}} \frac{\partial G^{(3D)}}{\partial \vec{n}} \mathrm{d}s$$

$$= -\sum_{i=1}^{N} \int_{S_i} G^{(3D)} \left(\Psi_3^{(3D)} \left[\vec{H}_{\mathrm{sp}}^{\mathrm{s}} - \nabla \phi_0^{\mathrm{sp}} \right] + \Psi_2^{(3D)} \left[-\nabla \phi_1^{\mathrm{sp}} \right] + \Psi_1^{(3D)} \left[-\nabla \phi_2^{\mathrm{sp}} \right] \right) \mathrm{d}s \tag{6.275}$$

where the operators $\Psi_m^{(2D)}$ and $\Psi_m^{(3D)}$, $m = 1, 2, 3$, are given in Equations 6.196 through 6.198 and Equations 6.199 through 6.201, respectively.

We now formulate the procedure for calculation of the distribution of the electromagnetic field over the conductor's surface, using Equation 6.26 as initial data. The formulation consists of the following three steps:

E–H and *A–K* formalisms (two-dimensional problems):

1. Obtain the space functions $v_m^{\mathrm{sp}}(\vec{r})$ and $u_m^{\mathrm{sp}}(\vec{r})$, $m = 0, 1, 2, 3$, by solving the surface integral equations (Equations 6.260 through 6.267).
2. Obtain the functions E^{s}, A^{s}, H_{ξ_2}, and K using the following formulae:

$$E^{\mathrm{s}} = \begin{cases} \mu \left(v_0^{\mathrm{sp}} + \sum_{m=1}^{3} v_m^{\mathrm{sp}} (j\omega)^{(1-m)/2} \right) \exp(j\omega t) & \text{Time-harmonic regime} \\ \mu \left(v_0^{\mathrm{sp}} \tilde{\theta}_0 + \sum_{m=1}^{3} v_m^{\mathrm{sp}} T_{m+1}^{\mathrm{b}} * \frac{\mathrm{d}\tilde{\theta}_0}{dt} \right) & \text{Pulsed regime} \end{cases} \tag{6.276}$$

$$A^{\mathrm{s}} = \begin{cases} -\left(v_0^{\mathrm{sp}} + \sum_{m=1}^{3} v_m^{\mathrm{sp}} (j\omega)^{-m/2} \right) \exp(j\omega t) & \text{Time-harmonic regime} \\ -\left(v_0^{\mathrm{sp}} \tilde{\theta}_0 + \sum_{m=1}^{3} v_m^{\mathrm{sp}} T_{m+1}^{\mathrm{b}} * \tilde{\theta}_0 \right) & \text{Pulsed regime} \end{cases} \tag{6.277}$$

$$H_{\xi_2} = K = \begin{cases} \left(u_0^{\mathrm{sp}} + \sum_{m=1}^{3} u_m^{\mathrm{sp}} (j\omega)^{-m/2} \right) \exp(j\omega t) & \text{Time-harmonic regime} \\ \left(u_0^{\mathrm{sp}} \tilde{\theta}_0 + \sum_{m=1}^{3} u_m^{\mathrm{sp}} T_{m+1}^{\mathrm{b}} * \tilde{\theta}_0 \right) & \text{Pulsed regime} \end{cases} \tag{6.278}$$

3. Obtain the functions E^{e} and A^{e} using Equations 3.132 and 3.118, respectively.

H–ϕ formalism (two- and three-dimensional problems):

1. Calculate the field $\vec{H}^{\mathrm{s}}_{\mathrm{sp}}$ using the formula:

$$\vec{H}^{\mathrm{s}}_{\mathrm{sp}} = \frac{1}{4\pi} \sum_{i=1}^{N} \int_{L_i} \vec{I}^{(i)}_{\mathrm{sp}}(\vec{r}\,') \times \frac{\vec{r} - \vec{r}\,'}{|\vec{r} - \vec{r}\,'|^3} \mathrm{d}l \tag{6.279}$$

2. Obtain the space functions $\phi^{\mathrm{sp}}_{m}(\vec{r}), m = 0, 1, 2, 3$, by solving the surface integral Equations 6.268 through 6.271 in the two-dimensional case or Equations 6.272 through 6.275 in the three-dimensional case.
3. Obtain the scalar potential using the formula:

$$\phi = \left(\phi^{\mathrm{sp}}_{0} \tilde{\theta}_0 + \sum_{m=1}^{3} \phi^{\mathrm{sp}}_{m} T^{\mathrm{b}}_{m+1} * \tilde{\theta}_0 \right) \tag{6.280}$$

At this point it is worth re-emphasizing the main advantages of the formulation developed:

1. The form of the integral equations, including the right-hand side, is independent of the time dependence of the incident current and is determined solely by the geometric parameters of the given system of conductors. Therefore, by solving the integral equations for space components just once for a given system of conductors, and multiplying the result by the corresponding time components we obtain solutions for any time dependence of the incident current.
2. The integral equations for the approximations differ only in the form of the right-hand side and can be solved by the same programmed routine; therefore, new computational complications do not arise beyond those involved in solving the problem in the well-known perfect electrical conductor limit.

6.6 Numerical Examples

To illustrate the theory discussed in the previous sections, consider the skin and proximity effect calculation in a system of two long parallel copper conductors of circular cross section, shown in Figure 6.1. As was already mentioned in Section 6.2, the problem can be considered as two-dimensional in the *xy*-plane of the conductor's cross sections. The characteristic scale, *D*, of the problem is the radius, *d*, of each conductor (equal to 21 mm) and the material of the conductor is copper. The distance between the centers of the conductors is the parameter of the problem and is denoted as L. In view of the double axial symmetry of the conductors system, only a half-contour of the cross section of one of the conductors need to be considered so that the coordinate, ξ, varies in the range between 0 and πD as shown in Figure 6.1.

Figures 6.2 through 6.5 show distributions of the functions $u_m^{sp}(\vec{r})$, given in Equations 6.260 through 6.267. To visualize the proximity effect, each of these figures contains two curves corresponding to L equal to $3D$ and $8D$. The function u_0^{sp} is the surface current density (or the tangential magnetic field) at the conductor's surface in the PEC limit. Due to the proximity effect, the surface current density is higher at the inner side of the conductors carrying oppositely directed currents. From Figure 6.2 it is clearly seen that the proximity effect is stronger when the conductors are located closer to each other. The PEC limit can be used unless the skin layer is very thin, i.e., in the initial moments of time, when the electromagnetic field diffusion into the conductor can be neglected. Expansion of the skin depth induces "redistribution" of the magnetic field "along" the conductor's surface. Obviously this phenomenon cannot be described within the PEC limit, but it is taken into account by the second term of expansion, u_1^{sp} (Leontovich's approximation). The effect of this process is to smooth out nonuniformities in the surface current density distribution so that u_1^{sp} is negative where u_0^{sp} is maximum and u_1^{sp} is positive where u_0^{sp} is minimum, as can be seen in Figure 6.3. It should also be noted that the magnitude of u_1^{sp} is higher in the case of strong proximity effect. From Figures 6.4 and 6.5, which show the distributions of

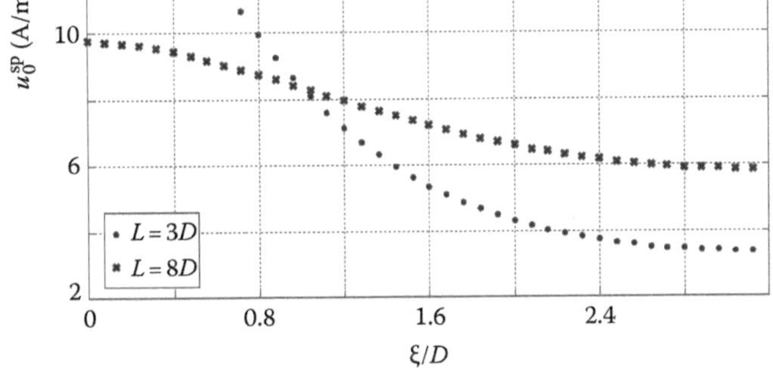

FIGURE 6.2
Distribution of the zero-order term u_0^{sp} over half the contour of the cross section of one conductor for $L = 2D$ and $L = 8D$.

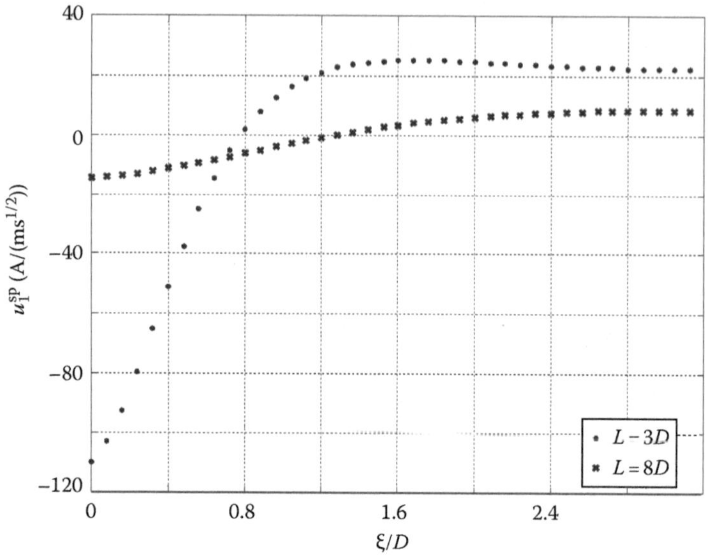

FIGURE 6.3
Distribution of the first-order term u_1^{sp} over half the contour of the cross section of one conductor for $L = 2D$ and $L = 8D$.

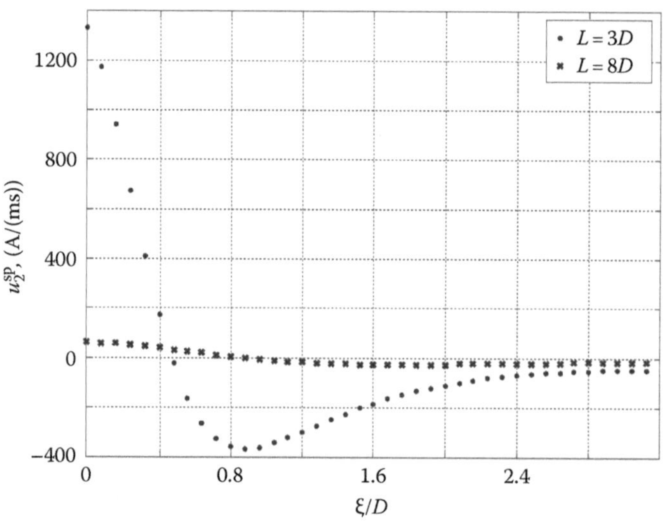

FIGURE 6.4
Distribution of the second-order term u_2^{sp} over half the contour of the cross section of one conductor for $L = 2D$ and $L = 8D$.

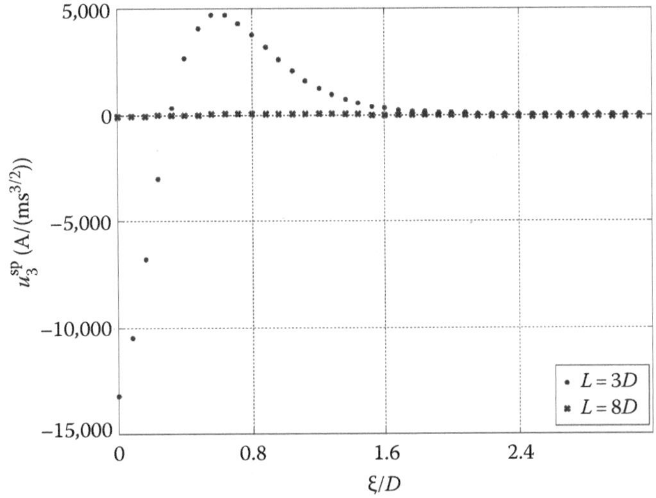

FIGURE 6.5
Distribution of the third-order term u_3^{sp} over half the contour of the cross section of one conductor for $L = 2D$ and $L = 8D$.

u_2^{sp} (Mitzner's approximation) and u_3^{sp} (Rytov's approximation), respectively, it follows that each coefficient of expansions tends to compensate the effect of the previous term. Another conclusion is that the magnitudes of the terms grow with the index number. This is not unusual in asymptotic expansions.

Next we demonstrate the evolution of the surface current density given by Equation 6.278 for the particular case of $D = 0.03$ m and $L = 0.12$ m (Figure 6.1). The system of conductors with these parameters will be used again in Chapter 9 for experimental validation of the SIBCs. The transient is obtained by closing a circuit breaker so that the resulting source current is exponential. Figure 6.6 shows the actual current pulse whereas the time-dependent functions T_k^b, $k = 0, 1, 2, 3$, are shown in Figure 6.7. Figure 6.8 shows the distributions of the surface current density K obtained using the PEC, Leontovich's, Mitzner's, and Rytov's boundary conditions compared with data obtained using a commercial FEM software for two instants of time: $t = 6 \cdot 10^{-3}$s ($\tilde{p} = 0.3$) (left-hand side figure) and $t = 12 \cdot 10^{-3}$s ($p = 0.43$) (right-hand side figure). From these results it can be concluded that the SIBC formulation allows an efficient and accurate simulation of the test case. The hypothesis of PEC gives definitely worse results, whereas increasing the order of the SIBC formulation is closer to the FEM solution and to the experimental measurements, considering the uncertainty in the latter. However, it is unclear *a priori*, until what times the BEM–SIBC formulation may be used. We will take up this issue in Chapter 9 where we develop an analytical formula that gives approximate limits of applicability of the surface impedance concept.

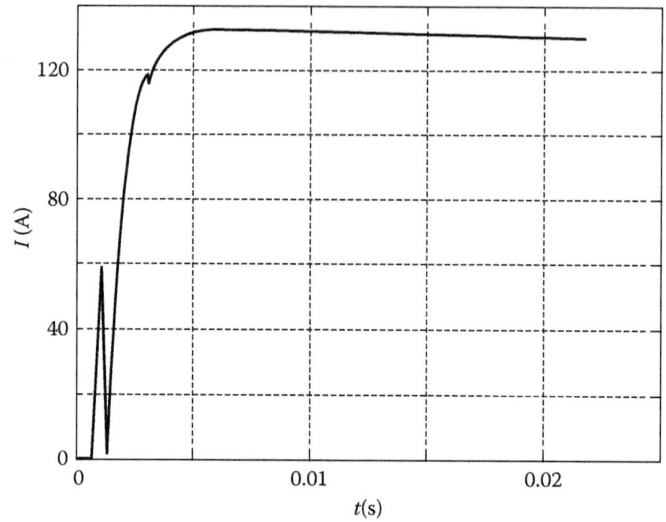

FIGURE 6.6
Current waveform.

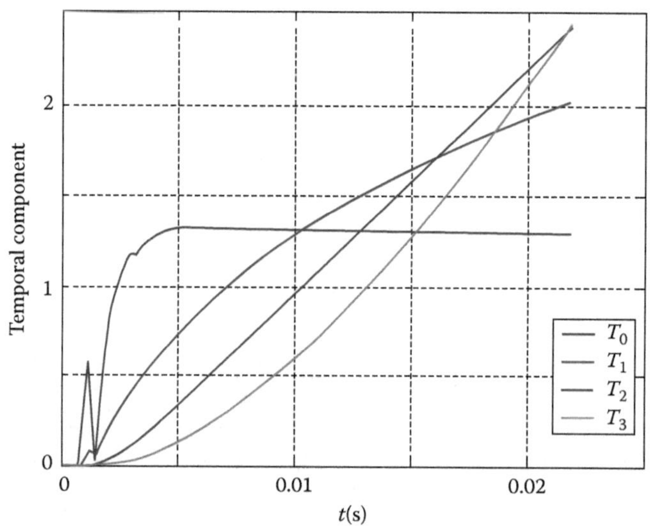

FIGURE 6.7
Time-dependent functions T_k^{b}, $k = 0, 1, 2, 3$, for the current pulse shown in Figure 6.6.

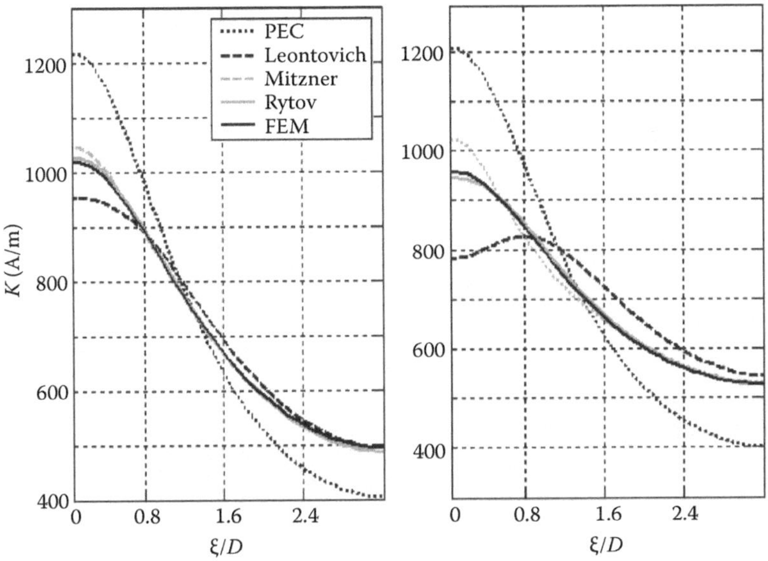

FIGURE 6.8
Distribution of the surface current density for $t = 6 \times 10^{-3}$ s ($\tilde{p} = 0.3$) (left) and $t = 12 \times 10^{-3}$ s ($\tilde{p} = 0.43$) (right).

6.7 Quasi-Three-Dimensional Integro-Differential Formulation for Symmetric Systems of Conductors

The two-dimensional formulations considered in Section 6.2 are valid only for very long conductors. In this section, we consider a system of N parallel finite length conductors with constant cross sections, connected by wires to a current source $I(t)$ [10]. We choose a fixed Cartesian coordinate system with the x- and y-axes in the plane of the conductor's cross sections and the z-axis parallel to the conductors' axes. The system is assumed to satisfy symmetry about the x- and y-axes (the simplest example of such a configuration is shown in Figure 6.9). The current distribution in the conductors is assumed to be symmetric about the y-axis and antisymmetric about the x-axis. This particular case is important from the methodological point of view since it gives another application of the perturbation technique in multiconductor skin effect problems.

As was shown in previous sections, the distribution of the magnetic scalar potential over the surface of the conductors, assuming small penetration depth of the field in the conductors, is described by an integral equation of the form

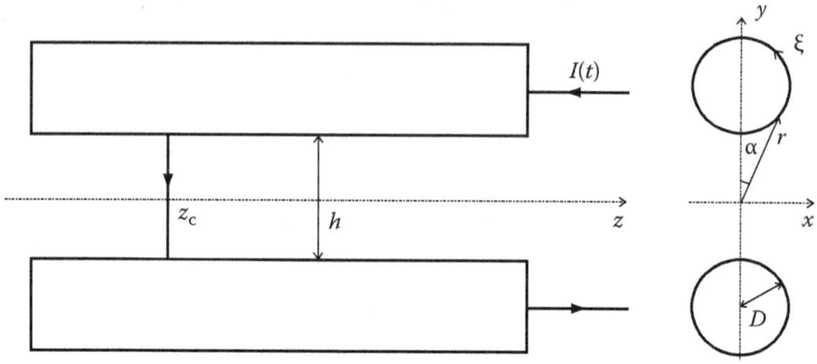

FIGURE 6.9
A system of two current-carrying conductors of circular cross sections connected in series by wires.

$$\tilde{c}\phi(s) + \sum_{i=1}^{N} \int_{S_i} \phi(s') \frac{\partial G^{(3D)}}{\partial \vec{n}'} ds' = X(s) \tag{6.281}$$

where
- X is the known right-hand side, which depends on the order of approximation
- $G^{(3D)}$ is given in Equation 6.161

The integral on the left-hand side of Equation 6.281 is composed of integrals over the ends and sides of the conductors. In the system considered here, the scalar potential distribution is symmetric about the axis of antisymmetry of the current distribution and antisymmetric about the axis of symmetry of the current. For the system shown in Figure 6.9, these conditions have the form:

$$\phi(-x, y, z) = \phi(x, y, z); \quad \phi(x, -y, z) = -\phi(x, y, z) \tag{6.282}$$

The integral over the lateral surfaces of the conductors can then be written in the form

$$\sum_{i=1}^{N} \int_{S_i} \phi \frac{\partial G^{(3D)}}{\partial \vec{n}} ds = \frac{1}{4\pi} \sum_{k=1}^{4} \int_{L} (-1)^{k+1} d\xi' \int_{z_1}^{z_2} \phi(\xi', z') \frac{\partial}{\partial \vec{n}} \left(\frac{1}{R_k^{(3D)}} \right) dz' \tag{6.283}$$

where the summation is over the quadrants of the xy-plane, L is the total path of integration, including all of the contours of the conductors cross sections in a given quadrant, the coordinate ξ runs along the contour (we assume that ξ is continuous and goes around all of the contours of the conductor cross sections in succession), z_1 and z_2 are the coordinates of the ends of the conductors, and the functions $R_k^{(3D)}$ are calculated as

$$R_k^{(3D)}(\xi',\xi,z',z) = \left[(z'-z)^2 + R_k^{(2D)}(\xi',\xi)\right]^{1/2} \tag{6.284}$$

where $R_k^{(2D)}(\xi',\xi)$ are the distances between the point ξ, which in view of the symmetry of the problem goes over only the part of L lying in the first quadrant of the xy-plane, and the points ξ' lying in the first, second, third, and fourth quadrants, respectively. To calculate these distances, let $r(\xi)$ be the radius vector to a point on the contour of the conductor cross section in the xy-plane and let $\alpha(\xi)$ be the angle between the radius vector and one of the coordinate axes, say y (Figure 6.9). The shape of the cross section is uniquely determined by these functions; therefore, assuming they are known, we have

$$\begin{cases} a_1 = \left(R_1^{(2D)}(\xi',\xi)\right)^2 = (r(\xi))^2 + (r(\xi'))^2 - 2r(\xi)r(\xi')\cos(\alpha(\xi') - \alpha(\xi)) \\ a_2 = \left(R_2^{(2D)}(\xi',\xi)\right)^2 = (r(\xi))^2 + (r(\xi'))^2 - 2r(\xi)r(\xi')\cos(\alpha(\xi') + \alpha(\xi)) \\ a_3 = \left(R_3^{(2D)}(\xi',\xi)\right)^2 = (r(\xi))^2 + (r(\xi'))^2 + 2r(\xi)r(\xi')\cos(\alpha(\xi') - \alpha(\xi)) \\ a_4 = \left(R_4^{(2D)}(\xi',\xi)\right)^2 = (r(\xi))^2 + (r(\xi'))^2 + 2r(\xi)r(\xi')\cos(\alpha(\xi') + \alpha(\xi)) \end{cases} \tag{6.285}$$

Differentiating the function $1/R^{(3D)}$ with respect to the normal to the lateral surface of the conductor and substituting the result into Equation 6.283, we obtain

$$\frac{1}{4\pi}\sum_{k=1}^{4}\int_L (-1)^{k+1} d\xi' \int_{z_1}^{z_2} \frac{b_k(\xi,\xi')\phi(\xi',z')}{\left((z'-z)^2 + a_k(\xi,\xi')\right)^{3/2}} dz' \tag{6.286}$$

where the functions $b_k = \partial a_k/\partial\vec{n}'$ are found directly from (Equation 6.286):

$$\begin{cases} b_1 = 2\left[r(\xi')\dfrac{\partial r(\xi')}{\partial\vec{n}'} - r(\xi)\left(\dfrac{\partial r(\xi')}{\partial\vec{n}'}\cos(\alpha(\xi') - \alpha(\xi)) - r(\xi')\sin(\alpha(\xi') - \alpha(\xi))\dfrac{\partial\alpha(\xi')}{\partial\vec{n}'}\right)\right] \\ b_2 = 2\left[r(\xi')\dfrac{\partial r(\xi')}{\partial\vec{n}'} - r(\xi)\left(\dfrac{\partial r(\xi')}{\partial\vec{n}'}\cos(\alpha(\xi') + \alpha(\xi)) - r(\xi')\sin(\alpha(\xi') + \alpha(\xi))\dfrac{\partial\alpha(\xi')}{\partial\vec{n}'}\right)\right] \\ b_3 = 2\left[r(\xi')\dfrac{\partial r(\xi')}{\partial\vec{n}'} + r(\xi)\left(\dfrac{\partial r(\xi')}{\partial\vec{n}'}\cos(\alpha(\xi') - \alpha(\xi)) - r(\xi')\sin(\alpha(\xi') - \alpha(\xi))\dfrac{\partial\alpha(\xi')}{\partial\vec{n}'}\right)\right] \\ b_4 = 2\left[r(\xi')\dfrac{\partial r(\xi')}{\partial\vec{n}'} + r(\xi)\left(\dfrac{\partial r(\xi')}{\partial\vec{n}'}\cos(\alpha(\xi') + \alpha(\xi)) - r(\xi')\sin(\alpha(\xi') + \alpha(\xi))\dfrac{\partial\alpha(\xi')}{\partial\vec{n}'}\right)\right] \end{cases} \tag{6.287}$$

Transforming to dimensionless variables using the characteristic linear dimension, D, of the cross section and the distance, Z, over which the magnetic field parameters vary significantly along the axis as the length scales in the ξ- and z-directions, respectively, the integral in Equation 6.286 takes the form:

$$\frac{1}{4\pi\tilde{\beta}}\sum_{k=1}^{4}\int_{L}(-1)^{k+1}b_k(\xi,\xi')d\xi'\int_{z_1}^{z_2}\frac{\phi(\xi',z')}{\left[\left(\frac{z'-z}{\tilde{\beta}}\right)^2+a_k(\xi,\xi')\right]^{3/2}}dz' \tag{6.288}$$

where

$$\tilde{\beta}=D/Z \tag{6.289}$$

is a small parameter.

The inner integral in Equation 6.288 is represented as an asymptotic series in $\tilde{\beta}$. Applying the method described in [11], and neglecting terms of order $O(\tilde{\beta}^3)$ and higher, we obtain the asymptotic results:

$$\int_{z_1}^{z_2}\frac{\phi(\xi',z')}{\left[\left(\frac{z'-z}{\tilde{\beta}}\right)^2+a_k(\xi,\xi')\right]^{3/2}}dz'=\sum_{n=0}^{2}\frac{\tilde{\beta}^{n+1}}{n!}\frac{\partial^n\phi}{\partial z^n}\int_{z_1}^{z_2}\left(\frac{z'-z}{\tilde{\beta}}\right)^n\frac{\phi(\xi',z')}{\left[\left(\frac{z'-z}{\tilde{\beta}}\right)^2+a_k(\xi,\xi')\right]^{3/2}}dz' \tag{6.290}$$

Evaluating the integrals analytically and substituting the results into Equation 6.288, we have

$$\begin{aligned}
&\frac{1}{4\pi}\int_{L}\phi d\xi'\sum_{k=1}^{4}(-1)^{k+1}\frac{b_k}{a_k}\left\{\frac{z-z_1}{\left[(z-z_1)^2+\tilde{\beta}^2a_k\right]^{1/2}}+\frac{z_2-z}{\left[(z_2-z)^2+\tilde{\beta}^2a_k\right]^{1/2}}\right\}\\
&-\frac{\tilde{\beta}^2}{4\pi}\int_{L}\frac{\partial\phi}{\partial z}d\xi'\sum_{k=1}^{4}(-1)^{k+1}\frac{b_k}{a_k}\left\{\frac{z-z_1}{\left[(z-z_1)^2+\tilde{\beta}^2a_k\right]^{1/2}}-\frac{z_2-z}{\left[(z_2-z)^2+\tilde{\beta}^2a_k\right]^{1/2}}\right\}\\
&-\frac{\tilde{\beta}^2}{8\pi}\int_{L}\frac{\partial^2\phi}{\partial z^2}d\xi'\sum_{k=1}^{4}(-1)^{k+1}b_k\ln a_k-\frac{\tilde{\beta}^2}{4\pi}\ln\tilde{\beta}\int_{L}\frac{\partial\phi}{\partial z}d\xi'\sum_{k=1}^{4}(-1)^{k+1}b_k\\
&-\frac{\tilde{\beta}^2}{8\pi}\int_{L}\frac{\partial^2\phi}{\partial z^2}d\xi'\sum_{k=1}^{4}(-1)^{k+1}b_k\left\{\frac{z-z_1}{\left[(z-z_1)^2+\tilde{\beta}^2a_k\right]^{1/2}}-\frac{z_2-z}{\left[(z_2-z)^2+\tilde{\beta}^2a_k\right]^{1/2}}\right\}\\
&+\frac{\tilde{\beta}^2}{8\pi}\int_{L}\frac{\partial^2\phi}{\partial z^2}d\xi'\sum_{k=1}^{4}(-1)^{k+1}b_k\ln\left\{\left[z-z_1+\left((z-z_1)^2+\tilde{\beta}^2a_k\right)^{1/2}\right]\right.\\
&\left.\times\left[z_2-z+\left((z_2-z)^2+\tilde{\beta}^2a_k\right)^{1/2}\right]\right\}
\end{aligned} \tag{6.291}$$

Taking the limit $\tilde{\beta} \to \infty$ under the integral sign, we obtain

$$\frac{1}{4\pi}\int_L \phi \mathrm{d}\xi' \sum_{k=1}^{4}(-1)^{k+1}\frac{b_k}{a_k} - \frac{\tilde{\beta}^2}{4\pi}\int_L \frac{\partial\phi}{\partial z}\left(\frac{1}{z_2 - z} - \frac{1}{z - z_1}\right)\mathrm{d}\xi' \sum_{k=1}^{4}(-1)^{k+1}b_k$$
$$-\frac{\tilde{\beta}^2}{8\pi}\int_L \frac{\partial^2\phi}{\partial z^2}\mathrm{d}\xi' \sum_{k=1}^{4}(-1)^{k+1}b_k \ln a_k - \frac{\tilde{\beta}^2}{4\pi}\ln\tilde{\beta}\int_L \frac{\partial^2\phi}{\partial z^2}\mathrm{d}\xi' \sum_{k=1}^{4}(-1)^{k+1}b_k$$
$$-\frac{\tilde{\beta}^2}{8\pi}\int_L \frac{\partial^2\phi}{\partial z^2}\mathrm{d}\xi' \sum_{k=1}^{4}(-1)^{k+1}b_k + \frac{\tilde{\beta}^2}{8\pi}\int_L \frac{\partial^2\phi}{\partial z^2}\ln[4(z_2 - z)(z - z_1)]\mathrm{d}\xi' \sum_{k=1}^{4}(-1)^{k+1}b_k \tag{6.292}$$

It follows from Equation 6.287 that

$$\sum_{k=1}^{4}(-1)^{k+1}b_k = 0 \tag{6.293}$$

Substituting Equation 6.293 into Equation 6.292, we finally have

$$\frac{1}{4\pi}\int_L \phi \mathrm{d}\xi' \sum_{k=1}^{4}(-1)^{k+1}\frac{b_k}{a_k} - \frac{\tilde{\beta}^2}{8\pi}\int_L \frac{\partial^2\phi}{\partial z^2}\mathrm{d}\xi' \sum_{k=1}^{4}(-1)^{k+1}b_k \ln a_k \tag{6.294}$$

We next consider the integral over the ends of the conductors. The normal to the surface in this case is along the z-axis and the integral in Equation 6.282 takes the form:

$$\sum_{i=1}^{N}\int_{S_i}\phi\frac{\partial G^{(3)}}{\partial \vec{n}}\mathrm{d}s = \frac{1}{4\pi}\sum_{i=1}^{N}\int_{S_i}(-1)^{k+1}\phi(s')\sum_{k=1}^{4}\frac{z^{\mathrm{end}} - z}{\left[(z^{\mathrm{end}} - z)^2 + a_k(s,s')\right]^{3/2}}\mathrm{d}s' \tag{6.295}$$

where z^{end} is z_1 or z_2. Introducing scale factors D and Z leads to the following form of the integral:

$$\frac{z^{\mathrm{end}} - z}{4\pi\tilde{\beta}}\sum_{i=1}^{N}\int_{S_i}(-1)^{k+1}\phi(s')\sum_{k=1}^{4}\frac{1}{\left[\left(\frac{z^{\mathrm{end}} - z}{\tilde{\beta}}\right)^2 + a_k(s,s')\right]^{3/2}}\mathrm{d}s' \tag{6.296}$$

Let us now estimate this integral. Depending on the distance between the end of the conductor and the observation point, the following cases are possible:

a. $|z^{\text{end}} - z| \sim \tilde{\beta}$: the observation point is close to the end of the conductor. Transforming to the variable $y = (z^{\text{end}} - z)/\tilde{\beta}$, it is not difficult to show that the integral in this case is of the order $O(1)$ (without taking symmetry into account).

b. $z^{\text{end}} - z$: the observation point is far from the end of the conductor. Let

$$\varpi(z) = \frac{1}{\tilde{\beta}} \sum_{k=1}^{4} \frac{(-1)^{k+1}}{\left[\left(\frac{z^{\text{end}} - z}{\tilde{\beta}}\right)^2 + a_k(s, s')\right]^{3/2}} \tag{6.297}$$

Factoring out the small parameter $\left|\tilde{\beta}^2 a_k (z^{\text{end}} - z)^{-2}\right| \gg \tilde{\beta}$ in Equation 6.297 and expanding $\varpi(z)$ in a power series in this parameter, we obtain

$$\varpi(z) = \tilde{\beta}^2 a_k (z^{\text{end}} - z)^{-3} \sum_{k=1}^{4} (-1)^{k+1} \left[1 - \frac{3}{2}\tilde{\beta}^2 a_k (z^{\text{end}} - z)^{-2} + O(\tilde{\beta}^4)\right] \tag{6.298}$$

It follows from Equation 6.287 that

$$\sum_{k=1}^{4} (-1)^{k+1} a_k = 0 \tag{6.299}$$

Therefore, the first two terms of expansion vanish in the sum over the quadrants as a result of symmetry, and $\varpi(z)$ and, hence, the entire integral equation (Equation 6.296) is of the order $O(\tilde{\beta}^6)$.

Therefore, end results are important only in a small region near the ends. Far from the ends only the integro-differential operator equation (Equation 6.294), which is of the order $O(\tilde{\beta}^3)$, has to be kept in Equation 6.281. It can be shown that

$$\frac{b_k}{a_k} = 4\pi \frac{\partial G^{(2D)}}{\partial \vec{n}'} \tag{6.300}$$

where $G^{(2D)}$ is given in Equation 6.17. Hence, the integral equation (Equation 6.281), describing the scalar potential distribution over the surfaces of the conductors can be written to within terms of the order $O(\tilde{\beta}^2)$ as

$$\tilde{c}\phi(\xi, z) + \sum_{i=1}^{N} \int_{L_i} \phi(\xi', z) \frac{\partial G^{(2D)}}{\partial \vec{n}'} d\xi' - \frac{\tilde{\beta}^2}{2\pi} \sum_{i=1}^{N} \int_{L_i} \frac{\partial^2 \phi(\xi', z)}{\partial z^2} \frac{\partial R^{(2D)}}{\partial \vec{n}'} R^{(2D)} \ln R^{(2D)} d\xi' = X(\xi, z) \tag{6.301}$$

The left-hand side of this equation differs from the left-hand sides of Equations 6.268 through 6.271 for the two-dimensional problem by the presence of the third term. Note that Equation 6.301 does not involve the end coordinates z_1 and z_2.

Since Equation 6.301 is an integro-differential equation, boundary conditions in the variable z must be specified. These conditions can be obtained by setting $z = z_1$ or $z = z_2$ in Equation 6.291 and repeating the subsequent operations. Neglecting terms of the order $O(\tilde{\beta}^3)$, we obtain

$$\tilde{c}\phi|_{z=z^{\text{end}}} + \sum_{i=1}^{N} \int_{L_i} \phi|_{z=z^{\text{end}}} \frac{\partial G^{(2D)}}{\partial \vec{n}'} d\xi' - \frac{\tilde{\beta}^2}{2\pi} \sum_{i=1}^{N} \int_{L_i} \left. \frac{\partial^2 \phi(\xi', z)}{\partial z^2} \right|_{z=z^{\text{end}}} \frac{\partial R^{(2D)}}{\partial \vec{n}'} R^{(2D)}$$
$$\times \ln R^{(2D)} d\xi' = X|_{z=z^{\text{end}}} \tag{6.302}$$

where $\phi_{z^{\text{end}}} = \phi(\xi, z^{\text{end}})$, $z^{\text{end}} = z_1, z_2$.

Therefore, for a symmetric system of long parallel conductors of finite length and constant cross section, we have obtained an integro-differential equation (Equation 6.301) containing only contour integrals and differing from the two-dimensional equation by the presence of an additional term, instead of the standard integral equation (Equation 6.281), where the integration is over the surfaces of the conductors.

To analyze Equation 6.301 we represent it in the following form using Equation 6.294:

$$\tilde{c}\phi(\xi, z) + \int_{L_i} \phi(\xi', z) E_0(\xi, \xi') d\xi' - \frac{\tilde{\beta}^2}{2\pi} \int_{L_i} \frac{\partial^2 \phi(\xi', z)}{\partial z^2} E_2(\xi, \xi') d\xi' = X(\xi, z) \tag{6.303}$$

where

$$E_0(\xi, \xi') = \frac{1}{4\pi} \sum_{k=1}^{4} (-1)^{k+1} \frac{b_k}{a_k}; \quad E_2(\xi, \xi') = \frac{1}{8\pi} \sum_{k=1}^{4} (-1)^{k+1} b_k \ln a_k \tag{6.304}$$

and the summation is, again, over the quadrants. Replacing Equations 6.303 and 6.304 by their discrete forms in ξ, we obtain

$$\Delta\xi \sum_{j=1}^{M_\xi} (E_0')_{ij} \phi_j(z) - \tilde{\beta}^2 \Delta\xi \sum_{j=1}^{M_\xi} \frac{\partial^2 \phi_j(z)}{\partial z^2} (E_2)_{ij} = X_i(z), \quad i = 1, \ldots, M_\xi \tag{6.305}$$

where

$$(E_0')_{ij} = \begin{cases} (E_0)_{ii} + c_i/\Delta\xi, & i = j \\ (E_0)_{ij}, & i \neq j \end{cases} \tag{6.306}$$

M_ξ is the number of segments in the contour L

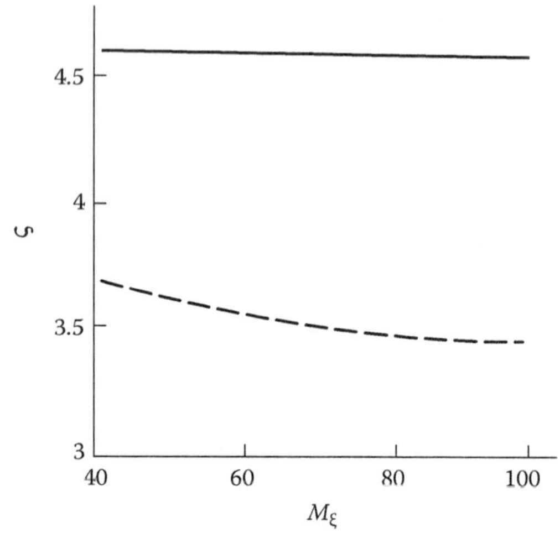

FIGURE 6.10
Dependence of the smallest eigenvalue of a pair of parallel conductors of circular (solid curve) and square (dashed curve) cross section on the number of nodes M_ξ of the contour ($D/h = 1$).

The general solution of Equation 6.306 has the form:

$$\phi_i(z) = g_i \exp(-\varsigma_i z/\tilde{\beta}) \qquad (6.307)$$

where

g_i are the functions of the point on the contour
ς_i are the eigenvalues of the matrix $\left(E_2^2\right)^{-1}E_0'$

The solution in Equation 6.307 describes the z dependence of the potential distribution resulting from the finite lengths of the conductors, and the smallest eigenvalue, $\varsigma_{\min}$, determines the degree of this dependence. We note that the eigenvalues depend only on the configuration of the cross sections of the conductors and not on their lengths. The dependence $\varsigma_{\min} = \varsigma_{\min}(\Delta\xi)$ is shown in Figure 6.10 for a pair of parallel conductors with circular cross section and a pair with square cross sections. We see that the curves have asymptotes in the limit $\Delta\xi \to 0$. The calculations show that for most systems of conductors the smallest eigenvalue lies between 1 and 10. Therefore, the function (Equation 6.307) is significantly different from zero only at small distances (of the order of a fraction of a diameter) from the ends. The dependence $\varsigma_{\min} = \varsigma_{\min}(h)$ is shown in Figure 6.11 for a pair of conductors of circular cross section, where h is the distance between the conductors. It follows from the graph that as h increases (and therefore as the interaction between the currents in the conductors decreases) the region where end effects are important widens, and $\varsigma_{\min}$ varies by about an order of magnitude between the limiting cases. Hence, the solution of the two-dimensional problem for the smallest eigenvalue of the matrix, $\left(E_2^2\right)^{-1}E_0'$, gives an estimate of the error in neglecting the third term involving the small parameter on the left-hand side of Equation 6.301.

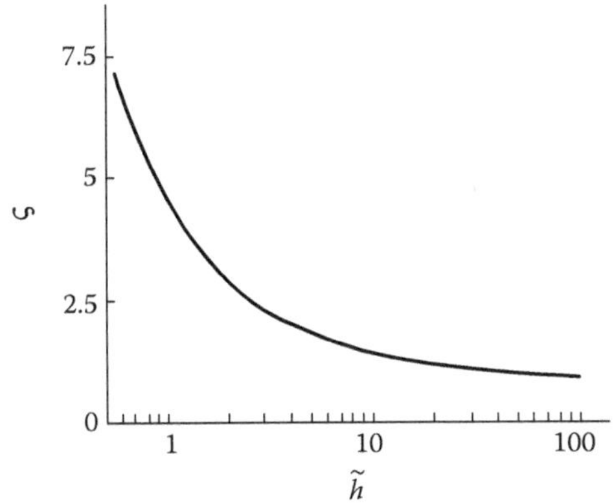

FIGURE 6.11
Dependence of the smallest eigenvalue of a pair of parallel conductors of circular cross section on the distance between them measured in cross-section radii.

As a conclusion we show the results of numerical implementation of the formulation (Equations 6.305 and 3.306) for a pair of parallel conductors of finite length connected in series by wires as shown in Figure 6.9. The wires are assumed to be connected to a current source which is so far from the conductors that it does not interact with them. The circuit is closed by a crosswire at $z = z_c$. Figures 6.12 and 6.13 show the distributions of the two components of the surface current density along the z-axis at the contour point $\xi = 0$, obtained using the quasi-three-dimensional formulation (Equations 6.305 and 3.306), a standard three-dimensional boundary element formulation, and a two-dimensional boundary element formulation using

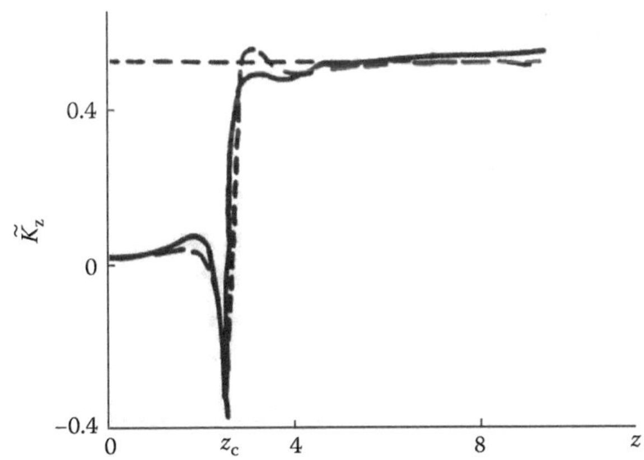

FIGURE 6.12
Longitudinal component of the surface current density as a function of z at the contour point $\xi = 0$ for a pair of parallel conductors shown in Figure 6.9. Solid curve: standard three-dimensional boundary element formulation; dashed curve: quasi-three-dimensional formulation (Equations 6.305 and 3.306); dashed line: two-dimensional boundary element formulation in the approximation of long parallel conductors.

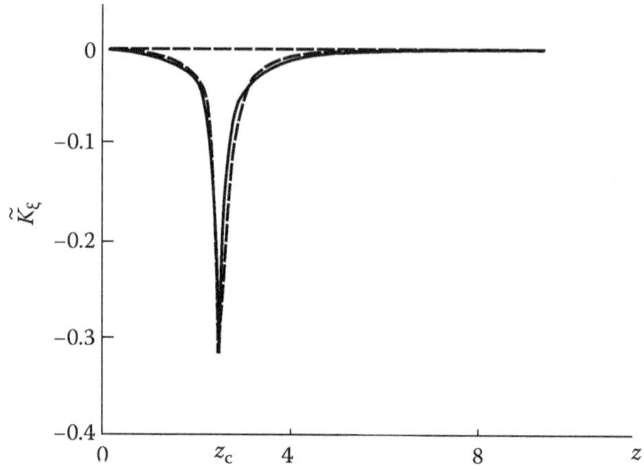

FIGURE 6.13
Transverse component of the surface current density as a function of z at the contour point $\xi = 0$ for a pair of parallel conductors shown in Figure 6.9. Solid curve: standard three-dimensional boundary element formulation; dashed curve: quasi-three-dimensional formulation (Equations 6.305 and 3.306); dashed line: two-dimensional boundary element formulation in the approximation of long parallel conductors.

the approximation of long parallel conductors. The crosswire is assumed to contact the conductors at a point whose contour coordinate is zero. We see that even in the most critical case, the quasi-three-dimensional formulation in Equations 6.305 and 3.306 qualitatively describes the correct behavior of the current and is much more accurate than the two-dimensional approximation.

References

1. C.A. Brebbia, *The Boundary Element Method for Engineers*, Pentech Press, London, 1978.
2. M.J. Tsuk and J.A. Kong, A hybrid method for the calculation of the resistance and inductance of transmission lines with arbitrary cross sections, *IEEE Transactions on Microwave Theory and Techniques*, 39(8), August 1991, 1338–1347.
3. R.B. Wu and J.C. Yang, Boundary integral equation formulation of skin effect problems in multiconductor transmission lines, *IEEE Transactions on Magnetics*, 25(4), July 1989, 3013–3015.
4. L. Di Rienzo, N. Ida, and S. Yuferev, Calculation of energy-related quantities of conductors using surface impedance concept, *IEEE Transactions on Magnetics*, 44(6), June 2008, 1322–1325.

5. S. Yuferev and N. Ida, Time domain surface impedance concept for low frequency electromagnetic problems—Part I: Derivation of high order surface impedance boundary conditions in the time domain, *IEE Proceedings—Science, Measurement and Technology*, 152(4), July 2005, 175–185.
6. S. Barmada, L. Di Rienzo, N. Ida, and S. Yuferev, The time domain surface impedance concept for low frequency electromagnetic problems—Part II: Application to transient skin and proximity effect problems in cylindrical conductors, *IEE Proceedings—Science, Measurement and Technology*, 152(5), September 2005, 207–216.
7. D. De Zutter, H. Rogier, L. Knockaert, and J. Sercu, Surface current modeling of the skin effect for on-chip interconnections, *IEEE Transactions on Advanced Packaging, B*, 30(2), May 2007, 342–349.
8. I. Mayergoyz, A new approach to the calculation of three-dimensional skin effect problems, *IEEE Transactions on Magnetics*, 19(5), September 1983, 2198–2200.
9. A.N. Tikhonov and V.Y. Arsenin, *Solution of Ill-Posed Problems*, Wiley, New York, 1977.
10. S. Yuferev and V. Yuferev, New quasi-three-dimensional integro-differential formulation of boundary integral equation method for the calculation of the skin and eddy currents in parallel conductors, *Journal of Technical Physics*, 39(1), 1994, 1–6.

7

Implementation of SIBCs for the Boundary Integral Equation Method: High-Frequency Problems

7.1 Introduction

In Chapter 6, we have shown how the SIBC concept can be coupled with boundary integral equations to form an efficient method of solution for low-frequency applications. The concepts developed are now extended to include high-frequency applications. We start, by reintroducing the displacement currents in Maxwell's equations. This means that the basic derivations in Chapter 3, including the surface impedance functions must be generalized to take into account the displacement currents in Maxwell's equations. Using the vector Green's function we first develop the electric field integral equation (EFIE) and the magnetic field integral equation (MFIE) representation of the fields based on the incident electric and magnetic fields. These equations are then coupled with the SIBCs previously developed to implement formulations of various orders of approximations in the time and frequency domains in a manner analogous to that in Chapter 6.

The shift from low-frequency to high-frequency carries both advantages and penalties. Of course, in conductors, the skin depths tend to be thinner, resulting in more accurate representation of fields. In that sense, one of the main differences is that at high frequencies, it is also possible to treat lossy dielectrics—that is, the surface impedance may now be applied at the interfaces between lossy dielectrics and dielectrics (such as free space). On the other hand, the introduction of displacement currents in Maxwell's equations results in negation of one of the most important of the advantages we found in the low-frequency transient representation—that of separation of variables. We saw in Chapter 6 that the variables in the surface impedance function given in Equations 3.82 through 3.84 admitted separation into time and space variables only by neglecting displacement currents and treating each term of Equations 3.82 through 3.84 independently. In the high-frequency case we cannot do that and the time convolution integrals cannot be precomputed and tabulated in advance. Yet, direct computation of the convolution integral in the time-domain integral equations is impractical due to the large computational time and storage requirements. To mitigate this difficulty we also include a discussion of methods for efficient implementation of the time convolution integrals in the formulations. More specifically, we show in Appendix 7.A.1 how Prony's method may be incorporated to approximate the impedance by decaying exponentials and by doing so we arrive at a simple recursive evaluation of the convolution integrals. As in previous chapters we use an example to illustrate the theory developed. In this case, the problem of scattering from a cylinder is used as an example.

7.2 Integral Representations of High-Frequency Electromagnetic Fields

Consider a homogeneous body V_c of finite conductivity ($\varepsilon_1 = \varepsilon, \mu_1 = \mu, \sigma_1 = \sigma$) surrounded by a nonconductive medium $V(\varepsilon_2 = \varepsilon_0, \mu_2 = \mu_0, \sigma_2 = 0)$. The distribution of the electromagnetic field in the dielectric space can be described by the frequency-domain Maxwell equations in the following form (with suppressed time variation $\exp(j\omega t)$):

$$\nabla \times \dot{\vec{E}} = -j\omega\mu_0\dot{\vec{H}} - \dot{\vec{J}}^{\text{magn}} \tag{7.1}$$

$$\nabla \times \dot{\vec{H}} = j\omega\varepsilon_0\,\dot{\vec{E}} + \dot{\vec{J}} \tag{7.2}$$

$$\nabla \cdot \dot{\vec{E}} = \rho^{\text{elec}}/\varepsilon_0 \tag{7.3}$$

$$\nabla \cdot \dot{\vec{H}} = \rho^{\text{magn}}/\mu_0 \tag{7.4}$$

where

$\dot{\vec{E}}$ and $\dot{\vec{H}}$ are the electric and magnetic field phasors
$\dot{\vec{J}}$ and $\dot{\vec{J}}^{\text{magn}}$ are the electric and magnetic current densities
ρ^{elec} and ρ^{magn} are the electric and magnetic charge densities

The laws of conservation of charge can be written in the form:

$$\nabla \cdot \dot{\vec{J}} = -j\omega\rho^{\text{elec}} \tag{7.5}$$

$$\nabla \cdot \dot{\vec{J}}^{\text{magn}} = -j\omega\rho^{\text{magn}} \tag{7.6}$$

Let V be a linear, homogeneous, isotropic medium so that ε_0 and μ_0 can be treated as scalars and the vector wave equations for $\dot{\vec{E}}$ and $\dot{\vec{H}}$ can be written as

$$\nabla \times \nabla \times \dot{\vec{E}} - k_\omega^2\,\dot{\vec{E}} = -j\omega\mu_0\dot{\vec{J}} - \nabla \times \dot{\vec{J}}^{\text{magn}} \tag{7.7}$$

$$\nabla \times \nabla \times \dot{\vec{H}} - k_\omega^2 \dot{\vec{H}} = -j\omega\mu_0 \dot{\vec{J}}^{\,\mathrm{magn}} + \nabla \times \dot{\vec{J}} \tag{7.8}$$

where k_ω is the wave number. Further transformations will be performed using the vector Green's theorem in the form:

$$\int_V \left(\vec{Q} \cdot \nabla \times \nabla \times \vec{P} - \vec{P} \cdot \nabla \times \nabla \times \vec{Q}\right) \mathrm{d}v = \int_S \left(\vec{P} \times \nabla \times \vec{Q} - \vec{Q} \times \nabla \times \vec{P}\right) \cdot \vec{n}_s \mathrm{d}s \tag{7.9}$$

where $\vec{Q}$ and $\vec{P}$ are vector functions of position with continuous first and second derivatives within and on S, where S is the boundary of V and the normal unit vector $\vec{n}_s$ is directed out of the dielectric space as shown in Figure 7.1. The boundary S can be decomposed into the surface S_c of the conducting body and the boundary S_i located far from the body.

We choose the vector function $\vec{Q}$ in the form:

$$\vec{Q} = \vec{Q}(\vec{r}, \vec{r}\,') = \frac{\vec{r} - \vec{r}\,'}{|\vec{r} - \vec{r}\,'|} \frac{\exp(-jk_\omega|\vec{r} - \vec{r}\,'|)}{|\vec{r} - \vec{r}\,'|^2} = \frac{\vec{r} - \vec{r}\,'}{|\vec{r} - \vec{r}\,'|} G(\vec{r}, \vec{r}\,') \tag{7.10}$$

$$G(\vec{r}, \vec{r}\,') = \frac{\exp(-jk_\omega|\vec{r} - \vec{r}\,'|)}{|\vec{r} - \vec{r}\,'|} \tag{7.11}$$

$$\nabla' G(\vec{r}, \vec{r}\,') = \frac{1 + jk_\omega|\vec{r} - \vec{r}\,'|}{|\vec{r} - \vec{r}\,'|} \frac{\exp(-jk_\omega|\vec{r} - \vec{r}\,'|)}{|\vec{r} - \vec{r}\,'|} \frac{\vec{r} - \vec{r}\,'}{|\vec{r} - \vec{r}\,'|} \tag{7.12}$$

Here $\vec{r}$ and $\vec{r}\,'$ are the observation and source point position vectors, respectively. It is easy to see that the function $\vec{Q}$ defined in Equation 7.10 meets the conditions of differentiability and continuity except in the case $\vec{r} = \vec{r}\,'$ because of the singularity in the Green function G. Substitution of the electric field

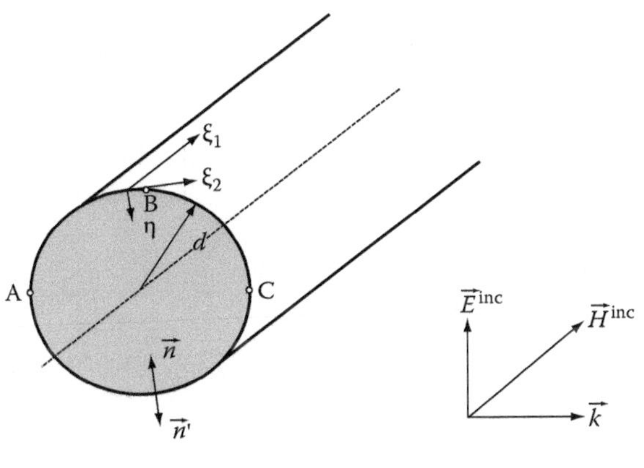

FIGURE 7.1
The problem of transient scattering from an infinitely long straight cylinder.

for $\vec{P}$ in Equation 7.9 leads to the following formula known as the Stratton–Chu formula [1]:

$$\int_V \left(j\omega\mu_0 \dot{\vec{J}} G + \dot{\vec{J}}^{\text{magn}} \times \nabla G - \frac{\rho^{\text{elec}}}{\varepsilon_0} \nabla G \right) dv$$

$$= \int_S \left[j\omega\mu_0 \left(\vec{n}_s \times \dot{\vec{H}} \right) G - \left(\vec{n}_s \times \dot{\vec{E}} \right) \times \nabla G - \left(\vec{n}_s \cdot \dot{\vec{E}} \right) \nabla G \right] ds \tag{7.13}$$

The vector operations indicated in Equation 7.13 are performed in the source coordinates.

Treatment of the singularity in G requires some derivations and results in the following equation [2]:

$$\dot{\vec{E}} = -\frac{1}{4\pi - \vartheta} \int_V \left(j\omega\mu_0 \dot{\vec{J}} G + \dot{\vec{J}}^{\text{magn}} \times \nabla' G - \frac{\rho^{\text{elec}}}{\varepsilon_0} \nabla' G \right) dv'$$

$$- \frac{1}{4\pi - \vartheta} \int_S \left[j\omega\mu_0 \left(\vec{n}' \times \dot{\vec{H}} \right) G - \left(\vec{n}' \times \dot{\vec{E}} \right) \times \nabla' G - \left(\vec{n}' \cdot \dot{\vec{E}} \right) \nabla' G \right] ds' \tag{7.14}$$

where

ϑ represents the absolute value of the solid angle subtended by S at $\vec{r}$ and the "prime" is used to indicate a normal defined at the source point

The surface integral in Equation 7.14 is the principal value integral over $S_i + S_c$. Invoking the duality of Maxwell's equations, the magnetic field can be written in a similar form:

$$\dot{\vec{H}} = \frac{1}{4\pi - \vartheta} \int_V \left(-j\omega\mu_0 \dot{\vec{J}}^{\text{magn}} G + \dot{\vec{J}} \times \nabla' G - \frac{\rho^{\text{magn}}}{\varepsilon_0} \nabla' G \right) dv'$$

$$+ \frac{1}{4\pi - \vartheta} \int_S \left[j\omega\varepsilon_0 \left(\vec{n}' \times \dot{\vec{E}} \right) G + \left(\vec{n}' \times \dot{\vec{H}} \right) \times \nabla' G + \left(\vec{n}' \cdot \dot{\vec{H}} \right) \nabla' G \right] ds' \tag{7.15}$$

A situation of great interest occurs when S_i recedes to infinity. For sources of bounded extent within V one finds that the radiation condition at infinity requires that the contribution from the integral over S_i be due entirely from sources outside S_i. We shall refer to this contribution as an incident field and subsequently write Equations 7.14 and 7.15 as

$$\dot{\vec{E}} = \frac{4\pi}{4\pi - \vartheta} \dot{\vec{E}}^{\text{inc}} - \frac{1}{4\pi - \vartheta} \int_V \left(j\omega\mu_0 \dot{\vec{J}} G + \dot{\vec{J}}^{\text{magn}} \times \nabla' G - \frac{\rho^{\text{elec}}}{\varepsilon_0} \nabla' G \right) dv'$$

$$- \frac{1}{4\pi - \vartheta} \int_S \left[j\omega\varepsilon_0 \left(\vec{n}' \times \dot{\vec{E}} \right) G + \left(\vec{n}' \times \dot{\vec{H}} \right) \times \nabla' G + \left(\vec{n}' \cdot \dot{\vec{H}} \right) \nabla' G \right] ds' \tag{7.16}$$

$$\dot{\vec{H}} = \frac{4\pi}{4\pi - \vartheta}\,\dot{\vec{H}}^{\text{inc}} + \frac{1}{4\pi - \vartheta}\int\limits_{V}\left(-j\omega\mu_0\dot{\vec{J}}^{\text{magn}}G + \dot{\vec{J}}\times\nabla' G - \frac{\rho^{\text{magn}}}{\varepsilon_0}\nabla' G\right)dv'$$
$$+\frac{1}{4\pi - \vartheta}\int\limits_{S_c}\left[j\omega\varepsilon_0\left(\vec{n}'\times\dot{\vec{E}}\right)G + \left(\vec{n}'\times\dot{\vec{H}}\right)\times\nabla' G + \left(\vec{n}'\cdot\dot{\vec{H}}\right)\nabla' G\right]ds' \tag{7.17}$$

To render the integral equations solvable with respect to the tangential components of the electric or the magnetic fields on the surface S_c, the observation and source points should be positioned on the surface of the conducting body. Let S_c be smooth and take into account the fact that $\vartheta = 2\pi$ for $\vec{r}$ on a smooth surface. Then, Equations 7.16 and 7.17 can be written in the forms:

$$\dot{\vec{E}}^{\text{b}} = 2\,\dot{\vec{E}}^{\text{inc}} - \frac{1}{2\pi}\int\limits_{V}\left(j\omega\mu_0\dot{\vec{J}}G + \dot{\vec{J}}^{\text{magn}}\times\nabla' G - \frac{\rho^{\text{elec}}}{\varepsilon_0}\nabla' G\right)dv'$$
$$-\frac{1}{2\pi}\int\limits_{S_c}\left[j\omega\mu_0\left(\vec{n}'\times\dot{\vec{H}}^{\text{b}}\right)G - \left(\vec{n}'\times\dot{\vec{E}}^{\text{b}}\right)\times\nabla' G - \left(\vec{n}'\cdot\dot{\vec{E}}^{\text{b}}\right)\nabla' G\right]ds' \tag{7.18}$$

$$\dot{\vec{H}}^{\text{b}} = 2\,\dot{\vec{H}}^{\text{inc}} + \frac{1}{2\pi}\int\limits_{V}\left(-j\omega\mu_0\dot{\vec{J}}^{\text{magn}}G + \dot{\vec{J}}\times\nabla' G - \frac{\rho^{\text{magn}}}{\varepsilon_0}\nabla' G\right)dv'$$
$$+\frac{1}{2\pi}\int\limits_{S_c}\left[j\omega\varepsilon_0\left(\vec{n}'\times\dot{\vec{E}}^{\text{b}}\right)G + \left(\vec{n}'\times\dot{\vec{H}}^{\text{b}}\right)\times\nabla' G + \left(\vec{n}'\cdot\dot{\vec{H}}^{\text{b}}\right)\nabla' G\right]ds' \tag{7.19}$$

where the vector $\vec{n}$ is directed into the dielectric space normal to the surface S_c.

Frequently, the only sources of the electromagnetic field in practical problems of electromagnetic scattering are incident fields E^{inc} or H^{inc}, and neither currents nor charges exist in the dielectric domain V. Thus the volume integrals vanish and Equations 7.18 and 7.19 reduce to the following forms:

$$\dot{\vec{E}}^{\text{b}} = 2\,\dot{\vec{E}}^{\text{inc}} - \frac{1}{2\pi}\int\limits_{S_c}\left[j\omega\mu_0\left(\vec{n}'\times\dot{\vec{H}}^{\text{b}}\right)G - \left(\vec{n}'\times\dot{\vec{E}}^{\text{b}}\right)\times\nabla' G - \left(\vec{n}'\cdot\dot{\vec{E}}^{\text{b}}\right)\nabla' G\right]ds' \tag{7.20}$$

$$\dot{\vec{H}}^{\text{b}} = 2\,\dot{\vec{H}}^{\text{inc}} + \frac{1}{2\pi}\int\limits_{S_c}\left[j\omega\varepsilon_0\left(\vec{n}'\times\dot{\vec{E}}^{\text{b}}\right)G + \left(\vec{n}'\times\dot{\vec{H}}^{\text{b}}\right)\times\nabla' G + \left(\vec{n}'\cdot\dot{\vec{H}}^{\text{b}}\right)\nabla' G\right]ds' \tag{7.21}$$

The time-domain analogs of Equations 7.20 and 7.21 can be obtained using Fourier transform techniques. The transform pair which is employed in this derivation is given by

$$\widehat{f}(\omega) = \int\limits_{-\infty}^{\infty} f(t)\exp(-j\omega t)dt; \quad f(t) = \Phi\left[\widehat{f}(\omega)\right] = \frac{1}{2\pi}\int\limits_{-\infty}^{\infty}\widehat{f}(\omega)\exp(j\omega t)d\omega \tag{7.22}$$

Since the interchange of the order of integration is valid, the operations Φ (Fourier transform) and $\int_{S_c} ds'$ can be interchanged. This leads to

$$\vec{E}^b(t) = 2\vec{E}^{inc}(t) - \frac{1}{2\pi}\int_{S_c}\left\{\mu_0\vec{n}' \times \Phi\left[j\omega\hat{\vec{H}}^b(\omega)\hat{G}\right] - \Phi\left[(\vec{n}' \times \hat{\vec{E}}^b(\omega)) \times \nabla'\hat{G}\right] - \Phi\left[(\vec{n}' \cdot \hat{\vec{E}}^b(\omega))\nabla'\hat{G}\right]\right\}ds' \tag{7.23}$$

$$\vec{H}^b(t) = 2\vec{H}^{inc}(t) + \frac{1}{2\pi}\int_{S_c}\left\{\varepsilon_0\vec{n}' \times \Phi\left[j\omega\hat{\vec{E}}^b(\omega)\hat{G}\right] + \Phi\left[(\vec{n}' \times \hat{\vec{H}}^b(\omega)) \times \nabla'\hat{G}\right] + \Phi\left[(\vec{n}' \cdot \hat{\vec{H}}^b(\omega))\nabla'\hat{G}\right]\right\}ds' \tag{7.24}$$

From Fourier transform theory it is known that if $f(t)$ and $d\,f(t)/dt$ approach zero as t approaches infinity, then

$$\Phi\left[j\omega\hat{f}(\omega)\right] = \frac{d}{dt}f(t) \tag{7.25}$$

Using Equation 7.20 and the identity

$$\Phi[\exp(-j\omega t')] = \delta(t - t') \tag{7.26}$$

we obtain

$$\Phi[G(\vec{r},\vec{r}',\omega)] = \Phi\left[\frac{\exp(-j(\omega/c)|\vec{r}-\vec{r}'|)}{|\vec{r}-\vec{r}'|}\right] = \delta\left(t - \frac{|\vec{r}-\vec{r}'|}{c}\right)\frac{1}{|\vec{r}-\vec{r}'|} \tag{7.27}$$

$$\Phi[\nabla'G(\vec{r},\vec{r}',\omega)] = \delta\left(t - \frac{|\vec{r}-\vec{r}'|}{c}\right)\frac{\vec{r}-\vec{r}'}{|\vec{r}-\vec{r}'|^3} + \frac{1}{c}\frac{\partial}{\partial t}\delta\left(t - \frac{|\vec{r}-\vec{r}'|}{c}\right)\frac{\vec{r}-\vec{r}'}{|\vec{r}-\vec{r}'|^2} \tag{7.28}$$

Substitution of Equations 7.27 and 7.28 into Equations 7.23 and 7.24 yields the time-domain EFIE and MFIE in the following forms:

Electric field integral equation:

$$\vec{E}^b = 2\vec{E}^{inc} - \frac{1}{2\pi}\int_{S_c}\left\{\frac{\mu_0}{R}\frac{\partial}{\partial t'}\left(\vec{n}' \times \vec{H}^b\right) - \left(\vec{n}' \times \vec{E}^b\right) \times \frac{\vec{R}}{R^3} - \frac{\partial}{\partial t'}\left(\vec{n}' \times \vec{E}^b\right)\right.$$
$$\left.\times\frac{\vec{R}}{cR^2} - \left(\vec{n}' \cdot \vec{E}^b\right)\frac{\vec{R}}{R^3} - \frac{\partial}{\partial t'}\left(\vec{n}' \cdot \vec{E}^b\right)\frac{\vec{R}}{cR^2}\right\}\Bigg|_{t'=t-R/c} ds' \tag{7.29}$$

Magnetic field integral equation:

$$\vec{H}^{\mathrm{b}} = 2\vec{H}^{\mathrm{inc}} + \frac{1}{2\pi}\int_{S_c}\left\{\frac{\varepsilon_0}{R}\frac{\partial}{\partial t'}\left(\vec{n}'\times\vec{E}^{\mathrm{b}}\right) + \left(\vec{n}'\times\vec{H}^{\mathrm{b}}\right)\times\frac{\vec{R}}{R^3} + \frac{\partial}{\partial t}\left(\vec{n}'\times\vec{H}^{\mathrm{b}}\right)\times\frac{\vec{R}}{cR^2}\right.$$
$$\left. + \left(\vec{n}'\cdot\vec{H}^{\mathrm{b}}\right)\frac{\vec{R}}{R^3} + \frac{\partial}{\partial t'}\left(\vec{n}'\cdot\vec{H}^{\mathrm{b}}\right)\frac{\vec{R}}{cR^2}\right\}\Bigg|_{t'=t-R/c} ds' \qquad (7.30)$$

where $\vec{R} = \vec{r} - \vec{r}\,'$.

Equations 7.29 and 7.30 can be represented in a more tractable form:

$$\vec{E}^{\mathrm{b}} = 2\vec{E}^{\mathrm{inc}} - \frac{1}{2\pi}\int_{S_c}\left\{\frac{\mu_0}{R}\frac{\partial}{\partial t'}\left(\vec{n}'\times\vec{H}^{\mathrm{b}}\right) - \Pi\left[\vec{n}'\times\vec{E}^{\mathrm{b}}\right]\times\frac{\vec{R}}{R} - \Pi\left[\vec{n}'\cdot\vec{E}^{\mathrm{b}}\right]\frac{\vec{R}}{R}\right\}\Bigg|_{t'=t-R/c} ds' \qquad (7.31)$$

$$\vec{H}^{\mathrm{b}} = 2\vec{H}^{\mathrm{inc}} + \frac{1}{2\pi}\int_{S_c}\left\{\frac{\varepsilon_0}{R}\frac{\partial}{\partial t'}\left(\vec{n}'\times\vec{E}^{\mathrm{b}}\right) + \Pi\left[\vec{n}'\times\vec{H}^{\mathrm{b}}\right]\times\frac{\vec{R}}{R} + \Pi\left[\vec{n}'\cdot\vec{H}^{\mathrm{b}}\right]\frac{\vec{R}}{R}\right\}\Bigg|_{t'=t-R/c} ds' \qquad (7.32)$$

Here the operator Π is defined as

$$\Pi[f(\vec{r}\,',t')] = \frac{f(\vec{r}\,',t')}{(R(\vec{r},\vec{r}\,'))^2} + \frac{1}{cR(\vec{r},\vec{r}\,')}\frac{\partial f(\vec{r}\,',t')}{\partial t'} \qquad (7.33)$$

7.3 SIBCs for Lossy Dielectrics

At very high frequencies the skin depth may be thin even when conductivity is poor. This means that the surface impedance concept can be applied to lossy dielectrics as well. However, in this case the displacement current cannot be neglected in the governing equations for conducting regions. Therefore, Equations 3.1 and 3.2 must be replaced by the complete Maxwell's equations in the form:

$$\nabla\times\vec{E} = -\mu\frac{\partial\vec{H}}{\partial t} \qquad (7.34)$$

$$\nabla\times\vec{H} = \sigma\vec{E} + \varepsilon\frac{\partial\vec{E}}{\partial t} \qquad (7.35)$$

We represent Equations 7.34 and 7.35 in local coordinates $\tilde{\xi}_1, \tilde{\xi}_2, \tilde{\eta}$ with nondimensional variables as follows:

$$\frac{\partial \tilde{E}_{\xi_k}}{\partial \tilde{\eta}} - \frac{\tilde{p}\tilde{d}_k}{\tilde{d}_k - \tilde{p}\tilde{\eta}} \frac{\partial \tilde{E}_\eta}{\partial \tilde{\xi}_k} - \frac{\tilde{E}_{\xi_\eta}}{\tilde{d}_k - \tilde{p}\tilde{\eta}} = (-1)^k \tilde{p} \frac{\partial \tilde{H}_{\xi_{3-k}}}{\partial \tilde{t}}, \quad k = 1, 2 \tag{7.36}$$

$$\sum_{i=1}^{2} (-1)^i \frac{\tilde{d}_i}{\tilde{d}_i - \tilde{p}\tilde{\eta}} \frac{\partial \tilde{E}_i}{\partial \tilde{\xi}_{3-i}} = -\frac{\partial \tilde{H}_\eta}{\partial \tilde{t}} \tag{7.37}$$

$$\frac{\partial \tilde{H}_{\xi_{3-k}}}{\partial \tilde{\eta}} - \tilde{p} \frac{\tilde{d}_{3-k}}{\tilde{d}_{3-k} - \tilde{p}\tilde{\eta}} \frac{\partial \tilde{H}_\eta}{\partial \tilde{\xi}_k} - \frac{\tilde{H}_{\xi_{3-k}}}{\tilde{d}_{3-k} - \tilde{p}\tilde{\eta}} = (-1)^k \left(\tilde{E}_{\xi_k} + \tilde{a} \frac{\partial \tilde{E}_{\xi_k}}{\partial \tilde{t}} \right), \quad k = 1, 2 \tag{7.38}$$

$$\tilde{p}^2 \sum_{i=1}^{2} (-1)^{3-k} \frac{\tilde{d}_i}{\tilde{d}_i - \tilde{p}\tilde{\eta}} \frac{\partial \tilde{H}_{\xi_{3-i}}}{\partial \tilde{\xi}_i} = \tilde{E}_\eta + \tilde{a} \frac{\partial \tilde{E}_\eta}{\partial \tilde{t}} \tag{7.39}$$

Here the parameter

$$\tilde{a}^{-1} = \sigma\tau/\varepsilon \tag{7.40}$$

can be viewed as a loss tangent. Substituting Equations 2.44 and 2.45 and transferring into the Laplace domain, we modify Equations 7.36 and 7.37 to the following forms:

$$\frac{\partial \tilde{\tilde{E}}_{\xi_k}}{\partial \tilde{\eta}} - \tilde{p} \left(\frac{\tilde{\tilde{E}}_{\xi_k}}{\tilde{d}_k} + \frac{\partial \tilde{\tilde{E}}_\eta}{\partial \tilde{\xi}_k} \right) - \tilde{p}^2 \tilde{\eta} \left(\frac{\tilde{\tilde{E}}_{\xi_k}}{\tilde{d}_k^2} + \frac{1}{\tilde{d}_k} \frac{\partial \tilde{\tilde{E}}_\eta}{\partial \tilde{\xi}_k} \right) = (-1)^k \tilde{p}\tilde{s}\tilde{\tilde{H}}_{\xi_{3-k}}, \quad k = 1, 2 \tag{7.41}$$

$$\sum_{i=1}^{2} (-1)^i \frac{\partial \tilde{\tilde{E}}_{\xi_i}}{\partial \tilde{\xi}_{3-i}} \left(\tilde{p} + \tilde{p}^2 \frac{\tilde{\eta}}{\tilde{d}_{3-i}} \right) = \tilde{p}\tilde{s}\tilde{\tilde{H}}_\eta \tag{7.42}$$

$$\frac{\partial \tilde{\tilde{H}}_{\xi_{3-k}}}{\partial \tilde{\eta}} - \tilde{p} \left(\frac{\tilde{\tilde{H}}_{\xi_{3-k}}}{\tilde{d}_{3-k}} + \frac{\partial \tilde{\tilde{H}}_\eta}{\partial \tilde{\xi}_{3-k}} \right) - \tilde{p}^2 \tilde{\eta} \left(\frac{\tilde{\tilde{H}}_{\xi_{3-k}}}{\tilde{d}_{3-k}^2} + \frac{1}{\tilde{d}_{3-k}} \frac{\partial \tilde{\tilde{H}}_\eta}{\partial \tilde{\xi}_{3-k}} \right) = (-1)^k \tilde{p}^{-1} (1 + \tilde{a}\tilde{s}) \tilde{\tilde{E}}_{\xi_k}, \quad k = 1, 2 \tag{7.43}$$

$$\sum_{i=1}^{2} (-1)^{3-i} \frac{\partial \tilde{\tilde{H}}_{\xi_{3-i}}}{\partial \tilde{\xi}_i} \left(\tilde{p} + \tilde{p}^2 \frac{\tilde{\eta}}{\tilde{d}_i} \right) = \tilde{p}^{-1} (1 + \tilde{a}\tilde{s}) \tilde{\tilde{E}}_\eta \tag{7.44}$$

Representing the electric and magnetic fields in the form of Equations 3.25 and 3.26 in the small parameter $\tilde{p}$ and transforming Equations 7.41 through 7.44 in a manner similar to Equations 3.16 through 3.19, the following relations between the terms of expansions are obtained:

PEC-limit:

$$\left(\tilde{\vec{\vec{E}}}_0\right)^{\mathrm{b}}_{\xi_{3-k}} = \left(\tilde{\vec{\vec{H}}}_0\right)^{\mathrm{b}}_{\eta} = 0 \tag{7.45}$$

First-order (Leontovich's) approximation:

$$\left(\tilde{\vec{\vec{E}}}_1\right)^{\mathrm{b}}_{\xi_{3-k}} = (-1)^k \tilde{p} \frac{\tilde{s}}{\sqrt{\tilde{s}+\tilde{a}\tilde{s}^2}} \left(\tilde{\vec{\vec{H}}}_0\right)^{\mathrm{b}}_{\xi_k}, \quad k = 1,2 \tag{7.46}$$

$$\left(\tilde{\vec{\vec{H}}}_1\right)^{\mathrm{b}}_{\eta} = \tilde{p} \frac{1}{\sqrt{\tilde{s}+\tilde{a}\tilde{s}^2}} \sum_{i=1}^{2} \frac{\partial \left(\tilde{\vec{\vec{H}}}_0\right)^{\mathrm{b}}_{\xi_i}}{\partial \tilde{\xi}_i} \tag{7.47}$$

Second-order (Mitzner's) approximation:

$$\left(\tilde{\vec{\vec{E}}}_2\right)^{\mathrm{b}}_{\xi_{3-k}} = (-1)^k \tilde{p} \frac{\tilde{s}}{\sqrt{\tilde{s}+\tilde{a}\tilde{s}^2}} \left(\left(\tilde{\vec{\vec{H}}}_1\right)^{\mathrm{b}}_{\xi_k} - \tilde{p} \frac{1}{\sqrt{\tilde{s}+\tilde{a}\tilde{s}^2}} \left(\tilde{\vec{\vec{H}}}_0\right)^{\mathrm{b}}_{\xi_k} \frac{\tilde{d}_k - \tilde{d}_{3-k}}{2\tilde{d}_k \tilde{d}_{3-k}} \right), \quad k = 1,2 \tag{7.48}$$

$$\left(\tilde{\vec{\vec{H}}}_2\right)^{\mathrm{b}}_{\eta} = \tilde{p} \frac{1}{\sqrt{\tilde{s}+\tilde{a}\tilde{s}^2}} \sum_{i=1}^{2} \frac{\partial}{\partial \tilde{\xi}_i} \left(\left(\tilde{\vec{\vec{H}}}_1\right)^{\mathrm{b}}_{\xi_i} - \tilde{p} \frac{1}{\sqrt{\tilde{s}+\tilde{a}\tilde{s}^2}} \left(\tilde{\vec{\vec{H}}}_0\right)^{\mathrm{b}}_{\xi_i} \frac{\tilde{d}_i - \tilde{d}_{3-i}}{2\tilde{d}_i \tilde{d}_{3-i}} \right) \tag{7.49}$$

Third-order (Rytov's) approximation:

$$\begin{aligned} \left(\tilde{\vec{\vec{E}}}_3\right)^{\mathrm{b}}_{\xi_{3-k}} = (-1)^k \tilde{p} \frac{\tilde{s}}{\sqrt{\tilde{s}+\tilde{a}\tilde{s}^2}} &\left[\left(\tilde{\vec{\vec{H}}}_2\right)^{\mathrm{b}}_{\xi_k} - \tilde{p} \frac{1}{\sqrt{\tilde{s}+\tilde{a}\tilde{s}^2}} \left(\tilde{\vec{\vec{H}}}_1\right)^{\mathrm{b}}_{\xi_k} \frac{\tilde{d}_k - \tilde{d}_{3-k}}{2\tilde{d}_k \tilde{d}_{3-k}} \right. \\ &\left. - \tilde{p}^2 \frac{1}{\tilde{s}+\tilde{a}\tilde{s}^2} \left(\left(\tilde{\vec{\vec{H}}}_0\right)^{\mathrm{b}}_{\xi_k} \frac{\tilde{d}_k^2 + 2\tilde{d}_k \tilde{d}_{3-k} - 3\tilde{d}_{3-k}^2}{8\tilde{d}_k^2 \tilde{d}_{3-k}^2} - \frac{1}{2} \frac{\partial^2 \left(\tilde{\vec{\vec{H}}}_0\right)^{\mathrm{b}}_{\xi_k}}{\partial \tilde{\xi}_k^2} + \frac{1}{2} \frac{\partial^2 \left(\tilde{\vec{\vec{H}}}_0\right)^{\mathrm{b}}_{\xi_k}}{\partial \tilde{\xi}_{3-k}^2} - \frac{\partial^2 \left(\tilde{\vec{\vec{H}}}_0\right)^{\mathrm{b}}_{\xi_{3-k}}}{\partial \tilde{\xi}_k \partial \tilde{\xi}_{3-k}} \right) \right] \end{aligned} \tag{7.50}$$

$$\begin{aligned} \left(\tilde{\vec{\vec{H}}}_3\right)^{\mathrm{b}}_{\eta} = \tilde{p} \frac{1}{\sqrt{\tilde{s}+\tilde{a}\tilde{s}^2}} \sum_{i=1}^{2} \frac{\partial}{\partial \tilde{\xi}_i} &\left[\left(\tilde{\vec{\vec{H}}}_2\right)^{\mathrm{b}}_{\xi_i} - \tilde{p} \frac{1}{\sqrt{\tilde{s}+\tilde{a}\tilde{s}^2}} \left(\tilde{\vec{\vec{H}}}_1\right)^{\mathrm{b}}_{\xi_i} \frac{\tilde{d}_i - \tilde{d}_{3-i}}{2\tilde{d}_i \tilde{d}_{3-i}} \right. \\ &\left. - \tilde{p}^2 \frac{1}{\tilde{s}+\tilde{a}\tilde{s}^2} \left(\left(\tilde{\vec{\vec{H}}}_0\right)^{\mathrm{b}}_{\xi_i} \frac{\tilde{d}_i^2 + 2\tilde{d}_i \tilde{d}_{3-i} - 3\tilde{d}_{3-i}^2}{8\tilde{d}_i^2 \tilde{d}_{3-i}^2} - \frac{1}{2} \frac{\partial^2 \left(\tilde{\vec{\vec{H}}}_0\right)^{\mathrm{b}}_{\xi_i}}{\partial \tilde{\xi}_i^2} + \frac{1}{2} \frac{\partial^2 \left(\tilde{\vec{\vec{H}}}_0\right)^{\mathrm{b}}_{\xi_i}}{\partial \tilde{\xi}_{3-i}^2} - \frac{\partial^2 \left(\tilde{\vec{\vec{H}}}_0\right)^{\mathrm{b}}_{\xi_{3-i}}}{\partial \tilde{\xi}_i \partial \tilde{\xi}_{3-i}} \right) \right] \end{aligned} \tag{7.51}$$

Returning to the frequency/time domains, we can represent Equations 7.46 through 7.51 in the forms:

$$\tilde{E}^{\mathrm{b}}_{\xi_{3-k}} = (-1)^{3-k}\frac{\partial}{\partial\tilde{t}}\tilde{F}\left[\tilde{H}^{\mathrm{b}}_{\xi_k}\right], \quad k=1,2 \tag{7.52}$$

$$\tilde{H}^{\mathrm{b}}_{\eta} = \sum_{i=1}^{2}\frac{\partial}{\partial\tilde{\xi}_i}\tilde{F}\left[\tilde{H}^{\mathrm{b}}_{\xi_i}\right] \tag{7.53}$$

Here, the surface impedance operator-function $\tilde{F}[\tilde{f}]$ generalizes Equations 3.86 through 3.88 and now covers both high- and low-frequency problems:

$$\tilde{F}^{\mathrm{Leontovich}}\left[\tilde{f}_{\xi_k}\right] = \begin{cases} \tilde{p}\dfrac{1}{\sqrt{\tilde{s}+\tilde{a}\tilde{s}^2}}\bar{\bar{\tilde{f}}}_{\xi_k} & \mathrm{LD} \\ \tilde{p}\dfrac{1-j}{2}\dfrac{1}{\sqrt{1+2j\tilde{a}}}\dot{\tilde{f}}_{\xi_k} & \mathrm{FD}, \quad k=1,2 \\ \tilde{p}\tilde{f}_{\xi_k} * \tilde{T}^{\mathrm{b}}_2 & \mathrm{TD} \end{cases} \tag{7.54}$$

$$\tilde{F}^{\mathrm{Mitzner}}\left[\tilde{f}_{\xi_k}\right] = \begin{cases} \tilde{p}\dfrac{1}{\sqrt{\tilde{s}+\tilde{a}\tilde{s}^2}}\bar{\bar{\tilde{f}}}_{\xi_k} - \tilde{p}^2\dfrac{1}{\tilde{s}+\tilde{a}\tilde{s}^2}\bar{\bar{\tilde{f}}}_{\xi_k}\dfrac{\tilde{d}_k-\tilde{d}_{3-k}}{2\tilde{d}_k\tilde{d}_{3-k}} & \mathrm{LD} \\ \tilde{p}\dfrac{1-j}{2}\dfrac{1}{\sqrt{1+2j\tilde{a}}}\dot{\tilde{f}}_{\xi_k} + \tilde{p}^2\dfrac{j}{2}\dfrac{1}{1+2j\tilde{a}}\dfrac{\tilde{d}_k-\tilde{d}_{3-k}}{2\tilde{d}_k\tilde{d}_{3-k}}\dot{\tilde{f}}_{\xi_k} & \mathrm{FD}, \quad k=1,2 \\ \tilde{p}\tilde{f}_{\xi_k} * \tilde{T}^{\mathrm{b}}_2 - \tilde{p}^2\dfrac{\tilde{d}_k-\tilde{d}_{3-k}}{2\tilde{d}_k\tilde{d}_{3-k}}\tilde{f}_{\xi_k} * \tilde{T}^{\mathrm{b}}_3 & \mathrm{TD} \end{cases} \tag{7.55}$$

$$\tilde{F}^{\mathrm{Rytov}}\left[\tilde{f}_{\xi_k}\right] = \begin{cases} \tilde{p}\left(\tilde{s}+\tilde{a}\tilde{s}^2\right)^{-1/2}\bar{\bar{\tilde{f}}}_{\xi_k} - \tilde{p}^2\left(\tilde{s}+\tilde{a}\tilde{s}^2\right)^{-1}\bar{\bar{\tilde{f}}}_{\xi_k}\dfrac{\tilde{d}_k-\tilde{d}_{3-k}}{2\tilde{d}_k\tilde{d}_{3-k}} \\ \quad -\tilde{p}^3\left(\tilde{s}+\tilde{a}\tilde{s}^2\right)^{-3/2}\left(\bar{\bar{\tilde{f}}}_{\xi_k}\dfrac{\tilde{d}_k^2+2\tilde{d}_k\tilde{d}_{3-k}-3\tilde{d}_{3-k}^2}{8\tilde{d}_k^2\tilde{d}_{3-k}^2} - \dfrac{1}{2}\dfrac{\partial^2\bar{\bar{\tilde{f}}}_{\xi_k}}{\partial\tilde{\xi}_k^2} + \dfrac{1}{2}\dfrac{\partial^2\bar{\bar{\tilde{f}}}_{\xi_k}}{\partial\tilde{\xi}_{3-k}^2} - \dfrac{\partial^2\bar{\bar{\tilde{f}}}_{\xi_{3-k}}}{\partial\tilde{\xi}_k\partial\tilde{\xi}_{3-k}}\right) & \mathrm{LD} \\ \tilde{p}\dfrac{1-j}{2}(1+2j\tilde{a})^{-1/2}\dot{\tilde{f}}_{\xi_k} + \tilde{p}^2\dfrac{j}{2}(1+2j\tilde{a})^{-1}\dfrac{\tilde{d}_k-\tilde{d}_{3-k}}{2\tilde{d}_k\tilde{d}_{3-k}}\dot{\tilde{f}}_{\xi_k} \\ \quad -\tilde{p}^3\dfrac{1+j}{4}(1+2j\tilde{a})^{-3/2}\left(\dot{\tilde{f}}_{\xi_k}\dfrac{\tilde{d}_k^2+2\tilde{d}_k\tilde{d}_{3-k}-3\tilde{d}_{3-k}^2}{8\tilde{d}_k^2\tilde{d}_{3-k}^2}\right. \\ \quad \left. -\dfrac{1}{2}\dfrac{\partial^2\dot{\tilde{f}}_{\xi_k}}{\partial\tilde{\xi}_k^2} + \dfrac{1}{2}\dfrac{\partial^2\dot{\tilde{f}}_{\xi_k}}{\partial\tilde{\xi}_{3-k}^2} - \dfrac{\partial^2\dot{\tilde{f}}_{\xi_{3-k}}}{\partial\tilde{\xi}_k\partial\tilde{\xi}_{3-k}}\right) & \mathrm{FD}, \quad k=1,2 \\ \tilde{p}\tilde{f}_{\xi_k} * \tilde{T}^{\mathrm{b}}_2 - \tilde{p}^2\dfrac{\tilde{d}_k-\tilde{d}_{3-k}}{2\tilde{d}_k\tilde{d}_{3-k}}\tilde{f}_{\xi_k} * \tilde{T}^{\mathrm{b}}_3 \\ \quad -\tilde{p}^3\left(\tilde{f}_{\xi_k}\dfrac{\tilde{d}_k^2+2\tilde{d}_k\tilde{d}_{3-k}-3\tilde{d}_{3-k}^2}{8\tilde{d}_k^2\tilde{d}_{3-k}^2} - \dfrac{1}{2}\dfrac{\partial^2\tilde{f}_{\xi_k}}{\partial\tilde{\xi}_k^2} + \dfrac{1}{2}\dfrac{\partial^2\tilde{f}_{\xi_k}}{\partial\tilde{\xi}_{3-k}^2} - \dfrac{\partial^2\tilde{f}_{\xi_{3-k}}}{\partial\tilde{\xi}_k\partial\tilde{\xi}_{3-k}}\right) * \tilde{T}^{\mathrm{b}}_4 & \mathrm{TD} \end{cases} \tag{7.56}$$

where the functions $\tilde{T}_m$ and $\tilde{\tilde{T}}_m$ were obtained using the inverse Laplace transform and written in the form [3]:

$$\frac{1}{\sqrt{\tilde{s}+\tilde{s}^2\tilde{a}}} \Leftrightarrow \tilde{a}^{-1/2} I_0\left(\frac{\tilde{t}}{2\tilde{a}}\right)\exp = \tilde{T}_2^{\mathrm{b}}(\tilde{t}) \tag{7.57}$$

$$\frac{\tilde{s}}{\sqrt{\tilde{s}+\tilde{s}^2\tilde{a}}} \Leftrightarrow \frac{1}{2\tilde{a}^{3/2}}\left[I_1\left(\frac{\tilde{t}}{2\tilde{a}}\right) - I_0\left(\frac{\tilde{t}}{2\tilde{a}}\right)\right]\exp\left(-\frac{\tilde{t}}{2\tilde{a}}\right) + U(\tilde{t})\tilde{a}^{-1/2} = \tilde{\tilde{T}}_2^{\mathrm{b}}(\tilde{t}) \tag{7.58}$$

$$\frac{1}{\tilde{s}+\tilde{s}^2\tilde{a}} \Leftrightarrow 1-\exp\left(-\frac{\tilde{t}}{\tilde{a}}\right) = \tilde{T}_3^{\mathrm{b}}(\tilde{t}) \tag{7.59}$$

$$\frac{\tilde{s}}{\tilde{s}+\tilde{s}^2\tilde{a}} \Leftrightarrow \frac{1}{\tilde{a}}\exp\left(-\frac{\tilde{t}}{\tilde{a}}\right) = \tilde{\tilde{T}}_3^{\mathrm{b}}(\tilde{t}) \tag{7.60}$$

$$\frac{1}{(\tilde{s}+\tilde{s}^2\tilde{a})^{3/2}} \Leftrightarrow 2\tilde{a}^{-1/2}\tilde{t} I_1\left(\frac{\tilde{t}}{2\tilde{a}}\right)\exp\left(-\frac{\tilde{t}}{2\tilde{a}}\right) = \tilde{T}_4^{\mathrm{b}}(\tilde{t}) \tag{7.61}$$

$$\frac{\tilde{s}}{(\tilde{s}+\tilde{s}^2\tilde{a})^{3/2}} \Leftrightarrow \tilde{a}^{-3/2}\tilde{t}\left[I_0\left(\frac{\tilde{t}}{2\tilde{a}}\right) - I_1\left(\frac{\tilde{t}}{2\tilde{a}}\right)\right]\exp\left(-\frac{\tilde{t}}{2\tilde{a}}\right) = \tilde{\tilde{T}}_4^{\mathrm{b}}(\tilde{t}) \tag{7.62}$$

where $I_n(\tilde{x})$ is the modified Bessel function of order n and $U(\tilde{t})$ is the Heaviside function. The relations between the functions $\tilde{T}_m^{\mathrm{b}}$ and $\tilde{\tilde{T}}_m^{\mathrm{b}}$ are written in the form:

$$\tilde{\tilde{T}}_3^{\mathrm{b}}(\tilde{t}) = \frac{d\tilde{T}_m^{\mathrm{b}}(t)}{d\tilde{t}} + \delta(\tilde{t})\tilde{T}_m^{\mathrm{b}}(0), \quad m = 2,3,4 \tag{7.63}$$

By setting $\tilde{a} \to 0$ in Equations 7.57, 7.59, and 7.61, the functions $\tilde{T}_m$ are reduced to their "low-frequency form" Equations 2.189 through 2.191, respectively.

Return to dimensional variables in Equations 7.52 through 7.56 finally yields

$$E_{\xi_{3-k}}^{\mathrm{b}} = (-1)^{3-k}\mu\frac{\partial}{\partial t}F\left[H_{\xi_k}^{\mathrm{b}}\right], \quad k = 1,2 \tag{7.64}$$

$$H_\eta^{\mathrm{b}} = \sum_{i=1}^{2}\frac{\partial}{\partial \xi_i}F\left[H_{\xi_i}^{\mathrm{b}}\right] \tag{7.65}$$

$$F^{\text{Leontovich}}\left[f_{\xi_k}\right] = \begin{cases} (j\omega\sigma\mu)^{-1/2}(1+j\omega\varepsilon/\sigma)^{-1/2}\dot{f}_{\xi_k} & \text{FD} \\ (\sigma\mu)^{-1/2}f_{\xi_k} * T_2^{\mathrm{b}} & \text{TD} \end{cases}, \quad k = 1,2 \tag{7.66}$$

$$F^{\text{Mitzner}}\left[f_{\xi_k}\right] = \begin{cases} (j\omega\sigma\mu)^{-1/2}(1+j\omega\varepsilon/\sigma)^{-1/2}\dot{f}_{\xi_k} - (j\omega\sigma\mu)^{-1}(1+j\omega\varepsilon/\sigma)^{-1}\dfrac{d_k - d_{3-k}}{2d_k d_{3-k}}\dot{f}_{\xi_k} & \text{FD} \\ (\sigma\mu)^{-1/2} f_{\xi_k} * T_2^{\text{b}} - (\sigma\mu)^{-1}\dfrac{d_k - d_{3-k}}{2d_k d_{3-k}} f_{\xi_k} * T_3^{\text{b}} & \text{TD}, \end{cases} \quad k = 1,2 \tag{7.67}$$

$$F^{\text{Rytov}}\left[f_{\xi_k}\right] = \begin{cases} (j\omega\sigma\mu)^{-1/2}\left(1+\dfrac{j\omega\varepsilon}{\sigma}\right)^{-1/2}\dot{f}_{\xi_k} - (j\omega\sigma\mu)^{-1}\left(1+\dfrac{j\omega\varepsilon}{\sigma}\right)^{-1}\dfrac{d_k - d_{3-k}}{2d_k d_{3-k}}\dot{f}_{\xi_k} \\ \quad -(j\omega\sigma\mu)^{-3/2}\left(1+\dfrac{j\omega\varepsilon}{\sigma}\right)^{-3/2}\left(\dot{f}_{\xi_k}\dfrac{d_k^2 + 2d_k d_{3-k} - 3d_{3-k}^2}{8d_k^2 d_{3-k}^2} - \dfrac{1}{2}\dfrac{\partial^2 \dot{f}_{\xi_k}}{\partial \xi_k^2}\right. \\ \quad \left. +\dfrac{1}{2}\dfrac{\partial^2 \dot{f}_{\xi_k}}{\partial \xi_{3-k}^2} - \dfrac{\partial^2 \dot{f}_{\xi_{3-k}}}{\partial \xi_k \partial \xi_{3-k}}\right) \quad \text{FD}, \ k = 1,2 \\ (\sigma\mu)^{-1/2} f_{\xi_k} * T_2^{\text{b}} - (\sigma\mu)^{-1}\dfrac{d_k - d_{3-k}}{2d_k d_{3-k}} f_{\xi_k} * T_3^{\text{b}} \\ \quad -(\sigma\mu)^{-3/2}\left(f_{\xi_k}\dfrac{d_k^2 + 2d_k d_{3-k} - 3\tilde{d}_{3-k}^2}{8d_k^2 d_{3-k}^2} - \dfrac{1}{2}\dfrac{\partial^2 f_{\xi_k}}{\partial \xi_k^2} + \dfrac{1}{2}\dfrac{\partial^2 f_{\xi_k}}{\partial \xi_{3-k}^2} - \dfrac{\partial^2 f_{\xi_{3-k}}}{\partial \xi_k \partial \xi_{3-k}}\right) * T_4^{\text{b}} \quad \text{TD} \end{cases} \tag{7.68}$$

where

$$T_2^{\text{b}}(t) = \sqrt{\frac{\sigma}{\varepsilon}} I_0\left(\frac{\sigma t}{2\varepsilon}\right)\exp\left(-\frac{\sigma t}{2\varepsilon}\right) \tag{7.69}$$

$$T_3^{\text{b}}(t) = 1 - \exp\left(-\frac{\sigma t}{\varepsilon}\right) \tag{7.70}$$

$$T_4^{\text{b}}(t) = 2t\sqrt{\frac{\sigma}{\varepsilon}} I_1\left(\frac{\sigma t}{2\varepsilon}\right)\exp\left(-\frac{\sigma t}{2\varepsilon}\right) \tag{7.71}$$

7.4 Direct Implementation of SIBCs into the Surface Integral Equations

Let us represent the frequency-domain equations (Equations 7.20 and 7.21) and time-domain equations (Equations 7.31 and 7.32) with local coordinates ξ_1, ξ_2, η:

Electric field integral equation:

$$\dot{E}_{\xi_k}^{\text{b}} = 2\dot{E}_{\xi_k}^{\text{inc}} + \frac{1}{2\pi}\int_{S_c}\left[j\omega\mu_0\left(\vec{e}_\eta \times \dot{\vec{H}}^{\text{b}}\right)_{\xi_k} G - \left(\left(\vec{e}_\eta \times \dot{\vec{E}}^{\text{b}}\right) \times \nabla' G\right)_{\xi_k} - \dot{E}_\eta^{\text{b}}(\nabla' G)_{\xi_k}\right] ds' \tag{7.72}$$

$$E^{\mathrm{b}}_{\xi_k} = 2E^{\mathrm{inc}}_{\xi_k} + \frac{1}{2\pi}\int\limits_{S_c}\left\{\frac{\mu_0}{R}\frac{\partial}{\partial t'}\left(\vec{e}_\eta \times \vec{H}^{\mathrm{b}}\right)_{\xi_k} - \left(\Pi\left[\vec{e}_\eta \times \vec{E}^{\mathrm{b}}\right] \times \frac{\vec{R}}{R}\right)_{\xi_k} - \Pi\left[E^{\mathrm{b}}_\eta\right]\frac{R_{\xi_k}}{R}\right\}\Bigg|_{t'=t-R/c} \mathrm{d}s' \tag{7.73}$$

Magnetic field integral equation:

$$\dot{H}^{\mathrm{b}}_{\xi_{3-k}} = 2\dot{H}^{\mathrm{inc}}_{\xi_{3-k}} - \frac{1}{2\pi}\int\limits_{S_c}\left[j\omega\varepsilon_0\left(\vec{e}_\eta \times \dot{\vec{E}}^{\mathrm{b}}\right)_{\xi_{3-k}} G + \left(\left(\vec{e}_\eta \times \dot{\vec{H}}^{\mathrm{b}}\right) \times \nabla' G\right)_{\xi_{3-k}} + \dot{H}^{\mathrm{b}}_\eta (\nabla' G)_{\xi_{3-k}}\right]\mathrm{d}s' \tag{7.74}$$

$$H^{\mathrm{b}}_{\xi_{3-k}} = 2H^{\mathrm{inc}}_{\xi_{3-k}} - \frac{1}{2\pi}\int\limits_{S_c}\left\{\frac{\varepsilon_0}{R}\frac{\partial}{\partial t'}\left(\vec{e}_\eta \times \vec{E}^{\mathrm{b}}\right)_{\xi_{3-k}} + \left(\Pi\left[\vec{e}_\eta \times \vec{H}^{\mathrm{b}}\right] \times \frac{\vec{R}}{R}\right)_{\xi_{3-k}} + \Pi\left[H^{\mathrm{b}}_\eta\right]\frac{R_{\xi_{3-k}}}{R}\right\}\Bigg|_{t'=t-R/c} \mathrm{d}s' \tag{7.75}$$

where

$$\vec{n}' \times \vec{H} = -\vec{e}_\eta \times \vec{H}; \quad \vec{n}' \cdot \vec{H} = -\vec{e}_\eta \cdot \vec{H} \tag{7.76}$$

The superscript "b" denoting quantities on the surface V will be omitted from here on. According to the vector identities

$$\left(\vec{e}_\eta \times \vec{f}\right)_{\xi_k} = (-1)^k f_{\xi_{3-k}}, \quad k = 1,2 \tag{7.77}$$

and

$$\left(\left(\vec{e}_\eta \times \vec{f}\right) \times \vec{g}\right)_{\xi_k} = (-1)^{3-k}\left(\left(\vec{e}_\eta \times \vec{f}\right)_{\xi_{3-k}} g_\eta - \left(\vec{e}_\eta \times \vec{f}\right)_\eta g_{\xi_{3-k}}\right) = (-1)^{3-k}\left(\vec{e}_\eta \times \vec{f}\right)_{\xi_{3-k}} g_\eta = f_{\xi_k} g_\eta \tag{7.78}$$

where $\vec{f}$ and $\vec{g}$ are any vectors. With the use of Equations 7.76 through 7.78 we represent Equations 7.72 through 7.75 in the following form:

Electric field integral equation:

$$\dot{E}_{\xi_k} = 2\dot{E}^{\mathrm{inc}}_{\xi_k} + \frac{1}{2\pi}\int\limits_{S_c}\left\{j\omega\mu_0(-1)^k \dot{H}_{\xi_{3-k}} G - \dot{E}_{\xi_k}(\nabla' G)_\eta - \dot{E}_\eta (\nabla' G)_{\xi_k}\right\}\mathrm{d}s' \tag{7.79}$$

$$E_{\xi_k} = 2E^{\mathrm{inc}}_{\xi_k} + \frac{1}{2\pi}\int\limits_{S_c}\left\{(-1)^k \frac{\mu_0}{R}\frac{\partial H_{\xi_{3-k}}}{\partial t'} - \Pi\left[E_{\xi_k}\right]\frac{R_\eta}{R} - \Pi\left[E_\eta\right]\frac{R_{\xi_k}}{R}\right\}\Bigg|_{t'=t-R/c} \mathrm{d}s' \tag{7.80}$$

Magnetic field integral equation:

$$\dot{H}_{\xi_{3-k}} = 2\dot{H}^{\mathrm{inc}}_{\xi_{3-k}} - \frac{1}{2\pi}\int\limits_{S_c}\left\{j\omega\varepsilon_0(-1)^{3-k}\dot{E}_{\xi_k}G + \dot{H}_{\xi_{3-k}}(\nabla' G)_{\eta} + \dot{H}_{\eta}(\nabla' G)_{\xi_{3-k}}\right\}\mathrm{d}s' \tag{7.81}$$

$$H_{\xi_{3-k}} = 2H^{\mathrm{inc}}_{\xi_{3-k}} - \frac{1}{2\pi}\int\limits_{S_c}\left\{(-1)^{3-k}\frac{\varepsilon_0}{R}\frac{\partial E_{\xi_k}}{\partial t'} + \Pi\left[H_{\xi_{3-k}}\right]\frac{R_{\eta}}{R} + \Pi\left[H_{\eta}\right]\frac{R_{\xi_{3-k}}}{R}\right\}\Bigg|_{t'=t-R/c}\mathrm{d}s' \tag{7.82}$$

The normal component of the electric field can be derived from Equation 7.2 and expressed in terms of the tangential components of the magnetic field:

$$\dot{E}_{\eta} = \frac{1}{j\omega\varepsilon_0}\vec{e}_{\eta}\cdot\left(\nabla'\times\dot{\vec{H}}\right) = \frac{1}{j\omega\varepsilon_0}\sum_{i=1}^{2}(-1)^{3-i}\frac{\partial\dot{H}_{\xi_{3-i}}}{\partial\xi_i} = \frac{j}{\omega\varepsilon_0}\sum_{i=1}^{2}(-1)^{i}\frac{\partial\dot{H}_{\xi_{3-i}}}{\partial\xi_i} \tag{7.83}$$

and

$$E_{\eta}(t) = -\frac{1}{\varepsilon_0}\int\limits_0^t\sum_{i=1}^{2}(-1)^{3-i}\frac{\partial\left(\vec{H}(t')\right)_{\xi_{3-i}}}{\partial\xi_i}\mathrm{d}t' = \frac{1}{\varepsilon_0}\int\limits_0^t\sum_{i=1}^{2}(-1)^{i}\frac{\partial\left(\vec{H}(t')\right)_{\xi_{3-i}}}{\partial\xi_i}\mathrm{d}t' \tag{7.84}$$

Substitution of Equations 7.83 and 7.84 into Equations 7.79 and 7.80 gives

$$\dot{E}_{\xi_k} = 2\dot{E}^{\mathrm{inc}}_{\xi_k} + \frac{1}{2\pi}\int\limits_{S_c}\left\{j\omega\mu_0(-1)^k\dot{H}_{\xi_{3-k}}G - \dot{E}_{\xi_k}(\nabla' G)_{\eta} - \frac{j}{\omega\varepsilon_0}(\nabla' G)_{\xi_k}\sum_{i=1}^{2}(-1)^{i}\frac{\partial\dot{H}_{\xi_{3-i}}}{\partial\xi_i}\right\}\mathrm{d}s' \tag{7.85}$$

$$E_{\xi_k} = 2E^{\mathrm{inc}}_{\xi_k} + \frac{1}{2\pi}\int\limits_{S_c}\left\{(-1)^k\frac{\mu_0}{R}\frac{\partial H_{\xi_{3-k}}}{\partial t'}\cdot\right.$$
$$\left.-\Pi\left[E_{\xi_k}\right]\frac{R_{\eta}}{R} - \frac{1}{\varepsilon_0}\frac{R_{\xi_k}}{R}\int\limits_0^t\sum_{i=1}^{2}(-1)^{i}\frac{\partial}{\partial\xi_i}\Pi\left[H_{\xi_{3-i}}\right]\mathrm{d}t'\right\}\Bigg|_{t'=t-R/c}\mathrm{d}s' \tag{7.86}$$

The EFIEs in Equation 7.86 or the MFIEs in Equations 7.81 and 7.82 still contain more than one unknown and require an additional equation that can be obtained by taking into account the electromagnetic properties of the conducting domain V_c. Under conditions of the skin effect, the SIBCs in Equations 7.52 and 7.53, giving the tangential components of the electric field and normal component of the magnetic field at the interface, can be substituted in Equations 7.85 and 7.86 and Equations 7.81 and 7.82 as follows:

Electric field integral equation:

$$(-1)^{3-k} j\omega\mu_0 F\left[\dot{H}_{\xi_{3-k}}\right] = 2\dot{E}^{\text{inc}}_{\xi_k} + j\frac{\mu_0}{2\pi}\int_{S_c}\left\{\omega(-1)^k \dot{H}_{\xi_{3-k}} G + (-1)^k \omega F\left[\dot{H}_{\xi_{3-i}}\right](\nabla' G)_\eta - \frac{1}{\omega\varepsilon_0\mu_0}(\nabla' G)_{\xi_k}\sum_{i=1}^{2}(-1)^i \frac{\partial \dot{H}_{\xi_{3-i}}}{\partial \xi_i}\right\}\mathrm{d}s' \quad (7.87)$$

$$(-1)^{3-k}\mu_0\frac{\partial}{\partial t}F[H_{\xi_{3-k}}] = 2E^{\text{inc}}_{\xi_k} + \frac{\mu_0}{2\pi}\int_{S_c}\left\{\frac{(-1)^k}{R}\frac{\partial H_{\xi_{3-k}}}{\partial t'} + (-1)^k\frac{R_\eta}{R}\frac{\partial}{\partial t'}F[\Pi[H_{\xi_{3-k}}]] - \frac{1}{\varepsilon_0\mu_0}\frac{R_{\xi_k}}{R}\int_0^t \sum_{i=1}^{2}(-1)^i\frac{\partial}{\partial \xi_i}\Pi[H_{\xi_{3-i}}]\mathrm{d}t'\right\}\Bigg|_{t'=t-R/c}\mathrm{d}s' \quad (7.88)$$

Magnetic field integral equation:

$$\dot{H}_{\xi_{3-k}} = 2\dot{H}^{\text{inc}}_{\xi_{3-k}} - \frac{1}{2\pi}\int_{S_c}\left\{-\omega^2\varepsilon_0\mu_0 F\left[\dot{H}_{\xi_{3-k}}\right]G + \dot{H}_{\xi_{3-k}}(\nabla' G)_\eta + (\nabla' G)_{\xi_{3-k}}\sum_{i=1}^{2}\frac{\partial}{\partial \xi_i}F\left[\dot{H}_{\xi_i}\right]\right\}\mathrm{d}s' \quad (7.89)$$

$$H_{\xi_{3-k}} = 2H^{\text{inc}}_{\xi_{3-k}} - \frac{1}{2\pi}\int_{S_c}\left\{\frac{\varepsilon_0\mu_0}{R}\frac{\partial^2}{\partial t'^2}F\left[H^{\mathrm{b}}_{\xi_{3-k}}\right] + \Pi[H_{\xi_{3-k}}]\frac{R_\eta}{R} + \frac{R_{\xi_{3-k}}}{R}\sum_{i=1}^{2}\frac{\partial}{\partial \xi_i}F[\Pi[H_{\xi_i}]]\right\}\Bigg|_{t'=t-R/c}\mathrm{d}s' \quad (7.90)$$

In the limiting case of a perfect electrical conductor, the tangential electric field on the surface S_c is assumed to be zero and Equations 7.87 through 7.90 reduce to the following equations:

Electric field integral equation:

$$-j\omega\frac{\mu_0}{2\pi}\int_{S_c}\left\{(-1)^k \dot{H}_{\xi_{3-k}} G - \frac{1}{\omega^2\varepsilon_0\mu_0}(\nabla' G)_{\xi_k}\sum_{i=1}^{2}(-1)^i\frac{\partial \dot{H}_{\xi_{3-i}}}{\partial \xi_i}\right\}\mathrm{d}s' = 2\dot{E}^{\text{inc}}_{\xi_k} \quad (7.91)$$

$$-\frac{\mu_0}{2\pi}\int_{S_c}\left\{\frac{(-1)^k}{R}\frac{\partial H_{\xi_{3-k}}}{\partial t'} - \frac{1}{\varepsilon_0\mu_0}\frac{R_{\xi_k}}{R}\int_0^t\sum_{i=1}^{2}(-1)^i\frac{\partial}{\partial \xi_i}\Pi\left[H_{\xi_{3-i}}\right]\mathrm{d}t'\right\}\Bigg|_{t'=t-R/c}\mathrm{d}s' = 2E^{\text{inc}}_{\xi_k} \quad (7.92)$$

Magnetic field integral equation:

$$\dot{H}_{\xi_{3-k}} + \frac{1}{2\pi}\int\limits_{S_c} \dot{H}_{\xi_{3-k}}(\nabla' G)_\eta \, ds' = 2\dot{H}^{\text{inc}}_{\xi_{3-k}} \tag{7.93}$$

$$H_{\xi_{3-k}} + \frac{1}{2\pi}\int\limits_{S_c} \Pi[H_{\xi_{3-k}}]\frac{R_\eta}{R}\bigg|_{t'=t-R/c} ds' = 2H^{\text{inc}}_{\xi_{3-k}} \tag{7.94}$$

Either of these equations can be used to solve for the tangential magnetic field. The principal difference between them is that the EFIEs in Equations 7.91 and 7.92 are Fredholm integral equations of the first kind whereas the MFIEs in Equations 7.93 and 7.94 are Fredholm integral equations of the second kind. The latter is generally preferable from the computational point of view [4]. However, the final choice is also governed by the geometry of the scatterer. For example, the EFIE is ideally suited for thin cylinders [2].

Frequently, simulations of antenna problems are performed using the equivalent surface current $\vec{J}^{\text{s}}$ defined as

$$\vec{J}^{\text{s}} = \vec{n} \times \vec{H}^{\text{b}} = -\vec{e}_\eta \times \vec{H}^{\text{b}} \left(J^{\text{s}}_{\xi_k} = H^{\text{b}}_{\xi_{3-k}}, \quad k = 1,2\right) \tag{7.95}$$

The EFIEs in Equations 7.91 and 7.92 and the MFIEs in Equations 7.93 and 7.94 for the PEC can be rewritten in terms of an equivalent surface current defined as follows:

Electric field integral equation:

$$-j\omega\frac{\mu_0}{2\pi}\int\limits_{S_c}\left\{(-1)^k \dot{J}^{\text{s}}_{\xi_k} G - \frac{1}{\omega^2\varepsilon_0\mu_0}(\nabla' G)_{\xi_k}\sum_{i=1}^{2}(-1)^i\frac{\partial \dot{J}^{\text{s}}_{\xi_k}}{\partial \xi_i}\right\} ds' = 2\dot{E}^{\text{inc}}_{\xi_k} \tag{7.96}$$

$$-\frac{\mu_0}{2\pi}\int\limits_{S_c}\left\{\frac{(-1)^k}{R}\frac{\partial J^{\text{s}}_{\xi_k}}{\partial t'} - \frac{1}{\varepsilon_0\mu_0}\frac{R_{\xi_k}}{R}\int\limits_0^t \sum_{i=1}^{2}(-1)^i\frac{\partial}{\partial \xi_i}\Pi\left[J^{\text{s}}_{\xi_k}\right]dt'\right\}\Bigg|_{t'=t-R/c} ds' = 2E^{\text{inc}}_{\xi_k} \tag{7.97}$$

Magnetic field integral equation:

$$\dot{J}^{\text{s}}_{\xi_k} + \frac{1}{2\pi}\int\limits_{S_c} \dot{J}^{\text{s}}_{\xi_k}(\nabla' G)_\eta ds' = 2\dot{H}^{\text{inc}}_{\xi_{3-k}} \tag{7.98}$$

$$J^{\text{s}}_{\xi_k} + \frac{1}{2\pi}\int\limits_{S_c} \Pi\left[J^{\text{s}}_{\xi_k}\right]\frac{R_\eta}{R}\bigg|_{t'=t-R/c} ds' = 2H^{\text{inc}}_{\xi_{3-k}} \tag{7.99}$$

Note that in general, the equivalent surface current is not the same as the surface current density $\vec{K}$. The latter can be related with $\vec{J}^s$ as follows:

$$K_{\xi_k} = \int_0^\infty J_{\xi_k}\mathrm{d}\eta = \int_0^\infty \left(\vec{n}\times\vec{H}\right)_{\xi_k}\mathrm{d}\eta = \int_0^\infty \left(\frac{\partial H_\eta}{\partial \xi_{3-k}} - \frac{\partial H_{\xi_{3-k}}}{\partial \eta}\right)\mathrm{d}\eta = H^{\mathrm{b}}_{\xi_{3-k}} + \int_0^\infty \frac{\partial H_\eta}{\partial \xi_{3-k}}\mathrm{d}\eta = J^{\mathrm{s}}_{\xi_k} + \int_0^\infty \frac{\partial H_\eta}{\partial \xi_{3-k}}\mathrm{d}\eta \tag{7.100}$$

The second term vanishes only in the particular case of the PEC-limit.

It is clear that the practical implementation of Equations 7.87 through 7.90 is a more complex task than Equations 7.91 through 7.94. In Section 7.5, we will modify Equations 7.87 through 7.90 to a form with the same structure of the left-hand side as in Equations 7.91 through 7.94 incorporating the PEC-condition.

7.5 Implementation Using the Perturbation Technique

Both the EFIEs (Equations 7.87 and 7.88) and the MFISs (Equations 7.89 and 7.90) can be brought into the following form by moving the incident fields and the terms containing the surface impedance function to the right-hand side:

Electric field integral equation:

$$-j\omega\mu_0\frac{1}{2\pi}\int_{S_c}\left\{(-1)^k\dot{H}_{\xi_{3-k}}G - \frac{1}{\omega^2\varepsilon_0\mu_0}(\nabla' G)_{\xi_k}\sum_{i=1}^{2}(-1)^i\frac{\partial \dot{H}_{\xi_{3-i}}}{\partial \xi_i}\right\}\mathrm{d}s'$$

$$= 2\dot{E}^{\mathrm{inc}}_{\xi_k} + (-1)^k j\omega\mu_0\left(F\left[\dot{H}_{\xi_{3-k}}\right] + \frac{1}{2\pi}\int_{S_c}F\left[\dot{H}_{\xi_{3-k}}\right](\nabla' G)_\eta\mathrm{d}s'\right) \tag{7.101}$$

$$-\mu_0\frac{1}{2\pi}\int_{S_c}\left\{\frac{(-1)^k}{R}\frac{\partial H_{\xi_{3-k}}}{\partial t'} - \frac{1}{\varepsilon_0\mu_0}\frac{R_{\xi_k}}{R}\int_0^t\sum_{i=1}^{2}(-1)^i\frac{\partial}{\partial \xi_i}\Pi[H_{\xi_{3-i}}]\mathrm{d}t'\right\}\Bigg|_{t'=t-R/c}\mathrm{d}s'$$

$$= 2E^{\mathrm{inc}}_{\xi_k} + (-1)^k\mu_0\left(\frac{\partial}{\partial t}F[H_{\xi_{3-k}}] + \frac{1}{2\pi}\int_{S_c}\frac{R_\eta}{R}\frac{\partial}{\partial t'}F[\Pi[H_{\xi_{3-k}}]]\Bigg|_{t'=t-R/c}\mathrm{d}s'\right) \tag{7.102}$$

Magnetic field integral equation:

$$\dot{H}_{\xi_{3-k}} + \frac{1}{2\pi}\int_{S_c} \dot{H}_{\xi_{3-k}}(\nabla' G)_\eta \mathrm{d}s' = 2\dot{H}^{\mathrm{inc}}_{\xi_{3-k}}$$

$$-\frac{1}{2\pi}\int_{S_c}\left\{-\omega^2\varepsilon_0\mu_0 F\left[\dot{H}_{\xi_{3-k}}\right]G + (\nabla' G)_{\xi_{3-k}}\sum_{i=1}^{2}\frac{\partial}{\partial\xi_i}F\left[\dot{H}_{\xi_i}\right]\right\}\mathrm{d}s' \quad (7.103)$$

$$H_{\xi_{3-k}} + \frac{1}{2\pi}\int_{S_c}\Pi[H_{\xi_{3-k}}]\frac{R_\eta}{R}\mathrm{d}s' = 2H^{\mathrm{inc}}_{\xi_{3-k}}$$

$$-\frac{1}{2\pi}\int_{S_c}\left\{\frac{\varepsilon_0\mu_0}{R}\frac{\partial^2}{\partial t'^2}F[H_{\xi_{3-k}}] + \frac{R_{\xi_{3-k}}}{R}\sum_{i=1}^{2}\frac{\partial}{\partial\xi_i}F[\Pi[H_{\xi_i}]]\right\}\Bigg|_{t'-t-R/c}\mathrm{d}s' \quad (7.104)$$

Introducing the scale factors into the operator Π, we obtain

$$\Pi\left[\tilde{f}f^*\right] = \frac{\tilde{f}f^*}{\tilde{R}^2D^2} + \frac{1}{c\tilde{R}D}\frac{\partial\left(\tilde{f}f^*\right)}{\partial\left(\tilde{t}'\tau\right)} = \frac{f^*}{D^2}\left(\frac{\tilde{f}}{\tilde{R}^2} + \frac{D}{c\tau}\frac{\partial\tilde{f}}{\partial\tilde{t}'}\right) = \frac{f^*}{D^2}\tilde{\Pi}\left[\tilde{f}\right] \quad (7.105)$$

where

$$\tilde{\Pi}\left[\tilde{f}\right] = \frac{\tilde{f}}{\tilde{R}^2} + \tilde{q}\frac{\partial\tilde{f}}{\partial\tilde{t}'} \quad (7.106)$$

$$\tilde{q} = \begin{cases} 2D/(c\omega) & \mathrm{FD} \\ D/(c\tau) & \mathrm{TD}\end{cases} \quad (7.107)$$

Here the parameter $\tilde{q}$ has a physical meaning as the electrical size of the problem.

Equations 7.101 through 7.104 can be represented with non-dimensional variables using the scale factors in Table 3.1 in the following forms:

Electric field integral equation:

$$-\frac{j}{2\pi}\int_{S_c}\left\{2(-1)^k\tilde{\dot{H}}_{\xi_{3-k}}\tilde{G} - (2\tilde{q})^{-2}(\nabla'\tilde{G})_{\xi_k}\sum_{i=1}^{2}(-1)^i\frac{\partial\tilde{\dot{H}}_{\xi_{3-i}}}{\partial\xi_i}\right\}\mathrm{d}\tilde{s}'$$

$$= 2\tilde{\dot{E}}^{\mathrm{inc}}_{\xi_k} + (-1)^k\left(\tilde{F}\left[\tilde{\dot{H}}_{\xi_{3-k}}\right] + \frac{1}{2\pi}\int_{S_c}\tilde{F}\left[\tilde{\dot{H}}_{\xi_{3-k}}\right](\nabla'\tilde{G})_\eta\mathrm{d}\tilde{s}'\right) \quad (7.108)$$

$$-\frac{1}{2\pi}\int\limits_{S_c}\left\{\frac{(-1)^k}{\tilde{R}}\frac{\partial\tilde{H}_{\xi_{3-k}}}{\partial\tilde{t}'}-\tilde{q}^{-2}\frac{\tilde{R}_{\xi_k}}{\tilde{R}}\int\limits_0^t\sum_{i=1}^{2}(-1)^i\frac{\partial}{\partial\tilde{\xi}_i}\tilde{\Pi}\left[\tilde{H}_{\xi_{3-i}}\right]\mathrm{d}\tilde{t}'\right\}\Bigg|_{\tilde{t}'=\tilde{t}-\tilde{R}/\tilde{c}}\mathrm{d}\tilde{s}'$$

$$=2\tilde{E}^{\mathrm{inc}}_{\xi_k}+(-1)^k\left(\frac{\partial}{\partial\tilde{t}}\tilde{F}\left[\tilde{H}_{\xi_{3-k}}\right]+\frac{1}{2\pi}\int\limits_{S_c}\frac{\tilde{R}_\eta}{\tilde{R}}\frac{\partial}{\partial\tilde{t}'}\tilde{F}\left[\tilde{\Pi}\left[\tilde{H}_{\xi_{3-k}}\right]\right]\Bigg|_{\tilde{t}'=\tilde{t}-\tilde{R}/\tilde{c}}\mathrm{d}\tilde{s}'\right)$$

(7.109)

Magnetic field integral equation:

$$\dot{\tilde{H}}_{\xi_{3-k}}+\frac{1}{2\pi}\int\limits_{S_c}\dot{\tilde{H}}_{\xi_{3-k}}(\nabla'\tilde{G})_\eta\mathrm{d}\tilde{s}'=2\dot{\tilde{H}}^{\mathrm{inc}}_{\xi_{3-k}}-\frac{1}{2\pi}\int\limits_{S_c}\left\{-2j\tilde{q}\tilde{F}\left[\dot{\tilde{H}}_{\xi_{3-k}}\right]\tilde{G}+(\nabla'\tilde{G})_{\xi_{3-k}}\sum_{i=1}^{2}\frac{\partial}{\partial\tilde{\xi}_i}\tilde{F}\left[\dot{\tilde{H}}_{\xi_i}\right]\right\}\mathrm{d}\tilde{s}' \quad (7.110)$$

$$\tilde{H}_{\xi_{3-k}}+\frac{1}{2\pi}\int\limits_{S_c}\tilde{\Pi}\left[\tilde{H}_{\xi_{3-k}}\right]\frac{\tilde{R}_\eta}{\tilde{R}}\mathrm{d}\tilde{s}'=2\tilde{H}^{\mathrm{inc}}_{\xi_{3-k}}-\frac{1}{2\pi}\int\limits_{S_c}\left\{\frac{\tilde{q}^2}{\tilde{R}}\frac{\partial^2}{\partial\tilde{t}'^2}\tilde{F}\left[\tilde{H}_{\xi_{3-k}}\right]+\frac{\tilde{R}_{\xi_{3-k}}}{\tilde{R}}\sum_{i=1}^{2}\frac{\partial}{\partial\tilde{\xi}_i}\tilde{F}\left[\tilde{\Pi}\left[\tilde{H}_{\xi_i}\right]\right]\right\}\Bigg|_{\tilde{t}'=\tilde{t}-\tilde{R}/\tilde{c}}\mathrm{d}\tilde{s}' \quad (7.111)$$

For further transformation we represent $\tilde{F}$ using the operators $\tilde{\Lambda}_m$, $m=1,2,3$, as was done in Equations 6.170 through 6.172:

$$\tilde{F}\left[\tilde{f}_{\xi_k}\right]\Big|^{\mathrm{Leontovich}}=\tilde{\Lambda}_1\left[\tilde{f}_{\xi_k}\right] \quad (7.112)$$

$$\tilde{F}\left[\tilde{f}_{\xi_k}\right]\Big|^{\mathrm{Mitzner}}=\tilde{\Lambda}_1\left[\tilde{f}_{\xi_k}\right]+\tilde{p}\tilde{\Lambda}_2\left[\tilde{f}_{\xi_k}\right] \quad (7.113)$$

$$\tilde{F}\left[\tilde{f}_{\xi_k}\right]\Big|^{\mathrm{Rytov}}=\tilde{\Lambda}_1\left[\tilde{f}_{\xi_k}\right]+\tilde{p}\tilde{\Lambda}_2\left[\tilde{f}_{\xi_k}\right]+\tilde{p}^2\tilde{\Lambda}_3\left[\tilde{f}_{\xi_k}\right] \quad (7.114)$$

where the operators $\tilde{\Lambda}_m$ in our high-frequency case are written in the form:

$$\tilde{\Lambda}_1\left[\tilde{f}_{\xi_k}\right]=\begin{cases}\dfrac{1-j}{2}\dfrac{1}{\sqrt{1+2j\tilde{a}}}\dot{\tilde{f}}^{\mathrm{b}}_{\xi_k} & \mathrm{FD}\\ \tilde{f}^{\mathrm{b}}_{\xi_k}*\tilde{T}^{\mathrm{b}}_2 & \mathrm{TD}\end{cases} \quad (7.115)$$

$$\tilde{\Lambda}_2\left[\tilde{f}_{\xi_k}\right]=\begin{cases}\dfrac{j}{2}\dfrac{1}{1+2j\tilde{a}}\dfrac{\tilde{d}_k-\tilde{d}_{3-k}}{2\tilde{d}_k\tilde{d}_{3-k}}\dot{\tilde{f}}^{\mathrm{b}}_{\xi_k} & \mathrm{FD}\\ -\dfrac{\tilde{d}_k-\tilde{d}_{3-k}}{2\tilde{d}_k\tilde{d}_{3-k}}\tilde{f}^{\mathrm{b}}_{\xi_k}*\tilde{T}^{\mathrm{b}}_3 & \mathrm{TD}\end{cases} \quad (7.116)$$

$$\tilde{\Lambda}_3\left[\tilde{f}_{\xi_k}\right] = S\begin{cases} -\frac{1+j}{4}(1+2j\tilde{a})^{-3/2}\left(\dot{\tilde{f}}^{\mathrm{b}}_{\xi_k}\frac{\tilde{d}_k^2+2\tilde{d}_k\tilde{d}_{3-k}-3\tilde{d}_{3-k}^2}{8\tilde{d}_k^2\tilde{d}_{3-k}^2}-\frac{1}{2}\frac{\partial^2\dot{\tilde{f}}^{\mathrm{b}}_{\xi_k}}{\partial\tilde{\xi}_k^2}+\frac{1}{2}\frac{\partial^2\dot{\tilde{f}}^{\mathrm{b}}_{\xi_k}}{\partial\tilde{\xi}_{3-k}^2}-\frac{\partial^2\dot{\tilde{f}}^{\mathrm{b}}_{\xi_{3-k}}}{\partial\tilde{\xi}_k\partial\tilde{\xi}_{3-k}}\right) & \mathrm{FD} \\ -\left(\tilde{f}^{\mathrm{b}}_{\xi_k}\frac{\tilde{d}_k^2+2\tilde{d}_k\tilde{d}_{3-k}-3\tilde{d}_{3-k}^2}{8\tilde{d}_k^2\tilde{d}_{3-k}^2}-\frac{1}{2}\frac{\partial^2\tilde{f}^{\mathrm{b}}_{\xi_k}}{\partial\tilde{\xi}_k^2}+\frac{1}{2}\frac{\partial^2\tilde{f}^{\mathrm{b}}_{\xi_k}}{\partial\tilde{\xi}_{3-k}^2}-\frac{\partial^2\tilde{f}^{\mathrm{b}}_{\xi_{3-k}}}{\partial\tilde{\xi}_k\partial\tilde{\xi}_{3-k}}\right)*\tilde{T}_4^{\mathrm{b}} & \mathrm{TD} \end{cases} \quad (7.117)$$

We now represent the magnetic field (unknown in both the EFIE and the MFIE) as power series in the small parameter $\tilde{p}$:

$$\tilde{H}_{\xi_k} = \sum_{m=0}^{\infty} \tilde{p}^m\left(\vec{\tilde{H}}_{\xi_k}\right)_m, \quad k = 1,2 \quad (7.118)$$

Substituting Equation 7.118 into Equations 7.108 through 7.111, equating coefficients at equal powers of $\tilde{p}$ and taking into account Equations 7.112 through 7.117, the following frequency- and time-domain equations for the terms of expansions are obtained:

Electric field integral equation:

$$-\frac{j}{2\pi}\int\limits_{S_c}\left\{2(-1)^k\left(\dot{\vec{\tilde{H}}}_m\right)_{\xi_{3-k}}\tilde{G}-(2\tilde{q})^{-2}(\nabla'\tilde{G})_{\xi_k}\sum_{i=1}^{2}(-1)^i\frac{\partial\left(\dot{\vec{\tilde{H}}}_m\right)_{\xi_{3-i}}}{\partial\tilde{\xi}_i}\right\}d\tilde{s}'$$
$$= \dot{\tilde{Y}}_m^{\mathrm{EFIE}}, \quad m = 0,1,2,3 \quad (7.119)$$

$$-\frac{1}{2\pi}\int\limits_{S_c}\left\{\frac{(-1)^k}{\tilde{R}}\frac{\partial\left(\vec{\tilde{H}}_m\right)_{\xi_{3-k}}}{\partial\tilde{t}'}-\tilde{q}^{-2}\frac{\tilde{R}_{\xi_k}}{\tilde{R}}\int\limits_0^t\sum_{i=1}^{2}(-1)^i\frac{\partial}{\partial\tilde{\xi}_i}\tilde{\Pi}\left[\left(\vec{\tilde{H}}_m\right)_{\xi_{3-i}}\right]d\tilde{t}'\right\}\Bigg|_{\tilde{t}'=\tilde{t}-\tilde{R}/\tilde{c}} d\tilde{s}'$$
$$= \tilde{Y}_m^{\mathrm{EFIE}}, \quad m = 0,1,2,3 \quad (7.120)$$

Here the right-hand sides $\dot{\tilde{Y}}_m^{\mathrm{EFIE}}$ and $\tilde{Y}_m^{\mathrm{EFIE}}$ are represented in the form:

PEC-limit ($m = 0$):

$$\dot{\tilde{Y}}_0^{\mathrm{EFIE}} = 2\dot{\tilde{E}}_{\xi_k}^{\mathrm{inc}}, \quad k = 1,2 \quad (7.121)$$

$$\tilde{Y}_0^{\mathrm{EFIE}} = 2\tilde{E}_{\xi_k}^{\mathrm{inc}}, \quad k = 1,2 \quad (7.122)$$

Leontovich's approximation ($m = 1$):

$$\dot{\tilde{Y}}_1^{\mathrm{EFIE}} = (-1)^k\left\{\tilde{\Lambda}_1\left[\left(\dot{\vec{\tilde{H}}}_0\right)_{\xi_{3-k}}\right]+\frac{1}{2\pi}\int\limits_{S_c}\tilde{\Lambda}_1\left[\left(\dot{\vec{\tilde{H}}}_0\right)_{\xi_{3-k}}\right](\nabla'\tilde{G})_\eta d\tilde{s}'\right\} \quad (7.123)$$

$$\tilde{Y}_1^{\mathrm{EFIE}} = (-1)^k \left\{ \frac{\partial}{\partial \tilde{t}} \tilde{\Lambda}_1 \left[\left(\vec{\tilde{H}}_0 \right)_{\xi_{3-k}} \right] + \frac{1}{2\pi} \int_{S_c} \frac{\tilde{R}_\eta}{\tilde{R}} \frac{\partial}{\partial \tilde{t}'} \tilde{\Lambda}_1 \left[\tilde{\Pi} \left[\left(\vec{\tilde{H}}_0 \right)_{\xi_{3-k}} \right] \right] \Bigg|_{\tilde{t}'=\tilde{t}-\tilde{R}/\tilde{c}} \mathrm{d}\tilde{s}' \right\} \tag{7.124}$$

Mitzner's approximation ($m = 2$):

$$\tilde{\tilde{Y}}_2^{\mathrm{EFIE}} = (-1)^k \left\{ \tilde{\Lambda}_1 \left[\left(\dot{\vec{\tilde{H}}}_1 \right)_{\xi_{3-k}} \right] + \tilde{\Lambda}_2 \left[\left(\dot{\vec{\tilde{H}}}_0 \right)_{\xi_{3-k}} \right] + \frac{1}{2\pi} \int_{S_c} \left(\tilde{\Lambda}_1 \left[\left(\dot{\vec{\tilde{H}}}_1 \right)_{\xi_{3-k}} \right] \right. \right.$$

$$\left. \left. + \tilde{\Lambda}_2 \left[\left(\dot{\vec{\tilde{H}}}_0 \right)_{\xi_{3-k}} \right] \right) (\nabla' \tilde{G})_\eta \mathrm{d}\tilde{s}' \right\} \tag{7.125}$$

$$\tilde{Y}_2^{\mathrm{EFIE}} = S(-1)^k \left\{ \frac{\partial}{\partial \tilde{t}} \left(\tilde{\Lambda}_1 \left[\left(\vec{\tilde{H}}_1 \right)_{\xi_{3-k}} \right] + \tilde{\Lambda}_2 \left[\left(\vec{\tilde{H}}_0 \right)_{\xi_{3-k}} \right] \right) \right.$$

$$\left. + \frac{1}{2\pi} \int_{S_c} \frac{\tilde{R}_\eta}{\tilde{R}} \frac{\partial}{\partial \tilde{t}'} \left(\tilde{\Lambda}_1 \left[\tilde{\Pi} \left[() \left(\vec{\tilde{H}}_1 \right)_{\xi_{3-k}} \right] \right] + \tilde{\Lambda}_2 \left[\tilde{\Pi} \left[\left(\vec{\tilde{H}}_0 \right)_{\xi_{3-k}} \right] \right] \right) \Bigg|_{\tilde{t}'=\tilde{t}-\tilde{R}/\tilde{c}} \mathrm{d}\tilde{s}' \right\} \tag{7.126}$$

Rytov's approximation ($m = 3$):

$$\tilde{\tilde{Y}}_3^{\mathrm{EFIE}} = (-1)^k \left\{ \tilde{\Lambda}_1 \left[\left(\dot{\vec{\tilde{H}}}_2 \right)_{\xi_{3-k}} \right] + \tilde{\Lambda}_2 \left[\left(\dot{\vec{\tilde{H}}}_1 \right)_{\xi_{3-k}} \right] + \tilde{\Lambda}_3 \left[\left(\dot{\vec{\tilde{H}}}_0 \right)_{\xi_{3-k}} \right] \right.$$

$$\left. + \frac{1}{2\pi} \int_{S_c} \left(\tilde{\Lambda}_1 \left[\left(\dot{\vec{\tilde{H}}}_2 \right)_{\xi_{3-k}} \right] + \tilde{\Lambda}_2 \left[\left(\dot{\vec{\tilde{H}}}_1 \right)_{\xi_{3-k}} \right] + \tilde{\Lambda}_3 \left[\left(\dot{\vec{\tilde{H}}}_0 \right)_{\xi_{3-k}} \right] \right) (\nabla' \tilde{G})_\eta \mathrm{d}\tilde{s}' \right\} \tag{7.127}$$

$$\tilde{Y}_3^{\mathrm{EFIE}} = (-1)^k \left\{ \frac{\partial}{\partial \tilde{t}} \left(\tilde{\Lambda}_1 \left[\left(\vec{\tilde{H}}_2 \right)_{\xi_{3-k}} \right] + \tilde{\Lambda}_2 \left[\left(\vec{\tilde{H}}_1 \right)_{\xi_{3-k}} \right] + \tilde{\Lambda}_3 \left[\left(\vec{\tilde{H}}_0 \right)_{\xi_{3-k}} \right] \right) \right.$$

$$+ \frac{1}{2\pi} \int_{S_c} \frac{\tilde{R}_\eta}{\tilde{R}} \frac{\partial}{\partial \tilde{t}'} \left(\tilde{\Lambda}_1 \left[\tilde{\Pi} \left[\left(\vec{\tilde{H}}_2 \right)_{\xi_{3-k}} \right] \right] + \tilde{\Lambda}_2 \left[\tilde{\Pi} \left[\left(\vec{\tilde{H}}_1 \right)_{\xi_{3-k}} \right] \right] \right.$$

$$\left. \left. + \tilde{\Lambda}_3 \left[\tilde{\Pi} \left[\left(\vec{\tilde{H}}_0 \right)_{\xi_{3-k}} \right] \right] \right) \Bigg|_{\tilde{t}'=\tilde{t}-\tilde{R}/\tilde{c}} \mathrm{d}\tilde{s}' \right\} \tag{7.128}$$

Magnetic field integral equation:

$$\left(\dot{\tilde{\vec{H}}}_m\right)_{\xi_{3-k}} + \frac{1}{2\pi}\int\limits_{S_c}\left(\dot{\tilde{\vec{H}}}_m\right)_{\xi_{3-k}}(\nabla'\tilde{G})_\eta d\tilde{s}' = \dot{\tilde{Y}}_m^{\mathrm{MFIE}}, \quad m = 0,1,2,3 \tag{7.129}$$

$$\left(\tilde{\vec{H}}_m\right)_{\xi_{3-k}} + \frac{1}{2\pi}\int\limits_{S_c}\tilde{\Pi}\left[\left(\tilde{\vec{H}}_m\right)_{\xi_{3-k}}\right]\frac{\tilde{R}_\eta}{\tilde{R}}d\tilde{s}' = \tilde{Y}_m^{\mathrm{MFIE}}, \quad m = 0,1,2,3 \tag{7.130}$$

Here the right-hand sides $\dot{\tilde{Y}}_m^{\mathrm{MFIE}}$ and $\tilde{Y}_m^{\mathrm{MFIE}}$ are represented in the forms:

PEC-limit ($m=0$):

$$\dot{\tilde{Y}}_0^{\mathrm{MFIE}} = 2\dot{\tilde{H}}_{\xi_{3-k}}^{\mathrm{inc}} \tag{7.131}$$

$$\tilde{Y}_0^{\mathrm{MFIE}} = 2\tilde{H}_{\xi_{3-k}}^{\mathrm{inc}} \tag{7.132}$$

Leontovich's approximation ($m=1$):

$$\dot{\tilde{Y}}_1^{\mathrm{MFIE}} = -\frac{1}{2\pi}\int\limits_{S_c}\left\{-2j\tilde{q}\tilde{\Lambda}_1\left[\left(\dot{\tilde{\vec{H}}}_0\right)_{\xi_{3-k}}\right]\tilde{G} + (\nabla'\tilde{G})_{\xi_{3-k}}\sum_{i=1}^{2}\frac{\partial}{\partial\tilde{\xi}_i}\tilde{\Lambda}_1\left[\left(\dot{\tilde{\vec{H}}}_0\right)_{\xi_i}\right]\right\}d\tilde{s}' \tag{7.133}$$

$$\tilde{Y}_1^{\mathrm{MFIE}} = -\frac{1}{2\pi}\int\limits_{S_c}\left\{\frac{\tilde{q}^2}{\tilde{R}}\frac{\partial^2}{\partial\tilde{t}'^2}\tilde{\Lambda}_1\left[\left(\tilde{\vec{H}}_0\right)_{\xi_{3-k}}\right] + \frac{\tilde{R}_{\xi_{3-k}}}{\tilde{R}}\sum_{i=1}^{2}\frac{\partial}{\partial\tilde{\xi}_i}\tilde{\Lambda}_1\left[\tilde{\Pi}\left[\left(\tilde{\vec{H}}_0\right)_{\xi_i}\right]\right]\right\}\Bigg|_{\tilde{t}'=\tilde{t}-\tilde{R}/\tilde{c}} d\tilde{s}' \tag{7.134}$$

Mitzner's approximation ($m=2$):

$$\dot{\tilde{Y}}_2^{\mathrm{MFIE}} = -\frac{1}{2\pi}\int\limits_{S_c}\left\{-2j\tilde{q}\left(\tilde{\Lambda}_1\left[\left(\dot{\tilde{\vec{H}}}_1\right)_{\xi_{3-k}}\right] + \tilde{\Lambda}_2\left[\left(\dot{\tilde{\vec{H}}}_0\right)_{\xi_{3-k}}\right]\right)\tilde{G}\right.$$
$$\left. + (\nabla'\tilde{G})_{\xi_{3-k}}\sum_{i=1}^{2}\frac{\partial}{\partial\tilde{\xi}_i}\left(\tilde{\Lambda}_1\left[\left(\dot{\tilde{\vec{H}}}_1\right)_{\xi_i}\right] + \tilde{\Lambda}_2\left[\left(\dot{\tilde{\vec{H}}}_0\right)_{\xi_i}\right]\right)\right\}d\tilde{s}' \tag{7.135}$$

$$\tilde{Y}_2^{\mathrm{MFIE}} = -\frac{1}{2\pi}\int\limits_{S_c}\left\{\frac{\tilde{q}^2}{\tilde{R}}\frac{\partial^2}{\partial\tilde{t}'^2}\left(\tilde{\Lambda}_1\left[\left(\tilde{\vec{H}}_1\right)_{\xi_{3-k}}\right] + \tilde{\Lambda}_2\left[\left(\tilde{\vec{H}}_0\right)_{\xi_{3-k}}\right]\right)\right.$$
$$\left. + \frac{\tilde{R}_{\xi_{3-k}}}{\tilde{R}}\sum_{i=1}^{2}\frac{\partial}{\partial\tilde{\xi}_i}\left(\tilde{\Lambda}_1\left[\tilde{\Pi}\left[\left(\tilde{\vec{H}}_1\right)_{\xi_i}\right]\right] + \tilde{\Lambda}_2\left[\tilde{\Pi}\left[\left(\tilde{\vec{H}}_0\right)_{\xi_i}\right]\right]\right)\right\}\Bigg|_{\tilde{t}'=\tilde{t}-\tilde{R}/\tilde{c}} d\tilde{s}' \tag{7.136}$$

Rytov's approximation ($m=3$):

$$\tilde{Y}_3^{\mathrm{MFIE}} = -\frac{1}{2\pi}\int\limits_{S_c}\Bigg\{-2j\tilde{q}\left(\tilde{\Lambda}_1\left[\left(\ddot{\tilde{\vec{H}}}_2\right)_{\xi_{3-k}}\right]+\tilde{\Lambda}_2\left[\left(\ddot{\tilde{\vec{H}}}_1\right)_{\xi_{3-k}}\right]+\tilde{\Lambda}_3\left[\left(\ddot{\tilde{\vec{H}}}_0\right)_{\xi_{3-k}}\right]\right)\tilde{G}$$
$$+(\nabla'\tilde{G})_{\xi_{3-k}}\sum_{i=1}^{2}\frac{\partial}{\partial\tilde{\xi}_i}\left(\tilde{\Lambda}_1\left[\left(\dot{\tilde{\vec{H}}}_2\right)_{\xi_i}\right]+\tilde{\Lambda}_2\left[\left(\dot{\tilde{\vec{H}}}_1\right)_{\xi_i}\right]+\tilde{\Lambda}_3\left[\left(\dot{\tilde{\vec{H}}}_0\right)_{\xi_{3-k}}\right]\right)\Bigg\}\mathrm{d}\tilde{s}' \qquad (7.137)$$

$$\tilde{Y}_3^{\mathrm{MFIE}} = -\frac{1}{2\pi}\int\limits_{S_c}\Bigg\{\frac{\tilde{q}^2}{\tilde{R}}\frac{\partial^2}{\partial\tilde{t}'^2}\left(\tilde{\Lambda}_2\left[\left(\tilde{\vec{H}}_1\right)_{\xi_{3-k}}\right]+\tilde{\Lambda}_2\left[\left(\tilde{\vec{H}}_1\right)_{\xi_{3-k}}\right]+\tilde{\Lambda}_3\left[\left(\tilde{\vec{H}}_0\right)_{\xi_{3-k}}\right]\right)$$
$$+\frac{\tilde{R}_{\xi_{3-k}}}{\tilde{R}}\sum_{i=1}^{2}\frac{\partial}{\partial\tilde{\xi}_i}\left(\tilde{\Lambda}_2\left[\tilde{\Pi}\left[\left(\tilde{\vec{H}}_1\right)_{\xi_i}\right]\right]+\tilde{\Lambda}_2\left[\tilde{\Pi}\left[\left(\tilde{\vec{H}}_1\right)_{\xi_i}\right]\right]\right.$$
$$\left.+\tilde{\Lambda}_3\left[\tilde{\Pi}\left[\left(\tilde{\vec{H}}_0\right)_{\xi_i}\right]\right]\right)\Bigg\}\Bigg|_{\tilde{t}'=\tilde{t}-\tilde{R}/\tilde{c}}\mathrm{d}\tilde{s}' \qquad (7.138)$$

It is well worth emphasizing here the following basic advantages of both the EFIE and MFIE formulations:

1. The integral equations for the terms of expansions differ only in the form of the right-hand side and can be solved by the same solution procedures. Therefore, new computational complications do not arise in comparison with solving the problem using the well-known perfect conductor limit ($m=0$).
2. The convolution integrals in the integral equations are on the right-hand sides only and can be calculated and tabulated before numerically solving the appropriate integral equation so that the computer resources required for the computation are greatly reduced as compared with the formulation developed in [5].

Representing the expansions in Equation 7.118 in dimensional form:

$$H = T\sum_{k=0}^{3}\tilde{p}^k\tilde{H}_k = \sum_{k=0}^{3}H_k \qquad (7.139)$$

where

$$H_k = \tilde{p}^k H^*\tilde{H}_k = \tilde{p}^k I D^{-1}\tilde{H}_k \qquad (7.140)$$

we get

$$\tilde{H}_k = \tilde{p}^{-k}(I)^{-1}DH_k = \delta^{-k}(I)^{-1}D^{k+1}H_k \qquad (7.141)$$

Substituting Equation 7.141 and other scale factors into Equations 7.119 through 7.138, we finally obtain the EFIE and MFIE formulations with dimensional variables:

Electric field integral equation:

$$-j\omega\mu_0\frac{1}{2\pi}\int\limits_{S_c}\left\{(-1)^k\left(\vec{H}_m\right)_{\xi_{3-k}}G-\frac{1}{\omega^2\varepsilon_0\mu_0}(\nabla'G)_{\xi_k}\sum_{i=1}^{2}(-1)^i\frac{\partial\left(\vec{H}_m\right)_{\xi_{3-i}}}{\partial\xi_i}\right\}ds'=\dot{Y}_m^{\mathrm{EFIE}} \tag{7.142}$$

$$-\mu_0\frac{1}{2\pi}\int\limits_{S_c}\left\{\frac{(-1)^k}{R}\frac{\partial\left(\vec{H}_m\right)_{\xi_{3-k}}}{\partial t'}-\frac{1}{\varepsilon_0\mu_0}\frac{R_{\xi_k}}{R}\int\limits_0^t\sum_{i=1}^{2}(-1)^i\frac{\partial}{\partial\xi_i}\Pi\left[\left(\vec{H}_m\right)_{\xi_{3-i}}\right]dt'\right\}\Bigg|_{t'=t-R/c}ds'=Y_m^{\mathrm{EFIE}} \tag{7.143}$$

Here the right-hand sides $\dot{Y}_m^{\mathrm{EFIE}}$ and Y_m^{EFIE} are represented in the forms:

PEC-limit ($m=0$):

$$\dot{Y}_0^{\mathrm{EFIE}}=2\dot{E}_{\xi_k}^{\mathrm{inc}},\quad k=1,2 \tag{7.144}$$

$$Y_0^{\mathrm{EFIE}}=2E_{\xi_k}^{\mathrm{inc}},\quad k=1,2 \tag{7.145}$$

Leontovich's approximation ($m=1$):

$$\dot{Y}_1^{\mathrm{EFIE}}=(-1)^k\left\{\Lambda_1\left[\left(\vec{H}_0\right)_{\xi_{3-k}}\right]+\frac{1}{2\pi}\int\limits_{S_c}\Lambda_1\left[\left(\vec{H}_0\right)_{\xi_{3-k}}\right](\nabla'G)_\eta ds'\right\} \tag{7.146}$$

$$Y_1^{\mathrm{EFIE}}=(-1)^k\left\{\frac{\partial}{\partial t}\Lambda_1\left[\left(\vec{H}_0\right)_{\xi_{3-k}}\right]+\frac{1}{2\pi}\int\limits_{S_c}\frac{R_\eta}{R}\frac{\partial}{\partial t'}\Lambda_1\left[\Pi\left[\left(\vec{H}_0\right)_{\xi_{3-k}}\right]\right]\Bigg|_{t'=t-R/c}ds'\right\} \tag{7.147}$$

Mitzner's approximation ($m=2$):

$$\dot{Y}_2^{\mathrm{EFIE}}=(-1)^k\left\{\Lambda_1\left[\left(\vec{H}_1\right)_{\xi_{3-k}}\right]+\Lambda_2\left[\left(\vec{H}_0\right)_{\xi_{3-k}}\right]+\frac{1}{2\pi}\int\limits_{S_c}\left(\Lambda_1\left[\left(\vec{H}_1\right)_{\xi_{3-k}}\right]\right.\right.$$
$$\left.\left.+\Lambda_2\left[\left(\vec{H}_0\right)_{\xi_{3-k}}\right]\right)(\nabla'G)_\eta ds'\right\} \tag{7.148}$$

$$Y_2^{\text{EFIE}} = (-1)^k \left\{ \frac{\partial}{\partial t} \left(\Lambda_1 \left[\left(\vec{H}_1 \right)_{\xi_{3-k}} \right] + \Lambda_2 \left[\left(\vec{H}_0 \right)_{\xi_{3-k}} \right] \right) \right.$$
$$\left. + \frac{1}{2\pi} \int_{S_c} \frac{R_\eta}{R} \frac{\partial}{\partial t'} \left(\Lambda_1 \left[\Pi \left[\left(\vec{H}_1 \right)_{\xi_{3-k}} \right] \right] + \Lambda_2 \left[\Pi \left[\left(\vec{H}_0 \right)_{\xi_{3-k}} \right] \right] \right) \Big|_{t'=t-R/c} ds' \right\} \tag{7.149}$$

Rytov's approximation ($m=3$):

$$\dot{Y}_3^{\text{EFIE}} = (-1)^k \left\{ \Lambda_1 \left[\left(\dot{\vec{H}}_2 \right)_{\xi_{3-k}} \right] + \Lambda_2 \left[\left(\dot{\vec{H}}_1 \right)_{\xi_{3-k}} \right] + \Lambda_3 \left[\left(\dot{\vec{H}}_0 \right)_{\xi_{3-k}} \right] \right.$$
$$\left. + \frac{1}{2\pi} \int_{S_c} \left(\Lambda_1 \left[\left(\dot{\vec{H}}_2 \right)_{\xi_{3-k}} \right] + \Lambda_2 \left[\left(\dot{\vec{H}}_1 \right)_{\xi_{3-k}} \right] + \Lambda_3 \left[\left(\dot{\vec{H}}_0 \right)_{\xi_{3-k}} \right] \right) (\nabla' G)_\eta ds' \right\} \tag{7.150}$$

$$Y_3^{\text{EFIE}} = (-1)^k \left\{ \frac{\partial}{\partial t} \left(\Lambda_1 \left[\left(\vec{H}_2 \right)_{\xi_{3-k}} \right] + \Lambda_2 \left[\left(\vec{H}_1 \right)_{\xi_{3-k}} \right] + \Lambda_3 \left[\left(\vec{H}_0 \right)_{\xi_{3-k}} \right] \right) \right.$$
$$+ \frac{1}{2\pi} \int_{S_c} \frac{R_\eta}{R} \frac{\partial}{\partial t'} + \left(\Lambda_1 \left[\Pi \left[\left(\vec{H}_2 \right)_{\xi_{3-k}} \right] \right] + \Lambda_2 \left[\Pi \left[\left(\vec{H}_1 \right)_{\xi_{3-k}} \right] \right] \right.$$
$$\left. \left. + \Lambda_3 \left[\Pi \left[\left(\vec{H}_0 \right)_{\xi_{3-k}} \right] \right] \right) \Big|_{t'=t-R/c} ds' \right\} \tag{7.151}$$

Magnetic field integral equation:

$$\left(\dot{\vec{H}}_m \right)_{\xi_{3-k}} + \frac{1}{2\pi} \int_{S_c} \left(\dot{\vec{H}}_m \right)_{\xi_{3-k}} (\nabla' G)_\eta ds' = \dot{Y}_m^{\text{MFIE}}, \quad m = 0, 1, 2, 3 \tag{7.152}$$

$$\left(\vec{H}_m \right)_{\xi_{3-k}} + \frac{1}{2\pi} \int_{S_c} \Pi \left[\left(\vec{H}_m \right)_{\xi_{3-k}} \right] \frac{R_\eta}{R} d\tilde{s}' = Y_m^{\text{MFIE}}, \quad m = 0, 1, 2, 3 \tag{7.153}$$

The right-hand sides $\dot{Y}_m^{\text{MFIE}}$ and Y_m^{MFIE} are represented in the forms:

PEC-limit ($m=0$):

$$\dot{Y}_0^{\text{MFIE}} = 2\dot{H}_{\xi_{3-k}}^{\text{inc}} \tag{7.154}$$

$$Y_0^{\text{MFIE}} = 2H_{\xi_{3-k}}^{\text{inc}} \tag{7.155}$$

Leontovich's approximation ($m=1$):

$$\dot{Y}_1^{\text{MFIE}} = -\frac{1}{2\pi}\int_{S_c}\left\{-\omega^2\varepsilon_0\mu_0\Lambda_1\left[\left(\vec{H}_0\right)_{\xi_{3-k}}\right]G + (\nabla' G)_{\xi_{3-k}}\sum_{i=1}^{2}\frac{\partial}{\partial\xi_i}\Lambda_1\left[\left(\vec{H}_0\right)_{\xi_i}\right]\right\}ds' \tag{7.156}$$

$$Y_1^{\text{MFIE}} = -\frac{1}{2\pi}\int_{S_c}\left\{\frac{\varepsilon_0\mu_0}{R}\frac{\partial^2}{\partial t'^2}\Lambda_1\left[\left(\vec{H}_0\right)_{\xi_{3-k}}\right] + \frac{R_{\xi_{3-k}}}{R}\sum_{i=1}^{2}\frac{\partial}{\partial\xi_i}\Lambda_1\left[\Pi\left[\left(\vec{H}_0\right)_{\xi_i}\right]\right]\right\}\Bigg|_{t'=t-R/c} ds' \tag{7.157}$$

Mitzner's approximation ($m=2$):

$$\begin{aligned}\dot{Y}_2^{\text{MFIE}} = -\frac{1}{2\pi}\int_{S_c}&\left\{-\omega^2\varepsilon_0\mu_0\left(\Lambda_1\left[\left(\vec{H}_1\right)_{\xi_{3-k}}\right] + \Lambda_2\left[\left(\vec{H}_0\right)_{\xi_{3-k}}\right]\right)G\right.\\ &\left. + (\nabla' G)_{\xi_{3-k}}\sum_{i=1}^{2}\frac{\partial}{\partial\xi_i}\left(\Lambda_1\left[\left(\vec{H}_1\right)_{\xi_{3-k}}\right] + \Lambda_2\left[\left(\vec{H}_0\right)_{\xi_{3-k}}\right]\right)\right\}ds'\end{aligned} \tag{7.158}$$

$$\begin{aligned}Y_2^{\text{MFIE}} = -\frac{1}{2\pi}\int_{S_c}&\left\{\frac{\varepsilon_0\mu_0}{R}\frac{\partial^2}{\partial t'^2}\left(\Lambda_1\left[\left(\vec{H}_1\right)_{\xi_{3-k}}\right] + \Lambda_2\left[\left(\vec{H}_0\right)_{\xi_{3-k}}\right]\right)\right.\\ &\left. + \frac{R_{\xi_{3-k}}}{R}\sum_{i=1}^{2}\frac{\partial}{\partial\xi_i}\left(\Lambda_1\left[\Pi\left[\left(\vec{H}_1\right)_{\xi_i}\right]\right] + \Lambda_2\left[\Pi\left[\left(\vec{H}_0\right)_{\xi_i}\right]\right]\right)\right\}\Bigg|_{t'=t-R/c} ds'\end{aligned} \tag{7.159}$$

Rytov's approximation ($m=3$):

$$\begin{aligned}\dot{Y}_3^{\text{MFIE}} = -\frac{1}{2\pi}\int_{S_c}&\left\{-\omega^2\varepsilon_0\mu_0\left(\Lambda_1\left[\left(\vec{H}_2\right)_{\xi_{3-k}}\right] + \Lambda_2\left[\left(\vec{H}_1\right)_{\xi_{3-k}}\right] + \Lambda_3\left[\left(\vec{H}_0\right)_{\xi_{3-k}}\right]\right)G\right.\\ &\left. + (\nabla' G)_{\xi_{3-k}}\sum_{i=1}^{2}\frac{\partial}{\partial\xi_i}\left(\Lambda_1\left[\left(\vec{H}_2\right)_{\xi_{3-k}}\right] + \Lambda_2\left[\left(\vec{H}_1\right)_{\xi_{3-k}}\right] + \Lambda_3\left[\left(\vec{H}_0\right)_{\xi_{3-k}}\right]\right)\right\}ds'\end{aligned} \tag{7.160}$$

$$Y_3^{\text{MFIE}} = -\frac{1}{2\pi}\int\limits_{S_c}\left\{\frac{\varepsilon_0\mu_0}{R}\frac{\partial^2}{\partial t'^2}\left(\Lambda_1\left[\left(\vec{H}_2\right)_{\xi_{3-k}}\right] + \Lambda_2\left[\left(\vec{H}_1\right)_{\xi_{3-k}}\right] + \Lambda_3\left[\left(\vec{H}_0\right)_{\xi_{3-k}}\right]\right)\right.$$
$$+\frac{R_{\xi_{3-k}}}{R}\sum_{i=1}^{2}\frac{\partial}{\partial\xi_i}\left(\Lambda_1\left[\Pi\left[\left(\vec{H}_2\right)_{\xi_i}\right]\right] + \Lambda_2\left[\Pi\left[\left(\vec{H}_1\right)_{\xi_i}\right]\right]\right.$$
$$\left.\left.+\Lambda_3\left[\Pi\left[\left(\vec{H}_0\right)_{\xi_i}\right]\right]\right)\right\}\Bigg|_{t'=t-R/c} \mathrm{d}s' \qquad (7.161)$$

where

$$\Lambda_1\left[f_{\xi_k}\right] = \begin{cases} (j\omega\sigma\mu)^{-1/2}(1+j\omega\varepsilon/\sigma)^{-1/2}\dot{f}_{\xi_k} & \text{FD} \\ (\sigma\mu)^{-1/2}f_{\xi_k} * T_2^{\text{b}} & \text{TD} \end{cases} \qquad (7.162)$$

$$\Lambda_2\left[f_{\xi_k}\right] = \begin{cases} -(j\omega\sigma\mu)^{-1}(1+j\omega\varepsilon/\sigma)^{-1}\dfrac{d_k-d_{3-k}}{2d_kd_{3-k}}\dot{f}_{\xi_k} & \text{FD} \\ -(\sigma\mu)^{-1}\dfrac{d_k-d_{3-k}}{2d_kd_{3-k}}f_{\xi_k} * T_3^{\text{b}} & \text{TD} \end{cases} \qquad (7.163)$$

$$\Lambda_3\left[f_{\xi_k}\right] = \begin{cases} -(j\omega\sigma\mu)^{-3/2}\left(1+\dfrac{j\omega\varepsilon}{\sigma}\right)^{-3/2}\left(\dot{f}_{\xi_k}\dfrac{d_k^2+2d_kd_{3-k}-3d_{3-k}^2}{8d_k^2d_{3-k}^2} - \dfrac{1}{2}\dfrac{\partial^2\dot{f}_{\xi_k}}{\partial\xi_k^2}\right. \\ \left.\qquad +\dfrac{1}{2}\dfrac{\partial^2\dot{f}_{\xi_k}}{\partial\xi_{3-k}^2} - \dfrac{\partial^2\dot{f}_{\xi_{3-k}}}{\partial\xi_k\partial\xi_{3-k}}\right) \quad \text{FD} \\ -(\sigma\mu)^{-3/2}\left(f_{\xi_k}\dfrac{d_k^2+2d_kd_{3-k}-3d_{3-k}^2}{8d_k^2d_{3-k}^2} - \dfrac{1}{2}\dfrac{\partial^2 f_{\xi_k}}{\partial\xi_k^2}\right. \\ \left.\qquad +\dfrac{1}{2}\dfrac{\partial^2 f_{\xi_k}}{\partial\xi_{3-k}^2} - \dfrac{\partial^2 f_{\xi_{3-k}}}{\partial\xi_k\partial\xi_{3-k}}\right) * T_4^{\text{b}} \quad \text{TD} \end{cases}$$

(7.164)

The principal difference between the practical implementations of time-domain formulations for low- and high-frequency problems is that in the latter case, separation of variables into space and time components is impossible. Thus the time convolution integrals cannot be precomputed and tabulated in advance. Direct computation of the convolution integral in the time-domain integral equations is impractical due to the large computational time and storage requirements [6]. Techniques for efficient implementation of the time convolution integrals in Equations 7.144 through 7.164 are considered in Appendix 7.A.1.

7.6 Numerical Example

To illustrate the foregoing theory, we consider the problem of transient scattering from an infinitely long straight cylinder of circular cross section illuminated by an incident Gaussian pulse. In this case, the coordinate ξ_1 is directed along the cylinder as shown in Figure 7.1 so that $d_1 = \infty$. Since the radius d of the cylinder cross section is constant, $d_2 = D$ and $\tilde{d}_2 = 1$. In the discussion of the numerical solution of the integral equations obtained above, we restrict ourselves by considering the transverse electric (TE) case. Under these conditions, only the circumferential ξ_2-component of $\vec{J}^s$ is nonzero and the single-component vector can be treated as a scalar.

Representing the equivalent surface current as an asymptotic expansion in the small parameter $\tilde{p}$:

$$\tilde{J}^s = \sum_{m=0}^{\infty} \tilde{p}\tilde{J}^s_m \tag{7.165}$$

we write the following time-domain MFIEs for the terms of expansion using Equations 7.130, 7.132, 7.134, 7.136, and 7.138:

$$\tilde{J}^s_m + \frac{1}{2\pi}\int_{S_c} \tilde{\Pi}\left[\tilde{J}^s_m\right]\frac{\tilde{R}_\eta}{\tilde{R}}\Bigg|_{\tilde{t}'=\tilde{t}-\tilde{q}\tilde{R}} = \tilde{Y}^{\text{MFIE}}_m, \quad m = 0, 1, 2, 3 \tag{7.166}$$

$$\tilde{Y}^{\text{MFIE}}_0 = 2\tilde{H}^{\text{inc}}_{\xi_2} \tag{7.167}$$

$$\tilde{Y}^{\text{MFIE}}_1 = -\frac{1}{2\pi}\int_{S_c} \frac{\tilde{q}^2}{\tilde{R}}\frac{\partial^2}{\partial \tilde{t}'^2}\left(\tilde{J}^s_0 * \tilde{T}^b_2\right)\Bigg|_{\tilde{t}'=\tilde{t}-\tilde{q}\tilde{R}} d\tilde{s}' - \frac{1}{2\pi}\frac{\partial}{\partial \xi_2}\int_{S_c} \tilde{\Pi}\left[\tilde{J}^s_0 * \tilde{T}^b_2\right]\frac{\tilde{R}_{\xi_2}}{\tilde{R}}\Bigg|_{\tilde{t}'=\tilde{t}-\tilde{q}\tilde{R}} d\tilde{s}' \tag{7.168}$$

$$\begin{aligned}\tilde{Y}^{\text{MFIE}}_2 = &-\frac{1}{2\pi}\int_{S_c} \frac{\tilde{q}^2}{\tilde{R}}\frac{\partial^2}{\partial \tilde{t}'^2}\left\{\tilde{J}^s_1 * \tilde{T}^b_2 + \frac{1}{2\tilde{d}_2}\tilde{J}^s_0 * \tilde{T}^b_3\right\}\Bigg|_{\tilde{t}'=\tilde{t}-\tilde{q}\tilde{R}} d\tilde{s}' \\ &- \frac{1}{2\pi}\frac{\partial}{\partial \xi_2}\int_{S_c} \left\{\tilde{\Pi}\left[\tilde{J}^s_1 * \tilde{T}^b_2\right] + \frac{1}{2\tilde{d}_2}\tilde{\Pi}\left[\tilde{J}^s_0 * \tilde{T}^b_3\right]\right\}\frac{\tilde{R}_{\xi_2}}{\tilde{R}}\Bigg|_{\tilde{t}'=\tilde{t}-\tilde{q}\tilde{R}} d\tilde{s}'\end{aligned} \tag{7.169}$$

$$\begin{aligned}\tilde{Y}^{\text{MFIE}}_3 = &-\frac{1}{2\pi}\int_{S_c} \frac{\tilde{q}^2}{\tilde{R}}\frac{\partial^2}{\partial \tilde{t}'^2}\left\{\tilde{J}^s_2 * \tilde{T}^b_2 + \frac{1}{2\tilde{d}_2}\tilde{J}^s_1 * \tilde{T}^b_3 + \frac{3}{8\tilde{d}_2^2}\tilde{J}^s_0 * \tilde{T}^b_4 + \frac{1}{2}\frac{\partial^2 \tilde{J}^s_0}{\partial \xi_2^2} * \tilde{T}^b_4\right\}\Bigg|_{\tilde{t}'=\tilde{t}-\tilde{q}\tilde{R}} d\tilde{s}' \\ &- \frac{1}{2\pi}\frac{\partial}{\partial \xi_2}\int_{S_c} \left\{\tilde{\Pi}\left[\tilde{J}^s_2 * \tilde{T}^b_2\right] + \frac{1}{2\tilde{d}_2}\tilde{\Pi}\left[\tilde{J}^s_1 * \tilde{T}^b_3\right] + \frac{3}{8\tilde{d}_2^2}\tilde{\Pi}\left[\tilde{J}^s_0 * \tilde{T}^b_4\right]\right. \\ &\left. + \frac{1}{2}\frac{\partial^2}{\partial \xi_2^2}\tilde{\Pi}\left[\tilde{J}^s_0 * \tilde{T}^b_4\right]\right\}\frac{\tilde{R}_{\xi_2}}{\tilde{R}}\Bigg|_{\tilde{t}'=\tilde{t}-\tilde{q}\tilde{R}} d\tilde{s}'\end{aligned} \tag{7.170}$$

From Equation 7.63 and Duhamel's theorem it follows that

$$\frac{\partial^2}{\partial t^2}\left(T_m^{\mathrm{b}}(t)*f(\vec{r},t)\right)=\frac{\mathrm{d}^2T_m^{\mathrm{b}}(t)}{\mathrm{d}t^2}*f(\vec{r},t)+\left.\frac{\mathrm{d}T_m^{\mathrm{b}}(t)}{\mathrm{d}t}\right|_{t=0}f(\vec{r},t)+T_m^{\mathrm{b}}(0)\frac{\partial f(\vec{r},t)}{\partial t} \tag{7.171}$$

Using Equations 7.57 through 7.62, we obtain

$$\frac{\mathrm{d}^2\tilde{T}_2^{\mathrm{b}}}{\mathrm{d}\tilde{t}^2}=\frac{1}{2\tilde{a}^{3/2}}\left\{\frac{1}{\tilde{a}}I_0\left(\frac{\tilde{t}}{2\tilde{a}}\right)-\frac{\tilde{a}+\tilde{t}}{\tilde{a}\tilde{t}}I_1\left(\frac{\tilde{t}}{2\tilde{a}}\right)\right\}\exp\left(-\frac{\tilde{t}}{2\tilde{a}}\right) \tag{7.172}$$

$$\frac{\mathrm{d}^2\tilde{T}_3^{\mathrm{b}}}{\mathrm{d}\tilde{t}^2}=-\frac{1}{\tilde{a}^2}\exp\left(-\frac{\tilde{t}}{\tilde{a}}\right) \tag{7.173}$$

$$\frac{\mathrm{d}^2\tilde{T}_4^{\mathrm{b}}}{\mathrm{d}\tilde{t}^2}=\frac{\tilde{a}-\tilde{t}}{\tilde{a}^{5/2}}I_0\left(\frac{\tilde{t}}{2\tilde{a}}\right)\exp\left(-\frac{\tilde{t}}{2\tilde{a}}\right) \tag{7.174}$$

Equations 7.166 through 7.170 were solved using the marching-on-time technique described in [2]. The scale factor $J_{\max}^{\mathrm{inc}}D$ was used for I where $J_{\max}^{\mathrm{inc}}$ is the maximum of J^{inc}. The width of the incident pulse was taken to equal the radius of the cylinder, i.e., $\tilde{q}=1$. The time $\tilde{t}=0$ corresponds to the time that the pulse reaches the body.

From the calculations it follows that two pulses, one on each side of the cylinder, travel while attenuating in amplitude, from the illuminated region to the shadow region of the scatterer [6]. Figure 7.2 shows the distribution

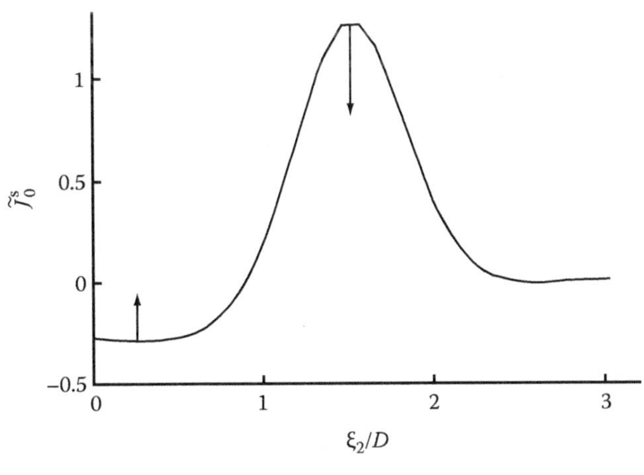

FIGURE 7.2
Distribution of the zero-order term $\tilde{J}_0^s$ of the expansion of the surface electric current over half the cross section of the cylinder.

of the function $\tilde{J}_0^s = \tilde{J}_0^s(\tilde{\xi}_2)$ obtained at time $\tilde{t} = 1$. The contour coordinate $\tilde{\xi}_2$ has its origin at point A and proceeds around the contour ending at point C (Figure 7.1). The function $\tilde{J}_0^s$ is the solution of the problem for a perfect conductor and cannot take into account the redistribution of the current induced by the field diffusion into the body. The effect of this process is to smooth out nonuniformities in the surface current distribution along the surface of the body. The surface current acts in the directions shown by arrows in Figure 7.2. In our model, this redistribution is described by the next terms of the expansions of the surface electric current. The distribution of the first-order term $\tilde{J}_1^s$ is shown in Figure 7.3 together with the function $\tilde{J}_0^s$. From Figure 7.3 it follows that the surface electric current will increase in the regions where the function $\tilde{J}_1^s$ is positive and decrease where the function $\tilde{J}_1^s$ is negative. The second-order term $\tilde{J}_2^s$ gives a further correction by taking into account the radius of curvature of the surface of the body. Figure 7.4 demonstrates the distributions of the terms $\tilde{J}_0^s$, $\tilde{J}_1^s$, and $\tilde{J}_2^s$. It is seen from Figure 7.4 that each of these terms tends to flatten the peak in the distribution of the previous term. The coefficients $\tilde{J}_1^s$ and $\tilde{J}_2^s$ demonstrate the redistribution of the field in the direction normal to the body's surface whereas the third-order coefficient $\tilde{J}_3^s$ reflects the redistribution of the field *along* the surface of the body. The function $\tilde{J}_3^s$ is shown in Figure 7.5.

Figures 7.6 through 7.8 show the time-distributions of the terms $\tilde{J}_0^s$, $\tilde{J}_1^s$, $\tilde{J}_2^s$, and $\tilde{J}_3^s$ at points A, B, and C, respectively. The final time-distributions of the surface electric current obtained in the PEC-approximation ($\tilde{J}^s = \tilde{J}_0^s$), in the Leontovich approximation ($\tilde{J}^s = \tilde{J}_0^s + \tilde{p}\tilde{J}_1^s$), and in the Rytov approximation ($\tilde{J}^s = \tilde{J}_0^s + \tilde{p}\tilde{J}_1^s + \tilde{p}^2\tilde{J}_2^s + \tilde{p}^3\tilde{J}_3^s$) at points A, B, and C are given in Figures 7.9 through 7.11, respectively. The parameter $\tilde{p}$ equals 1/2. From these figures

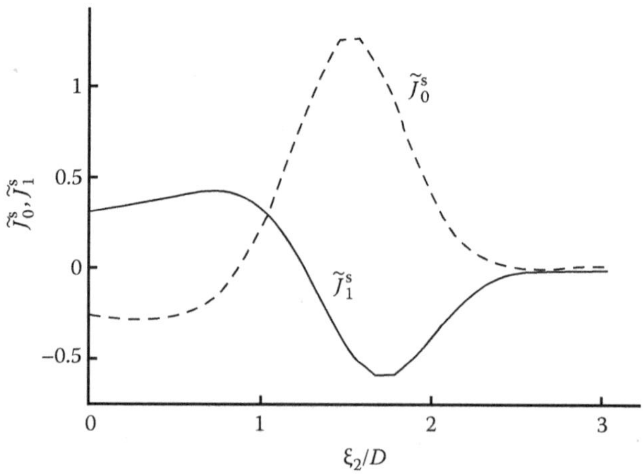

FIGURE 7.3
Distributions of the first-order term $\tilde{J}_1^s$ (solid line) and zero-order term $\tilde{J}_0^s$ (dashed line) of the expansion of the surface electric current over half the cross section of the cylinder.

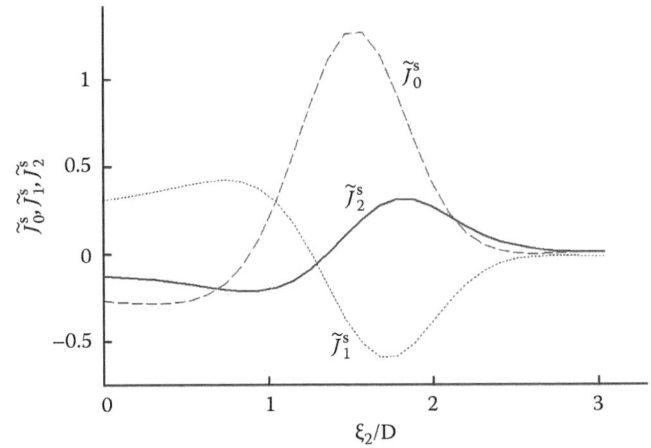

FIGURE 7.4
Distributions of the terms $\tilde{J}_2^s$ (solid line), $\tilde{J}_1^s$ (dotted line), and $\tilde{J}_0^s$ (dashed line) of the expansion of the surface electric current over half the cross section of the cylinder.

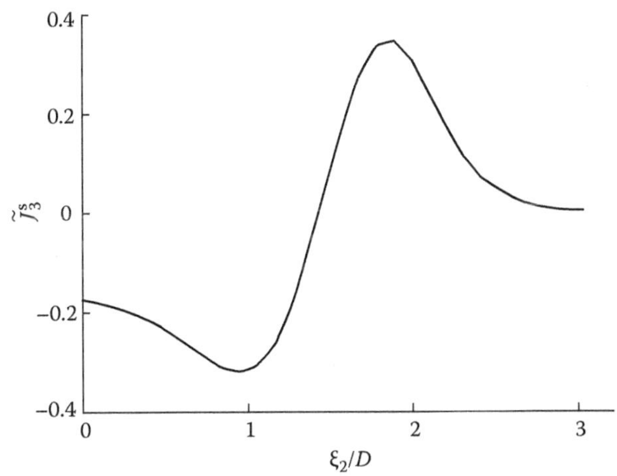

FIGURE 7.5
Distribution of the third-order term $\tilde{J}_3^s$ of the expansion of the surface electric current over half the cross section of the cylinder.

it follows that the proposed time-domain SIBC of the high order of approximation allows the accuracy of the calculations to be about 10%–20% higher compared with solving the problem using standard time-domain SIBC of the order of the Leontovich approximation.

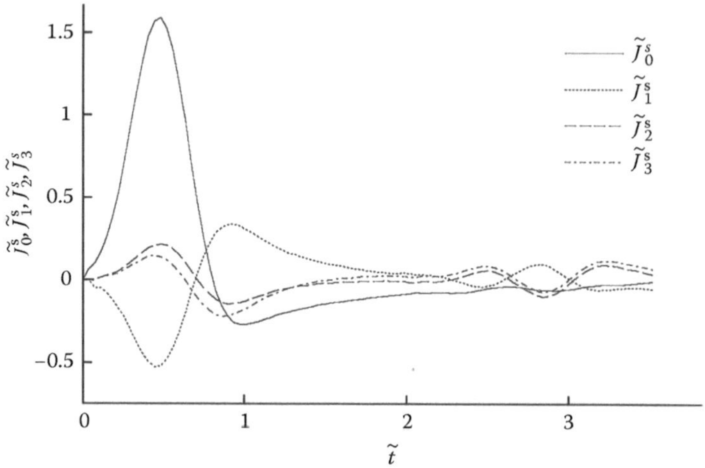

FIGURE 7.6
Time-distributions of the coefficients of expansion of the surface electric current at point *A* at the contour of the cylinder cross section.

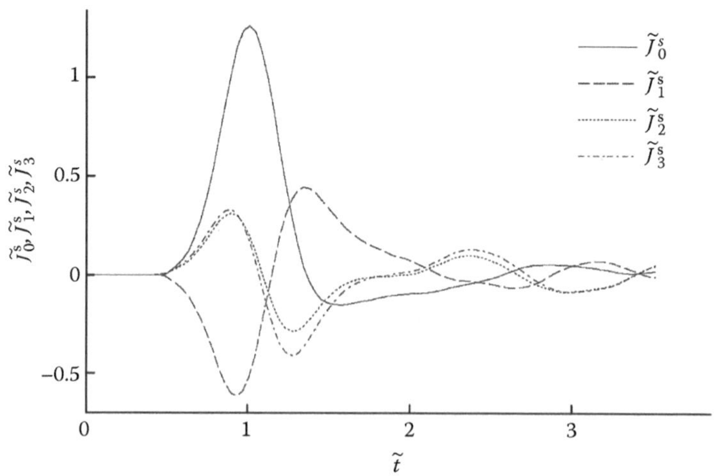

FIGURE 7.7
Time-distributions of the coefficients of expansion of the surface electric current at point *B* at the contour of the cylinder cross section.

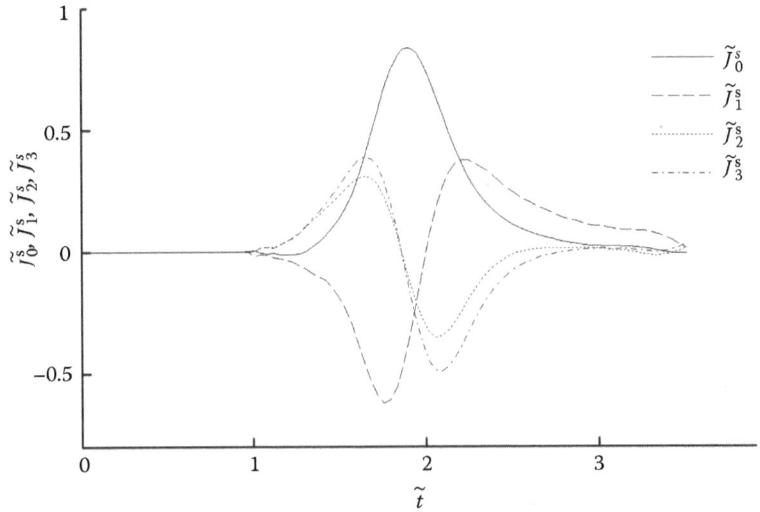

FIGURE 7.8
Time-distributions of the coefficients of expansion of the surface electric current at point C on the contour of the cylinder cross section.

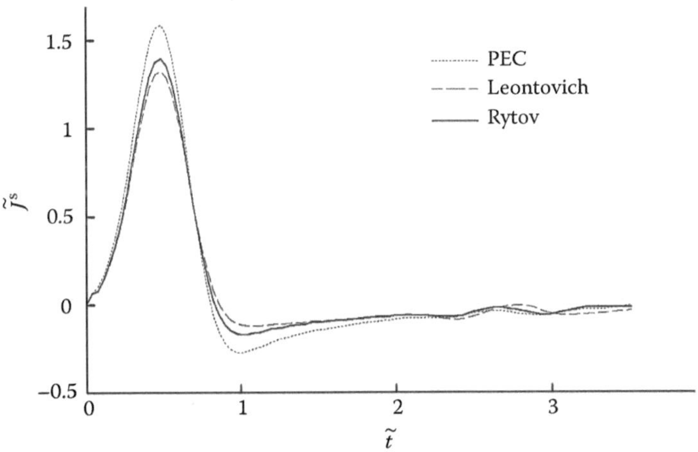

FIGURE 7.9
Time-distributions of the surface electric current, calculated in the perfect conductor limit, in the Leontovich approximation and Rytov's approximation, at point *A* on the contour of the cylinder cross section.

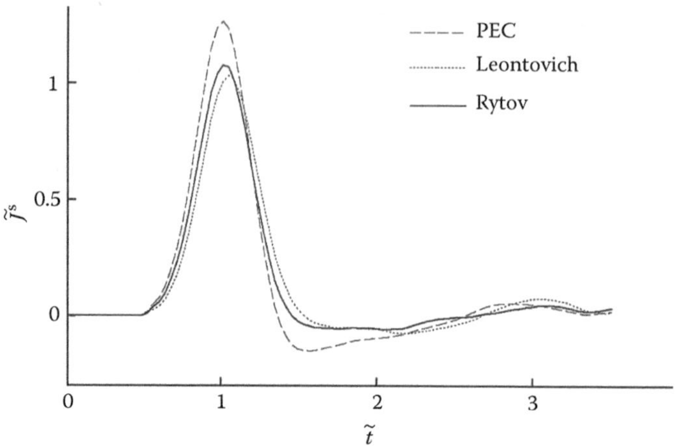

FIGURE 7.10
Time-distributions of the surface electric current, calculated in the perfect conductor limit, in the Leontovich approximation and Rytov's approximation, at point *B* on the contour of the cylinder cross section.

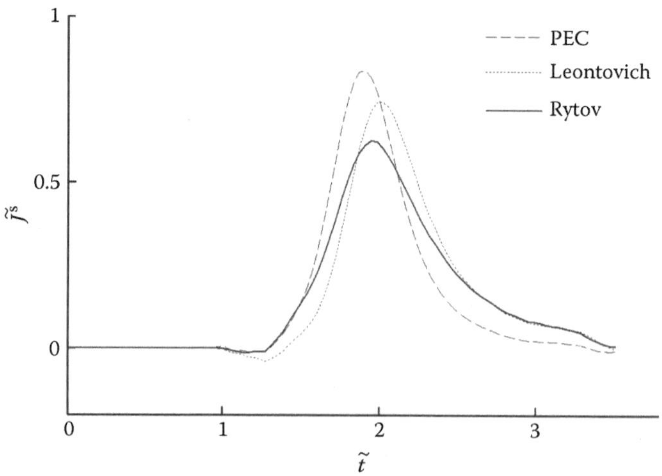

FIGURE 7.11
Time-distributions of the surface electric current, calculated in the perfect conductor limit, in the Leontovich approximation and Rytov's approximation, at point *C* on the contour of the cylinder cross section.

Appendix 7.A.1: Efficient Evaluation of Time Convolution Integrals

Consider again the problem of two-dimensional transient scattering from an infinitely long straight cylinder of circular cross section shown in Figure 7.1. In this case the time-domain SIBC (Equation 7.52) relating the tangential electric and magnetic fields can be written in the form (details are given in Section 7.6):

$$E_{\xi_1} = \hat{T}_2^{\mathrm{b}} * H_{\xi_2} - \hat{T}_3^{\mathrm{b}} * \frac{H_{\xi_2}}{d_2} - \hat{T}_4^{\mathrm{b}} * \left(\frac{H_{\xi_2}}{8d_2^2} + \frac{1}{2}\frac{\partial^2 H_{\xi_2}}{\partial \xi_2^2} \right) \tag{7.A.1}$$

where

$$\hat{T}_2^{\mathrm{b}}(t) = \left(\frac{\mu}{\varepsilon}\right)^{1/2} \left\{ \delta(t) + \frac{\sigma}{2\varepsilon}\left[I_1\left(\frac{\sigma t}{2\varepsilon}\right) - I_0\left(\frac{\sigma t}{2\varepsilon}\right)\right] \exp\left(-\frac{\sigma t}{2\varepsilon}\right)\right\} \tag{7.A.2}$$

$$\hat{T}_3^{\mathrm{b}}(t) = \varepsilon^{-1} \exp\left(-\frac{\sigma t}{\varepsilon}\right) \tag{7.A.3}$$

$$\hat{T}_4^{\mathrm{b}}(t) = t(\varepsilon\sqrt{\varepsilon\mu})^{-1}\left[I_0\left(\frac{\sigma t}{2\varepsilon}\right) - I_1\left(\frac{\sigma t}{2\varepsilon}\right)\right] \exp\left(-\frac{\sigma t}{2\varepsilon}\right) \tag{7.A.4}$$

Let us set $E = E_{\xi_1}$ and $H = H_{\xi_2}$ to simplify further derivations. Consider the implementation of the second term on the right-hand side.

$$\hat{T}_3^{\mathrm{b}} * H = \int_0^t H(t')\hat{T}_3^{\mathrm{b}}(t-t')\mathrm{d}t' = \varepsilon^{-1}\int_0^t H(t')\exp\left(-\frac{\sigma}{\varepsilon}(t-t')\right)\mathrm{d}t' \tag{7.A.5}$$

Assuming $H(t)$ is piecewise linear in time, Equation 7.A.5 in the discrete time domain with a time step Δt can be written as

$$\hat{T}_3^{\mathrm{b}} * H|_{n\Delta t} = \varepsilon^{-1}\sum_{l=0}^{n} \int_{(l-1)\Delta t}^{l\Delta t} \exp\left(-\frac{\sigma}{\varepsilon}(n\Delta t - t')\right)$$
$$\times \left[H((l-1)\Delta t) + (t' - (l-1)\Delta t)\frac{H(l\Delta t) - H((l-1)\Delta t)}{\Delta t}\right]\mathrm{d}t' \tag{7.A.6}$$

After integrating over t'

$$\hat{T}_3^b * H|_{n\Delta t} = \sigma^{-1} \sum_{l=0}^{n} \int_{(l-1)\Delta t}^{l\Delta t} \exp\left(-\frac{\sigma}{\varepsilon}(n\Delta t - t')\right) \times \left\{ H(l\Delta t)\left[1 + \frac{\varepsilon}{\sigma \Delta t}\left(\exp\left(-\frac{\sigma}{\varepsilon}\Delta t\right) - 1\right)\right] + H((l-1)\Delta t)\left[\frac{\varepsilon}{\sigma\Delta t} - \exp\left(-\frac{\sigma}{\varepsilon}\Delta t\right)\left(1 + \frac{\varepsilon}{\sigma \Delta t}\right)\right]\right\} \tag{7.A.7}$$

Now, a recursive form can be obtained by taking the nth term out of the summation and rewriting the remaining summation in terms of $\hat{T}_3^b * H|_{(n-1)\Delta t}$

$$\hat{T}_3^b * H|_{n\Delta t} = \hat{T}_3^b * H|_{(n-1)\Delta t} \exp\left(-\frac{\sigma}{\varepsilon}\Delta t\right) + \sigma^{-1}\left\{ H(n\Delta t)\left[1 + \frac{\varepsilon}{\sigma \Delta t}\left(\exp\left(-\frac{\sigma}{\varepsilon}\Delta t\right) - 1\right)\right] + H((n-1)\Delta t)\left[\frac{\varepsilon}{\sigma\Delta t} - \exp\left(-\frac{\sigma}{\varepsilon}\Delta t\right)\left(1 + \frac{\varepsilon}{\sigma \Delta t}\right)\right]\right\} \tag{7.A.8}$$

Equation 7.A.8 can be represented in another form:

$$\hat{T}_3^b * H|_{n\Delta t} = P_1 H(n\Delta t) + P_2 H((n-1)\Delta t) + P_3 \hat{T}_3^b * H|_{(n-1)\Delta t} \tag{7.A.9}$$

where

$$\begin{aligned} P_1 &= \sigma^{-1}\left[1 + \frac{\varepsilon}{\sigma\Delta t}\left(\exp\left(-\frac{\sigma}{\varepsilon}\Delta t\right) - 1\right)\right] \\ P_2 &= \sigma^{-1}\left[\frac{\varepsilon}{\sigma\Delta t} - \exp\left(-\frac{\sigma}{\varepsilon}\Delta t\right)\left(1 + \frac{\varepsilon}{\sigma\Delta t}\right)\right] \\ P_3 &= \exp\left(-\frac{\sigma}{\varepsilon}\Delta t\right) \end{aligned} \tag{7.A.10}$$

Therefore, if the surface impedance function is represented by a decaying exponential, the convolution summation can be recursively updated, avoiding the need for a complete time history of the field components. This provides an approach to implement the terms $\hat{T}_2^b * H|_{(n-1)\Delta t}$ and $\hat{T}_4^b * H|_{(n-1)\Delta t}$, namely approximate the functions $\hat{T}_2^b$ and $\hat{T}_4^b$ by a series of exponentials which in turns allows efficient evaluation of the convolution integral using recursion:

$$\hat{T}(t) \approx \sum_{i=1}^{Q^{(m)}} c_i^{(m)} \exp\left(\omega_i^{(m)} t\right), \quad m = 2, 4 \tag{7.A.11}$$

where $Q^{(m)}$ is the number of terms in the approximation of the function $\hat{T}_m^b$. One of the most accurate methods for obtaining an exponential approximation to an exact function or to a data set is Prony's method [7]. It has been used to extrapolate transmission line matrix (TLM) and finite-difference time-domain (FDTD) computed waveforms for microwave circuits by a number of investigators [8–10]. Implementation of time-domain SIBCs using Prony's method was also considered in [11–13]. In describing Prony's method we follow [12]. The method fits an exponential approximation of the form:

$$f(N) \approx \sum_{i=1}^{Q} c_i(\mu_i)^N \tag{7.A.12}$$

to a function $f(x)$ by sampling at $N_{\max}$ equally spaced points, $N = 0, 1, 2, \ldots, N_{\max} - 1$. This is a two-step procedure. The first step is to evaluate the values of μ. The set of equations:

$$\begin{array}{l} f_{Q+1} + f_Q\alpha_1 + f_{Q-1}\alpha_2 + \cdots + f_1\alpha_Q = 0 \\ \cdot \quad \cdot \quad \cdot \quad \cdot \quad \cdot \quad \cdot \\ \cdot \quad \cdot \quad \cdot \quad \cdot \quad \cdot \quad \cdot \\ \cdot \quad \cdot \quad \cdot \quad \cdot \quad \cdot \quad \cdot \\ f_{N_{\max}-1} + f_{N_{\max}-2}\alpha_1 + f_{N_{\max}-3}\alpha_2 + \cdots + f_Q\alpha_Q = 0 \end{array} \tag{7.A.13}$$

is solved using a least-square error algorithm for the coefficients α. The values of μ are then found as the roots of the polynomial:

$$\mu^Q + \alpha_1\mu^{Q-1} + \alpha_2\mu^{Q-2} + \cdots + \alpha_{Q-1}\mu + \alpha_Q = 0 \tag{7.A.14}$$

The second step is to find the coefficients c. The set of equations:

$$\begin{array}{l} c_1(\mu_1)^2 + c_2(\mu_2)^2 + \cdots + c_Q(\mu_Q)^2 = f_2 \\ \cdot \quad \cdot \quad \cdot \quad \cdot \quad \cdot \\ \cdot \quad \cdot \quad \cdot \quad \cdot \quad \cdot \\ \cdot \quad \cdot \quad \cdot \quad \cdot \quad \cdot \\ c_1(\mu_1)^{N_{\max}-1} + c_2(\mu_2)^{N_{\max}-1} + \cdots + c_Q(\mu_Q)^{N_{\max}-1} = f_{N_{\max}-1} \end{array} \tag{7.A.15}$$

is solved for the coefficients c using a least-square algorithm.

The functions $\hat{T}_2^b$ and $\hat{T}_4^b$ defined in Equations 7.A.2 and 7.A.4, respectively, depend on the medium properties and must be reapproximated every time the parameters of the medium are changed. We normalize these functions by introducing the variable:

$$\vartheta = \frac{\sigma}{2\varepsilon}t \tag{7.A.16}$$

So that

$$\hat{T}_2^{\text{norm}}(\vartheta) = (I_1(\vartheta) - I_0(\vartheta))\exp(-\vartheta) \tag{7.A.17}$$

$$\hat{T}_4^{\text{norm}}(\vartheta) = \vartheta(I_0(\vartheta) - I_1(\vartheta))\exp(-\vartheta) \tag{7.A.18}$$

The functions $\hat{T}_2^{\text{norm}}$ and $\hat{T}_4^{\text{norm}}$ are related with $\hat{T}_2^{\text{b}}$ and $\hat{T}_4^{\text{b}}$, respectively, as follows:

$$\hat{T}_2^{\text{b}}(t) = \left(\frac{\mu}{\varepsilon}\right)^{1/2}\left\{\delta(t) + \frac{\sigma}{2\varepsilon}\hat{T}_2^{\text{norm}}\left(\vartheta\frac{2\varepsilon}{\sigma}\right)\right\} \tag{7.A.19}$$

$$\hat{T}_4^{\text{b}}(t) = 2(\sigma\sqrt{\varepsilon\mu})^{-1}\hat{T}_4^{\text{norm}}\left(\vartheta\frac{2\varepsilon}{\sigma}\right) \tag{7.A.20}$$

Since the normalized functions in Equations 7.A.17 and 7.A.18 are independent of the properties of the medium, the rational approximation has to be performed only once. Application of Prony's method yields the coefficients $c_i^{(m)}$ and $\omega_i^{(m)}$: 6 terms to approximate $\hat{T}_2^{\text{norm}}$ and 20 terms to approximate $\hat{T}_4^{\text{norm}}$ [14]. Tables 7.A.1 and 7.A.2 show the coefficients $c_i^{(m)}$ and $\omega_i^{(m)}$ for $\hat{T}_m^{\text{norm}}$, $m = 2, 4$. The analytical and exponential approximations of $\hat{T}_2^{\text{norm}}$ and $\hat{T}_4^{\text{norm}}$ are shown in Figures 7.A.1 and 7.A.2, respectively.

Finally, evaluation of the integrals:

$$\hat{T}_m^{\text{norm}} * H = \sum_{i=1}^{Q^{(m)}} c_i^{(m)} e^{-\omega_i^{(m)}t} * H(t) = \sum_{i=1}^{Q^{(m)}} \int_0^t c_i^{(m)} \exp\left(-\omega_i^{(m)}(t-t')\right)H(t')\text{d}t', \quad m = 2,4 \tag{7.A.21}$$

is done by the following recursion:

$$\hat{T}_m^{\text{norm}} * H\big|_{n\Delta t} = \sum_{i=1}^{Q^{(m)}} \left(P_{1i}^{(m)}H(n\Delta t) + P_{2i}^{(m)}H((n-1)\Delta t) + P_{1i}^{(m)}\hat{T}_m^{\text{norm}} * H\Big|_{(n-1)\Delta t}\right), \quad m = 2,4 \tag{7.A.22}$$

TABLE 7.A.1

$\hat{T}_2^{\text{norm}}$ Exponential Approximation Coefficients

I	$\mathbf{c}_i^{(2)}$	$\boldsymbol{\omega}_i^{(2)}$
1	−2.46e-8	8.14e-6
2	−5.14e-6	3.71e-4
3	−3.04e-4	0.0648
4	−0.088	0.0686
5	−0.139	0.0481
6	−0.849	1.66

TABLE 7.A.2

$\hat{T}_4^{\text{norm}}$ Exponential Approximation Coefficients

I	$c_i^{(4)}$	$\omega_i^{(4)}$
1	1.01	0.60
2	−0.41	0.81
3	−6.35	1.01
4	−4.68	1.20
5	9.89	1.39
6	7.77	1.58
7	2.19	1.80
8	14.50	2.01
9	−9.98	2.21
10	−13.75	2.39
11	−2.12	2.60
12	4.34	2.79
13	−21.7	3.01
14	−28.36	3.21
15	24.68	3.38
16	29.03	3.61
17	8.81	3.82
18	−1.51	4.01
19	10.88	4.20
20	−24.31	4.41

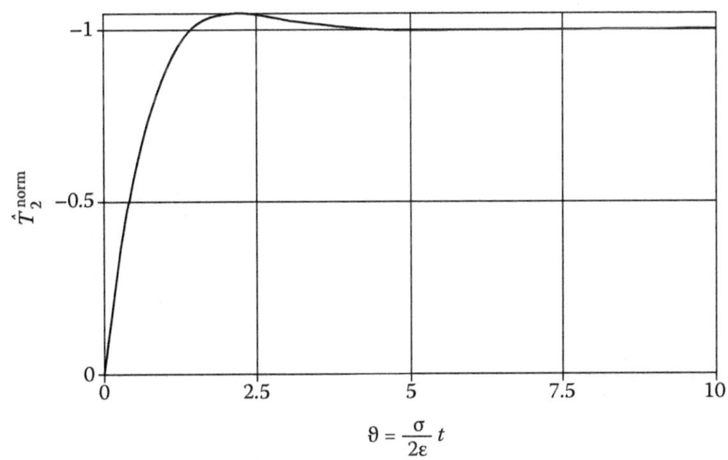

FIGURE 7.A.1
The $\hat{T}_2^{\text{norm}}$ function and its exponential approximation. Solid and dashed lines denote analytical solution and exponential approximation, respectively.

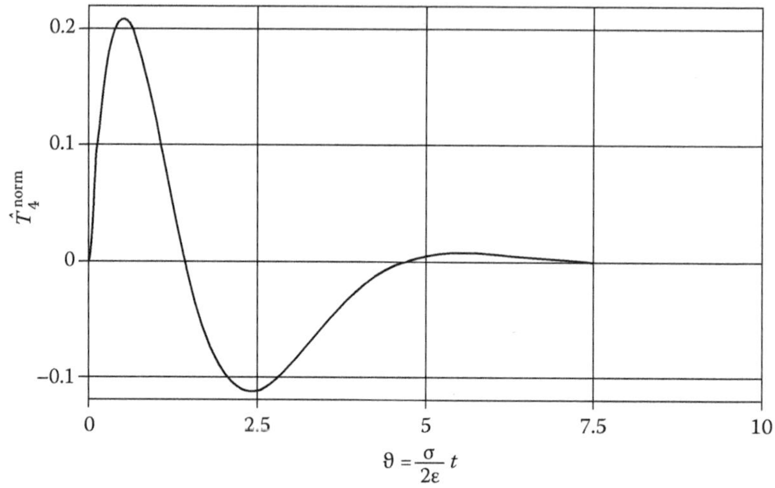

FIGURE 7.A.2
The $\hat{T}_4^{\text{norm}}$ function and its exponential approximation. Solid and dashed lines denote analytical solution and exponential approximation, respectively.

where

$$\begin{aligned} P_{1i}^{(m)} &= \frac{c_i^{(m)}}{\omega_i^{(m)}}\left[1+\frac{1}{\omega_i^{(m)}\Delta t}\left(\exp\left(-\omega_i^{(m)}\Delta t\right)-1\right)\right] \\ P_{2i}^{(m)} &= \frac{c_i^{(m)}}{\omega_i^{(m)}}\left[\frac{1}{\omega_i^{(m)}\Delta t}-\exp\left(-\omega_i^{(m)}\Delta t\right)\left(1+\frac{1}{\omega_i^{(m)}\Delta t}\right)\right] \\ P_{3i}^{(m)} &= \exp\left(-\omega_i^{(m)}\Delta t\right) \end{aligned} \tag{7.A.23}$$

Note that a different approach was considered in [15], namely: the normalized function $\hat{T}_2^{\text{norm}}$ is approximated in the Laplace domain using a series of first-order rational functions. Then, the inverse Laplace transform yields exponentials. Both approaches lead to similar recursions.

References

1. J.D. Stratton, *Electromagnetic Theory*, McGraw-Hill, New York, 1941.
2. A.J. Poggio and E.K. Miller, Integral equation solutions of three-dimensional problems, in R. Mittra (Ed.), *Computer Techniques for Electromagnetics*, Pergamon Press, Oxford, U.K., 1973.

3. M. Abramowitz and I.A. Stegun, *Handbook of Mathematical Functions*, Dover Publications, New York, 1972.
4. A.N. Tikhonov and V.Y. Arsenin, *Solution of Ill-Posed Problems*, Wiley, New York, 1977.
5. F.M. Tesche, On the inclusion of loss in time-domain solutions of electromagnetic interaction problems, *IEEE Transactions on Electromagnetic Compatibility*, 32(1), February 1990, 1–4.
6. C. Bennett, Jr. and W. Weeks, Transient scattering from conducting cylinders, *IEEE Transactions on Antennas and Propagation*, 18(5), September 1970, 627–633.
7. F.B. Hildebrand, *Introduction to Numerical Analysis*, Dover, New York, 1974, pp. 457–462.
8. J.L. Dubard, D. Pompei, J. Le Roux, and A. Papiernik, Characterization of microstrip antennas using the TLM simulation associated with a Prony-Pisrenko method, *International Journal of Numerical Modeling*, 3, 1990, 269–285.
9. W.L. Ko and R. Mitra, A combination of FDTD and Prony's methods for analyzing microwave integrated circuits, *IEEE Transactions Microwave Theory and Techniques*, 39, 1991, 2176–2181.
10. J.A. Pereda, L.A. Vielva, and A. Prieto, Computation of resonant frequencies and quality factors of open dielectric resonators by a combination of the finite-difference time-domain (FDTD) and Prony's methods, *IEEE Microwave and Guided Wave Letters*, 2, 1992, 431–433.
11. J.H. Beggs, R.J. Luebbers, K.S. Yee, and K.S. Kuz, Finite-difference time-domain implementation of surface impedance boundary conditions, *IEEE Transactions on Antennas and Propagation*, 40(1), 1992, 49–56.
12. J.G. Maloney and G.S. Smith, The use of surface impedance concepts in the finite-difference time-domain method, *IEEE Transactions on Antennas and Propagation*, 40(1), January 1992, 38–48.
13. A. Taflove and S.C. Hagness, *Computational Electrodynamics: The Finite Difference Time-domain Method*, 2nd edn., Artech House, Norwood, MA, 2000.
14. N. Farahat, The implementation of high order approximation SIBC for FDTD method, PhD dissertation, The University of Akron, Akron, OH, August 2000.
15. K.S. Oh and J.E. Schutt-Aine, An efficient implementation of surface impedance boundary conditions for the finite-difference time-domain method, *IEEE Transactions on Antennas and Propagation*, 43(7), 1995, 660–666.

8

Implementation of SIBCs for Volume Discretization Methods

8.1 Introduction

Following implementation of low- and high-order SIBCs for the boundary integral equations in Chapters 6 and 7 we now turn our attention to implementation of SIBC for volume discretization methods. We start with the FDTD and implement the SIBCs developed in Chapter 3 for high-frequency applications. The contour-path method, based on the integral forms of Faraday and Ampere's laws, is used for the implementation because it allows the grid to conform to the boundary and by doing so eliminates one of the main issues in use of the FDTD method—the use of regular grids. The fundamentals of the contour-path FDTD method are outlined separately in Appendix 8.A.1. This approach reduces the errors caused by discretization of boundaries and is well suited for incorporation of SIBCs. The conditions of applicability of the coupling between the FDTD and SIBC are discussed and examples to its use given to demonstrate the implementation.

The finite integration technique (FIT) is another volume discretization method related to the FDTD method. Unlike the FDTD method which solves for the electric and magnetic fields, the FIT method operates on circulations of the electric field (voltages), circulations of the magnetic field (magnetomotive forces) and fluxes of the electric and magnetic field. In that sense it may be viewed as a generalized FDTD method for the solution of Maxwell's equations in their integral form. As with the FDTD method, the FIT method is implemented for high-frequency problems using low- and high-order SIBCs in Cartesian and tetrahedral grids. The relations between the FIT and FDTD grids are discussed as well as the limits of applicability and the performance of the method is evaluated through an example.

Perhaps the best-known volumetric discretization method is the finite-element method (FEM) because of its extensive use in all engineering disciplines. It is however fundamentally different than the previous two methods in that Maxwell's equations must be transformed and adapted to the method before discretization and solution can commence. This in turn has implications on the implementation of the SIBCs. Also, the FEM has a diversity of element types and methods of formulation. For the sake of brevity we will discuss here only the formulation in the "weak" sense using tetrahedral edge elements although the method and approach are general enough to be useful for other types of elements and formulations. Numerical examples complete this chapter as well.

8.2 Statement of the Problem

Consider a nonconducting domain $V(\varepsilon_1, \mu_1, \sigma_1 = 0)$ surrounding a conducting body $V_c(\varepsilon_2, \mu_2, \sigma_2)$. Let S be the surface of V_c and, consequently, the interface between the conducting and nonconducting regions (Figure 8.1). The external boundary of the region V is denoted as S_{inf}. The distribution of the electromagnetic field in both regions can be described by Maxwell's equations in the following form:

Nonconducting region V:

$$\nabla \times \vec{H} = \varepsilon_1 \frac{\partial \vec{E}}{\partial t} \tag{8.1}$$

$$\nabla \times \vec{E} = -\mu_1 \frac{\partial \vec{H}}{\partial t} \tag{8.2}$$

Conducting region V_c:

$$\nabla \times \vec{H} = \sigma_2 \vec{E} + \varepsilon_2 \frac{\partial \vec{E}}{\partial t} \tag{8.3}$$

$$\nabla \times \vec{E} = -\mu_2 \frac{\partial \vec{H}}{\partial t} \tag{8.4}$$

We assume that the electromagnetic field in V is induced by external sources. Let the angular frequency ω or duration τ of the pulse of the incident field be such that the skin depth δ is much smaller than the characteristic size D of the

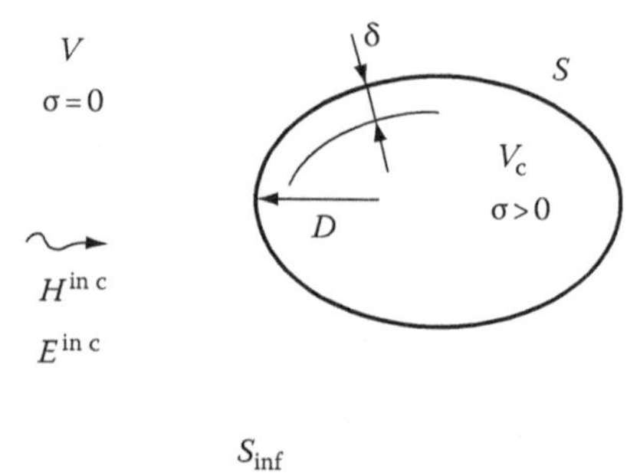

FIGURE 8.1
A general geometry. A conducting region surrounded by a dielectric.

conductor, i.e., the condition in Equation 2.1 is satisfied. Suppose our main interest is focused on the field in V. Then it is natural to replace Equations 8.3 and 8.4 by the SIBCs (Equation 3.132) applied on S and by doing so eliminate V_c from the numerical procedure so that only V has to be discretized.

8.3 Finite-Difference Time-Domain Method

Without any loss of generality we restrict ourselves to a two-dimensional problem in the xz-plane by assuming no variation along the y-axis (Figure 8.2). In this case, Equations 8.1 and 8.2 take the forms:

$$\frac{\partial H_x}{\partial t} = \frac{1}{\mu_1}\frac{\partial E_y}{\partial z} \tag{8.5}$$

$$\frac{\partial H_y}{\partial t} = \frac{1}{\mu_1}\left(\frac{\partial E_z}{\partial x} - \frac{\partial E_x}{\partial z}\right) \tag{8.6}$$

$$\frac{\partial H_z}{\partial t} = -\frac{1}{\mu_1}\frac{\partial E_y}{\partial x} \tag{8.7}$$

$$\frac{\partial E_x}{\partial t} = -\frac{1}{\varepsilon_1}\frac{\partial H_y}{\partial z} \tag{8.8}$$

$$\frac{\partial E_y}{\partial t} = \frac{1}{\varepsilon_1}\left(\frac{\partial H_x}{\partial z} - \frac{\partial H_z}{\partial x}\right) \tag{8.9}$$

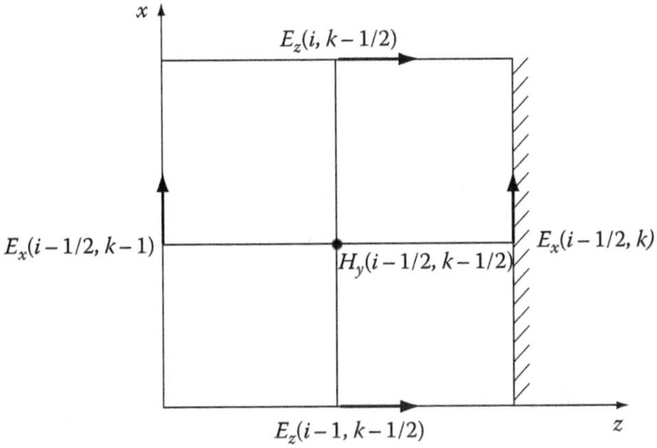

FIGURE 8.2
The original (Cartesian) FDTD boundary cell in two dimensions.

$$\frac{\partial E_z}{\partial t} = \frac{1}{\varepsilon_1} \frac{\partial H_y}{\partial x} \tag{8.10}$$

Equations 8.5 through 8.10 can be decoupled into transverse electric TE_y and transverse magnetic TM_y modes as follows:

TE_y mode:

$$\frac{\partial H_y}{\partial t} = \frac{1}{\mu_1} \left(\frac{\partial E_z}{\partial x} - \frac{\partial E_x}{\partial z} \right) \tag{8.11}$$

$$\frac{\partial E_x}{\partial t} = -\frac{1}{\varepsilon_1} \frac{\partial H_y}{\partial z} \tag{8.12}$$

$$\frac{\partial E_z}{\partial t} = \frac{1}{\varepsilon_1} \frac{\partial H_y}{\partial x} \tag{8.13}$$

TM_y mode:

$$\frac{\partial E_y}{\partial t} = \frac{1}{\varepsilon_1} \left(\frac{\partial H_x}{\partial z} - \frac{\partial H_z}{\partial x} \right) \tag{8.14}$$

$$\frac{\partial H_x}{\partial t} = \frac{1}{\mu_1} \frac{\partial E_y}{\partial z} \tag{8.15}$$

$$\frac{\partial H_z}{\partial t} = -\frac{1}{\mu_1} \frac{\partial E_y}{\partial x} \tag{8.16}$$

Independence of these modes allows one to decompose an arbitrary propagation in isotropic media into these two modes and solve them separately. Here we consider the TE_y case whereas the TM_y is analogous.

Discretizing the domain V with rectangular cells and applying Yee's algorithm [14], we write the following finite-difference approximations of Equations 8.11 through 8.13 for the ith cell in the x-direction and the kth cell in the z-direction at the nth time step:

$$\begin{aligned} &\frac{H_y^{n+1/2}(i-1/2,k-1/2) - H_y^{n-1/2}(i-1/2,k-1/2)}{\Delta t} \\ &= \frac{1}{\mu_1} \left(\frac{E_z^n(i,k-1/2) - E_z^n(i-1,k-1/2)}{\Delta x} - \frac{E_x^n(i-1/2,k) - E_x^n(i-1/2,k-1)}{\Delta z} \right) \end{aligned} \tag{8.17}$$

$$\begin{aligned} &\frac{E_x^{n+1}(i-1/2,k-1) - E_x^n(i-1/2,k-1)}{\Delta t} \\ &= -\frac{1}{\varepsilon_1} \frac{H_y^{n+1/2}(i-1/2,k-1/2) - H_y^{n+1/2}(i-1/2,k-3/2)}{\Delta z} \end{aligned} \tag{8.18}$$

$$\frac{E_z^{n+1}(i-1,k-1/2) - E_z^n(i-1,k-1/2)}{\Delta t}$$
$$= -\frac{1}{\varepsilon_1}\frac{H_y^{n+1/2}(i-1/2,k-1/2) - H_y^{n+1/2}(i-3/2,k-1/2)}{\Delta x} \tag{8.19}$$

where
Δx and Δy are the rectangular cell sizes
Δt is the time step

If the kth cell is a boundary cell adjacent to the interface (this case is shown in Figure 8.2), the electric field is determined at the interface so that the quantity $E_x^n(i-1/2,k)$ in Equation 8.17 is given by the SIBC (Equation 3.132). The simplest approximation of Equation 3.132 in the discrete space and time according to Yee's algorithm can be written in the form:

$$E_x^n(i-1/2,k) = \hat{F}\left[H_y^n(i-1/2,k)\right]$$
$$\approx \frac{\hat{F}\left[H_y^{n+1/2}(i-1/2,k-1/2)\right] + \hat{F}\left[H_y^{n-1/2}(i-1/2,k-1/2)\right]}{2} \tag{8.20}$$

Substituting Equations 8.20 into 8.17, we obtain

$$H_y^{n+1/2}(i-1/2,k-1/2) + \frac{\Delta t}{2\mu_1 \Delta z}\hat{F}\left[H_y^{n+1/2}(i-1/2,k-1/2)\right] = H_y^{n-1/2}(i-1/2,k-1/2)$$
$$+\frac{\Delta t}{\mu_1 \Delta x}\left(E_z^n(i,k-1/2) - E_z^n(i-1,k-1/2)\right) - \frac{\Delta t}{2\mu_1 \Delta z}\hat{F}\left[H_y^{n-1/2}(i-1/2,k-1/2)\right]$$
$$+\frac{\Delta t}{\mu_1 \Delta z}E_x^n(i\quad 1/2,k-1) \tag{8.21}$$

Since the high-order SIBC takes into account the curvature of the body we use the generalized FDTD approach for curved surfaces (contour-path method [7] described in Appendix 8.A.1). In this method, the boundary cells adjacent to the surface of the scatterer are distorted in order to conform to the arbitrary curved surface of the scatterer. Figure 8.3 shows a nonrectangular boundary cell for the generalized FDTD formulation. The magnetic field updating Equation 8.17 can be written in the following form:

$$H_y^{n+1/2}(i-1/2,k-1/2) = H_y^{n-1/2}(i-1/2,k-1/2)$$
$$+\frac{\Delta t}{\mu_1 A}\left(gE_z^n(i,k-1/2) - fE_z^n(i-1,k-1/2) - sE_{\tan}^n(i-1/2,k) + hE_x^n(i-1/2,k-1)\right) \tag{8.22}$$

where
g, f, s, h are the side lengths of the distorted cell
A is the area of the patch

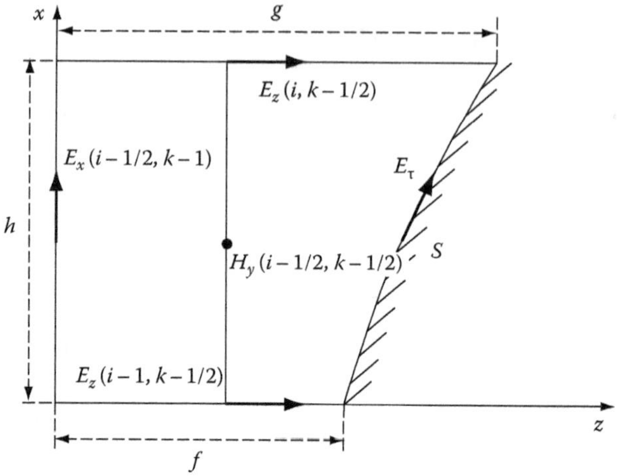

FIGURE 8.3
Contour-path FDTD boundary cell in two dimensions.

Replacing $E^n_{\tan}(i-1/2,k)$ by the SIBC in Equation 8.20, we obtain

$$H_y^{n+1/2}(i-1/2,k-1/2)+\frac{s\Delta t}{2\mu_1 A}\hat{F}\left[H_y^{n+1/2}(i-1/2,k-1/2)\right]=H_y^{n-1/2}(i-1/2,k-1/2)$$
$$+\frac{\Delta t}{\mu_1 A}\left(gE_z^n(i,k-1/2)-fE_z^n(i-1,k-1/2)-\frac{1}{2}s\hat{F}\left[H_y^{n-1/2}(i-1/2,k-1/2)\right]\right.$$
$$\left.+\,hE_x^n(i-1/2,k-1)\right) \qquad (8.23)$$

In our two-dimensional case, the time derivative $\hat{F}$ of the surface impedance function is written in the form:

$$\hat{F}\left[H_y\right]=\hat{T}_2^{\mathrm{b}}*H_y-\hat{T}_3^{\mathrm{b}}*\frac{H_y}{d}-\hat{T}_4^{\mathrm{b}}*\left(\frac{H_y}{8d^2}+\frac{1}{2}\frac{\partial^2 H_y}{\partial s^2}\right) \qquad (8.24)$$

where $\tilde{d}$ is the local radius of curvature of the interface and the time-dependent functions $T_m^{\mathrm{b}}, m=2,3,4$, are given by Equations 7.69 through 7.71. Substituting Equation 8.24 into Equation 8.23, we obtain

$$H_y^{n+1/2}(i-1/2,k-1/2)$$
$$+\frac{s\Delta t}{2\mu_1 A}\left\{H_y^{n+1/2}(i-1/2,k-1/2)*\left(\hat{T}_2^{\mathrm{b}}-\frac{1}{d}\hat{T}_3^{\mathrm{b}}-\frac{1}{8d^2}\hat{T}_4^{\mathrm{b}}\right)-\frac{1}{2}\frac{\partial^2 H_y^{n+1/2}(i-1/2,k-1/2)}{\partial s^2}*\hat{T}_4^{\mathrm{b}}\right\}$$
$$=H_y^{n-1/2}(i-1/2,k-1/2)+\frac{\Delta t}{\mu_1 A}\left\{gE_z^n(i,k-1/2)-fE_z^n(i-1,k-1/2)-hE_x^n(i-1/2,k-1)\right.$$
$$\left.-\frac{1}{2}sH_y^{n-1/2}(i-1/2,k-1/2)*\left(\hat{T}_2^{\mathrm{b}}-\frac{1}{d}\hat{T}_3^{\mathrm{b}}-\frac{1}{8d^2}\hat{T}_4^{\mathrm{b}}\right)+\frac{1}{4}s\frac{\partial^2 H_y^{n-1/2}(i-1/2,k-1/2)}{\partial s^2}*\hat{T}_4^{\mathrm{b}}\right\}$$
$$(8.25)$$

The following representation is used to approximate the term $\partial^2 H_y/\partial s^2$ in discrete space:

$$\frac{\partial^2 H_y^n}{\partial s^2} = \frac{\partial^2 H_y^n}{\partial x^2}\left(\frac{\partial s}{\partial x}\right)^{-1} + \frac{\partial^2 H_y^n}{\partial y^2}\left(\frac{\partial s}{\partial y}\right)^{-1} + \frac{\partial^2 H_y^n}{\partial x \partial y}\left[\left(\frac{\partial s}{\partial x}\right)^{-1} + \left(\frac{\partial s}{\partial y}\right)^{-1}\right]$$
$$- \left\{\frac{\partial H_y^n}{\partial x}\frac{\partial^2 s}{\partial x^2}\left(\frac{\partial s}{\partial x}\right)^{-2} + \frac{\partial H_y^n}{\partial y}\frac{\partial^2 s}{\partial x \partial y}\left(\frac{\partial s}{\partial y}\right)^{-2} + \frac{\partial H_y^n}{\partial x}\frac{\partial^2 s}{\partial x \partial y}\left(\frac{\partial s}{\partial x}\right)^{-2} + \frac{\partial H_y^n}{\partial y}\frac{\partial^2 s}{\partial y^2}\left(\frac{\partial s}{\partial y}\right)^{-2}\right\} \tag{8.26}$$

Although Equation 8.25 is suitable for computer implementation, it has the following disadvantages:

1. It cannot be solved with respect to $H_{k-1/2}^{n+1/2}$ by analytical methods so that the finite-difference algorithm becomes "implicit" for the "boundary" cell whereas it is "explicit" for other cells of the grid.
2. The full convolution with all past field components is included in Equation 8.25. This renders its evaluation for large problems impractical.

These difficulties can be overcome by application of recursive formulas (described in Appendix 7.A.1) for fast and economic computation of time-convolution integrals. In addition, the number of integrals on the left-hand side of Equation 8.25 can be reduced using the perturbation technique developed in the previous chapters. Using the scale factors introduced in Equation 3.52, we represent Equation 8.22 with nondimensional variables as follows:

$$\tilde{H}_y^{n+1/2}(i-1/2,k-1/2) = \tilde{H}_y^{n-1/2}(i-1/2,k-1/2) + \frac{\tilde{q}}{\tilde{Q}}\Big(\tilde{g}\tilde{E}_z^n(i,k-1/2)$$
$$-\tilde{f}\tilde{E}_z^n(i-1,k-1/2) - \tilde{s}\tilde{E}_{\tan}^n(i-1/2,k) + \tilde{h}\tilde{E}_x^n(i-1/2,k-1)\Big) \tag{8.27}$$

where
$\tilde{Q}$ is the Courant number
$\tilde{q}$ is the electric dimension of the body defined in Equation 7.107

The values g, f, s, and h in Equation 8.27 have been normalized by the scale factor A/Δ where

$$\Delta = \Delta x \Delta z \Big/ \sqrt{\Delta x^2 + \Delta y^2} \tag{8.28}$$

Switching in Equations 8.20 and 8.24 to nondimensional variables, we obtain the SIBC in the following form:

$$\tilde{E}_{\tan} = \tilde{\tilde{T}}_2^{\mathrm{b}} * \frac{\tilde{H}_y^{n+1/2} + \tilde{H}_y^{n-1/2}}{2} - \tilde{p}\frac{1}{\tilde{\tilde{d}}}\tilde{\tilde{T}}_3^{\mathrm{b}} * \frac{\tilde{H}_y^{n+1/2} + \tilde{H}_y^{n-1/2}}{2}$$
$$- \tilde{p}^2\tilde{\tilde{T}}_4^{\mathrm{b}} * \left(\frac{\tilde{H}_y^{n+1/2}}{16\tilde{d}^2} + \frac{\tilde{H}_y^{n-1/2}}{16\tilde{d}^2} + \frac{1}{4}\frac{\partial^2 \tilde{H}_y^{n+1/2}}{\partial \tilde{s}^2} + \frac{1}{4}\frac{\partial^2 \tilde{H}_y^{n-1/2}}{\partial \tilde{s}^2} \right) \tag{8.29}$$

We represent the magnetic field as power series in the small parameter $\tilde{p}$:

$$\tilde{H}_y^{n+1/2}(i-1/2, k-1/2) = \tilde{H}_{y1}^{n+1/2} + \tilde{p}\tilde{H}_{y2}^{n+1/2} + \tilde{p}^2\tilde{H}_{y3}^{n+1/2} \tag{8.30}$$

Substituting Equations 8.29 and 8.30 into Equation 8.27 and equating the coefficients of equal powers of $\tilde{p}$, provides the following equations for the terms of expansion:

$\tilde{p}^0$:

$$\tilde{H}_{y1}^{n+1/2} + \frac{\tilde{s}\tilde{q}}{2\tilde{Q}}\tilde{T}_2^{\mathrm{b}} * \tilde{H}_{y1}^{n+1/2} = \tilde{H}_y^{n-1/2}(i-1/2, k-1/2)$$
$$+ \frac{\tilde{q}}{\tilde{Q}}\Big[\tilde{g}\tilde{E}_z^n(i, k-1/2) - \tilde{f}\tilde{E}_z^n(i-1, k-1/2) - \frac{\tilde{s}}{2}\tilde{\tilde{T}}_2^{\mathrm{b}} * \tilde{H}_y^{n-1/2}$$
$$+ \tilde{h}\tilde{E}_x^n(i-1/2, k-1) \Big] \tag{8.31}$$

$\tilde{p}^1$:

$$\tilde{H}_{y2}^{n+1/2} + \frac{\tilde{s}\tilde{q}}{2\tilde{Q}}\tilde{\tilde{T}}_2^{\mathrm{b}} * \tilde{H}_{y2}^{n+1/2} = \frac{\tilde{s}\tilde{q}}{2\tilde{Q}}\frac{1}{\tilde{\tilde{d}}}\tilde{\tilde{T}}_3^{\mathrm{b}} * \left(\tilde{H}_{y1}^{n+1/2} + \tilde{H}_y^{n-1/2} \right) \tag{8.32}$$

$\tilde{p}^2$:

$$\tilde{H}_{y3}^{n+1/2} + \frac{\tilde{s}\tilde{q}}{2\tilde{Q}}\tilde{\tilde{T}}_2^{\mathrm{b}} * \tilde{H}_{y3}^{n+1/2} = \frac{\tilde{s}\tilde{q}}{2\tilde{Q}}\frac{1}{\tilde{\tilde{d}}}\tilde{\tilde{T}}_3^{\mathrm{b}} * \tilde{H}_{y2}^{n+1/2} - \frac{\tilde{s}\tilde{q}}{4\tilde{Q}}\tilde{\tilde{T}}_4^{\mathrm{b}} * \left(\frac{\tilde{H}_{y1}^{n+1/2}}{4\tilde{d}^2} + \frac{\tilde{H}_y^{n-1/2}}{4\tilde{d}^2} \right.$$
$$\left. + \frac{\partial^2 \tilde{H}_{y1}^{n+1/2}}{\partial \tilde{s}^2} + \frac{\partial^2 \tilde{H}_y^{n-1/2}}{\partial \tilde{s}^2} \right) \tag{8.33}$$

Returning to dimensional variables, the proposed formulation is written in the form:

$$H_y^{n+1/2}(i-1/2, k-1/2) = H_{y1}^{n+1/2} + H_{y2}^{n+1/2} + H_{y3}^{n+1/2} \tag{8.34}$$

$$H_{y1}^{n+1/2} + \frac{s\Delta t}{2\mu_1 A} T_2^{\mathrm{b}} * H_{y1}^{n+1/2} = H_y^{n-1/2}(i - 1/2, k - 1/2) + \frac{\Delta t}{\mu_1 A}\Big\{gE_z^n(i, k - 1/2) - fE_z^n(i - 1, k - 1/2) - \frac{s}{2}\hat{T}_2^{\mathrm{b}} * H_y^{n-1/2} + hE_x^n(i - 1/2, k - 1)\Big\} \tag{8.35}$$

$$H_{y2}^{n+1/2} + \frac{s\Delta t}{2\mu_1 A}\hat{T}_2^{\mathrm{b}} * H_2^{n+1/2} = \frac{s\Delta t}{2\mu_1 A}\frac{1}{d}\hat{T}_3^{\mathrm{b}} * \left(H_{y1}^{n+1/2} + H_y^{n-1/2}\right) \tag{8.36}$$

$$H_{y3}^{n+1/2} + \frac{s\Delta t}{2\mu_1 A}\hat{T}_2^{\mathrm{b}} * H_3^{n+1/2} = \frac{s\Delta t}{2\mu_1 A}\frac{1}{d}\hat{T}_3^{\mathrm{b}} * H_{y2}^{n+1/2} - \frac{s\Delta t}{4\mu_1 A}\hat{T}_4^{\mathrm{b}} * \left(\frac{H_{y1}^{n+1/2}}{4d^2} + \frac{H_y^{n-1/2}}{4d^2} + \frac{\partial^2 H_{y1}^{n+1/2}}{\partial s^2} + \frac{\partial^2 H_y^{n-1/2}}{\partial s^2}\right) \tag{8.37}$$

The first equation is the implementation of the Leontovich SIBC. The second and third equations provide corrections by taking into account the curvature of the body's surface and the field variation in the tangential direction. We emphasize that Equations 8.35 through 8.37 are different only in the right-hand sides.

Note that the left-hand sides of Equations 8.35 through 8.37 involve the time-convolution product, therefore, these equations cannot be solved with respect to $H_y^{n+1/2}(i - 1/2, k - 1/2)$ by analytical methods. This makes the scheme implicit for the "boundary" cell and reduces efficiency of Yee's algorithm (which is explicit for all other cells). However, Equations 8.35 through 8.37 admit further transformations due to the fact that the electric field in V decreases when we approach the conductor's surface.

Since $E_x^n(i - 1/2, k)$ is related to the interface whereas $E_x^n(i - 1/2, k - 1)$ is related to the dielectric medium, we can assume that

$$E_x^n(i - 1/2, k) \ll E_x^n(i - 1/2, k - 1) \tag{8.38}$$

Thus, Equation 8.23 contains terms of different orders of magnitude, which means we can transform this equation using the perturbation technique. As the first step we transfer to nondimensional variables by choosing appropriate scale factors.

Let E_v be the characteristic scale for variation of the electric field in all nodes of the boundary cell (located in the dielectric region V) except at the interface. Thus we can write

$$E_x^n(i - 1/2, k - 1) = E_v\tilde{E}_x^n(i - 1/2, k - 1) \tag{8.39}$$

$$E_z^n(i, k - 1/2) = E_v\tilde{E}_z^n(i, k - 1/2) \tag{8.40}$$

$$E_z^n(i-1,k-1/2) = E_v\tilde{E}_z^n(i-1,k-1/2) \tag{8.41}$$

As in previous chapters, the sign "$\sim$" denotes nondimensional values. Using Equation 8.39, the scale factor E_k for the electric field $E_x^n(i-1/2,k)$ at the interface can be represented in the form:

$$E_k^* = \tilde{\upsilon}E_v \tag{8.42}$$

$$\tilde{\upsilon} = E_k^*/E_v \ll 1 \tag{8.43}$$

$$E_x^n(i-1/2,k) = E_k^*\,\tilde{E}_x^n(i-1/2,k) = \tilde{\upsilon}E_v\tilde{E}_x^n(i-1/2,k) \tag{8.44}$$

where $\tilde{\upsilon}$ is the small parameter. Since the quantities $H_y^{n+1/2}(i-1/2,k-1/2)$ and $H_y^{n-1/2}(i-1/2,k-1/2)$ are of the same order of magnitude, they should have a common scale factor H_v. That scale factor can be defined from Equation 8.6 as follows:

$$H_v = E_v\Delta t/(\mu_1\Delta z) \tag{8.45}$$

Using Equation 8.45, the values $H_y^{n+1/2}(i-1/2,k-1/2)$ and $H_y^{n-1/2}$ $(i-1/2,$ $k-1/2)$ can be represented in the form:

$$\begin{aligned} H_y^{n+1/2}(i-1/2,k-1/2) &= H_v\tilde{H}_y^{n+1/2}(i-1/2,k-1/2) \\ &= \frac{E_v\Delta t}{\mu_1\Delta z}\tilde{H}_y^{n+1/2}(i-1/2,k-1/2) \end{aligned} \tag{8.46}$$

$$\begin{aligned} H_y^{n-1/2}(i-1/2,k-1/2) &= H_v\tilde{H}_y^{n-1/2}(i-1/2,k-1/2) \\ &= \frac{E_v\Delta t}{\mu_1\Delta z}\tilde{H}_y^{n-1/2}(i-1/2,k-1/2) \end{aligned} \tag{8.47}$$

With nondimensional variables, Equation 8.23 takes the form:

$$\begin{aligned} &\tilde{H}_y^{n+1/2}(i-1/2,k-1/2) + \tilde{\upsilon}\frac{1}{2}\tilde{\hat{F}}\left[\tilde{H}_y^{n+1/2}(i-1/2,k-1/2)\right] = \tilde{H}_y^{n-1/2}(i-1/2,k-1/2) \\ &+ \tilde{E}_z^n(i,k-1/2) - \tilde{E}_z^n(i-1,k-1/2) - \frac{1}{2}\tilde{\hat{F}}\left[\tilde{H}_y^{n-1/2}(i-1/2,k-1/2)\right] + \tilde{E}_x^n(i-1/2,k-1) \end{aligned} \tag{8.48}$$

We represent the function $\tilde{H}_y^{n+1/2}(i-1/2,k-1/2)$ for which the solution is sought in the form of expansions in the small parameter $\tilde{\upsilon}$:

$$\tilde{H}_y^{n+1/2}(i-1/2,k-1/2) = \tilde{h}_0 + \tilde{\upsilon}\tilde{h}_1 + O(\tilde{\upsilon}^2) \tag{8.49}$$

Substituting Equation 8.49 into Equation 8.48 and equating the coefficients of equal powers of $\tilde{\upsilon}$, we obtain the equations for $\tilde{h}_0$ and $\tilde{h}_1$:

$$\tilde{h}_0 = \tilde{H}_y^{n-1/2}(i-1/2,k-1/2) + \tilde{E}_z^n(i,k-1/2) - \tilde{E}_z^n(i-1,k-1/2) \\ -\frac{1}{2}\tilde{\hat{F}}\left[\tilde{H}_y^{n-1/2}(i-1/2,k-1/2)\right] + \tilde{E}_x^n(i-1/2,k-1) \tag{8.50}$$

$$\tilde{h}_1 = -\frac{1}{2}\tilde{\hat{F}}\left[\tilde{h}_0\right] \tag{8.51}$$

In effect, we developed a two-step technique whereby $\tilde{h}_0$ is calculated using Equation 8.50 and then $\tilde{h}_1$ is calculated using Equation 8.51.

In dimensional variables, Equations 8.50 and 8.51 take the forms:

$$h_0 = H_y^{n-1/2}(i-1/2,k-1/2) + \frac{\Delta t}{\mu_1 \Delta z}\left(E_z^n(i,k-1/2) - E_z^n(i-1,k-1/2)\right) \\ -\frac{\Delta t}{2\mu_1 \Delta z}\hat{F}\left[H_y^{n-1/2}(i-1/2,k-1/2)\right] + \frac{\Delta t}{\mu_1 \Delta z}E_x^n(i-1/2,k-1) \tag{8.52}$$

$$h_1 = -\frac{\Delta t}{2\mu_1 \Delta z}\hat{F}[h_0] \tag{8.53}$$

and

$$H_y^{n+1/2}(i-1/2,k-1/2) = h_0 + h_1 \tag{8.54}$$

Note that we do not actually know the exact values of $\tilde{\upsilon}$ and E_υ. Both of them are required only at the stage of derivation and are obviously not included in the dimensional formulation (Equations 8.52 through 8.54). All we really need to know is that the condition in Equation 8.38 is satisfied (and, consequently, Equation 8.43 holds).

The representation in Equations 8.52 through 8.54 has clear physical meaning, namely h_0 is the magnetic field calculated under the assumption that medium 2 is a perfect electrical conductor (PEC) and the electric field at the interface is zero; h_1 is the correction allowing for finite conductivity of medium 2.

Finally, combining Equations 8.34 through 8.37 and Equations 8.52 through 8.54, we obtain

$$H_y^{n+1/2}(i-1/2,k-1/2) = H_{y0}^{n+1/2} + H_{y1}^{n+1/2} + H_{y2}^{n+1/2} + H_{y3}^{n+1/2} \tag{8.55}$$

$$H_{y0}^{n+1/2} = H_y^{n-1/2}(i-1/2,k-1/2) + \frac{\Delta t}{\mu_1 A}\left\{gE_z^n(i,k-1/2) \right. \\ \left. - fE_z^n(i-1,k-1/2) + hE_x^n(i-1/2,k-1)\right\} \tag{8.56}$$

$$H_{y1}^{n+1/2} = -\frac{s\Delta t}{2\mu_1 A}\hat{T}_2^{b} * \left(H_{y0}^{n+1/2} + H_y^{n-1/2}(i-1/2,k-1/2)\right) \tag{8.57}$$

$$\begin{aligned} H_{y2}^{n+1/2} = &-\frac{s\Delta t}{2\mu_1 A}\hat{T}_2^{b} * \left(H_{y1}^{n+1/2} + H_y^{n-1/2}(i-1/2,k-1/2)\right) \\ &+\frac{s\Delta t}{2\mu_1 A}\frac{1}{d}\hat{T}_3^{b} * \left(H_{y0}^{n+1/2} + H_y^{n-1/2}(i-1/2,k-1/2)\right) \end{aligned} \tag{8.58}$$

$$\begin{aligned} H_{y3}^{n+1/2} = &-\frac{s\Delta t}{2\mu_1 A}\hat{T}_2^{b} * \left(H_{y2}^{n+1/2} + H_y^{n-1/2}(i-1/2,k-1/2)\right) \\ &+\frac{s\Delta t}{2\mu_1 A}\frac{1}{d}\hat{T}_3^{b} * \left(H_{y1}^{n+1/2} + H_y^{n-1/2}(i-1/2,k-1/2)\right) \\ &-\frac{s\Delta t}{4\mu_1 A}\hat{T}_4^{b} * \left(\frac{H_{y0}^{n+1/2}}{4d^2} + \frac{H_y^{n-1/2}}{4d^2} + \frac{\partial^2 H_{y0}^{n+1/2}}{\partial s^2} + \frac{\partial^2 H_y^{n-1/2}}{\partial s^2}\right) \end{aligned} \tag{8.59}$$

It is easy to see that Equations 8.55 through 8.59 are fully explicit because the term containing the impedance function is now on the right-hand sides of the equations.

Let us derive the conditions of applicability of the technique. We will perform the derivation in the frequency domain considering propagation of a uniform plane wave from region 1 ($z < 0$) to region 2 ($z > 0$) (one-dimensional problem). Because of reflection, the following waves travel in both regions:

Incident wave:

$$E_1^+ = E_{10}^+ \exp(-j\beta_1 z) \tag{8.60}$$

Reflected wave:

$$E_1^- = E_{10}^- \exp(j\beta_1 z) \tag{8.61}$$

Transmitted wave:

$$E_2^+ = E_{20}^+ \exp(-\gamma_2 z) \tag{8.62}$$

where

$$\beta_1 = \omega\sqrt{\mu_1\varepsilon_1} = 2\pi/\lambda; \quad \gamma_2 = \sqrt{(\sigma_2 + j\omega)j\omega\mu_2} \tag{8.63}$$

Since the total electric field is continuous at $z = 0$, the reflection coefficient can be represented in the form:

$$\Gamma = E_{10}^-/E_{10}^+ = \frac{\eta_2 - \eta_1}{\eta_2 + \eta_1} = \frac{\eta_2/\eta_1 - 1}{\eta_2/\eta_1 + 1} \tag{8.64}$$

$$\eta_1 = \sqrt{\mu_1/\varepsilon_1}; \quad \eta_2 = \sqrt{j\omega\mu_2/(\sigma_2 + j\omega\varepsilon_2)} \tag{8.65}$$

In the conducting region we have

$$\sigma_2 \gg \omega\varepsilon_2 \tag{8.66}$$

so that

$$\eta_2/\eta_1 \ll 1 \tag{8.67}$$

With these, the reflection coefficient can be represented in the form:

$$\Gamma \approx (s-1)(1-s) + O(s^2) = -1 + 2s + O(s^2) \tag{8.68}$$

where

$$s = \sqrt{\frac{j\omega\mu_2\varepsilon_1}{\sigma_2\mu_1}} = \sqrt{\frac{\mu_2\varepsilon_1}{\mu_1\varepsilon_2}}\sqrt{\frac{j\omega\varepsilon_2}{\sigma_2}} = \frac{1+j}{\sqrt{2}}\varsigma k_2^{-1} \ll 1 \tag{8.69}$$

$$k_2 = \sqrt{\frac{\sigma_2}{\omega\varepsilon_2}} = \sqrt{\tan\theta} \tag{8.70}$$

$$\varsigma = \sqrt{\frac{\mu_2\varepsilon_1}{\mu_1\varepsilon_2}} \tag{8.71}$$

and $\tan\theta$ is the loss tangent in medium 2.

Using Equation 8.68, we write the total electric field in medium 1 in the form:

$$\begin{aligned} E_1(z) &= E_{10}^{+}\exp(-j\beta_1 z) + E_{10}^{-}\exp(j\beta_1 z) = E_{10}^{+}(\exp(-j\beta_1 z) + \Gamma\exp(j\beta_1 z)) \\ &= E_{10}^{+}\left[\exp(-j\beta_1 z) + \left(-1 + \sqrt{2}\varsigma k_2^{-1} + j\sqrt{2}\varsigma k_2^{-1}\right)\exp(j\beta_1 z)\right] \\ &= E_{10}^{+}\sqrt{2}\varsigma k_2^{-1}\left\{[\cos\beta_1 z - \sin\beta_1 z] + j\left[\left(-\sqrt{2}\varsigma k_2 + 1\right)\sin\beta_1 z + \cos\beta_1 z\right]\right\} \end{aligned} \tag{8.72}$$

Setting $z = 0$ in Equation 8.72, we obtain the total field at the interface:

$$E_1(0) = E_{10}^{+}\sqrt{2}\varsigma k_2^{-1}(1+j) \tag{8.73}$$

The ratio of Equations 8.72 and 8.73 gives the variation of the field in medium 1 with respect to the field at the interface:

$$\frac{E_1(z)}{E_1(0)} = \left\{ \left[2\cos\frac{2\pi}{\lambda_1}z - \sqrt{2}\varsigma k_2 \sin\frac{2\pi}{\lambda_1}z \right] + j\left(-\sqrt{2}\varsigma k_2 + 2\right)\sin\frac{2\pi}{\lambda_1}z \right\} \tag{8.74}$$

Our goal is evaluation of the electric fields on the opposite sides of the "boundary" cell (see Figure 8.2). Usually, more than 10 cells per wavelength are taken in computations. Thus

$$2\pi z/\lambda_1 \ll 1 \tag{8.75}$$

Expanding $\sin\frac{2\pi}{\lambda_1}z$ and $\cos\frac{2\pi}{\lambda_1}z$ in a Taylor series and taking into account Equation 8.75, we represent Equation 8.74 in the form:

$$\frac{E_1(z)}{E_1(0)} \approx \left\{ \left[2 - \sqrt{2}\varsigma k_2 \frac{2\pi}{\lambda_1}z \right] + j\left(-\sqrt{2}\varsigma k_2 + 2\right)\frac{2\pi}{\lambda_1}z \right\} \tag{8.76}$$

Setting $z = \Delta z$, we obtain

$$\frac{E_1(\Delta z)}{E_1(0)} \gg 1 \quad \text{if} \quad Q = \sqrt{\tan\theta}\sqrt{\frac{\mu_1\varepsilon_2}{\mu_2\varepsilon_1}}\frac{2\pi}{\lambda_1}\delta z \gg 1 \tag{8.77}$$

The error ε due to the representation in Equation 8.49 is derived directly from the condition in Equation 8.77 and written in the form:

$$\varepsilon = Q^{-2} = (\tan\theta)^{-1}\frac{\mu_2\varepsilon_1}{\mu_1\varepsilon_2}\left(\frac{\lambda_1}{2\pi\Delta z}\right)^2 = \frac{\mu_2}{\omega\sigma_2(\mu_1\Delta z)^2} \tag{8.78}$$

Next we give a few numerical examples to illustrate the techniques. First consider the one-dimensional problem of normal incidence of a TM plane wave at the plane boundary of a lossy dielectric half-space with the following material parameters:

$$\varepsilon_1 = \varepsilon_2 = \varepsilon_0; \quad \mu_1 = \mu_2 = \mu_0; \quad \sigma_1 = 0; \quad \sigma_2 \neq 0 \tag{8.79}$$

The excitation is defined as follows:

$$\text{Excitation} = 1000\sin(2\pi t/5.56\text{E-8}), \quad 0 < t < (5.56\text{E-8})/2\text{s} \tag{8.80}$$

The parameters of the computational mesh are chosen as

$$\Delta x = \Delta z = 5\times 10^{-1}\,\text{m}; \quad \Delta t = 1.18\times 10^{-9}\,\text{s};$$
$$N_{z\text{-space steps}} = N_{x\text{-space steps}} = 100; \quad N_{\text{time steps}} = 300 \tag{8.81}$$

Figure 8.4 shows the distributions of the magnetic field at the interface obtained for a material with low loss tangent ($\tan\theta = 10$). The curves were

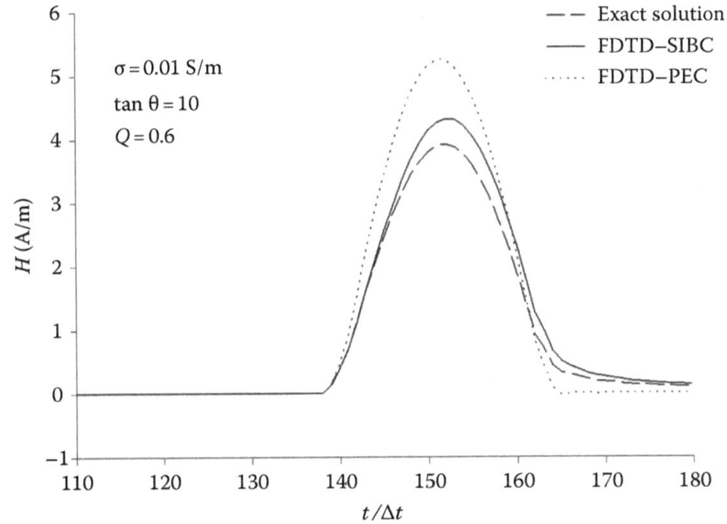

FIGURE 8.4
Distribution of the magnetic field at the interface for low-conductivity scatterer.

obtained using the PEC-condition (dotted line), exact formula (dashed line), and Equations 8.55 through 8.59 (solid line). The error is significant as predicted by Equation 8.78, ($Q < 1$). For a material with high loss tangent $\left(\tan\theta = 10^3\right)$ the actual error of 2.4% is in agreement with the estimate $\varepsilon = Q^{-2} = 2.2\%$ given by Equation 8.78 as is shown in Figure 8.5.

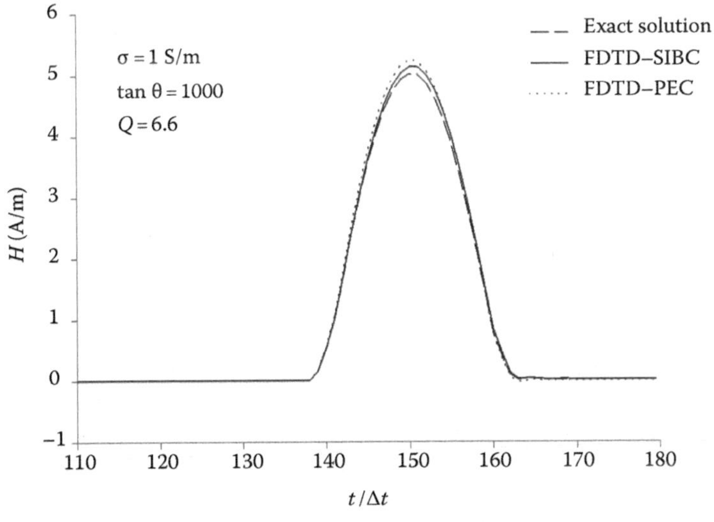

FIGURE 8.5
Distribution of the magnetic field at the interface for higher-conductivity scatterer.

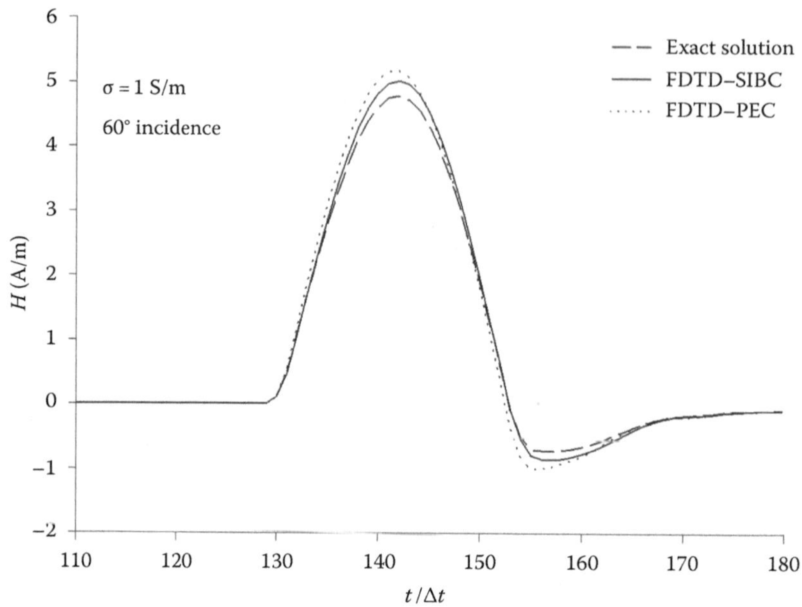

FIGURE 8.6
Distribution of the magnetic field at the interface for higher-conductivity scatterer at oblique incidence.

The problem becomes two-dimensional in the case of oblique incidence (let the incident angle be equal to 60°). The magnetic field was computed in the middle of the dielectric surface for both the exact and proposed methods. Figures 8.6 and 8.7 show the magnetic field computed in the middle of the dielectric surface for materials with low and high loss tangent, respectively, using low- and high-order SIBCs.

Another example is scattering of an incident TE_y wave by an infinitely long conducting cylinder as shown in Figure 7.1. The excitation function is the following:

$$\text{Excitation} = 1000 \times \exp\left(-16 \times \left(\frac{t}{7.071 \times 10^{-8}} - 1\right)^2\right) \tag{8.82}$$

The diameter of the cylinder is 12 cells which is almost one-third of the average wave length of the Gaussian pulse ($\lambda = 3 \times 10^8 \tau/2$). τ is the duration of the Gaussian pulse. Distorted cells are used adjacent to the cylinder surface to allow the FDTD cells to conform to the curved surface. Figures 8.8 and 8.9 show the total magnetic field at point A (in Figure 7.1) obtained using low-order (Leontovich's approximation) and high-order (Rytov's approximation) SIBCs for different values of the parameter $\tilde{p}$. From these figures it follows

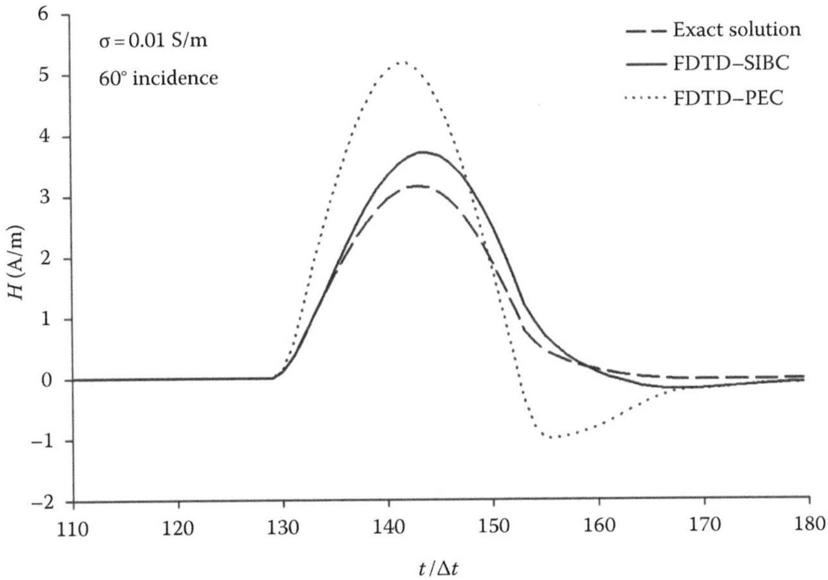

FIGURE 8.7
Distribution of the magnetic field at the interface for low-conductivity scatterer at oblique incidence.

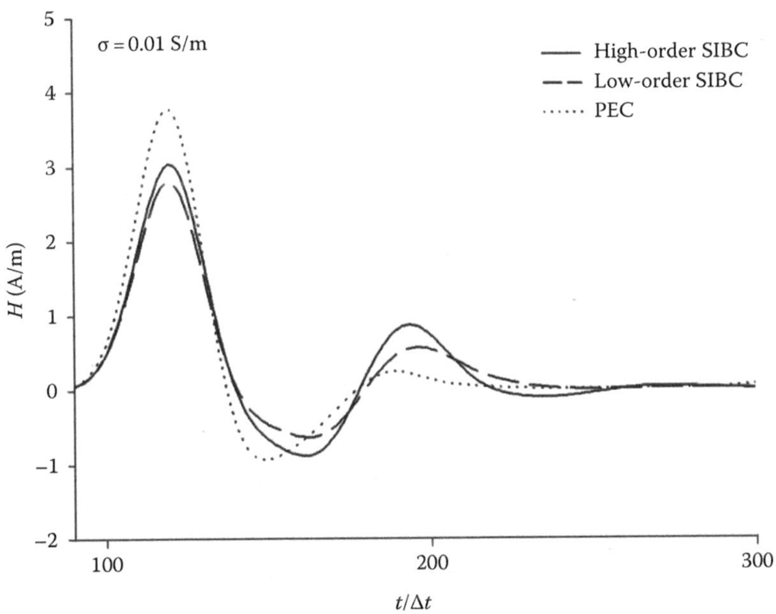

FIGURE 8.8
Magnetic field at point A, low conductivity.

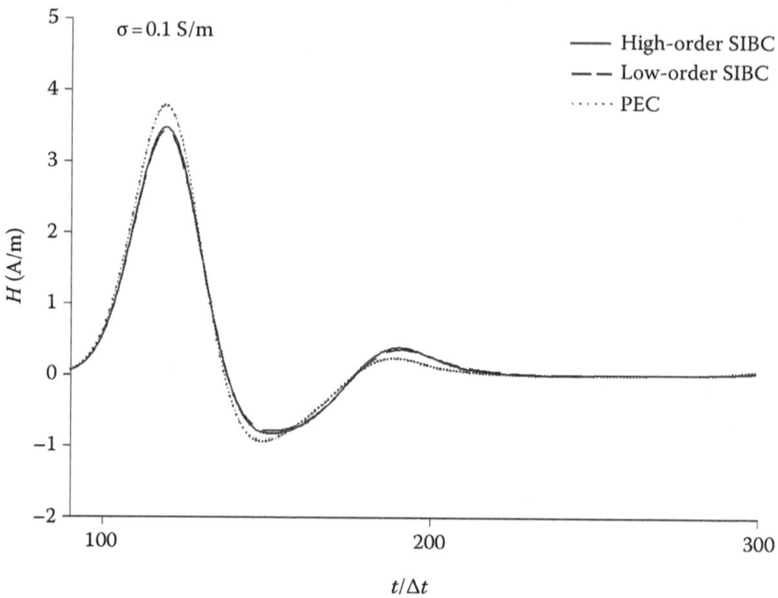

FIGURE 8.9
Magnetic field at point *A*, higher conductivity.

that the difference between results obtained using low- and high-order SIBCs is not visible at very small values of $\tilde{p}$.

8.4 Finite Integration Technique

A generalized FDTD scheme for the solution of Maxwell's equations in their integral form is known as the FIT [1]. In both methods, staggered dual grids are used for approximation of the electric- and magnetic-related parameters. An example of the orthogonal dual mesh used in FDTD and FIT is shown in Figure 8.10. However, the FDTD method operates in terms of the electric and magnetic fields whereas FIT employs circulations and fluxes as state variables. In FDTD, the nodes, where electric and magnetic fields are calculated, are located in the middle of edges and in the middle of facets, respectively. In FIT, the state variables are the so-called grid voltages and grid fluxes related to edges and facets [2]:

$$e = \int_{L_e} \vec{E} \cdot d\vec{l}; \quad h = \int_{L_h} \vec{H} \cdot d\vec{l}; \quad d = \int_{S_d} \vec{D} \cdot \vec{n} ds; \quad b = \int_{S_b} \vec{B} \cdot \vec{n} ds \tag{8.83}$$

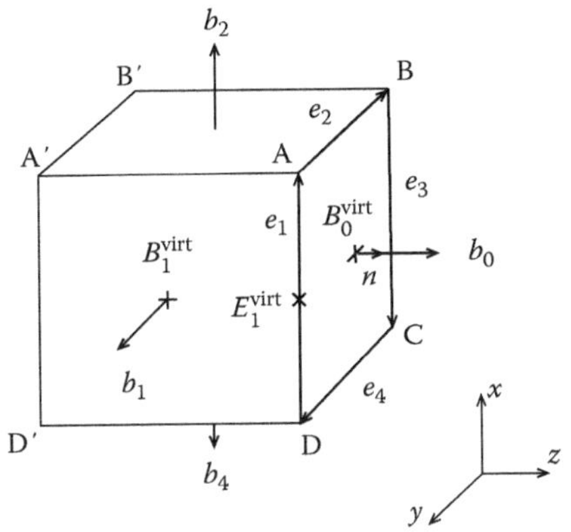

FIGURE 8.10
Cartesian computational cell used in the FIT and FDTD methods.

where

L_i and S_i are the cell's edge and facet, respectively
the subscript denotes the type of grid (primary or dual)
the vectors $d\vec{l}$ and $\vec{n}$ are directed along the edge and normal to the facet, respectively

Faraday's law in integral form applied to the facet *ABCD* of the computational cell shown in Figure 8.10 can be written with the variables in Equation 8.83 as follows:

$$\int_{L_e} \vec{E} \cdot d\vec{l} = -\int_{S_d} \frac{\partial \vec{B}}{\partial t} \cdot \vec{n} ds \quad \Leftrightarrow \quad e_1 + e_2 + e_3 + e_4 = db_0/dt \tag{8.84}$$

The left-hand side contains all voltage components related to edges adjacent to the considered facet, combined with proper signs ± 1 according to their orientation.

Material relations between e and d are obtained by introduction of a virtual continuous component E^{virt} at the intersection point of the dual and primary grids [1]. Since the normal vector to the dual facet is collinear with the direction of the intersecting primary edge, we obtain

$$e = \int_{L_e} \vec{E} \cdot d\vec{l} \approx E^{virt} L_e \tag{8.85}$$

$$d = \int_{S_d} \varepsilon \vec{E} \cdot \vec{n} ds \approx \varepsilon_{eff} E^{virt} A_d \quad \text{where} \quad \varepsilon_{eff} = \frac{1}{A_d} \int_{S_d} \varepsilon ds \tag{8.86}$$

The discrete permittivity expression can now be written in the form:

$$\frac{d}{e} = \frac{\int_{S_d} \varepsilon \vec{E} \cdot \vec{n} ds}{\int_{L_e} \vec{E} \cdot d\vec{l}} \approx \frac{\varepsilon_{eff} A_d}{L_e} \tag{8.87}$$

Similarly, a virtual magnetic flux density B^{virt} is allocated as a continuous normal component at the center of the primary facet so that

$$b = \int_{S_b} \vec{B} \cdot \vec{n} ds \approx B^{virt} S_b \tag{8.88}$$

$$h = \int_{L_h} \mu^{-1} \vec{B} \cdot d\vec{l} \approx \mu_{eff}^{-1} B^{virt} L_e \quad \text{where} \quad \mu_{eff}^{-1} = \frac{1}{L_h} \int_{L_h} \mu^{-1} dl \tag{8.89}$$

and

$$\frac{h}{b} = \frac{\int_{L_h} \mu^{-1} \vec{B} \cdot d\vec{l}}{\int_{S_b} \vec{B} \cdot \vec{n} ds} \approx \frac{\mu_{eff}^{-1} L_h}{A_b} \tag{8.90}$$

We now represent the SIBC relating the electric and magnetic fields in terms of the circulations and fluxes as state variables of the FIT. Suppose the facet *ABCD* of the Cartesian cell shown in Figure 8.10 belongs to the dielectric/conductor interface. To truncate the mesh, additional equations relating e_1, e_2, e_3, e_4, and b_0, and containing material properties of the conductor are needed. An SIBC relating the normal and tangential components of the magnetic flux at the interface is obtained directly from Equation 7.65 and written in the form:

$$B_z = B_\eta = \sum_{i=1}^{2} \frac{\partial}{\partial \xi_i} F\left[B_{\xi_i}\right] = \frac{\partial}{\partial x} F[B_x] + \frac{\partial}{\partial y} F\left[B_y\right] \tag{8.91}$$

Here the surface impedance function F is given by Equations 7.66 through 7.68 for the Leontovich, Mitzner, and Rytov approximations, respectively. Applying vector identities to Equation 7.65, we obtain

$$B_z = \nabla \cdot \left[\left(\vec{n} \times \vec{F}\right) \times \vec{n}\right] = \vec{n} \cdot \nabla \times \left(\vec{n} \times \vec{F}\right) \tag{8.92}$$

where the components of the vector $\vec{F}$ are defined as follows:

$$F_x = F[B_x]; \quad F_y = F\left[B_y\right] \tag{8.93}$$

Performing integration on both sides of Equation 8.91 over the facet *ABCD* and applying vector identities, we obtain

$$b_0 = \int\limits_{S_{b0}} \left(\vec{B}\right)_z \mathrm{d}s = \int\limits_{S_{b0}} \nabla \cdot \left[\left(\vec{n} \times \vec{F}\right) \times \vec{n}\right] \mathrm{d}s = -\int\limits_{S_{b0}} \vec{n} \cdot \nabla \times \left(\vec{n} \times \vec{F}\right) \mathrm{d}s \tag{8.94}$$

Application of Stoke's theorem to the last equation yields

$$b_0 = -\oint\limits_{\mathrm{L}} \left(\vec{n} \times \vec{F}\right) \cdot \mathrm{d}\vec{l} = -\sum_{k=1}^{4} \int\limits_{\mathrm{L}_k} \left(\vec{n} \times \vec{F}\right) \cdot \mathrm{d}\vec{l} \tag{8.95}$$

From Figure 8.10, it follows that

$$\vec{n} \times \vec{F}\Big|_{\mathrm{ABCD}} = -F_y \vec{a}_x + F_x \vec{a}_y = -F\left[B_y\right]\vec{a}_x + F[B_x]\vec{a}_y \tag{8.96}$$

where $\vec{a}_x$, $\vec{a}_y$, and $\vec{a}_z$ are unit vectors of the global Cartesian coordinates system shown in Figure 8.10 and $\vec{n} = \vec{a}_z$. Substituting Equation 8.96 into Equation 8.95 and taking into account that

$$\mathrm{d}\vec{l}_{\mathrm{DA}} = \mathrm{d}l_{\mathrm{DA}} \cdot \vec{a}_x; \quad \mathrm{d}\vec{l}_{\mathrm{AB}} = -\mathrm{d}l_{\mathrm{AB}} \cdot \vec{a}_y; \quad \mathrm{d}\vec{l}_{\mathrm{BC}} = -\mathrm{d}l_{\mathrm{BC}} \cdot \vec{a}_x; \quad \mathrm{d}\vec{l}_{\mathrm{CD}} = \mathrm{d}l_{\mathrm{CD}} \cdot \vec{a}_y \tag{8.97}$$

we obtain

$$\left.\begin{aligned}
&\int\limits_{L_{\mathrm{DA}}} \left(\vec{n} \times \vec{F}\right) \cdot \mathrm{d}\vec{l} = -\int\limits_{L_{\mathrm{DA}}} \left(\vec{F}_{\mathrm{DA}}\right)_y \mathrm{d}l_{\mathrm{DA}} = -F\left[B_1^{\mathrm{virt}}\right] L_{e_1} = -\frac{L_{e_1}}{S_{b_1}} F[b_1] \\
&\int\limits_{L_{\mathrm{AB}}} \left(\vec{n} \times \vec{F}\right) \cdot \mathrm{d}\vec{l} = -\int\limits_{L_{\mathrm{AB}}} \left(\vec{F}_{\mathrm{AB}}\right)_x \mathrm{d}l_{\mathrm{AB}} = -F\left[B_2^{\mathrm{virt}}\right] L_{e_2} = -\frac{L_{e_2}}{S_{b_2}} F[b_2] \\
&\int\limits_{L_{\mathrm{BC}}} \left(\vec{n} \times \vec{F}\right) \cdot \mathrm{d}\vec{l} = \int\limits_{L_{\mathrm{BC}}} \left(\vec{F}_{\mathrm{BC}}\right)_y \mathrm{d}l_{\mathrm{BC}} = -F\left[B_3^{\mathrm{virt}}\right] L_{e_3} = -\frac{L_{e_3}}{S_{b_3}} F[b_3] \\
&\int\limits_{L_{\mathrm{CD}}} \left(\vec{n} \times \vec{F}\right) \cdot \mathrm{d}\vec{l} = \int\limits_{L_{\mathrm{CD}}} \left(\vec{F}_{\mathrm{CD}}\right)_x \mathrm{d}l_{\mathrm{CD}} = -F\left[B_4^{\mathrm{virt}}\right] L_{e_4} = -\frac{L_{e_4}}{S_{b_4}} F[b_4]
\end{aligned}\right\} \tag{8.98}$$

where

$$L_{e1} = \text{DA};\quad L_{e2} = \text{AB};\quad L_{e3} = \text{BC};\quad L_{e4} = \text{CD}$$
$$S_{b1} = S_{\text{ADD}'\text{A}'};\quad S_{b2} = S_{\text{ABB}'\text{A}'};\quad S_{b3} = S_{\text{BCC}'\text{B}'};\quad S_{b4} = S_{\text{CDD}'\text{C}'}$$

Substitution of Equation 8.98 into Equation 8.96 yields the SIBC (Equation 8.91) in terms of the state variables of FIT:

$$b_0 = \sum_{k=1}^{4} \frac{L_{ek}}{S_{bk}} F[b_k] \tag{8.99}$$

Substituting Equation 8.99 into Faraday's law (Equation 8.84) we obtain the analog of Equation 7.64:

$$e_k = \frac{L_{ek}}{S_{bk}} \frac{\partial}{\partial t} F[b_k] \tag{8.100}$$

The relations in Equations 8.99 and 8.100 are the desired SIBCs in terms of circulations of the electric field and magnetic flux.

To illustrate and verify our results consider a canonical two-dimensional example, for which the analytical solution is known [3]: a line current $I(t)$ placed at point $(0, y_s)$ radiating over a half-space (Figure 8.11). Even though the high-order SIBC accurately models the curvature of the surface, the test case has been chosen with a planar surface in order not to introduce errors caused by the staircase effect (whereby the curved surface is modeled using a Cartesian grid). As a matter of fact, the modeling of curved surfaces by means of high-order SIBC and FIT would require a conformal scheme. In

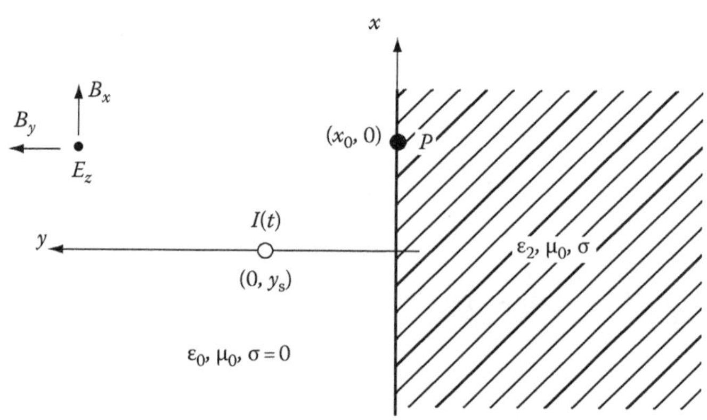

FIGURE 8.11
Geometry of the test problem.

the proposed example the improvement in accuracy provided by the SIBC of high order of approximation is only due to the modeling of the variation of the electromagnetic field along the surface, which is not modeled by low-order SIBC of the Leontovich type.

The computational domain (the dielectric half-space) is discretized into a 100×100 Cartesian grid made of one layer of three-dimensional Yee cubic cells with side length $\Delta = 0.015$ m. Mur's first-order absorbing boundary conditions are used at the other boundaries [4]. The following current pulse is considered (Figure 8.12):

$$I(t) = \frac{t - \tau_0}{\tau} e^{-\left(\frac{t-\tau_0}{\tau}\right)^2} \tag{8.101}$$

with $\tau = 40\Delta t, \tau_0 = 12\Delta t$, where the time step Δt was chosen as $\Delta t = \Delta x / 2c_0$.

The computed results were compared with the exact solution of the problem in the time domain, obtained from the solution in the frequency domain reported in [5] by means of the inverse fast Fourier transform algorithm. The Fourier transform of $I(t)$ of Equation 8.101 was derived analytically as

$$I(\omega) = -\frac{j\sqrt{\pi}}{2} \tau^2 \omega e^{-j\omega\tau_0} e^{-\frac{\omega^2\tau^2}{4}} \tag{8.102}$$

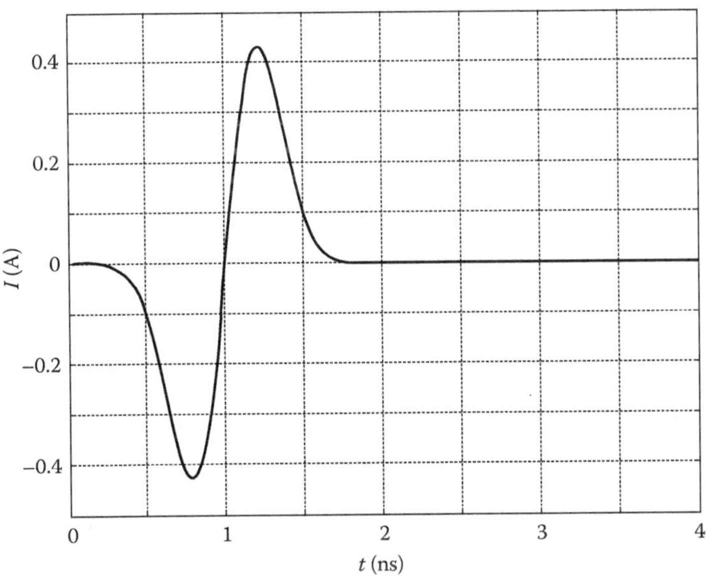

FIGURE 8.12
The current pulse $I(t)$.

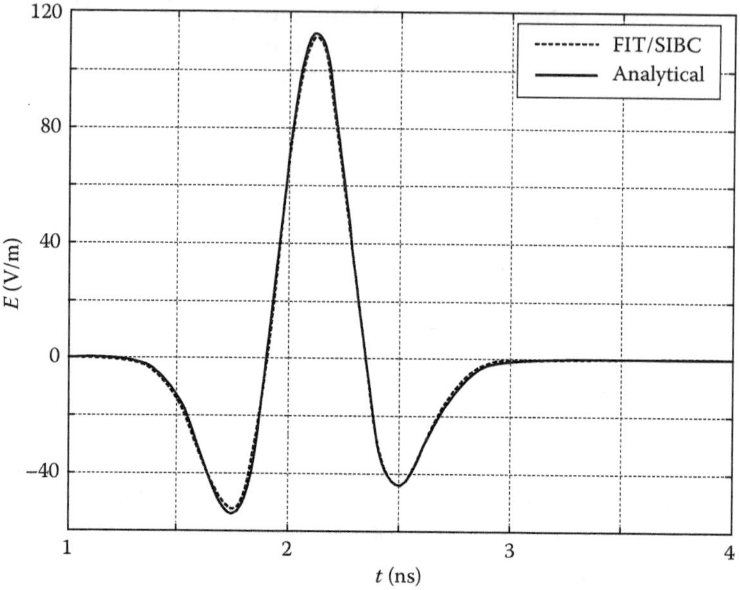

FIGURE 8.13
Electric field at the observation point for $y_s = 20\Delta$; $x_s = 10\Delta$; $\sigma = 10$ S/m.

and then substituted in the expression of the electric field at the observation point [5]:

$$E_z(\omega, x_0) = -\frac{c_0\mu_0 k_0}{2\pi} I(\omega) \int_0^{\infty} \frac{2}{\sqrt{1-\xi^2} + \sqrt{\varepsilon_{2r} - \xi^2}} \exp\left(-jk_0 y_s \sqrt{1-\xi^2}\right) \cos(k_0 x_0 \xi) d\xi \tag{8.103}$$

where $k_0 = \omega/c_0$.

The electric field at point $(x_0, 0)$ is computed using FIT with the low-order SIBC, FIT with the high-order SIBC, and the analytical solution (Figures 8.13 through 8.18). As can be seen from Figures 8.13 and 8.14, the low-order SIBC is sufficiently accurate when conductivity is relatively high ($\sigma = 10$ and 1 S/m). At low values of conductivity ($\sigma = 0.1$ S/m) and when the filamentary source is near the surface, the SIBC of low order does not provide accurate results and high-order SIBCs should be used (Figures 8.15 through 8.17). Figure 8.18 shows the particular case of $x_s = 0$ when the tangential derivatives are zero due to symmetry and, consequently, low- and high-order SIBCs give the same results.

As a conclusion of this section we show how the conditions in Equations 8.99 and 8.100 can be generalized for tetrahedral grids. Let the facet *ABC* of

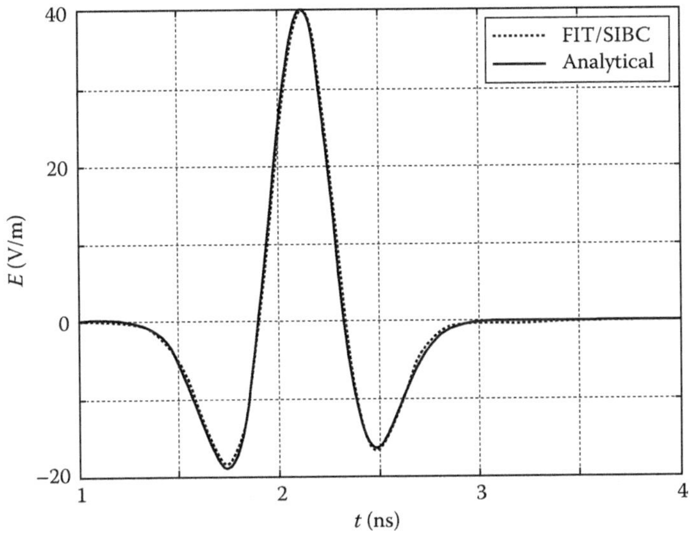

FIGURE 8.14
Electric field at the observation point for $y_s = 20\Delta$; $x_s = 10\Delta$; $\sigma = 1$ S/m.

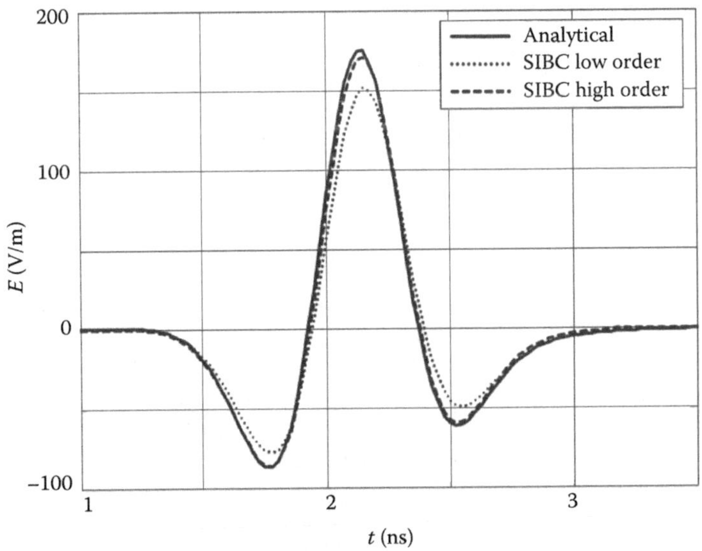

FIGURE 8.15
Electric field at the observation point for $y_s = 10\Delta$; $x_s = 20\Delta$; $\sigma = 0.1$ S/m.

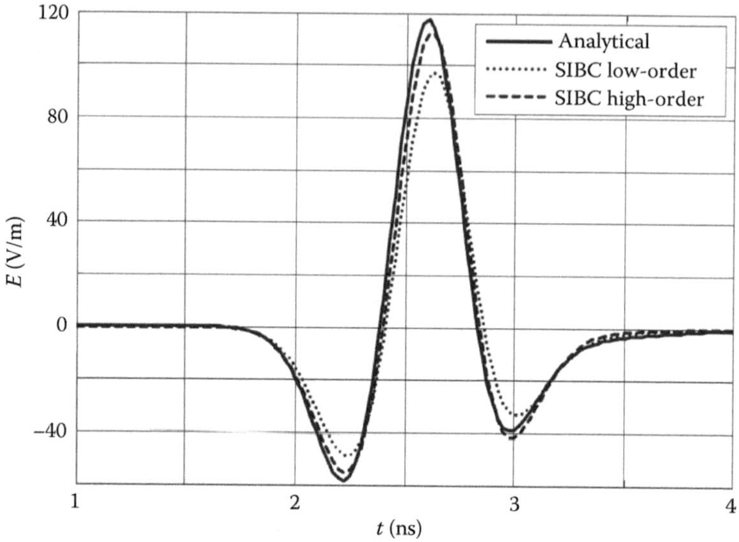

FIGURE 8.16
Electric field at the observation point for $y_s = 10\Delta$; $x_s = 30\Delta$; $\sigma = 0.1$ S/m.

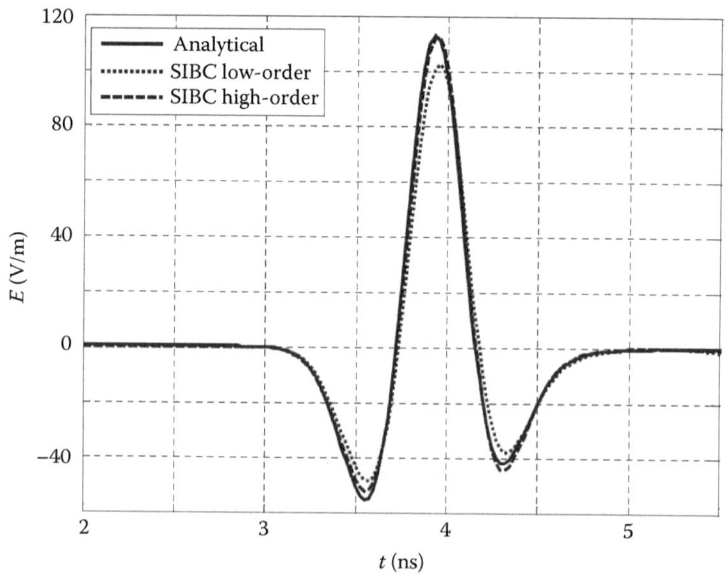

FIGURE 8.17
Electric field at the observation point for $y_s = 30\Delta$; $x_s = 50\Delta$; $\sigma = 0.1$ S/m.

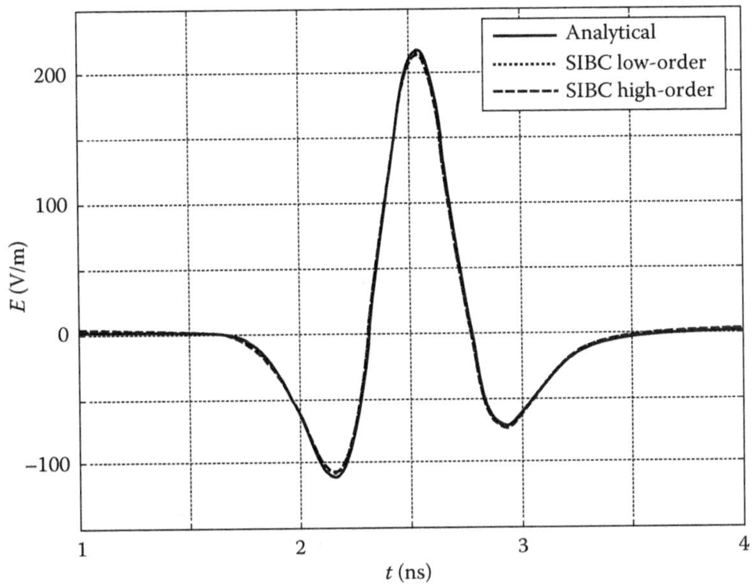

FIGURE 8.18
Electric field at the observation point for $y_s = 30\Delta$; $x_s = 0$; $\sigma = 0.1$ S/m.

the tetrahedron, shown in Figure 8.19, be a part of the conductor/dielectric interface. In this case the SIBCs can be written in the form:

$$b_{ABC} = \sum_{k=1}^{3} \frac{L_{ek}}{S_{bk}} F[b_k] \sin \zeta_k, \quad k = 1,2,3 \tag{8.104}$$

$$e_k = \frac{L_{ek}}{S_{bk}} \sin \zeta_k \frac{\partial}{\partial t} F[b_k] \tag{8.105}$$

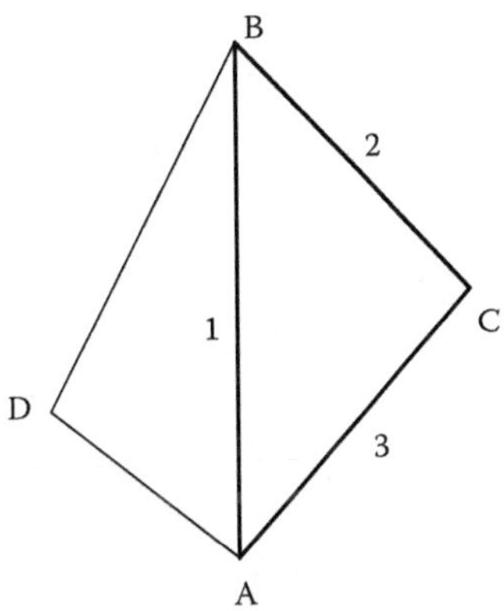

FIGURE 8.19
Tetrahedral computational cell.

where

$b_1 = b_{ABD}$
$b_2 = b_{BCD}$
$b_3 = b_{ACD}$
$L_{e1} = AB$
$L_{e2} = BC$
$L_{e3} = CD$
$S_{b1} = S_{ABD}$
$S_{b2} = S_{BCD}$
$S_{b3} = S_{ACD}$
ζ_1 is the angle between facets ABC and ABD
ζ_2 is the angle between facets ABC and BCD
ζ_3 is the angle between facets ABC and ACD

8.5 Finite-Element Method

The FEM also uses discretization of space in tetrahedral grids, but, unlike FDTD and FIT, the FEM requires that Maxwell's equations be transformed before discretization. Usually, the FEM is used in the frequency domain so we also will consider the case of single-frequency excitation. The governing equation for the magnetic field in the dielectric domain V (Figure 8.1) is obtained directly from Equations 8.1 and 8.2 and written in the form:

$$\nabla \times \left[(j\omega\varepsilon_1)^{-1} \nabla \times \vec{H} \right] + j\omega\mu\vec{H} = \nabla \times \vec{J}^{\text{inc}} \tag{8.106}$$

where $\vec{J}^{\text{inc}}$ denotes the incident current.

Following the method of weighted residuals, we multiply Equation 8.106 by a set of vector weighting functions $\vec{W}_m$, $m = 1, 2, \ldots, M$, and integrate over V as follows:

$$\int_V \nabla \times \left[(j\omega\varepsilon_1)^{-1} \nabla \times \vec{H} \right] \cdot \vec{W}_m dv + \int_V j\omega\mu\vec{H} \cdot \vec{W}_m dv = \int_V \nabla \times \vec{J}^{\text{inc}} \cdot \vec{W}_m dv \tag{8.107}$$

To represent Equation 8.107 in the "weak" form, we integrate by parts the first term on the left-hand side of Equation 8.107 and apply the vector identity:

$$\int_V \nabla \times \left[(j\omega\varepsilon_1)^{-1} \nabla \times \vec{H} \right] \cdot \vec{W}_m dv = \int_V (j\omega\varepsilon_1)^{-1} \left(\nabla \times \vec{H} \right) \cdot \left(\nabla \times \vec{W}_m \right) dv - \int_{S+S_{\text{inf}}} (j\omega\varepsilon_1)^{-1} \left[\left(\nabla \times \vec{H} \right) \times \vec{n} \right] \cdot \vec{W}_m ds \tag{8.108}$$

The surface integral in Equation 8.108 is written in the following form using Equation 8.1:

$$\int\limits_{S+S_{\text{inf}}} (j\omega\varepsilon_1)^{-1}\left[\left(\nabla \times \vec{H}\right) \times \vec{n}\right] \cdot \vec{W}_m \mathrm{d}s = -\int\limits_{S+S_{\text{inf}}} \left(\vec{n} \times \vec{E}\right) \cdot \vec{W}_m \mathrm{d}s \qquad (8.109)$$

Substituting Equation 8.109 into Equation 8.108, we obtain the "weak" formulation: find the magnetic field $\vec{H}$ such that

$$\int\limits_{V} \left\{\frac{1}{j\omega\mu} \nabla \times \vec{H} \cdot \left(\nabla \times \vec{W}_m\right) + j\omega\varepsilon\vec{H} \cdot \vec{W}_m\right\} \mathrm{d}v + \left\langle \left(\vec{n} \times \vec{E}\right), \vec{W}_m \right\rangle_S$$
$$= \int\limits_{V} \nabla \times \vec{J}^{\text{inc}} \cdot \vec{W}_m \mathrm{d}v - \left\langle \left(\vec{n} \times \vec{E}\right), \vec{W}_m \right\rangle_{S_{\text{inf}}} \qquad (8.110)$$

holds for square integrable test functions $\vec{W}_m$, and where

$$\left\langle \vec{u}, \vec{\lambda} \right\rangle_S = \int\limits_{S} \vec{u} \cdot \vec{\lambda} \mathrm{d}s \qquad (8.111)$$

A truncation technique must be applied to treat the second term on the right-hand side of Equation 8.110.

The magnetic field inside V can be expanded with a set of vector functions $\vec{W}_m$:

$$\vec{H} = \sum_{n=1}^{M} H_n^{(1)} \vec{W}_n \qquad (8.112)$$

where M is the number of degrees of freedom. Substitution of Equation 8.112 into Equation 8.110 and using the frequency domain SIBC in Equation 7.64 yields the following matrix equation for the expansions coefficients:

$$[A]\{H^{(1)}\} = \{J^{\text{inc}}\} \qquad (8.113)$$

$$A_{mn} = \int\limits_{V} \left\{\frac{1}{j\omega\mu} \nabla \times \vec{W}_n \cdot \left(\nabla \times \vec{W}_m\right) + j\omega\varepsilon\vec{W}_n \cdot \vec{W}_m\right\} \mathrm{d}v + \left\langle \vec{P}_n^{W}, \vec{W}_m \right\rangle_S \qquad (8.114)$$

$$J_m^{\text{inc}} = \int\limits_{V} \nabla \times \vec{J}^{\text{inc}} \cdot \vec{W}_m \mathrm{d}v; \quad m = 1, 2, \ldots, M, \quad n = 1, 2, \ldots, N \qquad (8.115)$$

where the following notation has been introduced:

$$\left(\vec{P}_i^{\mathrm{f}}\right)_{\xi_l} = -\frac{1+j}{2\theta}\mu\omega\delta\left\{\left(\vec{f}_i\right)_{\xi_l} + \frac{1-j}{4\theta}\delta\left(d_{3-l}^{-1} - d_l^{-1}\right)\left(\vec{f}_i\right)_{\xi_l} + \frac{\delta^2}{2j\theta}\frac{3d_{3-l}^2 - d_l^2 - 2d_l d_{3-l}}{8d_l^2 d_{3-l}^2}\left(\vec{f}_i\right)_{\xi_l}\right.$$
$$\left. + \frac{\delta^2}{2j\theta}\left(-\frac{\partial^2\left(\vec{f}_i\right)_{\xi_l}}{\partial\xi_{3-l}^2} + \frac{\partial^2\left(\vec{f}_i\right)_{\xi_l}}{\partial\xi_l^2} + 2\frac{\partial^2\left(\vec{f}_i\right)_{\xi_l}}{\partial\xi_l\partial\xi_{3-l}}\right)\right\} \tag{8.116}$$

The matrix equations for implementation of the Leontovich and Mitzner SIBCs can be easily obtained from Equations 8.113 through 8.115 by neglecting the terms containing δ^2 and δ^3, respectively.

Consider implementation of SIBCs using first-order edge finite elements [6,7]. The vector functions $\vec{W}_m$ and $\vec{F}_k$ have the forms:

$$\vec{W}_m = \lambda_1^{(m)}\nabla\lambda_2^{(m)} - \lambda_2^{(m)}\nabla\lambda_1^{(m)} \tag{8.117}$$

$$\vec{F}_k = \vartheta_1^{(k)}\nabla\vartheta_2^{(k)} - \vartheta_2^{(k)}\nabla\vartheta_1^{(k)} \tag{8.118}$$

The barycentric functions λ_i and ϑ_m can be represented in the form:

$$\lambda_i^{(m)} = (6V_{\mathrm{elem}})^{-1}\left(a_0^{(mi)} + \sum_{l=1}^{3} a_l^{(mi)} x_l\right), \quad i = 1,2 \tag{8.119}$$

$$\vartheta_j^{(k)} = (3S_{\mathrm{facet}})^{-1}\left(b_0^{(nj)} + \sum_{l=1}^{3} b_l^{(nj)} x_l\right), \quad j = 1,2 \tag{8.120}$$

where x_i are the global Cartesian coordinates that can be expressed in terms of the local coordinates ξ_k using the transformation matrix α:

$$x_i = x_i^{\mathrm{o}} + \sum_{j=1}^{3}\alpha_{ji}\xi_j, \quad i = 1,2,3 \tag{8.121}$$

where x_i^{o} are global coordinates of the origin of the local coordinate system. Thus Equations 8.119 and 8.120 can be represented in the forms:

$$\lambda_i^{(m)} = (6V_{\mathrm{elem}})^{-1}\left(\tilde{a}_0^{(mi)} + \sum_{l=1}^{3}\tilde{a}_l^{(mi)}\xi_l\right), \quad i = 1,2 \tag{8.122}$$

$$\vartheta_j^{(k)} = (3S_{\mathrm{facet}})^{-1}\left(\tilde{b}_0^{(nj)} + \sum_{l=1}^{3}\tilde{b}_l^{(nj)}\xi_l\right), \quad j = 1,2 \tag{8.123}$$

where

$$\tilde{a}_0^{(l)} = a_0^{(l)} + \sum_{k=1}^{3} a_k^{(l)} x_k^{\mathrm{o}}; \quad \tilde{b}_0^{(l)} = b_0^{(l)} + \sum_{k=1}^{3} b_k^{(l)} x_k^{\mathrm{o}}; \quad \tilde{a}_j^{(l)} = \sum_{k=1}^{3} \alpha_{jk} a_k^{(l)};$$

$$\tilde{b}_j^{(l)} = \sum_{k=1}^{3} \alpha_{jk} b_k^{(l)}, \quad l = 1,2,3 \tag{8.124}$$

Substitution of Equations 8.122 through 8.124 into Equations 8.117 and 8.118 allows representation of the functions $\vec{W}_m$ and $\vec{F}_k$ in the final forms:

$$\vec{W}_m = (6V_{\text{elem}})^{-2} \sum_{n=1}^{3} \vec{e}_n \left(A_{on}^{(m)} + \sum_{i=1}^{3} A_{in}^{(m)} \xi_i \right) \tag{8.125}$$

$$\vec{F}_k = (3S_{\text{facet}})^{-2} \sum_{l=1}^{3} \vec{e}_l \left(B_{ol}^{(k)} + \sum_{j=1}^{3} B_{jl}^{(k)} \xi_j \right) \tag{8.126}$$

where

$$A_{ij}^{(m)} = \tilde{a}_i^{(m1)} \tilde{a}_j^{(m2)} - \tilde{a}_j^{(m2)} \tilde{a}_i^{(m1)}; \quad B_{ij}^{(m)} = \tilde{b}_i^{(m1)} \tilde{b}_j^{(m2)} - \tilde{b}_j^{(m2)} \tilde{b}_i^{(m1)}, \quad i = 0,1,2,3, \quad j = 1,2,3 \tag{8.127}$$

The first, second, and third terms in Equation 8.116 are implemented in the same way. This means that implementation of Mitzner's SIBC does not lead to any additional computational difficulties compared to the standard Leontovich SIBC.

The fourth term in Equation 8.116 contains the integrals of the following kind:

$$\int_{S_{\text{fac}}} f_{\xi_i} \frac{\partial^2 f_{\xi_i}}{\partial \xi_i \partial \xi_j} \text{ds}; \quad \int_{S_{\text{fac}}} f_{\xi_i} \frac{\partial^2 f_{\xi_i}}{\partial \xi_i^2} \text{ds}; \quad \int_{S_{\text{fac}}} f_{\xi_i} \frac{\partial^2 f_{\xi_i}}{\partial \xi_j^2} \text{ds}, \quad i = 1,2, \quad j = 1,2 \tag{8.128}$$

where $\vec{f}$ denotes $\vec{W}_m$ and $\vec{F}_k$. From Equation 8.127 it directly follows that

$$A_{ll}^{(m)} = 0; \quad B_{ll}^{(m)} = 0, \quad l = 1,2,3 \tag{8.129}$$

By substituting Equation 8.129 into Equations 8.125 and 8.126, we obtain

$$\frac{\partial \left(\vec{W}_m \right)_{\xi_l}}{\partial \xi_l} = 0; \quad \frac{\partial \left(\vec{F}_k \right)_{\xi_l}}{\partial \xi_l} = 0, \quad l = 1,2 \tag{8.130}$$

It is easy to see that evaluation of the integrals in Equation 8.128 by taking into account Equation 8.130 leads to zero so that the last term in Equation 8.116 disappears. Thus, the use of the first-order vector approximating functions does not allow implementation of Rytov's SIBC accurately. That is, it does not take into account the tangential variation of the magnetic field over the surface of the conductor. Higher-order vector functions are needed for this purpose.

Appendix 8.A.1: Basics of Contour-Path FDTD Method

The Yee algorithm was originally formulated in rectangular Cartesian grids [14]. In these grids, curved boundaries must be approximated by a staircase model. In certain problems this approach leads to significant error. Many alternatives have been suggested and developed to modify the original FDTD formulation in order to accurately model curved boundaries [4,8–11]. Among them are orthogonal subgridding (use of fine cells near the boundary and coarse meshing elsewhere), general nonorthogonal gridding and local conformal grids. The contour-path FDTD method is one of the methods under the category of local conformal gridding. This method, developed by Jurgenns et al. [12] uses standard orthogonal Yee cells everywhere except in the vicinity of the curved boundaries. The cells immediately adjacent to the boundary of the scatterer are distorted to conform to the boundary. The method is based on Faraday's and Ampere's laws in integral forms instead of differential forms in the original FDTD. The basics of the method is briefly described in the following.

Faraday's law in integral form:

$$\mu \frac{\partial}{\partial t} \int_{s} \vec{H} \cdot \vec{n} ds = -\oint_{c} \vec{E} \cdot d\vec{l} \tag{8.A.1}$$

can be applied along a contour c as shown in Figure 8.A.1 assuming that the electric field is constant along each edge of the contour [13]:

$$\mu \frac{\partial}{\partial t} \int_{s} H_z\big|_{i,j,k} ds = -\left(l_1 E_x\big|_{i,j-1/2,k} + l_2 E_y\big|_{i+1/2,j,k} - l_3 E_x\big|_{i,j+1/2,k} - l_4 E_y\big|_{i-1/2,j,k}\right) \tag{8.A.2}$$

then with the assumption of the magnetic field being constant inside the surface s and applying the central differences for time derivatives:

$$\mu \left(\frac{H_z\big|_{i,j,k}^{n+1/2} - H_z\big|_{i,j,k}^{n-1/2}}{\Delta t}\right) \int_{s} ds = -\left(l_1 E_x\big|_{i,j-1/2,k} + l_2 E_y\big|_{i+1/2,j,k} - l_3 E_x\big|_{i,j+1/2,k} - l_4 E_y\big|_{i-1/2,j,k}\right) \tag{8.A.3}$$

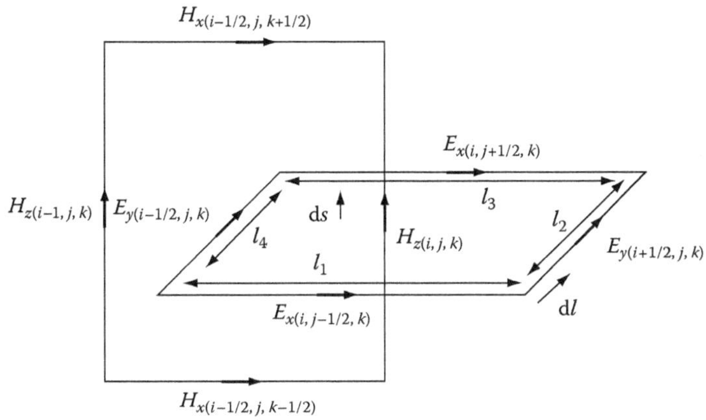

FIGURE 8.A.1
Faraday's law in integral form.

Rearranging Equation 8.A.3 gives

$$H_z\big|_{i,j,k}^{n+1/2} = H_z\big|_{i,j,k}^{n-1/2} - \frac{\Delta t}{\mu A}\left(l_1 E_x\big|_{i,j-1/2,k} + l_2 E_y\big|_{i+1/2,j,k} - l_3 E_x\big|_{i,j+1/2,k} - l_4 E_y\big|_{i-1/2,j,k}\right) \tag{8.A.4}$$

It is easy to see that Equation 8.A.4 reduces to the standard Cartesian Yee algorithm in the particular case of $l_1 = l_3 = \Delta x$, $l_2 = l_4 = \Delta y$, and $A = \Delta x \Delta y$.

Analogously, as shown in Figure 8.A.2, the integral form of Ampere's law

$$\varepsilon \frac{\partial}{\partial t} \int_S \vec{E} \cdot \vec{n} \mathrm{d}s = \int_C \vec{H} \cdot \mathrm{d}\vec{l} \tag{8.A.5}$$

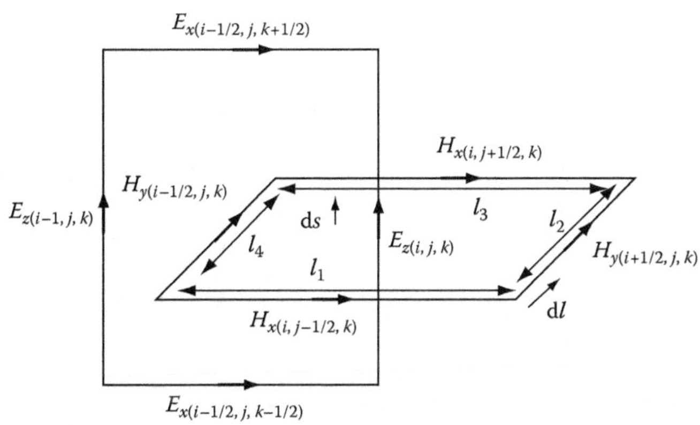

FIGURE 8.A.2
Ampere's law in integral form.

is applied to the contour c and the assumption of the magnetic field being constant along the contour yields

$$\varepsilon \frac{\partial}{\partial t} \int_{s} E_z\big|_{i,j,k} ds = l_1 H_x\big|_{i,j-1/2,k} + l_2 H_y\big|_{i+1/2,j,k} - l_3 H_x\big|_{i,j+1/2,k} - l_4 H_y\big|_{i-1/2,j,k} \quad (8.A.6)$$

Approximating the time derivatives with the central differences scheme finally gives

$$E_z\big|_{i,j,k}^{n+1} = E_z\big|_{i,j,k}^{n} + \frac{\Delta t}{\varepsilon A}\left(l_1 H_x\big|_{i,j-1/2,k}^{n+1/2} + l_2 H_x\big|_{i+1/2,j,k}^{n+1/2} - l_3 H_x\big|_{i,j+1/2,k}^{n+1/2} - l_4 H_y\big|_{i-1/2,j,k}\right) \quad (8.A.7)$$

References

1. T. Weiland, A discretization method for the solution of Maxwell's equations for six-component fields, *Electronics and Communications AEU*, 31(3), 1977, 116–120.
2. T. Weiland, Advances in FIT/FDTD modeling, *Proceedings of 18th Annual Review of Progress in Applied Computational Electromagnetics Conference* (*ACES*), Monterey, 2002, pp. 1–14.
3. G.S. Smith, On the skin effect approximation, *American Journal of Physics*, 58(10), 1990, 996–1002.
4. G. Mur, Absorbing boundary conditions for the finite-difference approximation of the time-domain electromagnetic field equations, *IEEE Transactions of Electromagnetic Compatibility*, EMC-23(4), November 1981, 377–382.
5. J.G. Maloney and G.S. Smith, The use of surface impedance concepts in the finite difference time-domain method, *IEEE Transactions on Antennas and Propagation*, 40(1), January 1992, 38–48.
6. M.L. Barton and Z.J. Cendes, New vector finite elements for three dimensional magnetic field computation, *Journal of Applied Physics*, 61(2), April, 1987, 3919–3921.
7. A. Bossavit, A rationale for edge elements in 3-D field computations, *IEEE Transactions on Magnetics*, MAG-24(1), January 1988, 74–79.
8. T.V. Yioultsis, N.V. Kantartzis, C.S. Antonopoulos, and T.D. Tsiboukis, A fully explicit Whithey elements—Time domain scheme with higher order vector finite elements for three-dimensional high frequency problems, *IEEE Transactions on Magnetics*, 34(5), September 1998, 191–202.
9. J.P. Berenger, A perfectly matched layer for the absorption of electromagnetic waves, *Journal of Computational Physics*, 114, 1994, 185–200.
10. K.S. Kunz and R.J. Luebbers, *The Finite Difference Time-Domain Method for Electromagnetics*, CRC Press, Boca Raton, FL, 1993.

11. C.S. Chan, H. Sangami, J.T. Elson, and R.F. Bowers, Conformal finite-difference time-domain methods, in *Time-Domain Methods for Microwave Structures*, T. Itoh and B. Houshmad, eds., IEEE Press, Piscataway, NJ, 1998, pp. 181–197.
12. T.G. Jurgens, A. Taflove, K. Umashankar, and T.G. Moore, Finite-difference time-domain modeling of curved surfaces, *IEEE Transactions on Antennas and Propagation*, 40(4), April 1992, 357–365.
13. A. Taflove, *Computational Electrodynamics, the Finite-Difference Time-Domain Method*, Artech House, Norwood, MA, 1995.
14. K.S. Yee, Numerical solution of initial boundary value problems involving Maxwell's equations in isotropic media, *IEEE Transactions on Antennas and Propagation*, 14(3), May 1996, 302–307.

9

Application and Experimental Validation of the SIBC Concept

9.1 Introduction

The purpose of this chapter is twofold. The first is to discuss the validation of the SIBCs developed in the previous chapters. Of particular interest here is the experimental validation of SIBCs because analytical and numerical validations have been incorporated in Chapters 4 through 8 as part of the formulation and development of the various SIBCs. The second, perhaps more important purpose is to describe a simple methodology for the selection of appropriate SIBCs for use with any particular problem the user may need to address and, indeed, to allow a simple decision process by which the user can first decide if SIBCs are in fact appropriate for the problem at hand and then to decide on the order of approximation needed. This should aid in the practical use of the foregoing SIBCs. We start with the second of these goals.

9.2 Selection of the Surface Impedance Boundary Conditions for a Given Problem

In the previous chapters, we classified the various SIBCs by their order of approximation. This classification afforded a certain understanding of their behavior, especially with regard to the ability of an SIBC to take into account curvatures and tangential variations of the fields. We said, and it is worth repeating here, that the zero-order approximation (PEC limit) allows no penetration of fields into the material, the first-order approximation (Leontovich condition) allows only perpendicular penetration into the conductor based on the skin depth condition, the second-order approximation (Mitzner's condition) allows for surface curvatures, and the third-order approximation (Rytov's condition) allows as well for tangential variations in the field. This classification clearly indicates that the errors in the Rytov approximation are fourth order in skin depth or, as we have indicated previously, in the small parameter $\tilde{p}$ (see Equation 2.19 as well as Table 3.1).

The correct choice of SIBCs for a given problem is a very important, practical consideration the user is often faced with. The proper selection affects not only the results obtained but also the effort and time involved in solving the problem. The required order is not always clear *a priori*, especially in transient applications. If the order of approximation of the SIBCs used in the computation is inadequate, the computational results will be inferior in that large errors will be present, something that is difficult to gauge when reliable experimental or numerical validation data are not available. On the other hand, application of high-order SIBCs for the calculation of very thin skin layers does not provide any gain in accuracy of the results and leads to useless computational expenses. Fortunately, the approach we have taken in the formulations of the previous chapters, especially the reliance on the Rytov expansion method and the reliance on the characteristic values of the physical problem, lends itself to a simple and reliable selection of the SIBC.

In this section, we look at some simple universal relationships between the characteristic values of the problem and the order of approximation so that the SIBCs best suited for a given problem can be selected in a simple and practical way. The methodology is applicable to all formulations and SIBCs described in the previous chapters including the time-domain, frequency-domain, linear, and nonlinear SIBCs. An example is included to illustrate the methodology.

9.2.1 Characteristic Values of the Problem

Any nonstatic electromagnetic problem involves two basic scales: the characteristic dimension D of the body's surface and the characteristic time τ. These are the required values in the methodology to follow since they affect the applicability of the SIBC, as was amply demonstrated in previous chapters. We use these values as the input to our methodology.

The characteristic time of a problem is defined as the ratio $2/\omega$ for time-harmonic incident fields or as the incident pulse duration τ for a pulsed source.

The characteristic dimension D is defined as (see Section 2.2 and Table 3.1):

$$D = \min(d_1, d_2) \tag{9.1}$$

where d_1 and d_2 are the radii of curvature of the surface coordinate lines in the case of smooth bodies. In the more general case of multiple conductors and sources, the condition in Equation 9.1 may be expanded as $D = \min(d_1, d_2, d_3, d_4)$, where d_3 might be the distance to the nearest source and d_4 the distance to the nearest conductor in the configuration. The main point is that D is the smallest "feature" of the physical configuration of the problem that may affect the fields on the surface of the conductors and, hence, in the skin layer.

Using D and τ, we defined the characteristic skin depth δ and characteristic dimension λ of the field variation along the body's surface as follows (see Equation 2.1):

$$\delta = \sqrt{\frac{\tau}{\mu\sigma}} \tag{9.2}$$

$$\lambda = c\tau \tag{9.3}$$

where c is the velocity of light.

If the material properties are nonlinear, the nonlinear characteristic permeability μ and conductivity σ must be used in Equation 9.2. This may be the case when a nonlinear BH-curve is given or when conductivity is temperature dependent.

The conditions of applicability of the surface impedance concept can now be written in terms of D, δ, and λ:

$$\lambda \ll D \quad \text{or} \quad \tilde{p} = \frac{\delta}{D} \ll 1 \tag{9.4}$$

$$\lambda \gg D \quad \text{or} \quad \tilde{q} = \frac{D}{\lambda} = \frac{D}{c\tau} \ll 1 \tag{9.5}$$

We now take $\tilde{p}$ and $\tilde{q}$ as the basic parameters of the problem (see also Chapter 2). The conditions (Equations 9.4 and 9.5) hold for all SIBCs. The various representations in the previous chapters were often written in terms of these parameters. Therefore, they are convenient in developing the methodology for SIBC selection.

9.2.2 Asymptotic Expansions

Although the arguments above and those to follow apply to all SIBCs we developed, and certainly to all formulations, we will nevertheless use as an example the E–H formulation from Section 3.3. Specifically, we can rewrite Equation 3.47, which gives the electric field in the frequency and time domains, in the Rytov approximation, as follows:

In the frequency domain,

$$\tilde{E}^{\mathrm{b}}_{\xi_{3-k}} = (-1)^k(1+j)\tilde{p}\left[\dot{\tilde{H}}^{\mathrm{b}}_{\xi_k} - \tilde{p}\frac{1-j}{2}\dot{\tilde{H}}^{\mathrm{b}}_{\xi_k}\frac{\tilde{d}_k - \tilde{d}_{3-k}}{2\tilde{d}_k\tilde{d}_{3-k}} + \tilde{p}^2\frac{j}{2}\left(\dot{\tilde{H}}^{\mathrm{b}}_{\xi_k}\frac{\tilde{d}_k^2 + 2\tilde{d}_k\tilde{d}_{3-k} - 3\tilde{d}_{3-k}^2}{8\tilde{d}_k^2\tilde{d}_{3-k}^2} - \frac{1}{2}\frac{\partial^2\dot{\tilde{H}}^{\mathrm{b}}_{\xi_k}}{\partial\tilde{\xi}_k^2} + \frac{1}{2}\frac{\partial^2\dot{\tilde{H}}^{\mathrm{b}}_{\xi_k}}{\partial\tilde{\xi}_{3-k}^2} - \frac{\partial^2\dot{\tilde{H}}^{\mathrm{b}}_{\xi_{3-k}}}{\partial\tilde{\xi}_k\partial\tilde{\xi}_{3-k}}\right)\right] + O(\tilde{p}^4), \quad k = 1,2 \tag{9.6}$$

In the time domain,

$$\tilde{E}^{\mathrm{b}}_{\xi_{3-k}} = (-1)^k \tilde{p}\left[\frac{\mathrm{d}}{\mathrm{d}\tilde{t}}\left(\tilde{H}^{\mathrm{b}}_{\xi_k} * \tilde{T}^{\mathrm{b}}_2\right) - \tilde{p}\frac{\tilde{d}_k - \tilde{d}_{3-k}}{2\tilde{d}_k\tilde{d}_{3-k}}\tilde{H}^{\mathrm{b}}_{\xi_k} * \tilde{T}^{\mathrm{b}}_1 - \tilde{p}^2\left(\tilde{H}^{\mathrm{b}}_{\xi_k}\frac{\tilde{d}_k^2 + 2\tilde{d}_k\tilde{d}_{3-k} - 3\tilde{d}_{3-k}^2}{8\tilde{d}_k^2\tilde{d}_{3-k}^2} - \frac{1}{2}\frac{\partial^2 \tilde{H}^{\mathrm{b}}_{\xi_k}}{\partial\tilde{\xi}_k^2}\right.\right.$$
$$\left.\left. + \frac{1}{2}\frac{\partial^2 \tilde{H}^{\mathrm{b}}_{\xi_k}}{\partial\tilde{\xi}_{3-k}^2} - \frac{\partial^2 \tilde{H}^{\mathrm{b}}_{\xi_{3-k}}}{\partial\tilde{\xi}_k\partial\tilde{\xi}_{3-k}}\right) * \tilde{T}^{\mathrm{b}}_2\right] + O(\tilde{p}^4), \quad k = 1, 2 \tag{9.7}$$

The approximation in Equations 9.6 or 9.7 allows us to replace the condition in Equation 9.4 by

$$\tilde{p}^4 \ll 1 \tag{9.8}$$

The representations in Equation 9.6 and 9.7 clearly show the various approximation orders. The zero-order (PEC) representation ($\tilde{p} = 0$) gives zero electric field, as required. The first-order approximation ($\tilde{p}^2 = 0$) gives the Leontovich approximation:

$$\tilde{E}^{\mathrm{b}}_{\xi_{3-k}} = (-1)^k(1 + j)\tilde{p}\dot{\tilde{H}}^{\mathrm{b}}_{\xi_k} + O(\tilde{p}^2) \tag{9.9}$$

$$\tilde{E}^{\mathrm{b}}_{\xi_{3-k}} = (-1)^k \tilde{p}\frac{\mathrm{d}}{\mathrm{d}\tilde{t}}\left(\tilde{H}^{\mathrm{b}}_{\xi_k} * \tilde{T}^{\mathrm{b}}_2\right) + O(\tilde{p}^2) \tag{9.10}$$

The second-order approximation ($\tilde{p}^3 = 0$) gives the Mitzner approximation:

$$\tilde{E}^{\mathrm{b}}_{\xi_{3-k}} = (-1)^k(1 + j)\tilde{p}\left[\dot{\tilde{H}}^{\mathrm{b}}_{\xi_k} - \tilde{p}\frac{1 - j}{2}\dot{\tilde{H}}^{\mathrm{b}}_{\xi_k}\frac{\tilde{d}_k - \tilde{d}_{3-k}}{2\tilde{d}_k\tilde{d}_{3-k}}\right] + O(\tilde{p}^3) \tag{9.11}$$

$$\tilde{E}^{\mathrm{b}}_{\xi_{3-k}} = (-1)^k(1 + j)\tilde{p}\left[\frac{\mathrm{d}}{\mathrm{d}\tilde{t}}\left(\tilde{H}^{\mathrm{b}}_{\xi_k} * \tilde{T}^{\mathrm{b}}_2\right) - \tilde{p}\frac{\tilde{d}_k - \tilde{d}_{3-k}}{2\tilde{d}_k\tilde{d}_{3-k}}\tilde{H}^{\mathrm{b}}_{\xi_k} * \tilde{T}^{\mathrm{b}}_1\right] + O(\tilde{p}^3) \tag{9.12}$$

The third-order (Rytov) approximation in Equations 9.6 and 9.7 clearly include the zeroth-, first-, and second-order approximations as well as the third-order corrections needed for the representation of tangential fields.

9.2.3 Methodology

Now we can evaluate the ranges of the characteristic values for which the SIBCs (Equations 9.6 through 9.12) are best applicable. The conditions (Equations 9.4 and 9.5) of applicability of the SIBCs involve the two characteristic scales (D and τ) and the two parameters of the problem ($\tilde{p}$ and $\tilde{q}$); therefore, the scales are "uniquely" expressed by these parameters.

From Equation 9.8 it follows that the approximation errors in the PEC limit, the Leontovich SIBCs, the Mitzner SIBCs, and the Rytov SIBCs are $\tilde{p}$, $\tilde{p}^2$, $\tilde{p}^3$, and $\tilde{p}^4$, respectively. Based on this observation, we can define approximate ranges of the parameter $\tilde{p}$, for which the SIBCs of these classes can be best applied:

1. For the PEC approximation,

$$\tilde{p} < 0.06 \tag{9.13}$$

2. For SIBCs in the Leontovich approximation,

$$\tilde{p} = 0.06 \div 0.25 \quad (\tilde{p}^2 \approx 0.003 \div 0.06) \tag{9.14}$$

3. For SIBCs in the Mitzner approximation,

$$\tilde{p} = 0.25 \div 0.4 \quad (\tilde{p}^3 \approx 0.02 \div 0.06) \tag{9.15}$$

4. For SIBCs in the Rytov approximation,

$$\tilde{p} = 0.4 \div 0.5 \quad (\tilde{p}^4 \approx 0.03 \div 0.06) \tag{9.16}$$

The range of the parameter $\tilde{q}$ can be defined as

$$\tilde{q} < 0.06 \tag{9.17}$$

With the definitions in Equations 9.13 through 9.17, the approximation error due to using the specific SIBCs will not exceed 6%.

These relations may be used as follows. From Equations 9.2, 9.4, and 9.5 we can write

$$D = \frac{\sqrt{\tau}}{\tilde{p}\sqrt{\sigma\mu}} \tag{9.18}$$

and

$$D = c\tau\tilde{q} \tag{9.19}$$

Now, by substituting the extreme values of the parameters $\tilde{p}$ and $\tilde{q}$ from Equations 9.13 through 9.17 into Equations 9.18 and 9.19, the desired ranges of the scales D and τ can be obtained for a given problem.

A small disadvantage of the relations in Equations 9.18 and 9.19 is that they are written in terms of specific material properties. To obtain more universal relations we introduce the following nondimensional variables:

$$\tilde{D} = \sigma\mu c D \tag{9.20}$$

$$\tilde{\tau} = \sigma\mu c^2 \tau \tag{9.21}$$

With the variables in Equations 9.20 and 9.21, the functions in Equations 9.18 and 9.19 can be written in the following forms:

$$\tilde{D} = \tilde{q}\tilde{\tau} \tag{9.22}$$

$$\tilde{D} = \frac{\sqrt{\tilde{\tau}}}{\tilde{p}} \tag{9.23}$$

These are more convenient than the relations in Equations 9.18 and 9.19 because they are nondimensional. Now we can plot these for the extreme values of $\tilde{p}$ and $\tilde{q}$ given in Equations 9.13 through 9.17. A representation of these values is shown in Figure 9.1.

Region 1a in Figure 9.1 shows the application area of the SIBCs in the PEC limit. Region 1b is the application range in the Leontovich approximation. Similarly, the Mitzner approximation and the Rytov approximation ranges are shown as 1c and 1d, respectively. If the point in the $\tilde{D}$–$\tilde{\tau}$ plane lies in regions 2 and 3, the surface impedance concept cannot be applied because the conditions in Equations 9.13 through 9.17 break down.

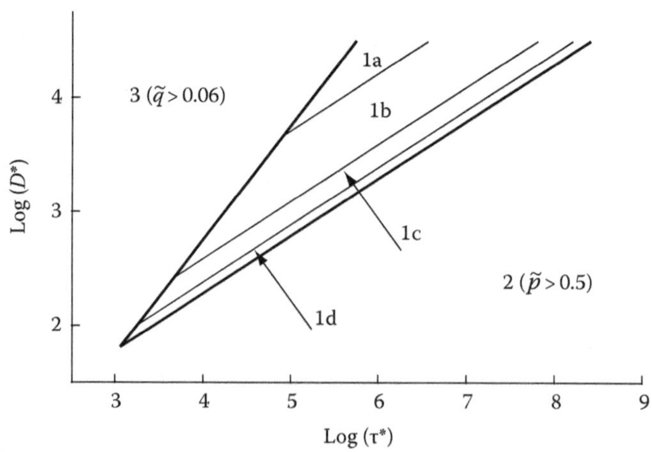

FIGURE 9.1
The $(\tilde{D})^* - (\tilde{\tau})^*$ plane.

Now we can summarize the methodology for selecting an appropriate SIBC for a given physical problem as follows:

1. Identify the characteristic values D and τ, and the material properties σ and μ.
2. Calculate or estimate the nondimensional values $\tilde{D}$ and $\tilde{\tau}$ using Equations 9.20 and 9.21.
3. Find the appropriate point in the $\tilde{D}-\tilde{\tau}$ plane.
4. If this point lies in the regions 1a through 1d, choose the corresponding SIBCs from Equation 9.6 or Equation 9.7 or the PEC boundary conditions by selecting those terms indicated in Equations 9.13 through 9.17.

It should be reemphasized that the functions shown in Figure 9.1 are "universal" because they do not depend on the properties of the conductors involved.

To demonstrate the application of the foregoing methodology, we consider here a problem in the time domain made of a pair of identical parallel copper conductors with a circular cross section in which equal and oppositely directed pulses of current of magnitude 1 A flow from an external source as shown in Figure 9.2. The radius of each conductor and the distance between the conductors were taken equal to 0.1 m (characteristic value $D = 0.1$ m). Under these conditions the current density has only one component directed along the conductors.

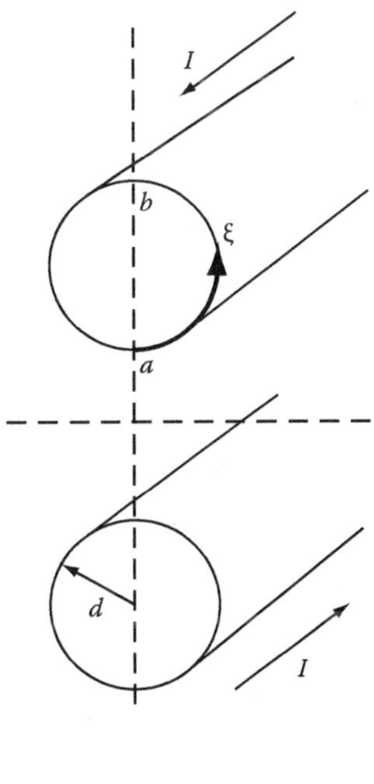

FIGURE 9.2
Physical geometry of the problem.

The SIBCs in Equations 9.10, 9.12, and 9.7 as well as the PEC condition were coupled with the surface integral equation using the formulation in Section 3.3 and solved using the boundary element method (BEM) as described in Chapter 6.

To illustrate the method, the distributions of the surface current density over one half of the cross section of one conductor were calculated for the following current pulses:

1. $\tau = 10^{-3}$ s ($\tilde{p} = 3.7 \times 10^{-2}$ and $\tilde{q} = 3.3 \times 10^{-7}$). From Equations 9.13 through 9.17 it follows that the PEC condition is suitable for this problem. Figure 9.3 shows that the use of an SIBC of the next order of approximation (Leontovich's SIBC) will not provide a significant increase in accuracy of the results (the difference between the curves does not exceed 4%).
2. $\tau = 10^{-2}$ s ($\tilde{p} = 1.2 \times 10^{-1}$ and $\tilde{q} = 3.3 \times 10^{-8}$). In this case, the Leontovich SIBC seems to be optimal. From Figure 9.4 it follows that the difference between the curves obtained using the PEC and Leontovich conditions is about 15% whereas application of the Mitzner SIBC increases the accuracy by only 2%.
3. $\tau = 10^{-1}$ s ($\tilde{p} = 3.7 \times 10^{-1}$ and $\tilde{q} = 3.3 \times 10^{-9}$). Figure 9.5 shows that the use of the Leontovich SIBC leads to unacceptable computational errors (about 18%). On the other hand, the difference between the curves obtained in the Mitzner and Rytov approximations does not exceed 3%. Therefore, in this problem it is necessary to use the Mitzner SIBC as the methodology predicts.

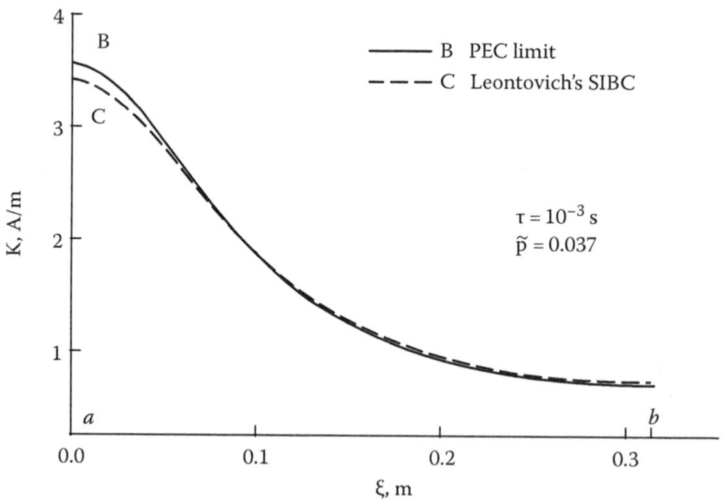

FIGURE 9.3
Distribution of the surface current density over the conductor's surface for $\tau = 10^{-3}$ s.

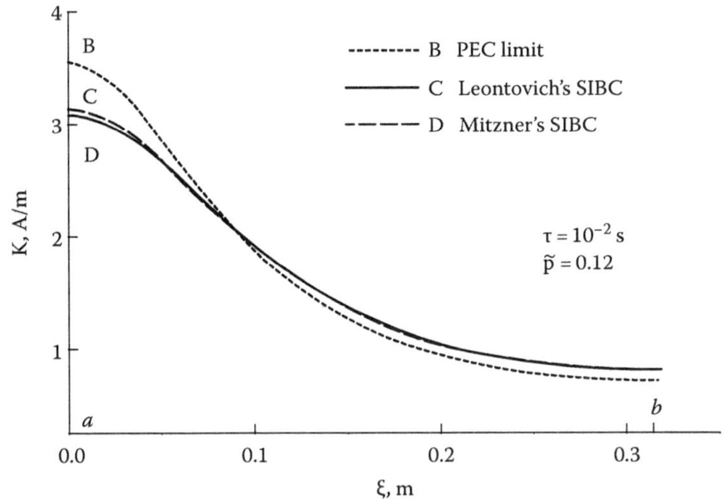

FIGURE 9.4
Distribution of the surface current density over the conductor's surface for $\tau = 10^{-2}$ s.

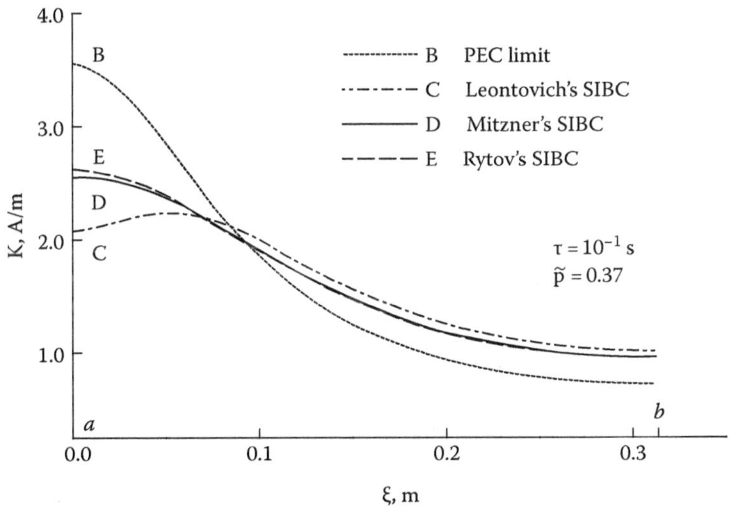

FIGURE 9.5
Distribution of the surface current density over the conductor's surface for $\tau = 10^{-1}$ s.

The methodology outlined above also facilitates the estimation of the approximation errors. Since the limit of applicability of a specific order of approximation is encapsulated in the condition in Equation 9.18 (or in Equation 2.1), we can rewrite this relation as follows:

$$\tilde{p}^m = \left[\tau/\left(\sigma\mu D^2\right)\right]^{m/2} \ll 1 \tag{9.24}$$

where the values $m=1,2,3$ correspond to the Leontovich SIBC, the Mitzner SIBC, and the Rytov SIBC, respectively, whereas $\tilde{p}=0$ corresponds to the PEC limit. We assume that the characteristics of the conductor's material are constant and Equation 9.24 contains two parameters: τ and D. Usually the duration of the pulse may vary whereas the radius of the conductor is constant. For such cases an approximation error ε_m can be estimated from Equation 9.24 as follows:

$$\varepsilon_m = \alpha^m \tau^{m/2}; \quad \alpha = \left(\sigma\mu D^2\right)^{-1/2}, \quad m = 1,2,3 \tag{9.25}$$

Note that the basic scale factors do not contain information about all the features of the problem. In particular, the definition of the time-scale factor as the duration of the pulse does not take into account the pulse shape. Another example is the space-scale factor D. It has been introduced as the characteristic size of the conductor's surface but it ignores the proximity effect in problems involving more than one conductor. In other words, according to Equation 9.25, the pulse shape and mutual configuration of the conductors do not have a significant effect on the approximation error of the SIBCs.

Consider again the example in Figure 9.2. We choose the duration τ of the incident pulse to equal 2 ms so that the small parameter $\tilde{p}$ is equal to 0.25, assuming D is the characteristic scale of the problem. To calculate the distributions of the surface current density for any instant of time using Equation 6.277, we need to define the shape of the current pulse (i.e., the function $\tilde{\theta}_0$). Calculations were performed for the following two different pulses $I_a(t)$ and $I_b(t)$ to investigate the role of the pulse shape:

$$I_a(t) = I_0\tilde{\theta}_a(t); \quad I_a(t) = I_0(t/\tau)^2 \tag{9.26}$$

$$I_b(t) = I_0\tilde{\theta}_b(t); \quad \tilde{\theta}_b = \sqrt{t/\tau} \tag{9.27}$$

Both pulses have the same duration, so that

$$I_a(0) = I_b(0); \quad I_a(\tau) = I_b(\tau) \tag{9.28}$$

The final distributions of the surface current density at $t=\tau$ for both pulses are shown in Figures 9.6 through 9.11. Figures 9.6 through 9.8 show $I_a(\tau)$ for $\mathrm{L}=2.5D$, $\mathrm{L}=3D$ and $\mathrm{L}=8D$ respectively whereas Figures 9.9 through 9.11 shown $I_b(\tau)$ for $\mathrm{L}=2.5D$, $\mathrm{L}=3D$ and $\mathrm{L}=8D$ respectively. For comparison purposes, each figure also shows the "reference data" obtained using the boundary element formulation based on the time-dependent fundamental solution [1] without the use of SIBCs (the so-called original BEM) and a commercial finite element software. It is clearly seen that application of the PEC limit when the skin depth is not thin ($\tilde{p}=0.25$) causes an error of up to 25% (Figures 9.6 and 9.9). Use of the Leontovich approximation reduces the

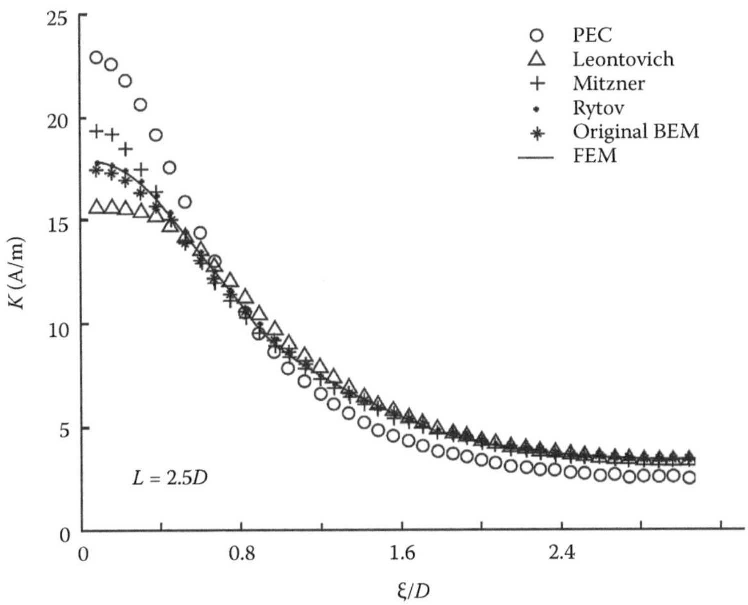

FIGURE 9.6
Surface current density for $I_a(t) = I_0\tilde{\theta}_a(t)$. The distance between the centers of the conductors is 2.5D.

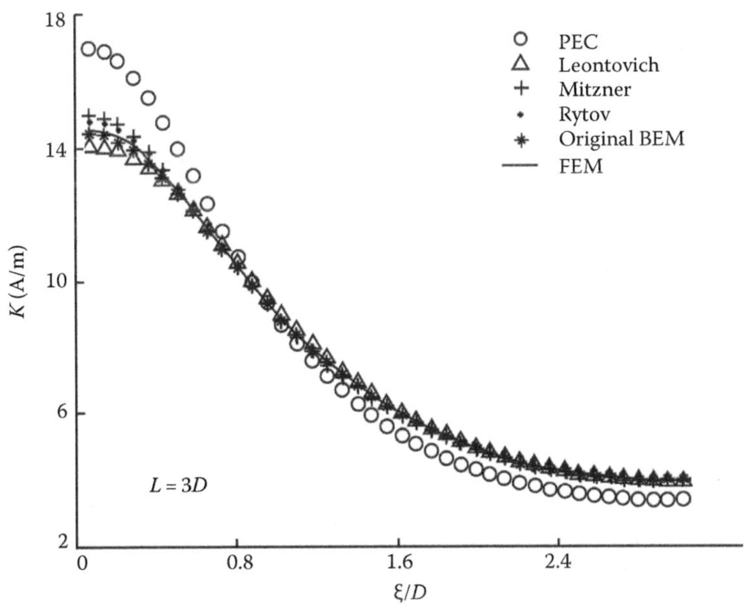

FIGURE 9.7
Surface current density for $I_a(t) = I_0\tilde{\theta}_a(t)$. The distance between the centers of the conductors is 3D.

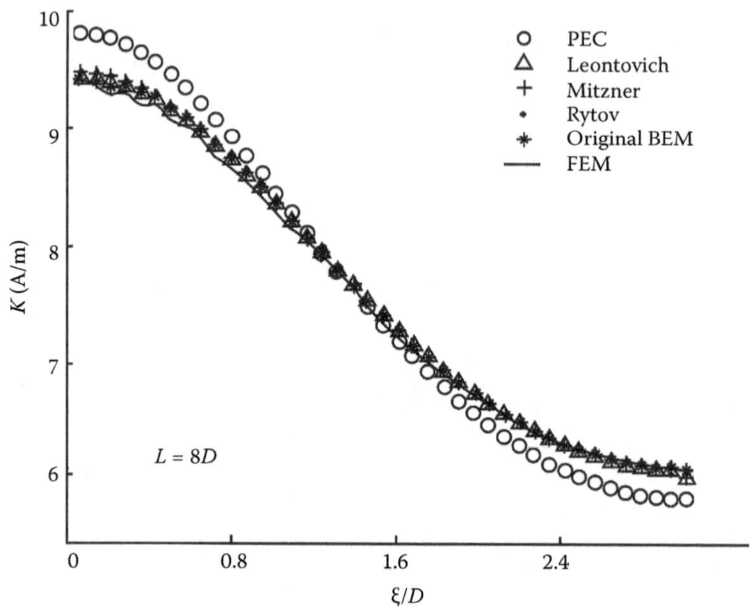

FIGURE 9.8
Surface current density for $I_a(t) = I_0\tilde{\theta}_a(t)$. The distance between the centers of the conductors is $8D$.

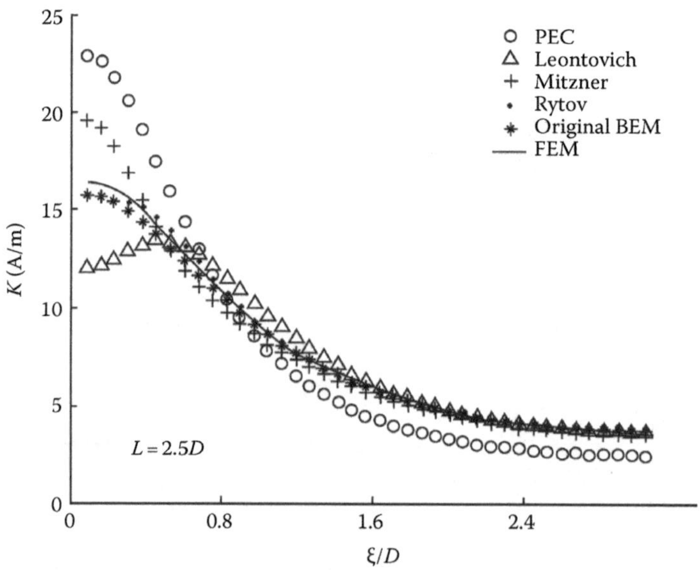

FIGURE 9.9
Surface current density for $I_b(t) = I_0\tilde{\theta}_b(t)$. The distance between the centers of the conductors is $2.5D$.

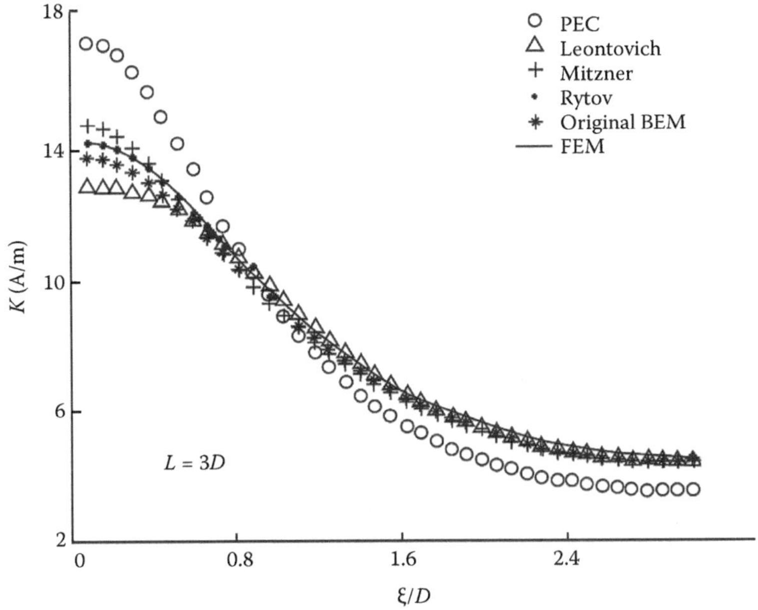

FIGURE 9.10
Surface current density for $I_b(t) = I_0\tilde{\theta}_b(t)$. The distance between the centers of the conductors is $3D$.

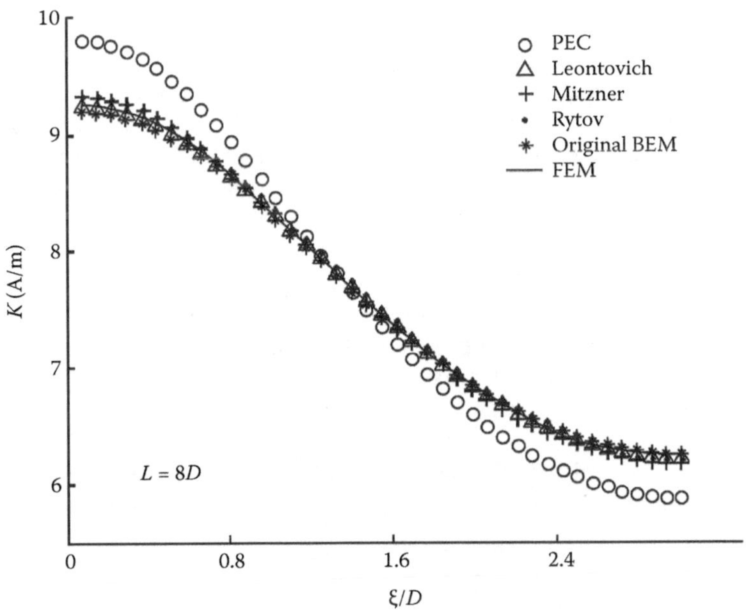

FIGURE 9.11
Surface current density for $I_b(t) = I_0\tilde{\theta}_b(t)$. The distance between the centers of the conductors is $8D$.

TABLE 9.1

Approximation Errors Calculated for the Problems Shown in Figures 9.6 through 9.11 and Given by Equation 9.25

	Figure 9.6	Figure 9.7	Figure 9.8	Figure 9.9	Figure 9.10	Figure 9.11	Equation 9.25
PEC (%)	26	17	4	34	23	6	
Leontovich's approximation (%)	11	3	<1	24	7	<1	6
Mitzner's approximation (%)	5	3	<1	7	4	<1	1.5
Rytov's approximation (%)	4	3	<1	5	4	<1	<1

error. However, it shows a nonphysical decay in the curve behavior (Figure 9.9). Accuracy is visibly improved by the use of Mitzner's approximation. Very good agreement between the BEM–SIBC formulation and the reference data in this example is achieved when the SIBC in Rytov's approximation is used.

Approximation errors calculated as the maximum relative error between BEM–SIBC and the "original BEM" solutions can be found in Table 9.1 together with results given by Equation 9.25. The accuracy of the "original BEM" solution is estimated to be within 1%.

The rather significant variations in the data shown in Table 9.1 are the results of the influences of two factors: the proximity effect and the symmetry of the system. The first factor is dominant when the conductors are located close to each other, whereas the second factor becomes important with an increase in the distance between the conductors.

Analysis of the results in Figures 9.6 through 9.11 shows that the values of the field at the surface of the conductor are highest when the proximity effect is strongest (Figures 9.6 and 9.9, $\xi = 0$). This leads to deeper penetration of the field inside the conductor at $\xi = 0$ and, consequently, to an increase in the role of the tangential derivatives that are neglected in the skin effect approximation. Thus the approximation error is highest in the cases of the strong proximity effect. On the other hand, when the conductors are located far from each other, the effect of symmetry may be significant, especially in the case of a circular cross section. It tends to smooth out nonuniformities in the field distribution around the conductor's surface so that the surface field values are lowest in Figures 9.8 and 9.11 resulting in a decrease in the approximation error.

The disagreement due to the proximity effect may be reduced if we redefine the space-scale factor D for the case of a multiconductor system as follows:

$$D = \min(d_{\min}, h_{\min}) \tag{9.29}$$

where $h_{\min}$ is the shortest distance between the conductors of the system. Application of Equation 9.29 to the problems shown in Figures 9.6 and 9.9

TABLE 9.2

Approximation Errors Calculated for the Problems Shown in Figures 9.6 and 9.9, Calculated from Equation 9.25 Using the Definition in Equation 9.29

	Figure 9.6	Figure 9.9	Formulas (Equations 9.25 and 9.29)
PEC (%)	26	34	
Leontovich's approximation (%)	11	24	25
Mitzner's approximation (%)	5	7	12
Rytov's approximation (%)	4	5	6

($L = 3d$) leads to a new value of D equal to 10.5 mm, which is half the distance between the conductors. Under these conditions Equation 9.25 yields more accurate estimates, given in Table 9.2.

From the results shown in Figures 9.6 through 9.11, it follows that the disagreement between the curves given by the BEM–SIBC and the "original BEM" starts earlier when the pulse I_b is used. The root of this phenomenon becomes clear if we consider the one-dimensional problem of electromagnetic field diffusion into a conducting half-space. The distribution of the magnetic field as a function of the distance from the surface for both pulses at $t = \tau$ is shown in Figure 9.12. With the pulse I_b the field penetrates deeper, and, consequently, the approximation error related to the thickness of the skin layer is higher than in the case of I_a. The increase in penetration depth causes an inductive process resulting in the redistribution of the magnetic

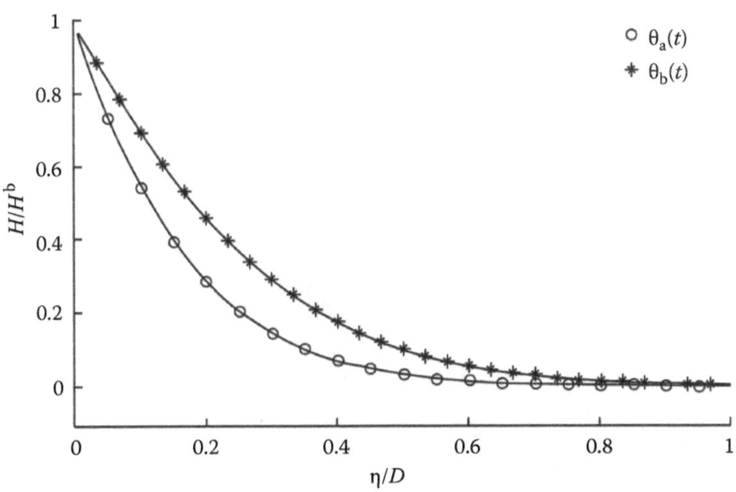

FIGURE 9.12
Distribution of the functions $\theta_a(t)$ and $\theta_b(t)$ inside the conductor at $t = \tau$.

field over the surface of the conductor and in an increase in the disagreement between the results obtained using the PEC limit and other approximations where the diffusion is taken into account. Therefore, if the incident pulse is such that the actual penetration depth is less than $\sqrt{\tau/(\sigma\mu)}$, the approximation error may be less than that predicted by Equation 9.25, but, at the same time, improvement of accuracy provided by the Leontovich and other approximations over the PEC limit may not be significant.

Since the representations in Equations 6.230 through 6.231 allow the function $\tilde{\theta}_0$ to be involved "explicitly" in the power series (Equation 6.30 or 6.76 and Equation 6.180), the role of the pulse shape in a given problem can be estimated by analyzing the functions $\tilde{\theta}_m/\tilde{\theta}_{m-1} = (\tilde{T}_m * \tilde{\theta}_0)/(\tilde{T}_{m-1} * \tilde{\theta}_0)$, $m = 1, 2, 3$. For instance, solution in the PEC limit would still be accurate if $\tilde{p} = 0.2$ and $\max(\tilde{\theta}_1/\tilde{\theta}_0) = 0.05$. Analysis of the distributions of the functions $\tilde{\theta}_m$ shown in Figure 6.12 for pulses I_a and I_b, respectively, leads to the following results:

$$\max(\tilde{\theta}_1/\tilde{\theta}_0)_a = 0.58; \quad \max(\tilde{\theta}_1/\tilde{\theta}_0)_b = 0.86 \tag{9.30}$$

$$\max(\tilde{\theta}_2/\tilde{\theta}_1)_a = 0.59; \quad \max(\tilde{\theta}_2/\tilde{\theta}_1)_b = 0.78 \tag{9.31}$$

$$\max(\tilde{\theta}_3/\tilde{\theta}_2)_a = 0.52; \quad \max(\tilde{\theta}_3/\tilde{\theta}_2)_b = 0.67 \tag{9.32}$$

Figures 9.13 and 9.14 show plots of the functions $\tilde{\theta}_m(\tilde{t}) = \tilde{T}_m(\tilde{t}) \tilde{*} \tilde{\theta}_0(\tilde{t})$ for both pulses.

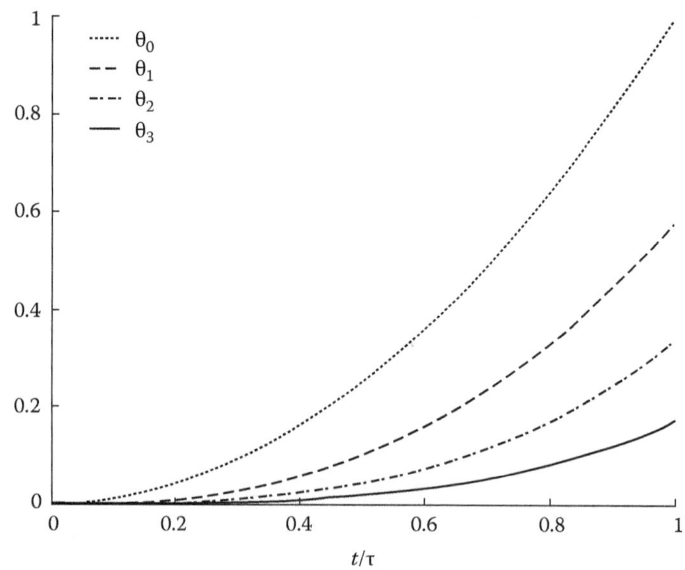

FIGURE 9.13
The functions $\tilde{\theta}_m$ for $I_a(t) = I_0\tilde{\theta}_a(t)$ ($\tilde{\theta}_0 = \tilde{\theta}_a$ and $\tilde{\theta}_m = \tilde{\theta}_a * \tilde{T}_m$, and $m = 1, 2, 3$).

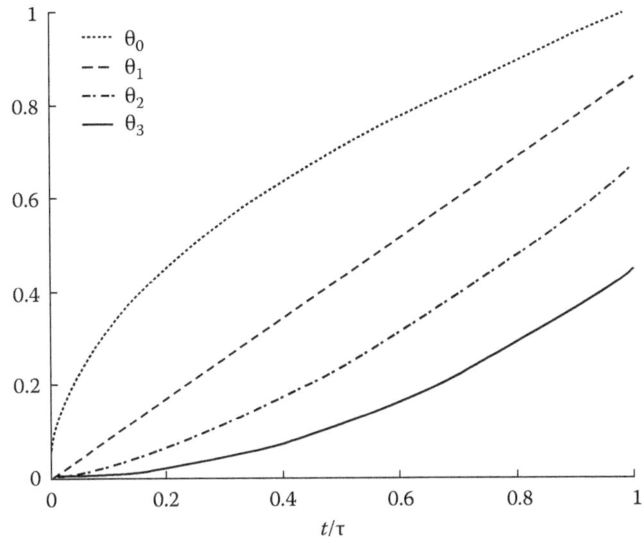

FIGURE 9.14
The functions $\tilde{\theta}_m$ for $I_b(t) = I_0\tilde{\theta}_b(t)$ $\left(\tilde{\theta}_0 = \tilde{\theta}_b \text{ and } \tilde{\theta}_m = \tilde{\theta}_b * \tilde{T}_m,\ m = 1, 2, 3\right)$.

9.3 Experimental Validation of SIBCs

Validation, especially as it relates to an algorithm, refers to the testing of that algorithm (or its implemented form in software) to ensure it satisfies its intended requirements. That is, validation is the process by which one ensures that the algorithm does what it is supposed to do. We will view validation in a slightly broader sense and will include in it the errors incurred and the consequences of the approximations involved. Indeed, since SIBCs are approximations, and since the exact results are often not known—otherwise there would be no need for the approximations—it is not practical to attempt to verify that the exact result is ever obtained. What is more practical is to attempt to ascertain that the results are a ''good'' approximation to the exact solution and to try to qualify how ''good'' the approximation is. Because SIBCs are intended to complement other numerical algorithms (such as FDTD, BEM, FEM, or FIT), the question of errors due to the SIBC as opposed to errors due to other approximations in the overall program must also be addressed.

In the previous chapters, we used specific examples to validate the results based on analytical and numerical results. Comparisons with data obtained from canonical problems (analytical validation) and from numerical methods are very useful validation tools and are routinely employed in the development of computational tools.

What we wish to present here is validation of SIBCs based on experimental data. This method of validation is often preferred in the sense that one compares numerical results with real data and, apart from experimental errors, the performance of the SIBC can be assessed. Nevertheless, experimental data are never exact and the errors from experimentation must be taken into account when analyzing the results.

As mentioned in the introduction, Chapters 4 through 8 offer both analytical and numerical validations of the SIBCs discussed in the context of a solution method such as the boundary element environment in Chapter 6 (low-frequency problems) or the solution in nonlinear media in Chapter 5. In many ways, analytical validations are easy as long as the analytical model is available or can be adapted for the problem at hand. Of course, only simple, typically highly idealized symmetric configurations (such as diffusion in the half-space used in Section 5.2 or the scattering from an infinite cylinder in Section 7.6) may be treated in this fashion. Numerical validation is wider in scope and typically consists of solving the problem by a different numerical method. In the context of SIBCs, the most logical approach is to compare results obtained with the SIBC formulated within a numerical technique such as the BEM with results obtained on the same problem without the use of the SIBC. An example to this approach may be found in Section 9.2.3, where the various order SIBCs (formulated with the BEM method) are compared with results from a straight BEM solution (no SIBCs) as well as results from the FEM, again without SIBCs. Naturally, for these comparisons to be valid, the BEM and FEM solutions must model the geometry properly, including the effects of skin depth.

Experimental validation is much more tedious and often requires complicated and expensive setups. Nevertheless, when experiments can be performed, they have the advantage of providing "absolute" verification—within the experimental error—of the correctness of SIBCs and, indeed, the numerical process.

The following sections discuss in some detail the experimental methods used to validate the formulations in the previous chapters and give examples of computations in various configurations. A main emphasis of the results given is the errors, or improvement in calculation due to the use of SIBCs.

9.3.1 Physical Configuration

The first experimental setup we use is shown in Figure 9.15. It consists of a pair of identical, long parallel aluminum ($\sigma = 3.82 \times 10^7 (\Omega/\text{m})$) conductors with a circular cross section of radii $D = 30$ mm. The distance between the centers of the conductors is 120 mm. The conductors are connected in series and the circuit is fed by a switched dc voltage source that provides equal and oppositely directed currents $I_1(t)$ and $I_2(t)$ flowing through the conductors. The transient is obtained by closing a circuit breaker so that the resulting

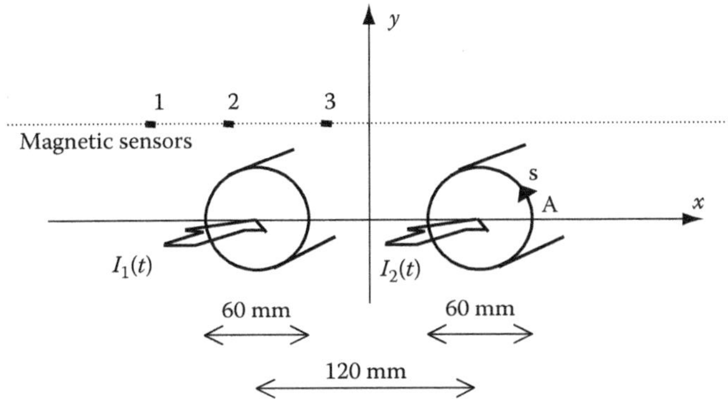

FIGURE 9.15
Two current carrying conductors used to calculate the currents from measured field data.

source current is exponential (see Figure 9.16). Three commercial magnetoresistive sensors (Philips KMZ10A) of nominal sensitivity $S = 80$ mV/(kA/m) are placed above the conductors, as described in Figure 9.15. Sensor 1 is in position (−121 mm, 54.2 mm), sensor 2 is in position (−73 mm, 54.2 mm), and sensor 3 is in position (−25 mm, 54.6 mm). The uncertainty in the measurement of the positions of the sensor was estimated at 0.5 mm. The sensors are oriented with their sensitivity axis parallel to the x-axis. Low-drift, high-accuracy instrument amplifiers (Burr Brown INA 128) were

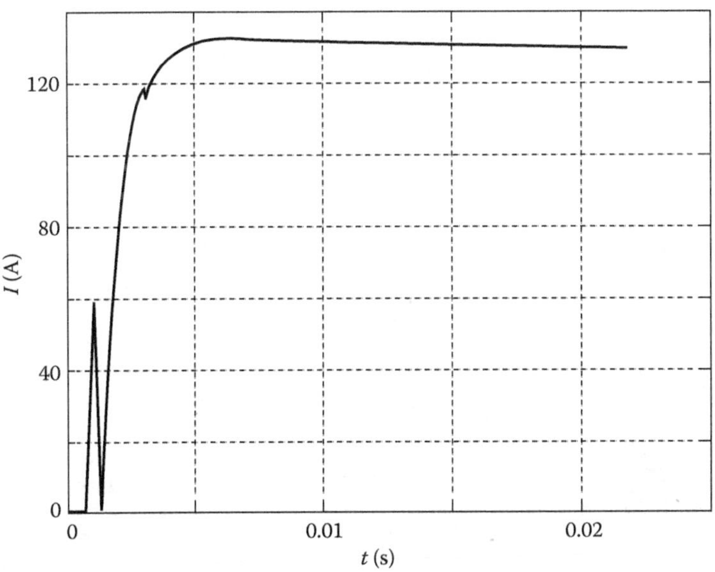

FIGURE 9.16
Current waveform $I_1(t)$.

employed in order to process and amplify the signals generated by each magnetoresistive sensor. The output signals were sampled and acquired by an eight-channel data acquisition system, with 12 bit resolution and 32 *k*samples/s rate per channel. The experimental data consist of the magnetic field at the sensors. This is then compared with the fields generated at the same location using the BEM employing a scalar potential formulation and time-domain SIBCs of various orders (see Section 3.4 for the scalar potential formulation and Chapter 6 for details of implementing of the BEM method).

To guarantee that the solution is valid, the duration of the source current pulse was chosen so that $\tau \ll \sigma\mu D^2$. The problem is two-dimensional in the plane of cross sections of the conductors [2].

Figures 9.17 through 9.19 show direct comparisons of the experimental results with those obtained using the PEC, Leontovich, Mitzner, and Rytov approximations as well as numerical results obtained using a commercial finite element (FEM) program [1]. These figures clearly show the value of high-order SIBCs in reducing errors in calculation. To quantify these errors better, consider Figures 9.20 through 9.22. In all of these, the absolute error in the Mitzner or Rytov approximation is everywhere less than 10 A/m; in most cases, it is less than about 3 A/m. This represents a maximum error of less than 6%. For larger times, say after 30 ms, the error is everywhere less than 4 A/m or an error of 2.5%. On the other hand, the PEC and Leontovich approximation

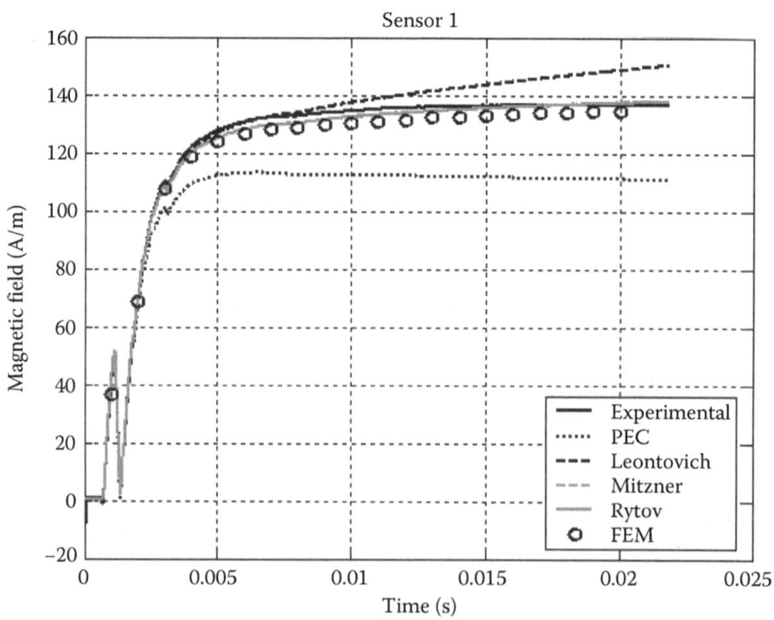

FIGURE 9.17
The magnetic field near sensor 1.

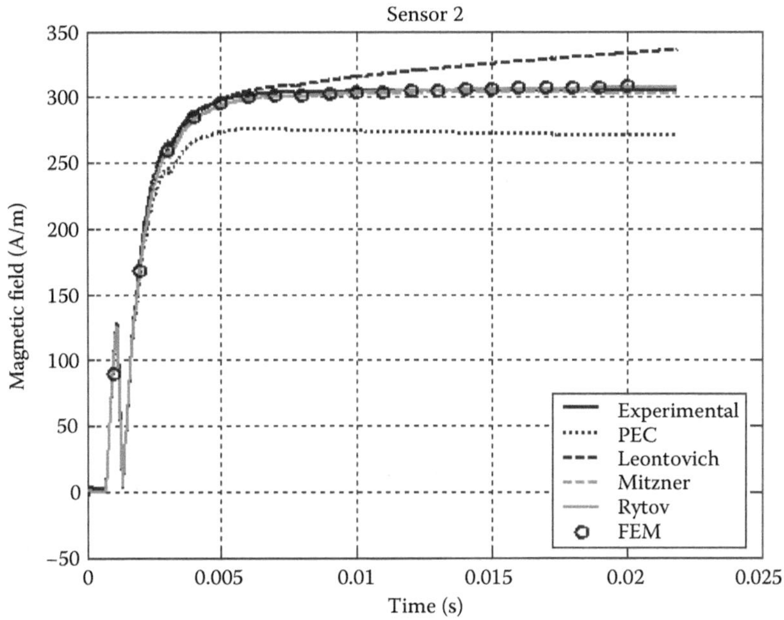

FIGURE 9.18
The magnetic field near sensor 2.

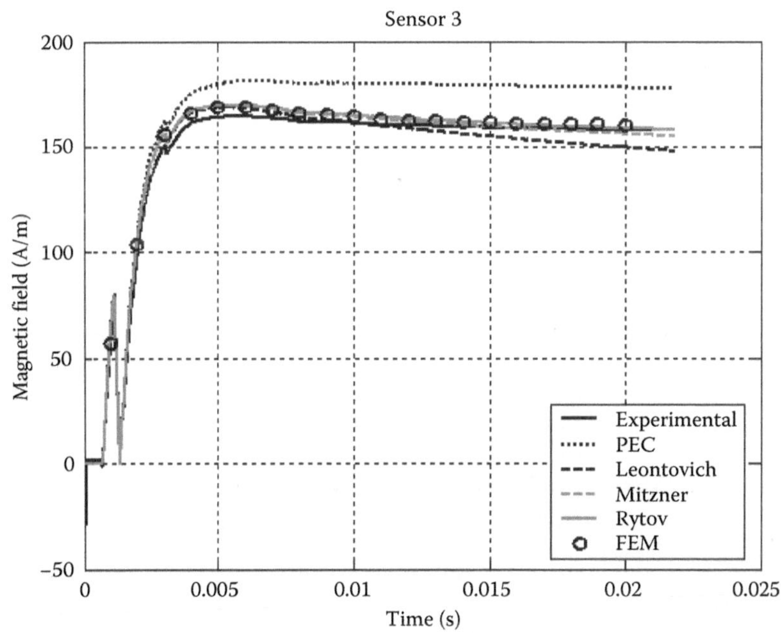

FIGURE 9.19
The magnetic field near sensor 3.

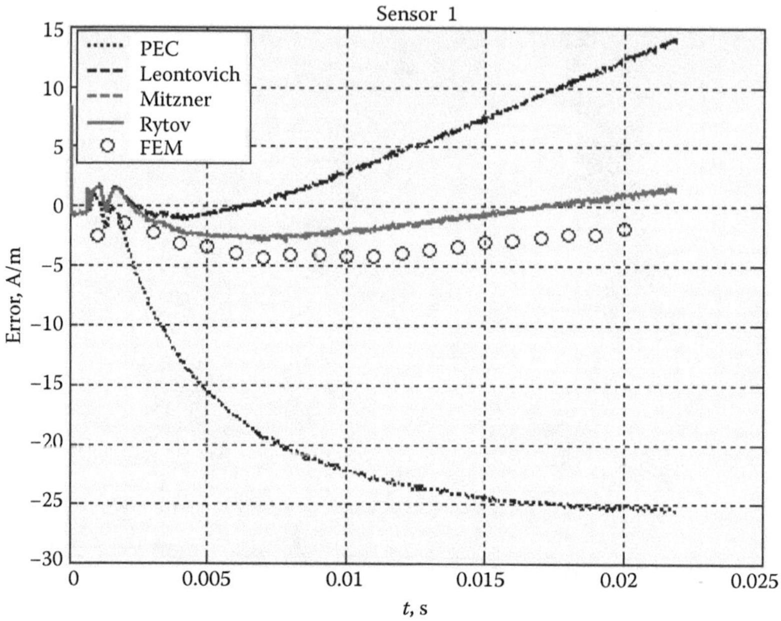

FIGURE 9.20
Errors in the magnetic field near sensor 1.

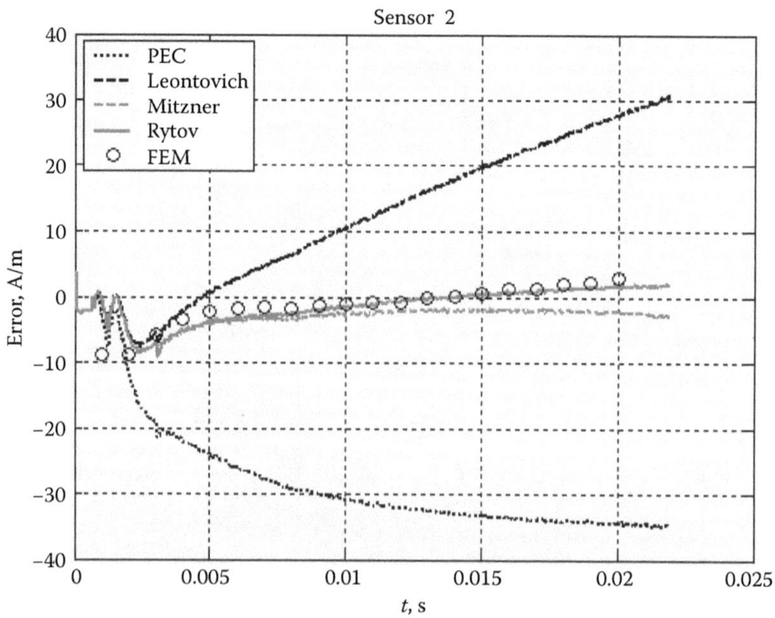

FIGURE 9.21
Errors in the magnetic field near sensor 2.

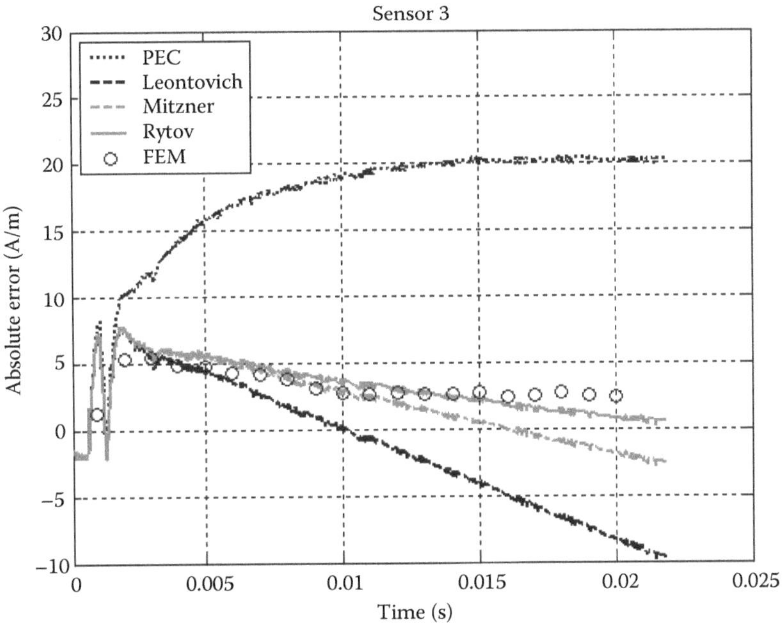

FIGURE 9.22
Errors in the magnetic field near sensor 3.

are much less accurate, with errors as high as 30 A/m (18%). Clearly, then, in this case the Mitzner approximation (second-order SIBC) is needed.

The results above also provide another way of looking at the issue of selection of SIBCs discussed in Section 9.3.1. Using the error estimate in Equation 9.25, we can plot the error as a function of the pulse duration. Figure 9.23 shows the distribution of the errors corresponding to the PEC limit, and Leontovich's, Mitzner's, and Rytov's approximations. For example, application of the Leontovich SIBC for simulations with a pulse duration equal to 0.0065 s leads to 10% error. Use of Rytov's SIBC allows simulations to be performed for a longer pulse duration equal to 0.021 s with the same error.

Note that the disagreement between the results obtained using the SIBCs and measured data is actually less than the error predicted by the formula in Equation 9.25 since it does not take into account such effects as symmetry of the problem, shape of the pulse, and the proximity effect. Nevertheless, Equation 9.25 gives a quick evaluation of the applicability of the surface impedance concept for a given problem and can be used for selecting the approximation order as an alternative form of Equations 9.13 through 9.17.

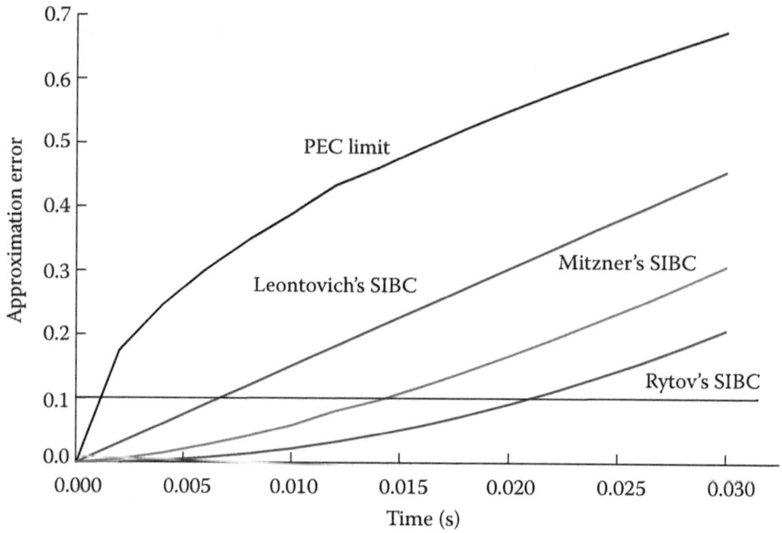

FIGURE 9.23
The approximation errors as a function of pulse duration.

9.3.2 Example: Reconstruction of Currents from Measured Magnetic Fields

To look at other sources of errors, including the effects of closely spaced conductors (proximity effect), we look now at a second problem, again in the time domain. The problem is that of reconstruction of transient currents in parallel conductors from measured magnetic data [3,4]. The physical configuration used is shown in Figure 9.24. It consists of three conductors

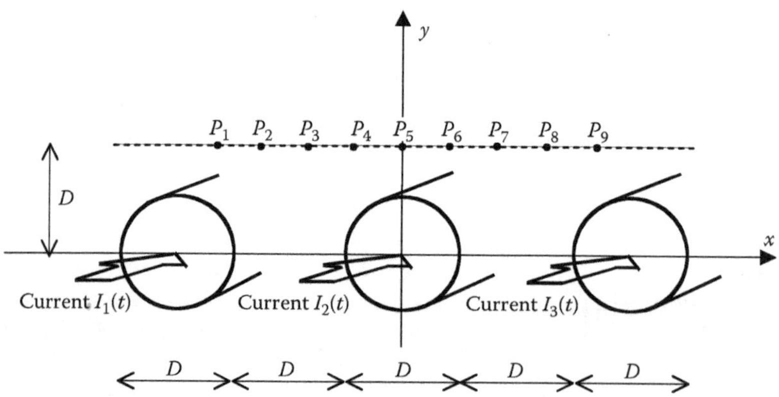

FIGURE 9.24
Geometry used for reconstruction calculations ($D = 50$ mm).

in which known, transient currents flow. The three conductors having a diameter $D = 50$ mm, are separated 100 mm apart (center to center) and are either copper (conductivity $\sigma = 5.8 \times 10^7$ S/m) or aluminum (conductivity $\sigma = 3.8 \times 10^7$ S/m). Nine magnetic sensors are placed at locations P_k, $k = 1, \ldots, 9$, with sensitivity vectors parallel to the x-axis to measure the field at their respective locations. Sensors P_1, P_5, and P_9 are placed exactly above the centers of conductors 1, 2, and 3, respectively. P_3 and P_7 are placed midway between conductors 1 and 2 and between conductors 2 and 3. P_2 and P_6 are above the right-hand edges of conductors 1 and 2 whereas P_4 and P_8 are above the left-hand edges of conductors 2 and 3.

The measured data at the sensors (which consists of the x-component of the magnetic field) are used as inputs to an inversion algorithm that computes the currents in the conductors [3]. The reconstructed currents are compared with the measured currents in the conductors and the errors are computed as the difference between the two values at all time steps between $t = 0$ and 20 ms. In this case, we started with known currents for the purpose of comparison and verification, but normally the currents would not be known and the method can then be used to measure the currents indirectly. The results that follow were generated for copper and aluminum conductors (the parameter $\tilde{p}$ is different for copper and aluminum because of the difference in conductivity).

To understand the results we first present the formulation and the inversion algorithm since the method given is both general and efficient and can be applied in other practical configurations. The problem is formulated here in terms of the scalar potential and the BEM as discussed in Chapters 3 and 6. However, whereas the general formulation is in three dimensions, the discussion that follows is in two dimensions.

For the formulation we use the scalar potential formulation introduced in Section 6.3, starting with the decomposition of the magnetic field in free space into its source magnetic field and that due to the scalar potential (see Equation 6.155):

$$\vec{H} = \vec{H}^{\mathrm{s}} - \nabla\varphi \tag{9.33}$$

where $\vec{H}^{\mathrm{s}}$ is the magnetic field created by the sources. In our case the sources are the conductors carrying currents I_i. To be able to calculate the field we will assume the current I_i flows through an imagined filament located at the position $\vec{r}_i^{\,\mathrm{fil}}$ inside the conductor [4]. The source field can then be obtained as the sum of the fields of all such filaments (three in our case, but we will leave the notation general by using N conductors):

$$\vec{H}^{\mathrm{s}} = \sum_{i=1}^{N} \vec{H}_i^{\mathrm{fil}} \tag{9.34}$$

The field in each filament $\vec{H}_i^{\text{fil}}$ is obtained from the Biot–Savart law or, in this case, because the conductors are long, from Ampere's law:

$$\vec{H}_{\text{fil}} = \frac{1}{2\pi}\sum_{i=1}^{N}\frac{I_i}{R_i} \tag{9.35}$$

These fields produce an output voltage in each of the magnetic sensors and the total voltage on each sensor is the sum of voltages produced by each filament. Each of the $M=9$ magnetic sensors produces an output voltage equal to

$$V_k(t) = S_k\left(\vec{s}_k \cdot \vec{H}\left(\vec{r}_k^{\,\text{sens}}, t\right)\right), \quad k = 1, \ldots, M \tag{9.36}$$

where
S_k is the sensor's sensitivity
$\vec{s}_k$ is a unit vector indicating the sensitivity direction of the sensor
$\vec{r}_k^{\,\text{sens}}$ is the position vector of sensor k

We assume the signals to be correlated in time so that the following decomposition can be done:

$$V_k(t) = V_k^{\text{s}} T_0(t), \quad k = 1, \ldots, M \tag{9.37}$$

where
V_k^{s}, $k = 1, \ldots, M$, are constant coefficients
$T_0(t)$ is a time-dependent function

By substituting Equations 9.33 and 9.34 into Equation 9.36 we get the time-dependent voltage on sensor k as

$$V_k(t) = \sum_{i=1}^{N} S_k \alpha_{ki} I_i(t) - S_k \vec{s}_k \cdot \nabla\phi\left(\vec{r}_k^{\,\text{sens}}, t\right), \quad m = 1, 2, \ldots, M \tag{9.38}$$

where α_{ki} is a coefficient depending only on the relative position of the kth sensor and the ith filamentary current [3]. Assuming a unit vector $\vec{e}_z$ along the conductors (the z-axis in Figure 9.24), the coefficient can be written as

$$\alpha_{ki} = \frac{\left(\vec{e}_z \times \vec{r}_k^{\,\text{sens}}\right) \cdot \vec{s}_k}{2\pi R_i} \tag{9.39}$$

Since the scalar potential in free space obeys the Laplace equation, the boundary integral equation method yields the following surface integral equation (see, for example, Equation 6.160 and the discussion that follows it):

$$\tilde{c}\phi + \sum_{i=1}^{N}\int_{L_i} \phi \frac{\partial G}{\partial \vec{n}}\, dl = \sum_{i=1}^{N}\int_{L_i} G \frac{\partial \phi}{\partial \vec{n}}\, dl \tag{9.40}$$

where

L_i is the contour of the cross section of conductor i

G is the fundamental solution of the two-dimensional Laplace equation in free space and the unit normal vector $\vec{n}$ is chosen inwards

Taking into account Equation 9.34, 9.40 can be rewritten in the form:

$$\tilde{c}\phi + \sum_{i=1}^{N} \int_{L_i} \phi \frac{\partial G}{\partial \vec{n}} \mathrm{d}l = \sum_{i=1}^{N} \int_{L_i} G\vec{n} \cdot \left(\vec{H}^{\mathrm{fil}} - \vec{H}\right) \mathrm{d}l \tag{9.41}$$

where

$$G(|\vec{r} - \vec{r}'|) = -(2\pi)^{-1} \ln(|\vec{r} - \vec{r}'|) \tag{9.42}$$

Equations 9.36, 9.38, and 9.41 contain four unknowns: I_i, ϕ, $\vec{H}_{\mathrm{fil}}$, and $\vec{n} \cdot \vec{H}$. The required additional relation between these quantities is supplied by the time-domain SIBC (see Sections 3.4 and 6.3 for the H–ϕ formulation of SIBCs):

$$\vec{n} \cdot \vec{H} = \frac{1}{\sqrt{\pi\sigma\mu t}} * \frac{\partial H_\xi}{\partial \xi} + \frac{1}{2d} \frac{U(t)}{\sigma\mu} * \frac{\partial H_\xi}{\partial \xi} + \sqrt{\frac{t}{\pi\sigma^3\mu^3}} * \left(\frac{3}{8d^2} \frac{\partial H_\xi}{\partial \xi} + \frac{1}{2} \frac{\partial^3 H_\xi}{\partial \xi^3}\right) \tag{9.43}$$

where

"$*$" denotes the time-convolution product

$U(t)$ is the unit step function

the coordinate ξ is directed along the contour of the conductor's cross section

d is the local radius of curvature corresponding to coordinate ξ

The expression in Equation 9.43 is much simpler than the general expression in Section 3.4 because there is only one radius of curvature.

The SIBC in Equation 9.43 is immediately recognized as the Rytov SIBC with the first term describing the diffusion in Leontovich's approximation, the second term providing the correction due to the curvature (Mitzner's approximation), and the third and fourth terms providing the corrections due to Rytov, which allow diffusion of the field tangential to the surface of the conductors.

Substituting Equation 9.43 into 9.41 leads to the following formulation:

$$\frac{\phi}{2} + \sum_{i=1}^{N} \int_{S_i} \left(\frac{\partial G}{\partial \vec{n}} \phi - G \sum_{m=1}^{3} \Lambda_m^{(2D)} [\nabla \phi]\right) ds = \sum_{i=1}^{N} \int_{S_i} G \left(H_\eta^s - \sum_{m=1}^{3} \Lambda_m^{(2D)} \left[H_\xi^s\right]\right) ds \tag{9.44}$$

Where the operators $\Lambda_m^{(2D)}$, $m = 1, 2, 3$, are given in Equation 6.193.

Equations 9.38, 9.35, and 9.44 are solved using the following iterative procedure.

Let $I_i^{(m)}$ be the total currents at step m of the iteration. $I_i^{(m+1)}$ are then obtained in the following way:

1. Calculate $\vec{H}_{\text{fil}}$ using Equation 9.35.
2. Solve Equation 9.44 to obtain $\phi^{(m+1)} = \phi^{(m+1)}(\vec{r}, t)$.
3. Calculate $\phi_k^{(m+1)}$, $k = 1, \ldots, M$, in the vicinity of the sensor k by moving the observation point in Equation 9.44.
4. Calculate $\nabla\phi^{(m+1)}$ at the location of each sensor.
5. Calculate $I_i^{(m+1)}$ using Equation 9.38.

The formulation used here is in the time domain. Therefore, the iterative procedure above must be repeated at each time step. It is also possible to formulate the problem in an invariant form because the formulation admits the separation of variables into spatial and time components. This was done in Section 6.5. A detailed formulation of the problem discussed here in time and invariant forms is given in [4] using the Leontovich SIBC. In the invariant formulation, the iterative procedure is only applied once. The results shown below were obtained with the time-dependent formulation discussed above.

The currents in the three conductors are set up as follows:

$$I_1(t) = 1 - \exp\left(-\left(\frac{t}{T}\right)^2\right) \tag{9.45}$$

$$I_2(t) = -\left(1 - \exp\left(-\left(\frac{2t}{T}\right)^2\right)\right) \tag{9.46}$$

$$I_3(t) = -\left(1 - \exp\left(-\left(\frac{3t}{T}\right)^2\right)\right) \tag{9.47}$$

with $T = 0.004$ s.

These currents are applied to the conductors, the fields are measured, and the iterative algorithm is used to reconstruct these currents as described above. This allows evaluation of errors against the applied currents, but one the value of the method in measuring currents indirectly from field data should be recognized apart from their use here for verification.

The inversion algorithm is applied in the interval $t = 0$ and $5T$ (20 ms). The BEM uses 40 elements on the surface of each conductor and the convolutions are computed using an inverse Fourier transform algorithm. 1000 time steps are required to assure stability in the numerical results.

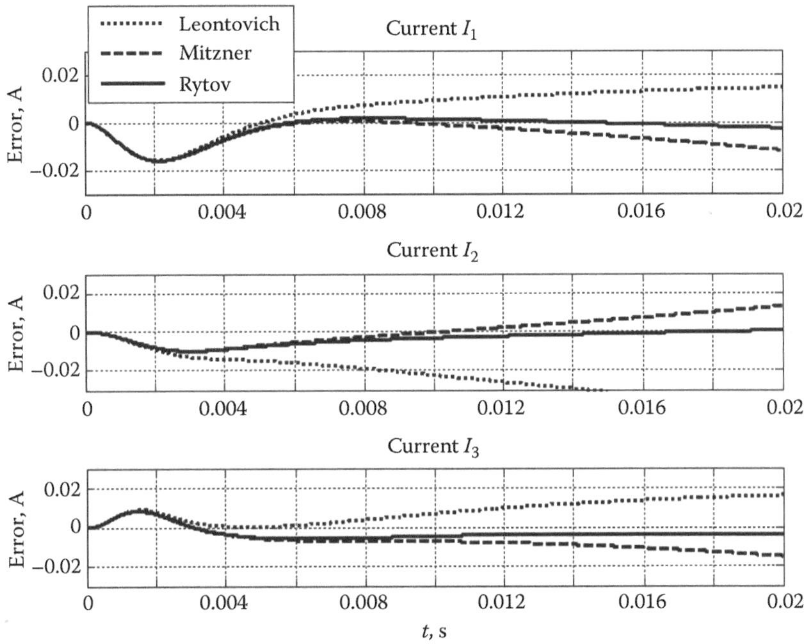

FIGURE 9.25
Absolute errors with nine sensors and copper conductors.

Figure 9.25 shows the errors in reconstructing the currents for copper conductors and nine magnetic sensors. The first thing to be noted is that the error in the Leontovich approximation starts increasing earlier in the timeline than the Mitzner and Rytov approximations, clearly indicating the effect of diffusion of the field in the conductors. The adoption of a higher-order approximation increases accuracy significantly. In particular, with the Rytov approximation, the error at $t = 20$ ms is still less than 0.25%. To see the effect of the number of input points to the algorithm, the number of sensors was reduced to five (P_1, P_3, P_5, P_7, P_9) and the problem solved again. The result is shown in Figure 9.26. Although the errors in the Leontovich limit have increased, those in the Mitzner and Rytov limits have not. This indicates both that a small number of sampling points is needed as well as the robustness of the inversion algorithm. Figure 9.27 shows the reconstructed currents in aluminum conductors using all nine sensors. Comparison with Figure 9.25 shows higher errors for all SIBCs due to the lower skin depth (and hence larger $\tilde{p}$). This is, of course, as expected.

To investigate the dependence of accuracy on the parameter $\tilde{p}$, the distance between conductors (center to centers) was reduced to 150 mm. This effectively increases $\tilde{p}$ since the skin depth remains unchanged but the characteristic

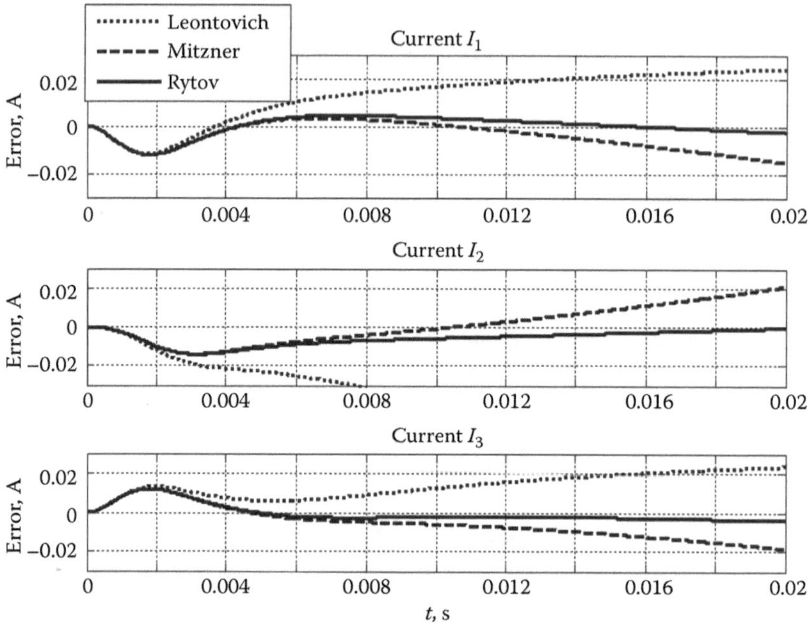

FIGURE 9.26
Absolute errors with five sensors and copper conductors.

dimension now is the distance between the conductor surfaces, which was reduced from 50 to 25 mm. This also gives rise to a fairly strong proximity effect, again accounting for an expected increase in errors. Figures 9.28 and 9.29 confirm this. They show the reconstructed currents for copper conductors with five and nine sensors, respectively. It should be immediately noticed that only the Rytov approximation provides acceptable results with a maximum error of less than 5%.

9.3.3 Example: Calculation of p.u.l. Parameters in Multiconductor Transmission Lines

As another example of the versatility of SIBCs, we consider here an example involving the calculation of per unit length (p.u.l.) parameters in multiconductor lossy transmission lines using low- and high-order SIBCs. This application in particular emphasizes the calculation of fields and related parameters within the skin layer as well as the postprocessing aspects of SIBCs. In addition, unlike previous examples, the calculations are performed in the frequency domain using the magnetic vector potential (MVP) formulation. The example also emphasizes the reach of high-order SIBCs into lower frequencies.

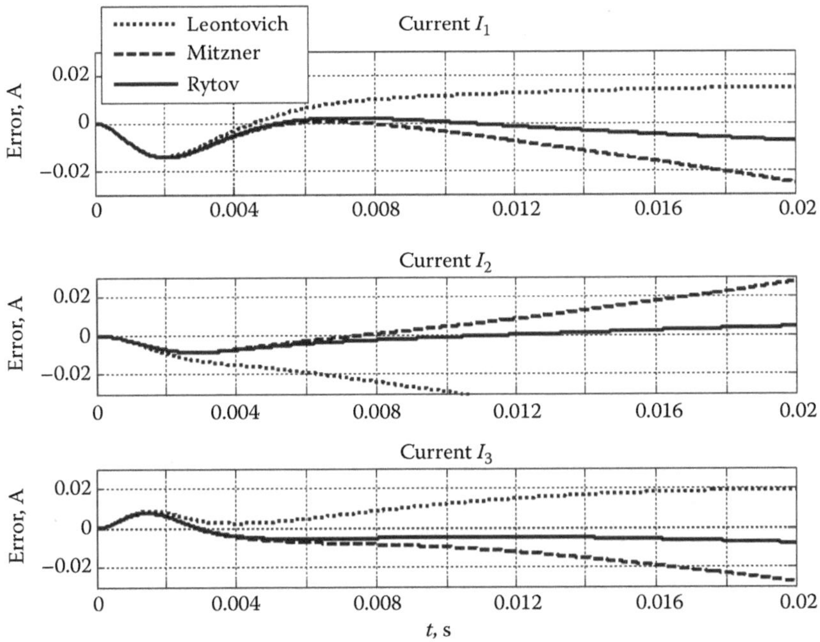

FIGURE 9.27
Absolute errors with nine sensors and aluminum conductors.

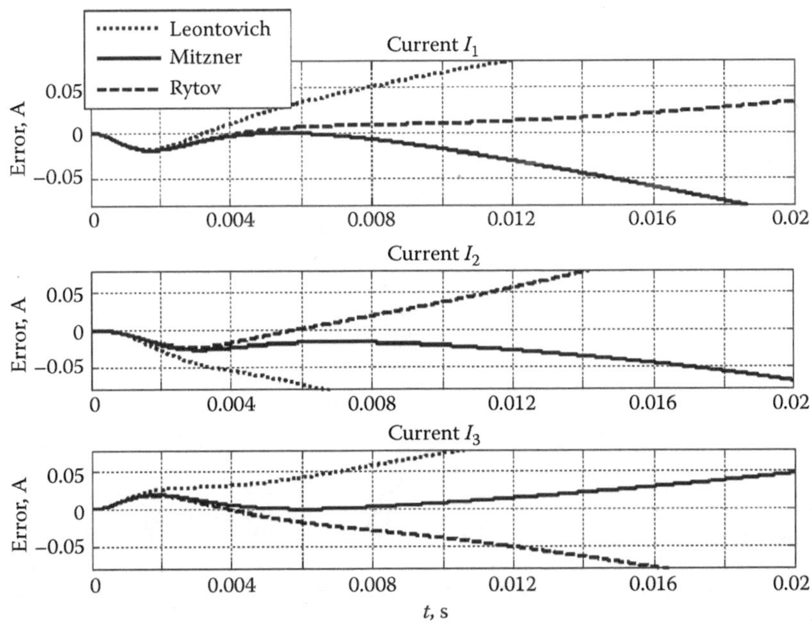

FIGURE 9.28
Absolute errors with five sensors (P_1, P_3, P_5, P_7, P_9) and copper conductors.

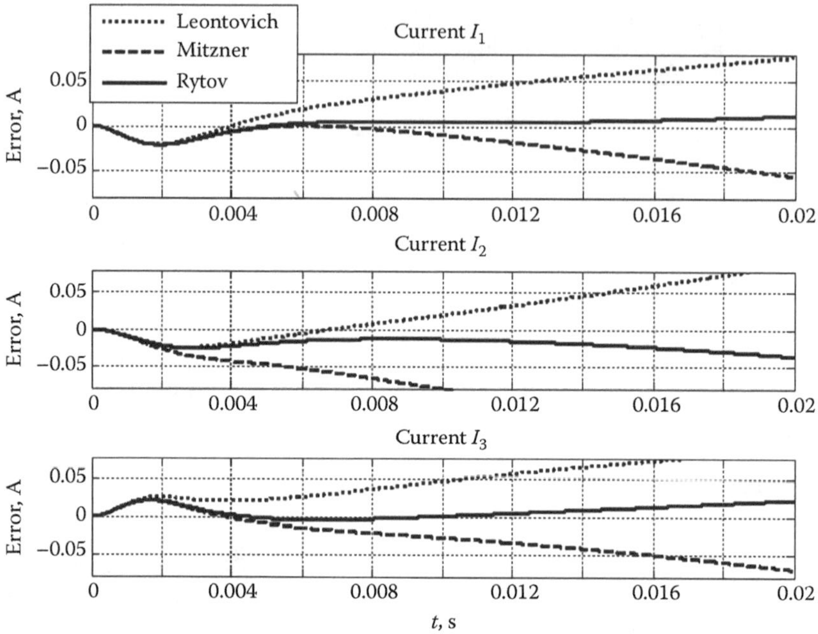

FIGURE 9.29
Absolute errors with nine sensors ($P_1, \ldots, P_9$) and copper conductors.

The problem to be solved is shown in Figure 9.30. Of interest are the self- and mutual resistances and inductances (p.u.l.) of the configuration.

The formulation: As with any problem to be solved, we must start with an appropriate formulation. In this case we select the MVP formulation as

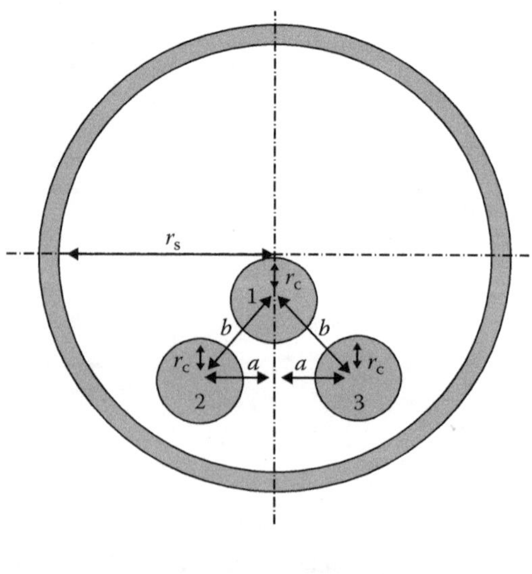

FIGURE 9.30
Simulated three-phase power cable ($r_s =$ 100 mm, $r_c = 17$ mm, $a = 1.5r_c$, $b = 3r_c$; copper conductors with $\sigma_c = 5.8 \cdot 10^7$ S/m, aluminum shield with $\sigma_a = 3.8 \cdot 10^7$ S/m; all media have relative permeability $\mu_r = 1$).

described in Section 3.5 (but modified for time-harmonic fields) and implemented in the context of the BEM in Section 6.2.2. Although we do not repeat here the formulation, it is worth pointing out a number of steps as they affect understanding of the implementation and the results. Following the standard process of writing Maxwell's equations in terms of the MVP (see, e.g., Section 3.5), we split the MVP into its source and eddy components:

$$\vec{A}(\vec{r},t) = \vec{A}^{s}(\vec{r},t) + \vec{A}^{e}(\vec{r},t) \tag{9.48}$$

The source component $\vec{A}^{s}$ is defined to satisfy the following condition:

$$j\omega\vec{A}^{s} = -\nabla V \tag{9.49}$$

Substitution of Equation 9.53 into the expression for the electric field (Equation 3.92) yields a relation between the electric field and the eddy component of the MVP:

$$\vec{E} = -j\omega\vec{A}^{e} \tag{9.50}$$

The present problem is two-dimensional in the cross section of the cable with the MVP as well as the electric field entirely in the direction of the conductors and the magnetic field having two components in the transverse directions (in the cross-section plane). With these we obtain a formulation in terms of $\vec{A}^{s}$ and the surface current K from which we can then calculate the eddy component of A as well [5,6]. It should also be noted here that the formulation is independent of frequency, which means that once a solution is obtained, the results for any frequency (within the limits of the approximation) may be obtained without the need to solve the problem again.

The p.u.l. parameters may be calculated using the multiconductor transmission line (MTL) impedance matrix:

$$\frac{\mathrm{d}}{\mathrm{d}x}\mathbf{V}(x) = -\mathbf{Z}\cdot\mathbf{I}(x) \tag{9.51}$$

The coefficients of the impedance matrix are calculated as follows.

From Equation 9.53 we have for each conductor

$$j\omega\vec{A}^{s} = -\frac{\partial V}{\partial x} \tag{9.52}$$

We assume a current I_i flows in the ith conductor and returns in the reference conductor, with the sum of currents I_j of all other conductors ($j \neq i$) equal to zero. From Equations 9.51 and 9.52 it can be seen that once the source component A_i^s has been computed, the line self-impedance Z_{ii} and consequently the p.u.l. resistance R_{ii} and the p.u.l. inductance L_{ii} can be easily calculated as

$$Z_{ii} = R_{ii} + j\omega L_{ii} = \sum_{k=0,i} \frac{j\omega A_k^s}{I_i} \tag{9.53}$$

where the sum over k takes into account the signal conductor i and the reference conductor 0, ($I_0 = -I_i$). The mutual impedance between conductors i and j can be calculated using a similar expression:

$$Z_{ij} = R_{ij} + j\omega L_{ij} = \sum_{k=0,j} \frac{j\omega A_k^s}{I_i} \tag{9.54}$$

These expressions can also be obtained from energy considerations [5,6] using the values of A^s and A^e calculated above. This involves calculation of the energy stored and power dissipated to obtain the self- and mutual inductances and resistances of the MTL.

In the example being considered here there are three conductors and a common, outer "ground" conductor. The results obtained using the formulation above are tabulated in Tables 9.3 through 9.5. Table 9.3 shows

TABLE 9.3

Per Unit Length Self-Resistance and Self-Inductance of Conductor 1 for the Three-Phase System of Figure 9.30

	FEM		BEM–SIBC	
Frequency (Hz)	$R_{11}(\Omega/m)$	$L_{11}(H/m)$	$R_{11}(\Omega/m)$	$L_{11}(H/m)$
50	3.2534×10^{-5}	3.8605×10^{-7}	3.3674×10^{-5}	3.8828×10^{-7}
100	4.5510×10^{-5}	3.6337×10^{-7}	4.6398×10^{-5}	3.6302×10^{-7}
200	6.4437×10^{-5}	3.4360×10^{-7}	6.4894×10^{-5}	3.4340×10^{-7}
300	7.8817×10^{-5}	3.3449×10^{-7}	7.9249×10^{-5}	3.3437×10^{-7}
500	1.0174×10^{-4}	3.2524×10^{-7}	1.0214×10^{-4}	3.2515×10^{-7}
10^3	1.4516×10^{-4}	3.1525×10^{-7}	1.4440×10^{-4}	3.1574×10^{-7}
10^4	4.5792×10^{-4}	3.0011×10^{-7}	4.5762×10^{-4}	3.0004×10^{-7}
10^5	1.4812×10^{-3}	2.9511×10^{-7}	1.4492×10^{-3}	2.9506×10^{-7}

TABLE 9.4

Per Unit Length Self-Resistance and Self-Inductance of Conductor 2 for the Three-Phase System of Figure 9.30

	FEM		BEM–SIBC	
Frequency (Hz)	$R_{22}(\Omega/m)$	$L_{22}(H/m)$	$R_{22}(\Omega/m)$	$L_{22}(H/m)$
50	3.3590×10^{-5}	3.0116×10^{-7}	3.3399×10^{-5}	3.1097×10^{-7}
100	4.6042×10^{-5}	2.7958×10^{-7}	4.6380×10^{-5}	2.8175×10^{-7}
200	6.4571×10^{-5}	2.6033×10^{-7}	6.4738×10^{-5}	2.6109×10^{-7}
300	7.8540×10^{-5}	2.5144×10^{-7}	7.8825×10^{-5}	2.5194×10^{-7}
500	1.0077×10^{-4}	2.4245×10^{-7}	1.0116×10^{-4}	2.4276×10^{-7}
10^3	1.4684×10^{-4}	2.3311×10^{-7}	1.4221×10^{-4}	2.3352×10^{-7}
10^4	4.4752×10^{-4}	2.1810×10^{-7}	4.4537×10^{-4}	2.1826×10^{-7}
10^5	1.4954×10^{-3}	2.1324×10^{-7}	1.4036×10^{-5}	2.1344×10^{-7}

TABLE 9.5

Per Unit Length Mutual Resistance and Mutual Inductance between Conductors 1 and 2 for the Three-Phase System of Figure 9.30

	FEM		BEM–SIBC	
Frequency (Hz)	$R_m(\Omega/m)$	$L_m(H/m)$	$R_m(\Omega/m)$	$L_m(H/m)$
50	6.8090×10^{-6}	1.1826×10^{-7}	7.5031×10^{-6}	1.1623×10^{-7}
100	1.0478×10^{-5}	1.0954×10^{-7}	1.0978×10^{-5}	1.0929×10^{-7}
200	1.5447×10^{-5}	1.0384×10^{-7}	1.6024×10^{-5}	1.0392×10^{-7}
300	1.9360×10^{-5}	1.0134×10^{-7}	1.9937×10^{-5}	1.0146×10^{-7}
500	2.5608×10^{-5}	9.8796×10^{-8}	2.6177×10^{-5}	9.8942×10^{-8}
10^3	3.8502×10^{-5}	9.6235×10^{-8}	3.7691×10^{-5}	9.6378×10^{-8}
10^4	1.2288×10^{-5}	9.1939×10^{-8}	1.2305×10^{-4}	9.2099×10^{-8}
10^5	4.2558×10^{-5}	9.0575×10^{-8}	3.9325×10^{-4}	9.0740×10^{-8}

the self-inductance and self-resistance of conductor 1. Table 9.4 shows the self-inductance and self-resistance for conductor 2. Table 9.5 shows the mutual inductances and resistances between conductors 1 and 2. In all cases the frequency is varied from 50 Hz to 100 kHz and the results are compared with numerical results obtained using an FEM solution [7].

Figures 9.31 through 9.33 display these data, clearly showing the increase in self- and mutual resistance and decrease in self- and mutual inductance with frequency.

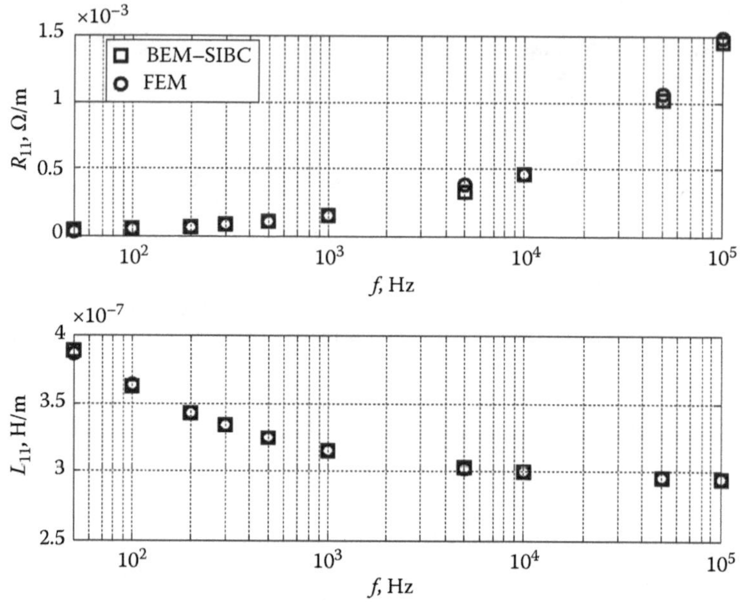

FIGURE 9.31
Per unit length self-resistance and self-inductance of conductor 1 versus frequency, for the three-phase power cable of Figure 9.30 (see Table 9.3).

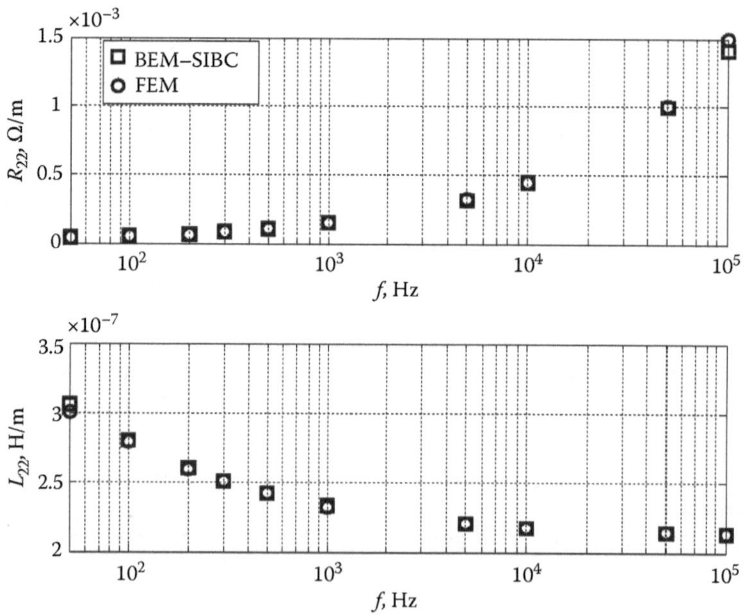

FIGURE 9.32
Per unit length self-resistance and self-inductance of conductor 2 versus frequency, for the three-phase power cable of Figure 9.30 (see Table 9.4).

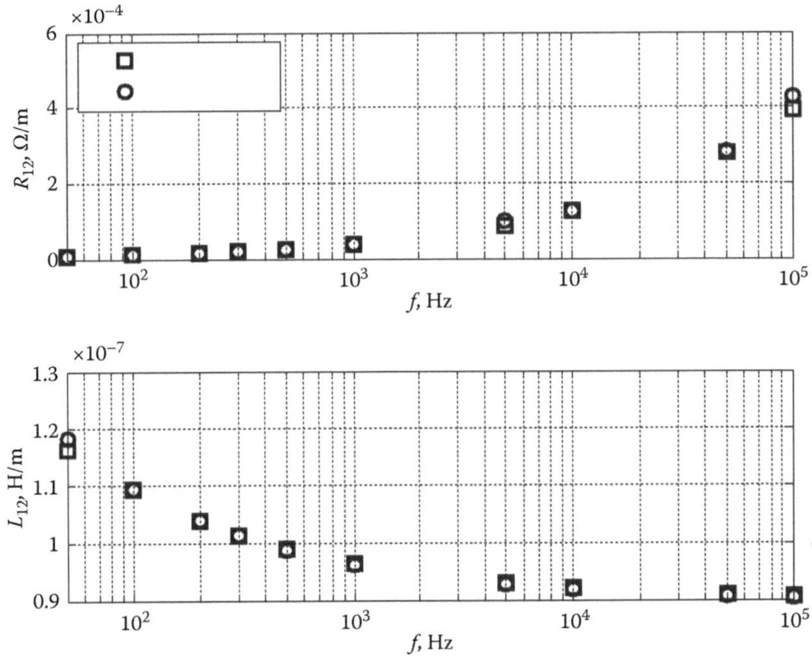

FIGURE 9.33
Per unit length mutual resistance and mutual inductance between conductors 1 and 2 versus frequency, for the three-phase power cable of Figure 9.30 (see Table 9.5).

References

1. Maxwell 2D Transient, Ansoft Corporation.
2. S. Barmada, L. Di Rienzo, N. Ida, and S. Yuferev, The use of surface impedance boundary conditions in time domain problems: Numerical and experimental validation, *ACES Journal*, 19(2), July 2004, 76–83.
3. L. Di Rienzo, S. Yuferev, N. Ida, and C.M. Arturi, Reconstruction of transient currents from magnetic data, *COMPEL*, 24(4), October 2005, 1200–1212.
4. L. Di Rienzo, N. Ida, and S. Yuferev, Application of surface impedance concept to inverse problems of reconstructing transient currents, *IEEE Transactions on Magnetics*, 39(3), May 2003, 1626–1629.
5. L. Di Rienzo, S. Yuferev, and N. Ida, Calculation of energy-related quantities of conductors using surface impedance concept, *IEEE Transactions on Magnetics*, 44(6), June 2008, 1322–1325.
6. L. Di Rienzo, S. Yuferev, and N. Ida, Computation of the impedance matrix of multiconductor transmission lines using high-order surface impedance boundary conditions, *IEEE Transactions on Electromagnetic Compatibility*, 50(4), November 2008, 974–984.
7. Maxwell 2D, Ansoft Corporation, Version 11 [Online] Available: http://www.ansoft.com/products/em/max2d/

Appendix A: Review of Numerical Methods

A.1 Introduction

The purpose of this appendix is to review the numerical methods with which the SIBCs described in previous chapters can be used. It is certainly too ambitious to try to discuss all numerical methods in a limited space. Many books have been written on any one of the methods and a sampling of these are listed in the reference section of this appendix. Our goal is much more modest than that. We only discuss representative methods of computation and only the basic principles involved are outlined. Nevertheless, this short review should be useful in placing the SIBCs in proper context.

A cursory review of the literature will reveal a long and varied list of numerical methods, all useful and important, with applications that span the whole range of engineering and science disciplines. Some of the more familiar methods are the various finite-element methods (FEMs), finite-difference methods (FDMs), boundary-integral methods (BIMs), boundary-element methods (BEMs), method of moments (MOM), transmission line matrix methods (TLMs), and many more, each with a seemingly infinite number of variants. Then there are hybrid methods that combine properties of two or more methods to overcome specific problems in computation. Examples are the FEM–FDM methods in which finite differences are used to handle time variations whereas finite elements still handle the spatial variations, the BEM–wavelet transform (WT) methods in which the WT is employed to reduce the size of the computational problem, and many other interesting and useful hybrid methods.

In light of this variety, we present here in some detail three methods that serve to represent all of the above, at least to a certain extent. The three methods are

Finite-element method

Finite difference time-domain (FDTD) method

Boundary-element method

The choice here is clear. The FEM method represents those methods that are essentially volumetric and differential in nature—those that are based on computation in the volume. The FDTD method represents finite-difference-based techniques and in particular those that operate in the time domain. For example, the TLM is fairly close in principle and in implementation to the FDTD method, and there is little to be gained by discussing it separately. The same applies to the finite-integration techniques (FIT) which may be viewed as a generalization of the FDTD method. The BEM computes variables on a boundary (such as equivalent sources or potentials) from which fields may be computed outside the volumes and, sometimes, depending on the method used, inside the volume, and hence represents other methods that take a similar approach such as the MOM. But the choice is not only representative. It is also based on the relative importance of these specific methods in computational engineering and, in particular, in computational electromagnetics.

A.2 Finite-Element Method

The FEM dates to the late 1950s [1–5] and early 1960s although the mathematical principles on which it is based are much older. The principles had to wait for the development of computers before the method could take off. Of course that is the case with all numerical methods. In particular, electrical engineering applications of the FEM came later and only dates to 1970 [6].

The FEM attempts to approximate the solution of a continuous problem in a rather straightforward way. The steps involved are the following [1,3,7,8]:

1. Establishment of the physical equations: The starting point is the physics of the problem. Usually, once the decision to use numerical methods has been made, one must establish the physical equations that will properly and accurately solve the problem.
2. Discretization: The solution space is discretized into finite elements. Finite elements are either lines, surfaces, or volumes, depending on dimensionality of the problem and the needs of the solution. Line, surface, and volume elements can be combined in the solution of a problem.
3. Approximation: An approximation over a finite element is defined. The approximation can be very simple (say, a linear function) or much more complex. It must conform to requirements of continuity and geometry. The choice of approximation depends on a variety of issues associated with the type of problem and the needs of the solution.
4. Minimization: A method of minimizing the error due to the discretization must be included. The discrete finite elements impose a solution that is only valid in the element itself. The minimization process ensures that the global solution is the "best" that can be obtained under the given conditions of minimization. A principle such as energy minimization or other equivalent principles are used for this purpose.

5. Solution: The previous steps result in a system of equations (linear or nonlinear) that must be solved.
6. Postprocessing of the solution may be required to obtain derived quantities.

The FEM is based on spatial discretization of the solution domain. A given finite element has associated with it physical dimensions of space. Therefore, it is not suited for solution in the time domain. While it is possible to define a finite element to approximate a solution in time, this is not normally done since finite differences are much more efficient for this purpose. When time-dependent problems must be solved, the time variable is handled either by a combined finite-element/finite-difference technique or, if possible, by transforming the time-dependent variables into the frequency domain (using the phasor transformation or, in more general terms, using the Fourier transform) and solving the problem in the frequency domain.

A.2.1 Physical Equations

The governing equations will depend on the application although the starting point in the electromagnetic case is obviously Maxwell's equations. But that is not usually sufficient. More often than not, we will introduce approximation as well as auxiliary functions, such as potentials, for the purpose of simplifying some aspects of the overall process. What is most convenient for an analytical solution is not always the best approach for a numerical solution and what fits one numerical method does not fit another. Even within the FEM, various formulations exist or can be obtained, each with its advantages and disadvantages. Also, the formulation is often influenced by the type of finite elements used and vice versa. Thus, no general, single form of the governing equations can or should be obtained. In any case, the general approach is to obtain the simplest form that adequately describes the physics of the problem at hand and to use judicious approximations in all aspects of the formulation and solution.

To give concrete examples, suppose we need to solve an eddy current problem in a general space that includes conducting and nonconducting media, including ferromagnetic materials. We have a number of options but first we must decide on a number of issues:

- Is the geometry nonlinear or can it be approximated as linear?
- Is the source transient or is it at steady state (single frequency)?
- Can the problem be solved in two-dimensional space or must it be three-dimensional?
- What quantities need to be calculated?
- What are the valid approximations that can be made?

Example

Suppose we decided that the source is transient and the problem is essentially two-dimensional, forcing us to start with the time-domain Maxwell's equations.

Eddy currents will exist in conducting media. Therefore, Ampere's law will be written as

$$\nabla \times \vec{H} = \vec{J} + \sigma\vec{E} \tag{A.1}$$

In two dimensions, in particular, it is convenient to use the magnetic vector potential, since then, by assuming the current to be perpendicular to the geometry, it is one-dimensional whereas the magnetic field is two-dimensional, distributed perpendicular to the current. With the magnetic vector potential we write

$$\vec{B} = \mu\vec{H} = \nabla \times \vec{A}; \quad \vec{E} = -\frac{\partial \vec{A}}{\partial t} - \nabla V \tag{A.2}$$

where V is the electric potential. At low frequencies, we can remove the potential by invoking the Coulomb gauge and get

$$\nabla \times \left(\nabla \times \vec{A}\right) + \sigma\frac{\partial \vec{A}}{\partial t} - \vec{J}_s = 0 \tag{A.3}$$

The triple product may be written in 2D as $-\nabla \cdot (\nu\nabla A)$, and we get

$$\nabla \cdot (\nu\nabla A) - \sigma\frac{\partial A}{\partial t} + J_s = 0 \tag{A.4}$$

Here the vector notation was removed since $\vec{A}$ has a single component. The time dependence in the eddy current term can be approximated as a finite difference approximation. Choosing a backward difference scheme,

$$\frac{\partial A}{\partial t} = \frac{A^i - A^{i-1}}{\Delta t} \tag{A.5}$$

where

- i indicates the current value of A
- $i-1$ is the value at the previous time step
- Δt is the time step chosen for the solution

With this we obtain the governing equation:

$$\nabla \cdot \left(\nu\nabla A^i\right) - \frac{\sigma A^i}{\Delta t} + \frac{\sigma A^{i-1}}{\Delta t} + J_s = 0 \tag{A.6}$$

This equation represents the fields in a nonlinear domain for a transient source, J_s. In addition, we must define a set of boundary conditions for the solution domain since, by necessity, we must limit the solution domain. Typically, the boundary conditions will be specified values of the variable (Dirichlet boundary conditions), Neumann boundary conditions (zero gradient of A across the boundary), or a more complex boundary condition such an absorbing boundary condition [8]. The boundary conditions are incorporated as part of the formulation in conjunction with the finite-element discretization.

Example

The same basic problem as previously but now we assume linear materials and a single-frequency source. Under these conditions, the fields may be represented in a time-harmonic form through the use of phasors. Assuming the source is $J_s(t) = J_s\cos(\omega t)$, we can write

$$J_s(t) = \mathrm{Re}\{\tilde{J}_s e^{j\omega t}\} \tag{A.7}$$

where $\tilde{J}_s e^{j\omega t}$ is the phasor form of the current density. Similarly, we can write with the help of the phasor form of $A(\tilde{A}e^{j\omega t})$:

$$A(t) = \mathrm{Re}\{\tilde{A}e^{j\omega t}\} \tag{A.8}$$

Substituting the phasor forms in Equation A.4 we have

$$\nabla \cdot (\nu \nabla \tilde{A}) - \sigma \frac{\partial \tilde{A}}{\partial t} + \tilde{J}_s = 0 \tag{A.9}$$

or

$$\nabla \cdot (\nu \nabla \tilde{A}) - j\omega\sigma\tilde{A} + \tilde{J}_s = 0 \tag{A.10}$$

Equation A.10 now needs to be solved in the FEM context. We note first that the variables now are complex and the equation is linear. Because of that, and assuming isotropy in permeability, we could also write it as

$$\nabla^2 \tilde{A} - j\omega\mu\sigma\tilde{A} + \mu\tilde{J}_s = 0 \tag{A.11}$$

Although the two examples above were written in two dimensions, they are completely general. Their extension to three dimensions involves little more than retaining the vector notation.

There are very many other forms and variations that can and are being used; a sampling of these can be found in Refs. [4,7–9].

A.2.2 Discretization

The discretization process in the FEM is simple. At its most basic form, the solution space is divided into a set of subdomains—volumes, surfaces, or lines. For example, one may use tetrahedra in three dimensions or triangular simplexes in two dimensions. Other shapes are equally possible but the choice of elements has consequences in terms of the steps that must necessarily follow. In turn, the choice of elements must take into account these steps. For example, a tetrahedral element is probably the most common element used in three dimensions. It, however, imposes specific possible approximations within the element. Other shapes (prisms, hexahedra, etc.) impose different constraints and produce different results with different levels of accuracy. There are a few constraints on the discretization itself. Some are geometrical: the elements must be capable of accurately describing the geometry and, as a rule, elements should not have angles that are too sharp because of errors these angles introduce, and elements cannot overlap geometrically. There are also special elements for specific applications. One example is a thin, flat element used for air gaps in electric machines. A geometry may combine various types of elements provided they are compatible—their shared edges, faces, and nodes must be uniquely defined in both elements. In addition, there are the so-called adaptive elements. Some, such as the Delauney elements [10] are designed to "optimize" the discretization in the sense of producing the best-quality elements possible. Others are adaptive in that finer discretization is possible by further subdivision of elements (*n*-adaptivity) [11–14] or the level of approximation of the element is adaptive (*p*-adaptivity) [14–17] and a combination of the two in *h–p* adaptivity [18–20], all in a way that is compatible with the requirements imposed by the software. These elements can produce automatic or manual refinement of the discretization of the geometry to improve accuracy of results. The whole process of discretization produces a mesh of elements, called a finite-element mesh. It should be mentioned here that there are specialized and general purpose software packages whose whole purpose is the geometrical division of a solution domain into finite elements with and without adaptive options [21–24].

In Section A.2.3, we discuss the process of approximation on a simple finite element, a process that will explain much of what has been said in this section.

A.2.3 Approximation

Once the finite elements have been defined geometrically, an approximation over the finite element must be defined. This approximation describes the variations of the variables within the finite element. Again, there is a large variety of approximations that may be defined within the element. In its simpler form, the variation will be linear in all dimensions of the element or it may be any higher-order polynomial that can be fitted on the element. The coefficients of the approximation are derived from the geometrical properties of the element. The approximation will also depend on how coefficients are defined. For example, the approximation in the volume of the element may be defined based on the vertices of a simplex. These are termed "nodal elements." Another way is to define them in terms of the edges of the

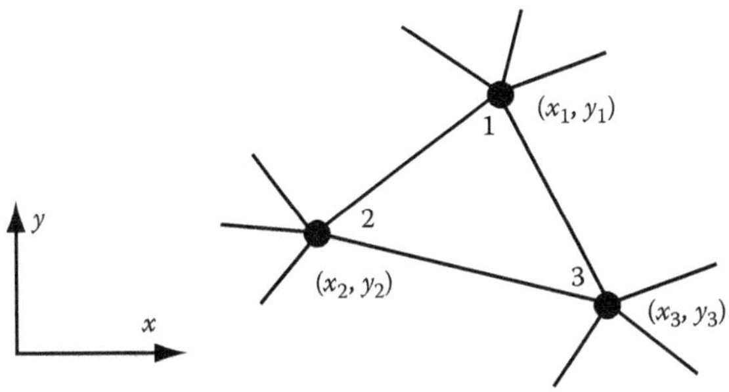

FIGURE A.1
A triangular finite element and notation used to define shape functions.

element (edge elements). Surface (facet) elements exist as well, all with specific properties and ideally suited for specific applications. To exemplify the method, we will start with nodal elements as these are the easiest to understand and commonly used.

Consider, for example, the triangular element shown in Figure A.1. We assume a linear approximation within the element:

$$V(x,y) = a_1 + a_2x + a_3y \tag{A.12}$$

Clearly this is a linear approximation in both x and y. $V(x,y)$ is the unknown value to be evaluated (e.g., the magnetic potential). To evaluate the coefficients a_1, a_2, and a_3, we use the unknown values at the three nodes of the element:

$$V_i = a_1 + a_2x_i + a_3y_i \tag{A.13}$$

Solving the three equations obtained using this relation, we get the general approximation in the element as

$$V(x,y) = \sum_{i=1}^{3} \left[\frac{1}{2D}(p_i + q_ix_i + r_iy)\right] V_i \tag{A.14}$$

where D is the area of the triangle and

$$p_i = x_{i+1}y_{i+2} - x_{i+2}y_{i+1}; \quad q_i = y_{i+1} - y_{i+2}; \quad r_i = x_{i+2} - x_{i+1} \tag{A.15}$$

where cyclical permutations are used for subscripts larger than 3. For example, $x_{3+1} = x_1$, $y_{3+2} = y_2$, and so on.

Returning now to Equation A.14, we note the following:

1. The value of V depends on the nodal values, V_i, and the geometrical functions in the square brackets.
2. V_i are the unknown values to be solved for. These are called degrees of freedom because, in the solution process, these are "adjusted" to

obtain the best possible solution using a minimization process to be discussed next.

3. The geometric functions in the square brackets are called shape functions and are dependent only on the shape and coordinates of the element. These shape functions have a few simple properties [1,3], the most important of which is that their sum equals 1 everywhere within and on the boundaries of the element.
4. The element described here is linear (sometimes called bilinear because it is linear in both dimensions). Higher-order elements with higher-order approximations can be obtained by either using geometries with more nodes or adding nodes to the triangular element. Of course, the same considerations apply to other types of elements.

Each element is defined entirely separately from any other element in the finite element mesh. Only in the minimization process are the elements linked. The process above can be extended to any element of any shape and any number of nodes, either in two or three dimensions. The approximation in Equation A.14 must of course contain as many coefficients as there are nodes in the element and the approximation may not be linear. In fact, in two dimensions, if there are more than three nodes, the polynomial used for approximation must contain terms with x^2 and/or y^2 or higher, depending on the exact number of nodes. In three dimensions, a tetrahedral element will have four nodes as a minimum, in which case the approximation will be linear in x, y, and z (trilinear).

However, there are many other methods that can be used to obtain approximations. Perhaps the best known of these is the isoparametric method [3,4].

The nodal elements described above are not the only elements possible. To demonstrate a totally different approach, consider the so-called edge or vector elements. In these, the shape functions are associated with both a magnitude and a direction in space and are particularly well suited for certain vector field applications in which the circulations along edges rather than potentials at nodes become the degrees of freedom [8,25,26]. A rectangular finite element is shown in Figure A.2. On this element, the

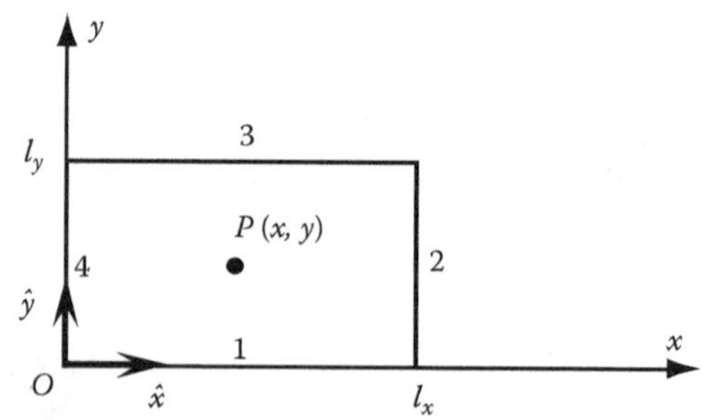

FIGURE A.2
A rectangular edge element.

approximation is similar to that in Equation A.14 but now we assume vectors for the degrees of freedom. Since the element has four edges, we write the approximation as

$$\vec{A}(x, y) = \sum_{i=1}^{4} \vec{w}_i A_i \qquad \text{(A.16)}$$

where $\vec{w}_i$ is the shape function related to the ith edge of the element.

The shape functions for the element are

$$\vec{w}_1 = \hat{x}\left[1 - \frac{y}{l_y}\right]; \quad \vec{w}_2 = \hat{y}\left[\frac{x}{l_x}\right]; \quad \vec{w}_3 = \hat{x}\left[\frac{y}{l_y}\right]; \quad \vec{w}_4 = \hat{y}\left[1 - \frac{x}{l_x}\right] \qquad \text{(A.17)}$$

where l_x and l_y are the lengths of the edges parallel to the x- and y-axes, respectively. These shape functions have a direction ($\hat{x}$ or $\hat{y}$, depending on the edge) in addition to a ''placement'' or ''position'' expression. The magnitude has been chosen to satisfy the basic requirement that the sum of the four shape functions on any edge be equal to 1.

The degrees of freedom, A_i in Equation A.16 represent the projections of the vector $\vec{A}$ on the ith edge of the element. The same process can be performed on any element including triangular and tetrahedral elements. In fact the more common method is to base the shape functions on the so-called Whitney complex [25–29] but this simple example is much more explicit and suffices to introduce the method.

A.2.4 Minimization

So far, each finite element has been treated as an independent entity without any attempt to take into account the degrees of freedom in neighboring elements. Obviously that cannot produce an acceptable solution unless we can find a method through which the degrees of freedom can be linked together. This process is called a minimization process. In effect we look for a principle which, when applied, forces all degrees of freedom to adjust to satisfy this principle. The simplest method is minimization of energy in the system. Even when no energy *per se* is minimized, the function to minimize is energy related in some sense. The normal terminology is ''energy functional minimization.'' The process by which the functional is obtained and minimized depends very much on the problem at hand but there are a few common methods that can be applied for this purpose. Of these the Rayleigh–Ritz method, Galerkin's method, and the variational method are the most commonly employed [3,8] but, as with all aspects of the FEM, considerable variations from the basic principles exist and many useful and ingenious modifications are being used. In all of these methods, a weighting function is identified, the weighting functions as well as the shape functions are applied in an appropriate functional, followed by its minimization on an element by element basis, yielding in the end a system of linear equations to be solved.

The variational method, while not always the choice method is, nevertheless the most instructive and therefore is worth discussing it here.

Suppose that a functional F exists, and that it is a function of a variable P and of its partial derivatives $P'_x = \partial P/\partial x$, $P'_y = \partial P/\partial y$, and $P'_z = \partial P/\partial z$. Then

$$F = \int_v f\left(P, P'_x, P'_y, P'_z\right) dv \tag{A.18}$$

where v is the volume in which F is defined.

The necessary and sufficient condition for F to be stationary (minimum condition) is that for any small variation δP of the independent variable P, the corresponding variation δF is zero:

$$\delta F = \int_v \left(\frac{\partial f}{\partial P} \delta P + \frac{\partial f}{\partial P'_x} \delta P'_x + \frac{\partial f}{\partial P'_y} \delta P'_y + \frac{\partial f}{\partial P'_z} \delta P'_z \right) dv = 0 \tag{A.19}$$

where $\delta P'_x = \delta(\partial P/\partial x) = \partial/\partial x(\delta P)$ (and similarly for $\delta P'_y$ and $\delta P'_z$). Denoting

$$\hat{x}\frac{\partial f}{\partial P'_x} + \hat{y}\frac{\partial f}{\partial P'_y} + \hat{z}\frac{\partial f}{\partial P'_z} = \vec{g}$$

$$\hat{x}\frac{\partial}{\partial x}\delta P + \hat{y}\frac{\partial}{\partial y}\delta P + \hat{z}\frac{\partial}{\partial z}\delta P = \nabla(\delta P) \tag{A.20}$$

With these, Equation A.19 becomes

$$\delta F = \int_v \left(\frac{\partial f}{\partial P} \delta P + \vec{g} \cdot \nabla(\delta P) \right) dv = 0 \tag{A.21}$$

Applying the divergence theorem to the second term in the integrand and using the vector identity $\nabla \cdot \left(c\vec{d}\right) = c\nabla \cdot \vec{d} + \vec{d} \cdot \nabla c$ gives

$$\delta F = \int_v \left(\frac{\partial f}{\partial P} - \nabla \cdot \vec{g} \right) \delta P dv + \oint_S \delta P \vec{g} \cdot \hat{n} ds = 0 \tag{A.22}$$

where

S is the surface enclosing the volume v

$\hat{n}$ is the normal unit vector at any point on S

Because the two integrals are independent, for their sum to be zero the two terms must satisfy the following:

$$\int_v \left(\frac{\partial f}{\partial P} - \nabla \cdot \vec{g}\right) \delta P dv = 0 \tag{A.23}$$

and

$$\oint_S \delta P \vec{g} \cdot \hat{n} ds = 0 \tag{A.24}$$

Since δP is arbitrary, and since Equation A.23 must be zero, we get

$$\frac{\partial f}{\partial P} - \nabla \cdot \vec{g} \tag{A.25}$$

or

$$\frac{\partial f}{\partial P} - \frac{\partial}{\partial x}\left(\frac{\partial f}{\partial P'_x}\right) - \frac{\partial}{\partial x}\left(\frac{\partial f}{\partial P'_y}\right) - \frac{\partial}{\partial x}\left(\frac{\partial f}{\partial P'_z}\right) = 0 \tag{A.26}$$

Equation A.26 is known as "Euler's equation." For the functional F to be valid, it must have an integrand that satisfies Equation A.26. Equation A.24, which relates to the integration on the surface S, is now examined separately. For this equation to be zero, either δP or $(\vec{g} \cdot \hat{n})$ is zero. We can, a priori, divide the surface S into two parts S_1 and S_2 (one of these surfaces can be nonexistent, or have zero area):

- S_1 is that part of the boundary on which $\delta P = 0$; on this part the potential is imposed as a boundary condition, called a "Dirichlet condition." Since P is fixed, its variation δP is zero.
- $S_2 = S - S_1$ is the part of the boundary on which $\vec{g} \cdot \hat{n} = 0$; on this boundary we have the so-called Neumann boundary condition and no condition is imposed. The Neumann boundary condition is natural and is an intrinsic property of this formulation. On S_2, we can write

$$\vec{g} \cdot \hat{n} = \frac{\partial f}{\partial P'_x} n_x + \frac{\partial f}{\partial P'_y} n_y + \frac{\partial f}{\partial P'_z} n_z = 0 \tag{A.27}$$

The value of P minimizing the functional F entails:

- Establishment of the functional F whose integrand satisfies Euler's equation.

- Application of boundary conditions, pertinent to the problem, where either P is fixed (Dirichlet condition), or Equation A.27 (Neumann condition) is satisfied.

For example, assuming an electrostatic problem, with charge density ρ in a dielectric medium the energy-related functional is

$$F = \int_v f\left(V, V'_x, V'_y\right) dv = \int_v \left(\frac{1}{2}\varepsilon E^2 - \rho V\right) dv \tag{A.28}$$

where the integrand is the volume energy density.

Similarly, assuming a magnetostatic problem with nonlinear media and the use of the magnetic vector potential as the variable leads to the following energy-related functional

$$F = \int_v \left[\int_0^B (H dB - JA)\right] dv \tag{A.29}$$

where $\vec{H} = \vec{B}/\mu$. Again, the energy relation is unmistakable.

We now calculate the terms of Euler's equation, for a two-dimensional application using Equation A.29 observing that here $P = A$, $P'_x = A'_x$, and $P'_y = A'_y$.

$$\frac{\partial F}{\partial A} = -J \tag{A.30}$$

Since

$$A'_x = \frac{\partial A}{\partial x} = -B_y \quad \text{and} \quad A'_y = \frac{\partial A}{\partial y} = -B_x \tag{A.31}$$

we have

$$\frac{\partial f}{\partial A'_y} = -\frac{\partial}{\partial B}\left[\left[\int_0^B (H dB - JA)\right]\right]\frac{\partial B}{\partial B_y} = -H\frac{\partial}{\partial B_y}\left(B_x^2 + B_y^2\right)^{1/2} \tag{A.32}$$

or, after evaluation,

$$\frac{\partial f}{\partial A'_y} = -H\frac{B_y}{B} = -\frac{B_y}{\mu} = -H_y \tag{A.33}$$

and similarly for the x-component of H:

$$\frac{\partial f}{\partial A'_y} = \frac{\partial f}{\partial B_x} = \frac{\partial f}{\partial B}\frac{\partial B}{\partial B_x} = H_x \tag{A.34}$$

Substituting Equations A.30, A.33, and A.34 into Euler's equation (Equation A.26) gives

$$-J - \frac{\partial}{\partial x}(-H_y) + \frac{\partial}{\partial y}(H_x) = 0 \tag{A.35}$$

This is equivalent to

$$\nabla \times \vec{H} = \vec{J} \tag{A.36}$$

which is the equation to be solved. To incorporate the Dirichlet boundary condition, we use the fact that on S_1, A is imposed and constant, and the field is tangent to the surface. For the Neumann boundary condition, Equation A.27 must be satisfied

$$\vec{g} \cdot \hat{n} = \frac{\partial f}{\partial P'_x} n_x - \frac{\partial f}{\partial P'_y} n_y = 0 \tag{A.37}$$

Using Equations A.33 and A.34, we get

$$-H_y n_x + H_x n_y = 0 \tag{A.38}$$

or, in general terms,

$$\vec{H} \times \hat{n} = 0 \tag{A.39}$$

meaning that the vectors $\vec{H}$ and $\hat{n}$ are colinear.

The variational method is both simple and instructive but it relies on an appropriate functional which, once established, leads to a solution process. It is instructive to follow the finite element process a step further and apply the principles and results above to a specific finite element example, using the triangular element discussed above (a three-dimensional example would be much more involved without adding much to the understanding of the process):

The discretized form of the functional in Equation A.29 is

$$F_i = \int\limits_{S_i} \left[\int\limits_{0}^{B} H \, \mathrm{d}B - JA \right] \mathrm{d}s \tag{A.40}$$

The term $\partial F_i/\partial A_k$ is obtained by the sum of the partial results of the two terms of Equation A.40. Noting that $\vec{H} = \nu\vec{B}$, where $\nu = 1/\mu$, we have

$$S(n,k) = \frac{\nu}{2D}(q_n q_k + r_n r_k) \tag{A.41}$$

The term

$$\frac{\partial}{\partial A_k}\int\limits_{S_i}(-JA)\mathrm{d}s \tag{A.42}$$

gives

$$-\frac{D}{6}JA \tag{A.43}$$

Collecting these in matrix form for a single element:

$$\begin{bmatrix} \dfrac{\partial F_i}{\partial A_1} \\ \dfrac{\partial F_i}{\partial A_2} \\ \dfrac{\partial F_i}{\partial A_3} \end{bmatrix} = \begin{bmatrix} S_{1,1} & S_{1,2} & S_{1,3} \\ S_{2,1} & S_{2,2} & S_{2,3} \\ S_{3,1} & S_{3,2} & S_{3,3} \end{bmatrix}\begin{bmatrix} A_1 \\ A_2 \\ A_3 \end{bmatrix} - \frac{D}{6}\begin{bmatrix} J \\ J \\ J \end{bmatrix} \tag{A.44}$$

This elemental matrix is now written independently for every element, and assembled into a global matrix which contains the coefficients for all elements. The boundary conditions are then applied to the global matrix (but see below) and the system so obtained is solved using any convenient method of solving a fairly large system of linear equations.

Unfortunately, finding an appropriate functional for general applications is either difficult or, sometimes not possible or practical. For that reason, other methods, of which the Galerkin method is perhaps the best known, have been established. However, unlike the variational method which can be established independently of finite elements, the Galerkin method is intrinsically linked with the finite elements. To demonstrate the method, we will use here an eddy current formulation which does not lend itself to application of the variational method.

Starting first with the general principle of weighted residuals and assuming that a function, $f(v, \Delta v) = 0$ which depends on a variable v and its derivatives Δv has an exact solution v and an approximate solution u. Then, substituting both solutions into the function a residual is obtained as follows:

$$R = f(v) - f(u) = -f(u) \tag{A.45}$$

Now, minimization of this residual guarantees the "best" solution to the problem at hand. In the context of finite elements, the following notation may be used:

$$W_k\, Rd\, D = 0, \quad 1 \leq k \leq K \tag{A.46}$$

where W_k is the weighting function defined for the K unknown nodes of the domain D. Galerkin's method is a particular form of the weighted residual method in which the weighting function and the approximating function are the same.

As an example of the use of the Galerkin method, we analyze here the two-dimensional problem described in Equation A.6 assuming general media (conducting, nonconducting, and nonlinear ferromagnetic media):

$$\nabla \cdot (\nu \nabla A) - \frac{\sigma A^i}{\Delta t} + \frac{\sigma A^{i-1}}{\Delta t} + J_s = 0 \tag{A.47}$$

A full three-dimensional formulation follows similar steps but requires additional considerations including gauging and use of the vector form of the operator (which retains the curl in Equation A.3) However these complications, while necessary do not add to the instructiveness of the application of Galerkin's method and therefore the choice of using a simplified two-dimensional formulation for this purpose.

Now we apply the Galerkin procedure to Equation A.47, where the weighting functions are the shape functions N given in the previous sections (for example, N may be the shape functions in Equation A.14 or the functions $\vec{w}$ in Equation A.17 if vector elements are used but may be any other properly defined shape functions on a finite element). By applying the weights in a weighted residual method, the Galerkin procedure gives

$$\int_S N^t \nabla \cdot (\nu \nabla A) \mathrm{ds} - \int_S N^t \frac{\sigma A^i}{\Delta t} \mathrm{ds} + \int_S N^t \frac{\sigma A^{i-1}}{\Delta t} \mathrm{ds} + \int_S N^t J_s \mathrm{ds} = 0 \tag{A.48}$$

It should be remembered that the approximation for A will also be in terms of the shape functions, which in any element is

$$A = \sum_n N_n A_n \tag{A.49}$$

Using the identity $U \nabla \cdot \vec{A} = \nabla \cdot U\vec{A} - \vec{A} \cdot \nabla U$, Equation A.48 becomes

$$\int_\Gamma N^t \nu \nabla A \cdot \vec{n} \mathrm{d}\Gamma - \int_S \nu \nabla A \cdot \nabla N^t \mathrm{ds} + \frac{\sigma}{\Delta t} \int_S N^t A^i \mathrm{ds} + \frac{\sigma}{\Delta t} \int_S N^t A^{i-1} \mathrm{ds} + \int_S N^t J_s = 0 \tag{A.50}$$

The first term is the contribution due to the boundary of the problem. This term is zero as can be seen by direct evaluation of the term $\nabla A \cdot \vec{n}$ on the boundary Γ. This, of course means that Neumann boundary conditions are implicit in the formulation (need not be specified) as was the case with the variational formulation. To obtain the elemental matrix, we substitute Equation A.49 into Equation A.50:

$$-\int_S \nabla N^t \cdot \nu \nabla N A_n \mathrm{ds} + \frac{\sigma}{\Delta t}\int_S N^t N A_n^i \mathrm{ds} + \frac{\sigma}{\Delta t}\int_S N^t N A_n^{i-1} \mathrm{ds} + \int_S N^t J_s = 0 \quad \text{(A.51)}$$

The subscript n indicates this is an elemental matrix whereas the superscript i indicates the time step. For an element with n nodes, this produces n equations. Note that in the form in Equation A.51 the reluctance ν has been retained as nonlinear and hence this formulation is equally suitable for linear and nonlinear applications.

From here on, the solution proceeds as previously by evaluating the coefficients of the elemental system in Equation A.51, incorporating these into a global system and then solving that system taking into account nonlinearity (using, for example a Newton–Raphson iterative method [4] at each time step) and stepping in time as with any finite difference scheme.

A.2.5 Solution

Whatever the method by which the final system of equations was obtained, these equations must be solved. If the problem is nonlinear, an iterative process is established and the system of equations is solved at each iteration. Similarly, in time-domain problems, the system of equations must often be solved at each time step (there are exceptions to this). Although the solution is straight forward, it is often important that efficient solution methods be employed. In most cases, a Gaussian elimination process or one of its variants is sufficient especially since the system of equations is almost always symmetric, positive definite. Other methods that have been employed successfully are various forms of the Choleski decomposition, with or without preconditioning [30–33]. The incomplete Choleski conjugate gradient (ICCG) in its various forms has been particularly useful in reducing solution time or, more often, allowing larger problems to be solved in reasonable times. These methods are also responsible, to a large extent for the success in the use of personal computers to solution of significant problems in computational electromagnetics and other areas of inquiry. Nevertheless, most users of the FEM will tend to view the solution as a necessary block in the overall process and often give it but scant attention.

A.2.6 Postprocessing

A necessary step in any FEM computation, and indeed in any numerical method, is postprocessing of the data obtained in the solution step. Sometimes all that is necessary is the field distribution but often other parameters such as forces, torques, power, and energy and the like are needed. Even calculation of fields will often require additional steps since the FEM program is likely to produce auxiliary functions such as potentials rather than fields. Usually the postprocessing is done separately, especially in commercial programs, allowing the user to analyze the solution.

A.3 Finite-Difference Time-Domain Method

The FDTD method [34,35] takes a very different approach to modeling; it approximates the field equations directly without the need for intermediate steps such as the approximation and minimization processes that are so important in FEM methods. To understand how this is done, consider Ampere's and Faraday's laws in the following forms [36,37]:

$$\frac{\partial \vec{H}}{\partial t} = -\frac{1}{\mu}\left(\nabla \times \vec{E}\right) \tag{A.52}$$

$$\frac{\partial \vec{E}}{\partial t} = \frac{1}{\varepsilon}\left(\nabla \times \vec{H}\right) - \frac{\sigma}{\varepsilon}\vec{E} \tag{A.53}$$

The curl terms are first written explicitly. For example, for the x-components:

$$\frac{\partial E_x}{\partial t} = \frac{1}{\varepsilon}\left(\frac{\partial H_z}{\partial_y} - \frac{\partial H_y}{\partial_z}\right) - \frac{\sigma}{\varepsilon}E_x \tag{A.54}$$

$$\frac{\partial H_x}{\partial t} = -\frac{1}{\mu}\left(\frac{\partial E_z}{\partial_y} - \frac{\partial E_y}{\partial_z}\right) \tag{A.55}$$

Similar terms can be written for the other components. Note however that source current densities are not included. These as well as boundary conditions are specified separately although sources can be included directly in the formulation that follows.

The derivatives in these two relations are approximated directly by finite differences in time and space. Because the formulation links the magnetic field and electric field in each of Equations A.54 and A.55, the calculation of the magnetic or electric field in one equation requires the calculation of the magnetic field from the second equation and vice versa. For this reason, the calculation proceeds in a leapfrog fashion by which the electric field is calculated at a certain time step and location in space, assuming the magnetic field is known. Then, the magnetic field at a different time step and different location in space is calculated based on the now known electric field. To do so, a grid is defined over the space of the solution domain as shown in Figure A.3. The spatial distribution in each dimension is assumed to be the same, denoted by h but that is only a convenience to simplify the relations. Based on this basic cell, Equations A.54 and A.55 are discretized as follows. First, a finite difference scheme is defined for space and for time discretization:

$$\frac{\partial E^m_{i,j,k}}{\partial x} = \frac{E^m_{i+1/2,j,k} - E^m_{i-1/2,j,k}}{h} \tag{A.56}$$

$$\frac{\partial E^m_{i,j,k}}{\partial t} = \frac{E^{m+1/2}_{i,j,k} - E^{m-1/2}_{i,j,k}}{\Delta t} \tag{A.57}$$

where

m indicates a point in time
i, j, k is a point in space
h is the cell size
Δt is the time step

Clearly these are central difference schemes. The notation 1/2 indicates a half step in time and/or space. Applying these approximations to Equations A.54 and A.55 on the grid in Figure A.3:

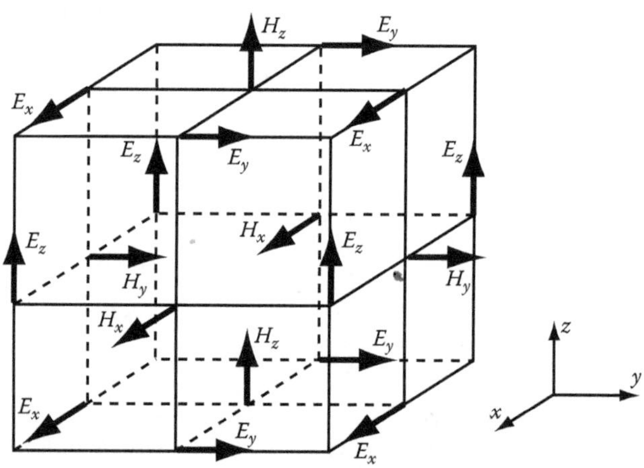

FIGURE A.3
Spatial relation between the magnetic and electric field in FDTD. The electric and magnetic fields are calculated at different locations on the grid.

$$H_x^{m+1/2}(i,j+1/2,k+1/2) = H_x^{m-1/2}(i,j+1/2,k+1/2) + \frac{\Delta t}{\mu(i,j+1/2,k+1/2)h}\Big[E_y^m(i,j+1/2,k+1)$$
$$-E_y^m(i,j+1/2,k) + E_y^m(i,j,k+1/2) - E_y^m(i,j+1/2,k+1/2)\Big] \quad \text{(A.58)}$$

$$E_x^{m+1}(i+1/2,j,k) = \left(1 - \frac{\sigma(i+1/2,j,k)}{\varepsilon(i+1/2,j,k)h}\right)E_x^m(i+1/2,j,k)$$
$$+\frac{\Delta t}{\varepsilon(i+1/2,j,k)h}\Big[H_y^{m+1/2}(i+1/2,j+1/2,k) - H_y^{m+1/2}(i+1/2,j-1/2,k)$$
$$+H_y^{m+1/2}(i+1/2,j-1/2,k) - H_y^{m+1/2}(i+1/2,j,k+1/2)\Big] \quad \text{(A.59)}$$

Similar expressions are obtained for the other components of the electric and magnetic fields in Equations A.54 and A.55. The most important observation here is that the electric and magnetic fields are interlaced and calculation of one field can only occur after the second has been evaluated.

The cell spacing and time step must satisfy the following stability criterion:

$$\Delta t \le \frac{1}{\sqrt{(\Delta x)^2 + (\Delta y)^2 + (\Delta z)^2 v_\mathrm{p}}} \quad \text{or} \quad \Delta t \le \frac{h}{\sqrt{3}v_\mathrm{p}} \quad \text{for } \Delta x = \Delta y = \Delta z = h \quad \text{(A.60)}$$

where v_p is the maximum phase velocity in the discretization space. From a practical point of view one would prefer that Δt be as large as possible and hence, h should be as large as possible. The latter however is constrained by the accuracy required and is usually taken as $h < \lambda/10$.

As with any numerical method many more variations and modifications are possible, including definition of grids on cylindrical geometries, inclusion of anisotropic materials, and others. In all cases, however it is important to note that for the solution to start, the initial conditions for the fields must be available (zero values are often used for this purpose). In other cases, perhaps a wave front is available which can serve as initial condition. Also to be noted is that material properties are defined point by point. Therefore anisotropic media are easily taken into account. Finally, the absolute simplicity of this formulation should be again indicated.

To obtain a solution it is also necessary to properly terminate the space discretization through imposition of boundary conditions and termination in time based on some appropriate criterion. Sometimes boundary conditions are imposed by the geometry. A PEC is an example of this type of boundary condition. More often, especially with propagating waves some type of radiation or absorption boundary condition is necessary. These allow close truncation of the solution domain so that only regions of interest are included thereby reducing the time needed for solution. Here again the FDTD method holds a clear advantage in the simplicity with which boundary conditions are taken into account. These boundary conditions are incorporated directly into the formulation. Because of this ease of implementation some very useful radiation/absorption boundary conditions have been introduced [38–40] allowing the FDTD to work effectively in open geometry configurations. Other methods include the development of artificial absorbers (perfectly matched layes [PMLs]) [41–44] and surface impedance boundary conditions [45–47]. Because of these important developments and its direct approach to modeling, the FDTD has been the method of choice in high-frequency applications for many years. Nevertheless, it can also be used at low frequencies, either through use of modified radiation boundary conditions [48] or, indeed, through the use of the SIBCs described in the present work.

In addition to boundary conditions, the initial conditions needed to start the solution must be specified. This may take the form of sources at nodes or may take the form of, say, a plane wave, propagating into the solution domain.

There are however some basic limitation in the use of the FDTD. A basic one, which is shared by other methods such the TLM, is discretization of nonrectangular boundaries. The basic solution to this difficulty is the staircase discretization of such boundaries. If the staircase is done properly (i.e., sufficiently fine), then this method is acceptable. Others have adapted the FDTD for use in non-Cartesian grids [49–52] and for nonuniform grids, all in attempts to extend the applicability of the method. Another limitation is the stability criterion defined in Equation A.60. It is always possible to find a combination of time/space steps that will satisfy these conditions but the time or space steps may have to be so small as to make the solution either slow and difficult or, indeed impractical.

A.4 Boundary-Element Method

In the previous two sections we discussed two fundamental methods for solution of engineering problems based on solution of differential equations. Whereas these methods, and those related to them account for many important applications, they are not the only methods applicable. In particular, there is a large class of numerical methods based on formulation and solution of integral equations. The class of methods called BEMs is based on integral equations formulation of boundary value problems in conjunction with numerical methods. Unlike methods such as the FEM or FDM methods, the discretization is on the boundaries of the solution domain rather than in the volume. Hence one of the most important characteristics of the method—that of reduction of the dimensionality of the problem by one. This characteristic accounts for the method's importance and widespread use as well as its limitations. Because of this fundamental characteristic, the BEM preceded the FEM in many applications and remains useful to this day, especially in problems with unbounded domains. For example, one of the most important numerical methods for high-frequency application is the MOM. It can be used as a differential method but most of its applications are in its integral formulation.

Another characteristic of integral equations is that the boundary conditions to the problem are incorporated within the integral equations. Therefore, unlike differential formulations, in which boundary conditions are applied externally, the integral formulation of a boundary value problem includes both the differential equation and its boundary conditions.

The BEM method has the advantage of simplicity in implementation in computer codes but has been hampered in tackling nonlinear problems and some problems that can be handled easily with methods like the FEM. Nevertheless, in many applications it is the method of choice primarily because of the reduced dimensionality of the problem.

There are two fundamental methods of obtaining a BEM formulation. An indirect method of formulation consists of use of the physical aspects of the problem to define equivalent sources on the boundaries of the problem. These sources are first calculated and then the fields due to the sources are evaluated in the region of interest. This method is commonly used to find radiation fields due to antennas and other configurations. For example, the work of Mitzner in defining SIBCs [53] was based on exactly this form of integral equations (see also Chapter 1). However, many others have found the indirect method useful for a variety of important applications [37,54].

A second method, which is also common in BEM formulations is the so-called direct method, which is based on the fact that if the Green's function of a given differential equation together with prescribed boundary conditions on a geometry is known, then the solution is also known in the form of a differential equation, which can be solved numerically. The numerical formulation is a weak-type variational or weighted residual formulation in which the prescribed boundary conditions are used to link the unknown boundary values with the known values. This approach has found much in common with methods such as the FEM and draws on work and advances obtained in work with FEM.

To see how the integral equations are obtained we use the direct method (see Chapter 1 for the method used to obtain the MFIE and EFIE using the indirect method). We will use the Poisson equation as an example for solution of the general class of potential problems, following the derivation in [9,54]. Other formulations are available and are equally valid, each with its own advantages. We assume that Poisson's equation for a potential, U, applies in a domain Ω, bounded by a boundary Γ as in Figure A.4. U itself may be any function, including, of course vector functions but to simplify things it is best to keep to a scalar function:

$$\nabla \cdot (\nu \nabla U) = -R \tag{A.61}$$

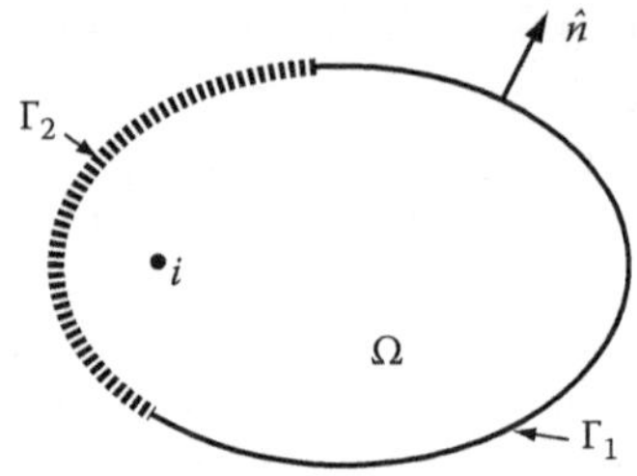

FIGURE A.4
Solution domain and notation for BEM formulation.

The applied and natural boundary conditions on $\Gamma = \Gamma_1 + \Gamma_2$:

$$U = \bar{U} \quad \text{on } \Gamma_1 \tag{A.62}$$

and

$$\frac{\partial U}{\partial n} = \frac{\partial \bar{U}}{\partial n} \quad \text{on } \Gamma_2 \tag{A.63}$$

Applying the method of weighted residual (or, for that matter, the Galerkin method described above):

$$\int_\Omega w_i[\nabla \cdot (\nu \nabla U) + R]\mathrm{d}\Omega + \int_{\Gamma_1} \bar{w}_i[U - \bar{U}]\mathrm{d}\Gamma + \int_{\Gamma_2} \bar{\bar{w}}_i \left[\frac{\partial U}{\partial n} - \frac{\partial \bar{U}}{\partial n}\right] \mathrm{d}\Gamma = 0 \tag{A.64}$$

In this form the weighting functions w_i, $\bar{w}_i$, and $\bar{\bar{w}}_i$ are arbitrary and will be chosen judiciously.

The first term is now transformed using Green's second theorem by first taking the term ν out of the divergence (and hence assuming isotropic media) as follows:

$$\nu \int_\Omega w_i \nabla^2 U \mathrm{d}\Omega = \nu \int_\Omega U \nabla^2 w_i \mathrm{d}\Omega + \int_\Gamma w_i \frac{\partial U}{\partial n} \mathrm{d}\Gamma - \int_\Gamma U \frac{\partial w_i}{\partial n} \mathrm{d}\Gamma \tag{A.65}$$

Now by choosing the following weights:

$$\bar{w}_i = \nu \frac{\partial w_i}{\partial n} \quad \text{and} \quad \bar{\bar{w}}_i = -\nu w_i \tag{A.66}$$

and substituting these as well as Equation A.65 into Equation A.64 we get

$$\int_\Omega \nabla^2 w_i \mathrm{d}\Omega - \nu \int_\Omega w_i \frac{R}{\nu} \mathrm{d}\Omega = -\int_{\Gamma_2} w_i \frac{\partial \bar{U}}{\partial n} \mathrm{d}\Gamma - \int_{\Gamma_1} w_i \frac{\partial U}{\partial n} \mathrm{d}\Gamma + \int_{\Gamma_2} \frac{\partial w_i}{\partial n} U \mathrm{d}\Gamma + \int_{\Gamma_1} \frac{\partial w_i}{\partial n} \bar{U} \mathrm{d}\Gamma \tag{A.67}$$

The net effect is that the unknown functions U are integrated on the boundary while the weighting functions are differentiated. It is however necessary to choose the functions w_i such that the first integral on the left-hand side can be easily evaluated. This is done by choosing w_i to be the fundamental solution corresponding to the Laplacian operator:

$$\nabla^2 w_i + \delta_i = 0 \tag{A.68}$$

The fundamental solutions are

$$w = \frac{1}{2\pi r} \text{ in 3D} \quad \text{and} \quad w = \frac{1}{2\pi} \ln\left(\frac{1}{r}\right) \text{ in 2D} \tag{A.69}$$

These solutions represent the potential due to a unit source placed at a point i. The weighting functions are then composed of a series of these fundamental solutions applied over the boundary.

If, for example, w_i is the potential of a point charge at the origin, then we can write

$$\nu \int_{\Omega} U \nabla^2 w_i \mathrm{d}\Omega = -\nu U_i \tag{A.70}$$

With these, the solution in Equation A.67 in the interior of the domain becomes

$$U_i = -\int_{\Omega} w_i \frac{R}{\nu} d\Omega + \int_{\Gamma_1} w_i \frac{\partial U}{\partial n} \mathrm{d}\Gamma - \int_{\Gamma_1} \frac{\partial w_i}{\partial n} \bar{U} \mathrm{d}\Gamma + \int_{\Gamma_2} w_i \frac{\partial \bar{U}}{\partial n} \mathrm{d}\Gamma - \int_{\Gamma_2} \frac{\partial w_i}{\partial n} U \mathrm{d}\Gamma \tag{A.71}$$

This equation calculates the potential U at point i in the interior of the domain Ω from the values of U, $\bar{U}$, $\partial U/\partial n$, and $\partial \bar{U}/\partial n$ on the boundaries of the domain. The index i on w indicates these to be functions of both the source point and the field point positions as indicated in Equation A.69.

For a BEM formulation it is necessary to take the point i to be on the boundary. On the boundaries themselves, the solution must be modified since the fundamental solutions are singular on the boundary. The singular integrals need to be evaluated first. On the boundary, either U or $\partial U/\partial n$ must be specified. In addition, the boundary must be divided into elements of some predetermined shape and size. The necessary steps are as follows:

1. The singular integrals are evaluated on the boundary of the domain. This integral will depend on the dimensionality of the solution and on the type of problem being solved and type of boundary.

It typically adds a quantity to U_i in Equation A.71 modifying it to cU_i. For example, in two or three dimensions, on a smooth boundary $c=1/2$.

2. A set of finite elements of a predetermined shape and physical dimensions is defined on the boundary of the problem. In two dimensions these may be simple line segments (linear elements) or they may be higher-order polynomials on lines (curved elements). In three dimensions they may be triangles (linear elements) or perhaps higher-order elements (curved elements).
3. A decision on the type of formulation to pursue. Specifically, the question that remains is what type of basis functions to use for the approximation for U. The common approach is to use what amounts to constant basis functions within each element and perform collocation at the center of the element. This is usually used in two and three dimensions when low-order elements are used. This avoids the need for integration over the operator and simplifies the setup. In this approach, the values of U and $\partial U/\partial n$ are assumed to be constant within each element and represented at the center of the element.

With these, Equation A.71 becomes

$$cU_i + \int_\Omega w_i \frac{R}{\nu} \mathrm{d}\Omega + \int_\Gamma U_j \frac{\partial w}{\partial n} \mathrm{d}\Gamma = \int_\Gamma \frac{\partial U_j}{\partial n} w \mathrm{d}\Gamma \tag{A.72}$$

It should be noted here that the different indices for the calculation point i and the index j under the integral point indicates these are calculated at different locations. Also, $\Gamma = \Gamma_1 + \Gamma_2$ so that the boundary conditions $U = \bar{U}$ will be applied on Γ_1 and $\partial U/\partial n = \partial \bar{U}/\partial n$ will be applied on Γ_2 as required.

Using the constant approximation as discussed above we have

$$cU + B_i + \int_\Omega w_i \frac{R}{\nu} \mathrm{d}\Omega + \sum_{j=1}^{n} U_j \int_\Gamma \frac{\partial w}{\partial n} \mathrm{d}\Gamma = \sum_{j=1}^{n} \frac{\partial U_j}{\partial n} \int_\Gamma w \mathrm{d}\Gamma \tag{A.73}$$

where B_i is the volume integral over the domain for each node i of the elements. This is now performed on all elements on the boundary (the integration is only on the element) to obtain a system matrix to be solved. The integrals then become the coefficients of the system of equations:

$$B_i + \sum_{j=1}^{n} H_{i,j} U_j = \sum_{j=1}^{n} \frac{\partial U_j}{\partial n} G_{i,j}; \quad H_{i,j} = \int_{\Gamma_j} \frac{\partial w}{\partial n} \mathrm{d}\Gamma + c\delta_{i,j}; \quad G_{i,j} = \int_{\Gamma_j} w \mathrm{d}\Gamma \tag{A.74}$$

This system is written as a full, nonsymmetric matrix which must be solved. This point is in fact a major problem associated with BEM methods—the solution of the resulting large, dense matrix is slow, limiting the size of problems that can be solved.

Once the system has been solved, the values of U in the domain can be calculated using the following (which is essentially Equation A.71):

$$U_i = \sum_{j=1}^{n} \frac{\partial U_j}{\partial n} G_{i,j} - \sum_{j=1}^{n} H_{i,j} U_j \tag{A.75}$$

If the constant approximation is not appropriate, and a higher-order approximation is necessary, Equation A.73 must be modified to include the sum in the integral sign, essentially returning to Equation A.72.

The formulation above of course is for Poisson's equation. Nevertheless, it can be easily modified for other equations including the Helmholtz equation. What changes, of course, is the operator and the fundamental solutions. In all other respects the formulation is similar. The basic steps in the formulation of the Helmholtz equation are as follows. Starting with the Helmholtz equation:

$$\nabla^2 U + \lambda^2 U = 0 \tag{A.76}$$

With the boundary conditions in Equations A.62 and A.63, the weighted residual equation now becomes

$$\int_{\Omega} U \nabla^2 w_i \mathrm{d}\Omega - \int_{\Omega} U \lambda^2 w_i \mathrm{d}\Omega = -\int_{\Gamma_1} w_i \frac{\partial U}{\partial n} \mathrm{d}\Gamma + \int_{\Gamma_2} \frac{\partial w_i}{\partial n} U \mathrm{d}\Gamma + \int_{\Gamma_2} w_i \frac{\partial \bar{U}}{\partial n} \mathrm{d}\Gamma - \int_{\Gamma_1} \frac{\partial w_i}{\partial n} \bar{U} \mathrm{d}\Gamma \tag{A.77}$$

The fundamental solutions are now the solutions to

$$\nabla^2 w + \lambda^2 w + \delta_i = 0 \tag{A.78}$$

which give

$$w = \frac{1}{4\pi r} e^{-j\lambda r} \tag{A.79}$$

Again, after integrating the singularity on the boundary and writing c for the contribution of the singularity ($c = 1/2$ for a smooth boundary):

$$cU_i + \int_{\Gamma} U_j \frac{\partial w}{\partial n} \mathrm{d}\Gamma = \int_{\Gamma} \frac{\partial U_j}{\partial n} w \mathrm{d}\Gamma \tag{A.80}$$

as previously.

Whatever the method used to formulate the problem, the method ends up in a system of equations that must be solved. It should be noted however that the assembly of the final matrix can be a slow and tedious task. If the terms of the system of equations are obtained by integration (this is the case in all but the constant weighting functions) the time and effort in assembling the matrix represents a significant part of the total effort. The system of equations is dense and results in long and slow solution times as was indicated above.

The methods discussed in this appendix should be viewed as an introduction to the subject. More advanced treatment of specific aspects of the numerical methods can be found in Chapters 6 through 8 and, of course in some of the references to this appendix. We have kept the discussion in this appendix separate from that in previous chapters on purpose and have rather used the common notations in the literature to facilitate further reading on the subject. The references included here are obviously a small sampling of the literature. We tried to include those that are significant and likely to be useful to the reader.

References

1. K.H. Huebner, *The Finite Element Method for Engineers*, Wiley Interscience, New York, 1975.
2. J.R. Brauer, Ed., *What Every Engineer Should Know About Finite Element Analysis*, 2nd edn., Marcel Dekker, New York, 1993.
3. O.C. Zienkiewicz and R.L. Taylor, *The Finite Element Method*, 4th edn., McGraw-Hill, London, 1994.
4. N. Ida and J.P.A. Bastos, *Electromagnetics and Calculation of Fields*, 2nd edn., Springer, New York, 1997.
5. B. Szabo and I. Babuska, *Finite Element Analysis*, Wiley, New York, 1991.
6. P.P. Silvester and M.V.K. Chari, Finite element solution of saturable magnetic field problems, *IEEE Transactions on Power Apparatus and Systems*, 89, 1970, 1642–1651.
7. P.P. Silvester and R.P. Ferrari, *Finite Elements for Electrical Engineers*, 2nd edn., Cambridge University Press, Cambridge, 1990.
8. J. Jin, *The Finite Element Method in Electromagnetics*, Wiley, New York, 1993.
9. K.J. Binns, P.J. Lawrenson, and C.W. Trowbridge, *The Analytical and Numerical Solution of Electric and Magnetic Fields*, Wiley, Chichester, U.K., 1992.
10. N.N. Shenton and Z.J. Cendes, Three-dimensional finite element mesh generation using Delaunay tesselation, *IEEE Transactions on Magnetics*, MAG-21(6), November 1985, 2535–2538.
11. I. Babuska and A.D. Miller, The post-processing approach in the finite element method, III: A posteriori error estimation and adaptive mesh selection, *International Journal for Numerical Methods in Engineering*, 20, 1984, 2311–2324.
12. O.C. Zienkiewicz and J.Z. Zhu, Adaptivity and mesh generation, *International Journal for Numerical Methods in Engineering*, 32, 1991, 783–810.

13. D.K. Sun, Z. Cendes, and J.F. Lee, Adaptive mesh refinement, *h*-version, for solving multi-port microwave devices in three dimensions, *IEEE Transactions on Magnetics*, 36, July 2000, 1596–1599.
14. W. Bangerth and R. Rannacher, *Adaptive Finite Element Methods for Solving Differential Equations*, Birkhauser Verlag, Basel, 2003.
15. I. Babuška, B. Szabo, and I.N. Katz, The *p*-version of the finite element method, *SIAM Journal of Numerical Analysis*, 18, 1981, 515–545.
16. J.P. Webb, *P*-adaptive methods for electromagnetic wave problems using hierarchal tetrahedral edge elements, *Electromagnetics*, 22(5), July 2002, 443–451.
17. P. Ladeveze, G. Goffignal, J.P. Pelle, I. Babuska, et al., *Accuracy Estimates and Adaptive Refinements in Finite Element Computation*, John Wiley & Sons, Chichester, U.K., 1986, pp. 181–203.
18. A.W. Craig, M. Ainsworth, Z.J. Zhu, and O.C. Zienkiewicz, *h* and *h-p* version error estimation and adaptive procedures from theory to practice, *Engineering with Computers*, 5, 1989, 221–234.
19. O.C. Zienkiewicz, J.Z. Zhu, and N.G. Gong, Effective and practical *h-p* version adaptive analysis procedures for the finite element method, *International Journal for Numerical Methods in Engineering*, 28, 1989, 879–891.
20. J.T. Oden, L. Demkowitz, W. Rachowitz, and T.A. Westermannm, Towards a universal *h-p* adaptive finite element strategy, part 2, a-posteriori error estimation, *Computer Methods in Applied Mechanics and Engineering*, 77, 1989, 113–180.
21. M.S. Shephard, Automatic and adaptive mesh generation, *IEEE Transactions on Magnetics*, MAG-21(6), November 1985, 2484–2489.
22. N.N. Shenton and Z.J. Cendes, Three-dimensional finite element mesh generation using Delaunay tesselation, *IEEE Transactions on Magnetics*, MAG-21(6), November 1985, 2535–2538.
23. A.R. Pinchuk and P.P. Silvester, Error estimation for automatic adaptive finite element mesh generation, *IEEE Transactions on Magnetics*, MAG-21(6), November 1985, 2551–2554.
24. S. Pisanetzky, KUBIK: An automatic three-dimensional finite element mesh generator, *International Journal for Numerical Methods in Engineering*, 17, 1981, 255–269.
25. M.L. Barton and Z.J. Cendes, New vector finite elements for three dimensional magnetic field computation, *Journal of Applied Physics*, 61(2), April, 1987, 3919–3921.
26. A. Bossavit, A rationale for edge elements in 3-D field computations, *IEEE Transactions on Magnetics*, MAG-24(1), January 1988, 74–79.
27. A. Bossavit, Whitney forms: A class of finite elements for three dimensional computations in electromagnetism, *IEE Proceedings*, 135-A(3), March 1988, 179–187.
28. G. Mur, A mixed finite element method for computing three dimensional time domain electromagnetic fields in strongly inhomogeneous media, *IEEE Transactions on Magnetics*, MAG-26(2), March 1990, 674–677.
29. G. Mur, Edge elements, their advantages and their disadvantages, *IEEE Transactions on Magnetics*, 30(5), September 1994, 3552–3557.
30. D.M. Young, *Iterative Solution of Large Linear Systems, Computer Science and Applied Mathematics*, Academic Press, New York, 1971.
31. J.K. Reid, Ed., *Large Sparse Sets of Linear Equations*, Academic Press, New York, 1971.
32. V.A. Barker, Ed., *Lecture Notes in Mathematics 572*, Springer-Verlag, Berlin, 1976.

33. D.J. Rose and R.A. Willoughby, Eds., *Sparse Matrices and Their Applications*, Plenum Press, New York, 1972.
34. K.S. Yee, Numerical solution of initial boundary value problems involving Maxwell's equations in isotropic media, *IEEE Transactions on Antenna and Propagation*, AP-14, May 1966, 302–307.
35. A. Taflove and S.C. Hagness, *Computational Electrodynamics: The Finite-Difference Time-Domain Method*, 3rd edn., Artech House, Norwood, MA, 2005.
36. K.S. Kunz and R.J. Luebbers, *The Finite Difference Time Domain Method for Electromagnetics*, CRC Press, Boca Raton, FL, 1993.
37. M.N.O. Sadiku, *Numerical Techniques in Electromagnetics*, CRC Press, Boca Raton, FL, 2000.
38. G. Mur, Absorbing boundary conditions for the finite difference approximation in the time domain electromagnetic field equations, *IEEE Transactions of Electromagnetic Compatibility*, EMC-23(4), November 1981, 377–382.
39. G.A. Kriegsmann, A. Taflove, and K.R. Umashankar, A new formulation of electromagnetic wave scattering using an on-surface radiation boundary condition approach, *IEEE Transactions on Antennas and Propagation*, 35, February 1987, 153–161.
40. T.G. Moore, J.G. Blaschak, A. Taflove, and G.A. Kriegsmann, Theory and application of radiation boundary operators, *IEEE Transactions on Antennas and Propagation*, 36, December 1988, 1797–1812.
41. J.-P. Berenger, A perfectly matched layer for the absorption of electromagnetic waves, *Journal of Computational Physics*, 114(1), 1994, 185–200.
42. W.C. Chew and W.H. Weedon, A 3D perfectly matched medium from modified Maxwell's equations with stretched coordinates, *Microwave and Optical Technology Letters*, 7, September 1994, 599–604.
43. J.-P. Berenger, Three-dimensional perfectly matched layer for the absorption of electromagnetic waves, *Journal of Computional Physics*, 127, 1996, 363–379.
44. J.-P. Berenger, Improved PML for the FDTD solution of wave-structure interaction problems, *IEEE Transactions on Antennas and Propagation*, 45, March 1997, 466–473.
45. T.B.A. Senior and J.L.O. Volakis, *Approximate Boundary Conditions in Electromagnetics* (*IEE Electromagnetic Waves Series*), IEE, London, 1995.
46. S.A. Tretyakov, Generalized impedance boundary conditions for isotropic multilayers, *Microwave and Optical Technology Letters*, 17(4), 1998, 262–265.
47. X. Antoine and H. Barucq, Approximation by generalized impedance boundary conditions of a transmission problem in acoustic scattering, *Mathematical Modeling and Numerical Analysis*, 39(5), 2005, 1041–1059.
48. M.E. Lee, S.I. Hariharan, and N. Ida, Solving general time dependent two dimensional eddy current problems, *Journal of Computational Physics*, 89(2), August 1990, 319–348.
49. S.G. Jurgens, A. Taflove, K.R. Umashankar, and T.G. Moore, Finite-difference time-domain modeling of curved surfaces, *IEEE Transactions on Antennas and Propagation*, 40, April 1992, 357–366.
50. K.S. Yee, J.S. Chen, and A.H. Chang, Conformal finite-difference time-domain (FDTD) with overlapping grids, *IEEE Transactions on Antennas and Propagation*, 40, September 1992, 1068–1075.
51. P. Mezzanotte, L. Roselli, and R. Sorrentino, A simple way to model curved metal boundaries in FDTD algorithm avoiding staircase approximation, *IEEE Microwave Guided Wave Letters*, 5, August 1995, 267–269.

52. T.G. Jurgens, A. Taflove, K. Umashankar, and T.G. Moore, Finite-difference time-domain modeling of curved surfaces, *IEEE Transactions on Antennas and Propagation*, 40, April 1992, 357–366.
53. K.M. Mitzner, An integral equation approach to scattering from a body of finite conductivity, *Radio Science*, 2(12), 1967, 1459–1470.
54. C.A. Brebbia and S. Walker, *Boundary Element Techniques in Engineering*, Newness-Butterworths, London, 1980.

Index

T

U

V

W

FSC
www.fsc.org
MIX
Papier | Fördert gute Waldnutzung
FSC® C083411

Kolibri 360 GmbH
Ferdinand-Jühlke-Straße 7
99095 Erfurt, Deutschland
produktsicherheit@kolibri360.de